D1208518

Introduction to Biological Membranes

SECOND EDITION

Mahendra Kumar Jain
Department of Chemistry
University of Delaware
Newark, Delaware

A WILEY-INTERSCIENCE PUBLICATION
JOHN WILEY & SONS
NEW YORK CHICHESTER BRISBANE TORONTO SINGAPORE

To My Mother
who once wondered
why the evening news always
lasts thirty minutes

Library of Congress Cataloging-in-Publication Data:

Jain, Mahendra K., 1938–
Introduction to biological membranes / Mahendra Kumar Jain.—2nd ed.

p. cm.
"A Wiley-Interscience publication."
Bibliography: p.
Includes index.
ISBN 0-471-84471-3: $39.95 (est.)
1. Membranes (Biology) I. Title.
QH601.J36 1988
574.87′5—dc19 87-18914
CIP

Printed in the United States of America

10 9 8 7 6 5 4 3 2 1

Preface

Experts do not
give objective
opinion, they
give their opinion.
Morarji Desai

Considerable progress has been made during the last two decades toward a molecular description of membrane processes. The study of membranes has the distinction of not being part of any one of the classical disciplines; therefore, an introductory text on membranes must transcend the boundaries of most classical disciplines while permitting the reader to draw information and inspiration from diverse areas of investigation. Although the perspectives from the various disciplines are neither unified nor always coincident, the sophistication of techniques and articulation of models have reached a stage at which, with modest imagination, we can begin to appreciate the functional complexity of membranes in the form of an organizational hierarchy. We have reached a descriptive level at which we can write schemes and often dissect the rate constants, and the stage is set to elaborate on the molecular dynamics and the underlying mechanisms.

In the hierarchy of biological organization, the structure and function of membranes lies somewhere between macromolecules and cells. As noncovalently aggregated macroscopic structures arising from amphipathic molecules, membranes in general have their unique characteristics that are not shared by other biomolecules and their aggregates. This book is an attempt to capture at present a theme that underlies the phenomenon of biological membranes, to provide a conceptual context, and to guide the novice to the specialized literature. Whenever possible, along with the phenomenology, a qualitative description of the underlying biophysical concepts is presented. This is not an authoritative treatise nor a critical review, although I have tried to approximate the general consensus on the state-of-the-art. Since this book is intended for a wide variety of readers, primary and secondary key references leading to the various levels of complexity are included. No attempt has been made to provide an extensive bibliography, although references as a general guide for the background material are also given.

I have greatly benefited from the thoughtful comments and reprints provided by a large number of membranologists. In the light of such suggestions I have tried to strike a balance while maintaining a conceptual continuity. I have refrained from summaries that require oversimplifications and unwarranted generalizations. I deeply appreciate informal comments from friends and strangers, and I look forward to suggestions from novices as well as experts, which I hope to incorporate in future versions.

MAHENDRA KUMAR JAIN

Newark, Delaware

Contents

1. **Introduction** **1**
2. **Components of Biological Membranes** **10**
   Isolation 10
   Characterization of Isolated Membranes 17
   Components of Biomembranes 22
   Conformation of Phospholipids 44
   Synthesis of Phospholipids 48
3. **Self-Association of Phospholipids** **51**
   Polymorphism of Lipid–Water Aggregates 54
   Monolayers 59
   Micelles 69
   Lamellar Phase 80
   Hydration of the Interface 84
4. **Properties of Bilayers** **86**
   Liposomes 86
   Electrical Double Layer at the Interface 98
   Transbilayer Movement of Phospholipids 105
   Lateral or Translational Diffusion 108
   Exchange of Phospholipids Between Bilayers 112
   Transverse Diffusion of Solutes Across Bilayers 115
5. **Order and Dynamics in Bilayers** **122**
   Thermotropic Changes 126
   Bilayers of Mixture of Lipids 137

6. Solutes in Bilayers 147

Partition Coefficient 148
Effects of Solutes on Bilayers 157
Biomembranes as Targets for Drugs 162

7. Lipid–Protein Interactions in Membranes 166

Solubilization of Membrane Proteins 167
Characterization of Isolated Membrane Proteins 171
Role of Lipids 185

8. Glycoproteins and Glycolipids 194

Identification of Glycoconjugates 195
Glycoproteins in Membranes 204

9. Ionophores 213

Carrier-Mediated Transport 215
Ion Transport Through Channels 238
Channels of Gramicidin A 241
Voltage-Gated Channels 254

10. Facilitated Transport 257

Transporter for Glucose in Erythrocytes 263
Anion-Exchanger of Erythrocytes 270
ADP–ATP Exchanger of Mitochondria 274
Ionic Channels in Epithelial Cells 275
Porins 277
Channels from Toxins 278

11. Coupled Transport 281

Cotransport 282
Phosphotransferase System 290

12. Gated-Channels 294

Acetylcholine Receptor 296
Action Potential 307
Miscellaneous Gated Channels 326

13. Energy Transduction 330

Bacteriorhodopsin 333
Oxidative Phosphorylation 336
The Calcium Pump 340
The Sodium Pump 346
The Proton Pump 350
Sensors of Proton Gradients 353

14. Transbilayer Response of Signals 356

Adenylated Cyclase Stimulated by Hormones 357
Phosphoinositol Cycle 365
Sodium-Proton Exchanger 368
Phototransduction in Vision 370

15. Bulk Transport by Fusion and Secretion 374

Fusion 376
Endocytosis 378
Liposomes as Drug Carriers 387

Bibliography 389

Index 417

1 Introduction

. . . liquids pose a great problem of packaging, which every experienced chemist knows. And it was well known to God Almighty, who solved it brilliantly, as he is wont to, with cellular membranes, egg shells, the multiple peels of oranges, and our own skins, because after all we too are liquids. Now at that time, there did not exist polyethylene, which would have suited me perfectly since it is flexible, light, and splendidly impermeable: but it is also a bit too incorruptible, and not by chance God Almighty himself, although he is a master of polymerization, abstained from patenting it: He does not like incorruptible things.

PRIMO LEVI
The Periodic Table*

Containing, compartmentalizing, and regulating the transfer of metabolites and macromolecules in a living organism are the primary functions of membranes. In nucleated cells, for example, plasma membranes surround an impressive array of cytoplasmic organelles, each with its own membrane as shown schematically in Fig. 1-1. The identity and the very existence of a cell and its organelles are not possible without membranes which separate the internal *milieu* from the surrounding environment. Functional specialization of structures enclosed with membranes, like organelles, is due to certain distinctive characteristics of their membranes.

Membranes provide a biophysical basis for the very concept of the cell (for an introductory text, see Alberts et al., 1983) as a unifying theme in the biosphere. It is not easy to conceive how life in all its manifestations, as we know it, could have evolved without a membrane-like structure that isolates the internal microenvironment from the variability and fluctuations of its surroundings. The problem of the evolution of membrane is multifaceted. The concept of a two-dimensional continuous molecular matrix is intrinsic to the concept of membrane-enclosed structures and it is necessary for compartmentalization. However, compared with the behavior of linear biological polymeric macromolecules like proteins and nucleic acids, the noncovalent supramolecular organization of biomembranes poses unique questions about

* This dilemma was encountered by the author, while in a German concentration camp, trying to steal in order to stave off starvation

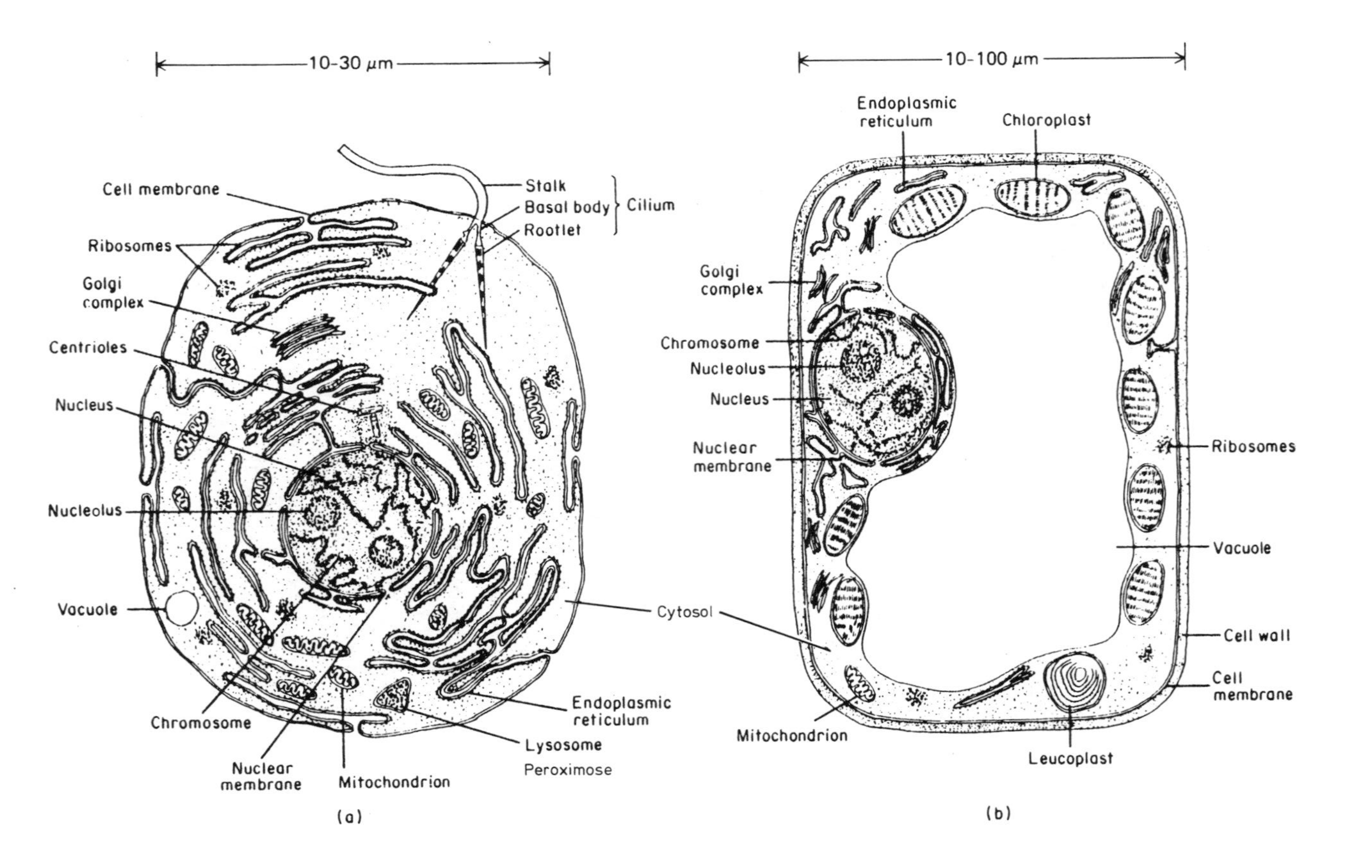
10-30 μm
Cell membrane
Stalk
Basal body
Rootlet
Cilium
Ribosomes
Golgi complex
Centrioles
Nucleus
Nucleolus
Vacuole
Cytosol
Chromosome
Nuclear membrane
Mitochondrion
Endoplasmic reticulum
Lysosome
Peroximose
(a)
10-100 μm
Endoplasmic reticulum
Chloroplast
Golgi complex
Chromosome
Nucleolus
Nucleus
Nuclear membrane
Ribosomes
Vacuole
Cell wall
Cell membrane
Mitochondrion
Leucoplast
(b)

thermodynamics and kinetics of assembly, organization, repair, material balance, cooperative interactions, asymmetry, and degrees of motional freedom.

The ubiquity of membranes in living systems implies their antiquity and functional diversity in the evolutionary sense. Our concept of evolution depends upon the spontaneous evolution of a cell. This cannot be achieved without membranes, although genes and proteins are also necessary, but not sufficient. As one would also expect from the diversity of life forms, membranes have evolved to perform diverse tasks (Fig. 1-2). For example, simple membranes such as those of the egg yolk sac essentially act as barriers; the plant cell membranes permit passage of certain solutes but not others; bacterial membranes tend to accumulate nutrients from the surroundings against concentration gradients; chloroplast membranes convert light energy into the energy of the proton gradient; and nerve cell membranes transmit coded electrical signals in response to specific stimuli. The underlying biophysical processes, which range from osmosis, to gated channels ultimately form the molecular basis for understanding such complex organismic functions as blood-brain-barrier (BBB), respiration, motility, and memory. Thus, for example, the BBB is made up of epithelial cells, and the passage of solutes through this complex barrier could occur via intercellular channels, through solubility-diffusion or facilitated transport mechanism, by transbilayer *flip-flop*, or by endocytosis.

Fig. 1-1. Caricature of the thin section of a generalized animal (a) and of a higher plant cell (b) illustrating relationships between different compartments created by the plasma membrane and the membranes of organelles. Membranes are shown as a pair of lines separated by a light interzone. The invaginations of the cell surface (especially in the animal cell) are indicated in several areas; some of these endoplasmic reticular structures extend for a considerable distance into the cell. The nuclear membrane is composed of flattened sacs of the endoplasmic reticulum (ER), and the enclosed space is in continuity with the cytoplasm. Golgi complex is shown here as modified ER but separated from cellular membrane. Mitochondria are shown with their cristae formed by invagination of their inner membrane. Chloroplasts have a similar, but not identical, structure. Lysosomes represent a class of organelles whose condition within a tissue varies under different stimuli. Organelles in general vary in form, size, and location within the cell. Also the number of organelles in a given type of cell may vary several hundredfold. Extracellular matrix (a "fuzz" always seen on the outer surface of a cell) and filamentous cytoskeleton are not shown. In animal cells the cytoskeleton is often organized in areas near the nucleus that contain the cell's pair of centrioles. The cytoskeletal filaments are made up of arrays of proteins that seem to hold organelles in the cytosol, give membrane its shape, and anchor certain membrane proteins. Three kinds of filaments that are often found are microtubules (diameter 25 nm), actin filaments (7 nm), and intermediate filaments (10 nm). For further details of the various organelles and morphological compartments see Fig. 2-1.

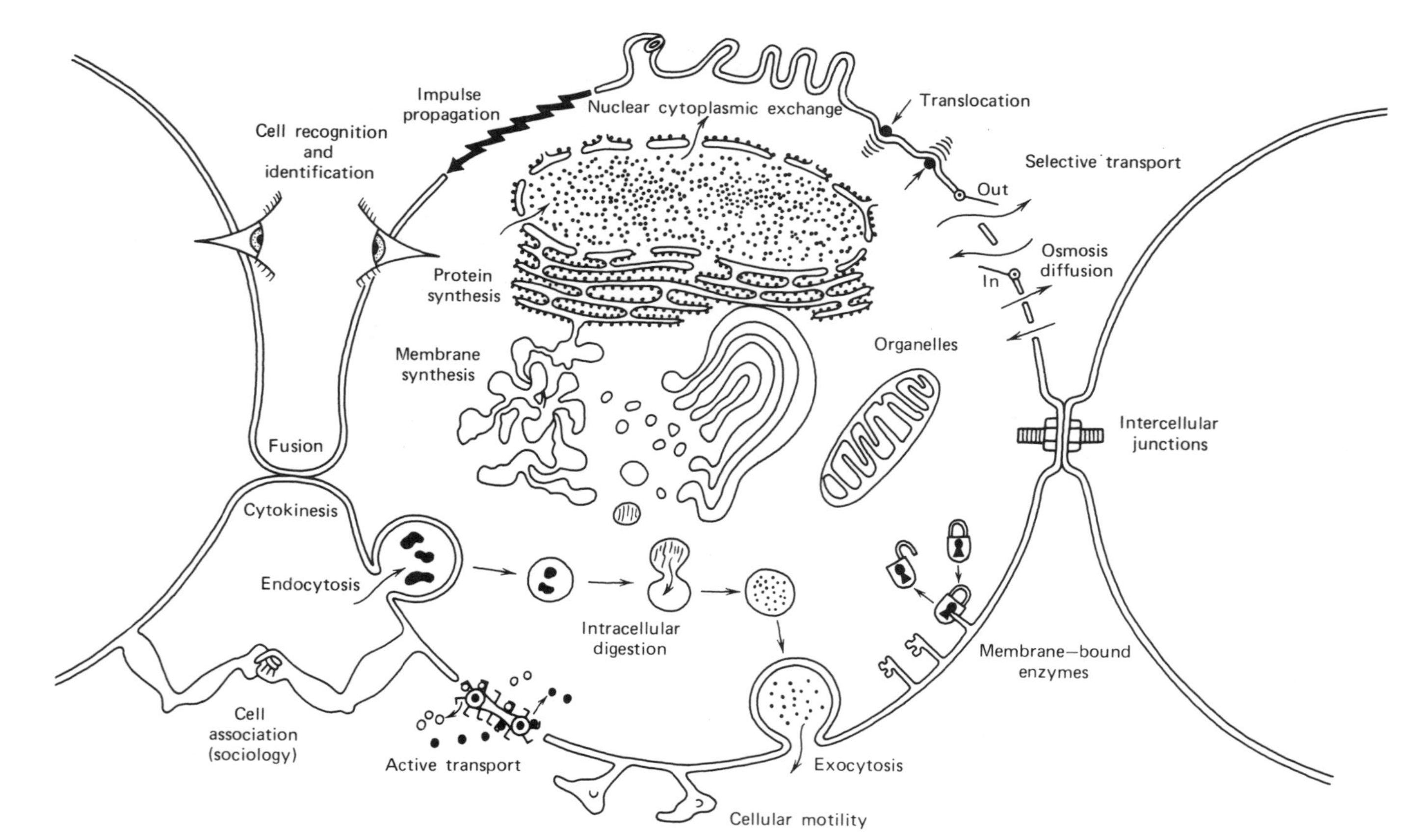

Fig. 1-2. A cartoon of a cell, emphasizing the various processes and functions (individual and social) mediated and modulated by the plasma membrane. (Courtesy of R. Wagner.)

In spite of the staggering diversity of form and function of organelles and cells, some first-order generalizations can be outlined as a basis for a discussion of the general functions of membranes. These include:

Compartmentalization provides morphological identity to the cell and organelles.

Selective barrier properties ultimately control the internal milieu.

The bilayer matrix provides a surface for specific distribution, orientation, and sidedness for a variety of functional molecules.

Communication and stimulus–response coupling across and along the membrane provide a basis for functions such as excitability, adhesion, immune response, and hormone action.

To mediate such functions, lipids act as barriers, solvents, anchors, activators, and conformational stabilizers for proteins that carry out specific catalytic and translocation functions. How such apparently diverse functions are mediated and regulated by and through membranes is an area of active investigation. To a first approximation, the diversity of functions arises from qualitative and quantitative differences in their composition that ultimately give rise to heterogeneity in lateral and transbilayer organization. Therefore, it is not very meaningful to surmise about the general structure and organization of membranes as they relate to specific functions. The underlying diversity of composition and organization must be kept in perspective as one often does, albeit more so, while talking about the structure of proteins and nucleic acids.

The following heuristic conceptual generalizations have been useful in stimulating thinking about membrane structure and organization (Overton, 1899; Gorter and Grendel, 1925; Danielli and Davson, 1935; Singer and Nicholson, 1972; Jain and White, 1977):

1. Biological membranes essentially consist of a two-dimensional matrix made up of a phospholipid bilayer interrupted and coated by proteins (Fig. 1-3). Thus, the interior of the membrane is much less polar than the interfacial region. Such an organization is a direct consequence of the **hydrophobic effect**, whereby the apolar acyl chains of lipids and the side-chains of the nonpolar amino acid residues in proteins tend to be *squeezed away* from the aqueous phase.
2. The components in a bilayer matrix are held together largely by noncovalent forces. The hydrophobic effect accounts for most of the interaction energy that stabilizes the bilayer organization; however, hydro-

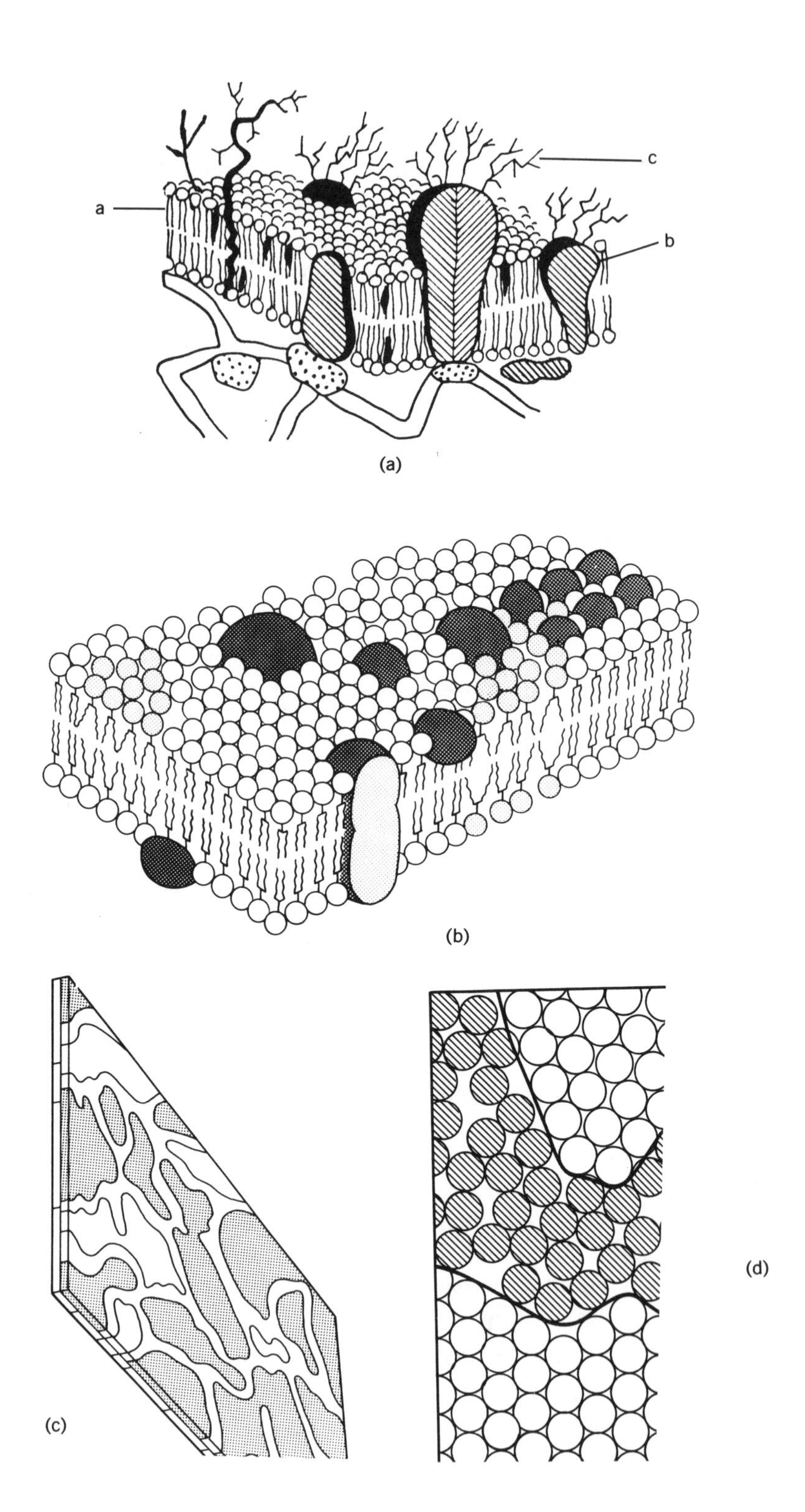
a
b
c
(a)
(b)
(c)
(d)

gen bonding and electrostatic interactions are significant in the polar group region.

3. The uncatalyzed exchange of components from one interface of the bilayer to the other is very slow for lipids and probably nonexistent for proteins. **Transbilayer movement** of amphipathic solutes is energetically unfavorable because it requires insertion of the polar groups into nonpolar regions and exposure of the apolar groups to the polar regions. Compositional **asymmetry** gives rise to morphological and functional asymmetry. The composition of the aqueous environment at the two interfaces is also always asymmetrical. A loss of such asymmetries is lethal to cells.
4. Specific interactions between the membrane components lead to **selective orientation** and **segregation** of the components in the plane of the membrane.
5. Several types of molecular motions are experienced by the components within the constraints of the bilayer organization (Fig. 1-4). **Rotation** of molecules along their axes perpendicular to the plane of the membrane occurs every 0.1–100 nsec for lipids and 0.01–100 msec for proteins; **segmental** motion of acyl chains (0.01–1 nsec) gives rise to an increased disorder toward the center of the membrane; **translational** motion of molecules in the plane of the membrane occurs with a **lateral diffusion** coefficient of 10^{-13} to 10^{-8} $cm^2 \cdot sec^{-1}$. These orientational and motional parameters for the components in the membrane differ more than what would be expected only on the basis of the size of these components. It should be emphasized that not all the molecules of the same type in the same membrane necessarily have the same motional properties.
6. The two-dimensional matrix of a biomembrane probably consists of patches of phospholipid molecules in different degrees of conformational disorder. Under certain conditions, the bilayer organization can

---

Fig. 1-3. Organizational and structural framework of biomembranes in various degrees of schematization. (A) A general patchwork representation of the organization of the components emphasizing lipids (a), proteins (b), carbohydrates (c), bilayer organization, asymmetry, channel, orientation, and anchoring of proteins. Additional points to be emphasized are long-range order and defects. As shown in B, both lipids and proteins may form distinct patches due to specific intermolecular interactions within the plane of the membrane. A nonrandom topological view of the membrane is exaggerated in C, where regions of differing composition and organization may coexist along with the intervening regions of relative disorder. Such regions of relative disorder are emphasized in D, where only the top view of the lipid molecules in a bilayer is shown.

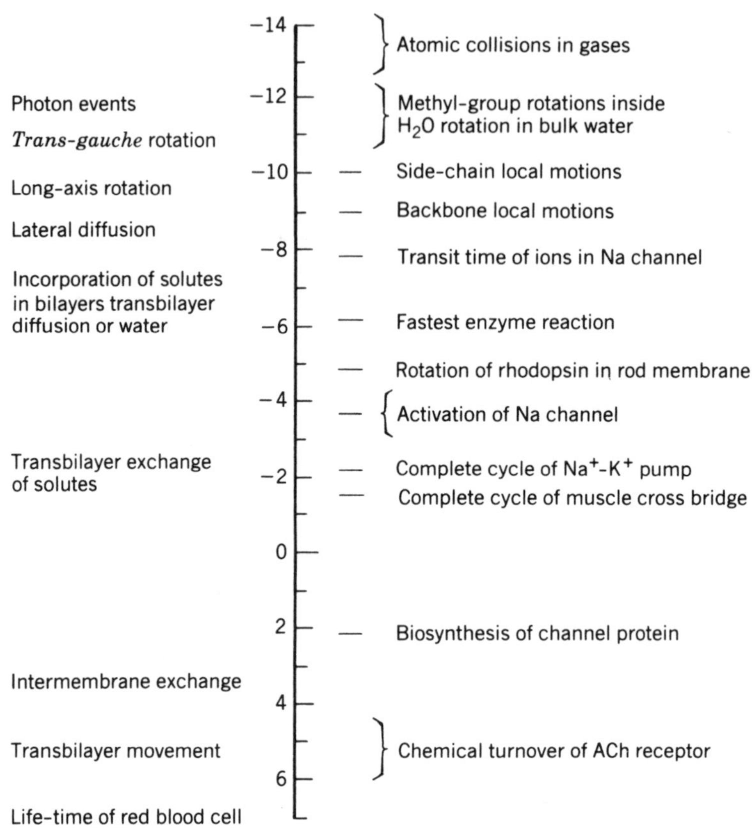

Fig. 1-4. The time-scale of events in membranes. The vertical axis is logarithmic in seconds (i.e., 1 ns = −9).

be interrupted by nonbilayer phases as well as by bilayer phases of differing compositions, and by the regions of mismatch between coexisting phases. Such features within the gross organization of a membrane can have different life-times, and may be induced in response to environmental and metabolic perturbations. Such organizational defects probably act as sites for a variety of functions that require topological discontinuities in the plane of the bilayer.

A description of biological membranes that emerges from these generalizations also has some broad consequences. The functional consequences of the organizational and motional constraints imposed upon the matrix of a biomembrane by virtue of its constituents and environment are elaborated in the following chapters. The hydrophobic forces lead to energetically favorable conformations and self-aggregation of phospholipid molecules in a bilayer matrix. This matrix can accommodate the wide range of constraints required for an interface between the relatively stable internal *milleu* of a cell and its changing external environment. This is accomplished with relatively few types of molecules, whose self-assembly is a spontaneous process not directly requiring genetic information. However, the overall process does make an economical use of the genetic information, in the sense that the properties of the membrane are modulated generally by altering the lipid composition rather than the structure of specific lipid components. Similarly, the ability of a large variety of relatively simple molecules to form bilayer organization is consistent with the suggestion that the spontaneous formation of a membranous structure from the primordial soup did not require any elaborate mechanism at a stage of evolution where the very identity of living forms was yet to be established.

2 Components of Biological Membranes

The harmonious cooperation of all beings arose, not from the orders of a superior authority external to themselves, but from the fact that they were all parts in a hierarchy of whole forming a cosmic pattern, and what they obeyed were the internal dictates of their own natures.

CHUNG TZU
Fourth Century BC

There are over 200 different types of cells in the human body. These are assembled into a variety of tissues (epithelia, muscle, nervous tissue), each containing a mixture of cells. A tissue may contain several types of cells, and each cell has several membrane-bound organelles. On the other hand, prokaryotes and single-celled eukaryotes have only a few types of membranes. The functional variety of biomembranes arises primarily from the diversity of their composition. Although the organizational principle underlying all these membranes is essentially the same, membranes do differ morphologically, physiologically, biochemically, and compositionally. The composition of membranes *in vivo* is preserved metabolically by biosynthesis and degradation, as well as by fusion and recycling.

ISOLATION

Several types of membranes are usually present in homogenized cells or tissues, therefore isolation of pure membrane fractions and characterization of their components usually involves several steps (DeDuve, 1964; Wallach and Lin, 1973; Fleischer and Packer, 1974). Three major steps that are usually necessary for isolation of members are:

1. Preparation of a cell population that is homogeneous with regard to specific functions under consideration. If preparation involves disrup-

tion of tissues into single cells by treatment with hydrolytic enzymes like collagenase, some modification of the plasma membrane is inevitable. As shown in Fig. 2-1 many types of cells contain elaborate envelopes that protect the plasma membrane against a variety of assaults. Disruption of the envelope is necessary for the isolation of the plasma membranes. So far, no general method has evolved for doing this.

2. Homogenization of cells and tissues is achieved by methods involving: shearing force between two low-clearance surfaces as in Potter-Elvehjem or Dounce homogenizer; by high-speed rotating blades as in blenders; by osmotic shock; by shaking cells with glass beads; by freezing-thawing cycles; by gas ebullition involving decompression of nitrogen that has been solubilized at high pressure; and by ultrasonic radiation. The thermodynamics of membrane organization are such that the disrupted membranes or membrane fragments are spontaneously vesiculated to form closed structures (Table 2-1) so as to minimize the interactions between water and the hydrophobic portions of a membrane. The osmolarity and the ionic composition of the medium are also important in regulating the size dispersion and sometimes the composition of the fragmented vesicles. Inert and impermeable solutes

TABLE 2-1 Vesicular Structures Derived from Cells and Organelles

Vesicles	Characteristics
Ghosts	Derived from the plasma membrane of the whole cell by osmotic lysis; they have appearance of empty sacks with slight "crumpling" of the surface and the same dimensions as intact cells
Microsomes (sarcoplasmic and endoplasmic)	Formed by revesiculation of the fragmented plasma or reticular membranes; they are only a fraction of the size of the original membrane
Myelin	Derived from pinched-off myelinated axons
Protoplasts[a]	Cell wall-less membranes of microorganisms (e.g., yeast, molds, gram-positive bacteria) and some higher plants; they can be prepared from bacteria in the presence of penicillin or cell wall-degrading enzymes
Spheroplasts	Membrane-derived gram-negative bacteria by removal of the cell wall by alkali or lysozyme treatment accompanied by EDTA
Submitochondrial particles (SMP)	Derived from inner mitochondrial membranes by ultrasonic disruption; they have inside-out geometry
Synaptosomes	Vesicles derived from pinched-off nerve ends and synaptic junctions

[a] Some microorganisms such as *Mycoplasma* exist normally in a wall-less state, while others assume the protoplast form by mutation (L forms).

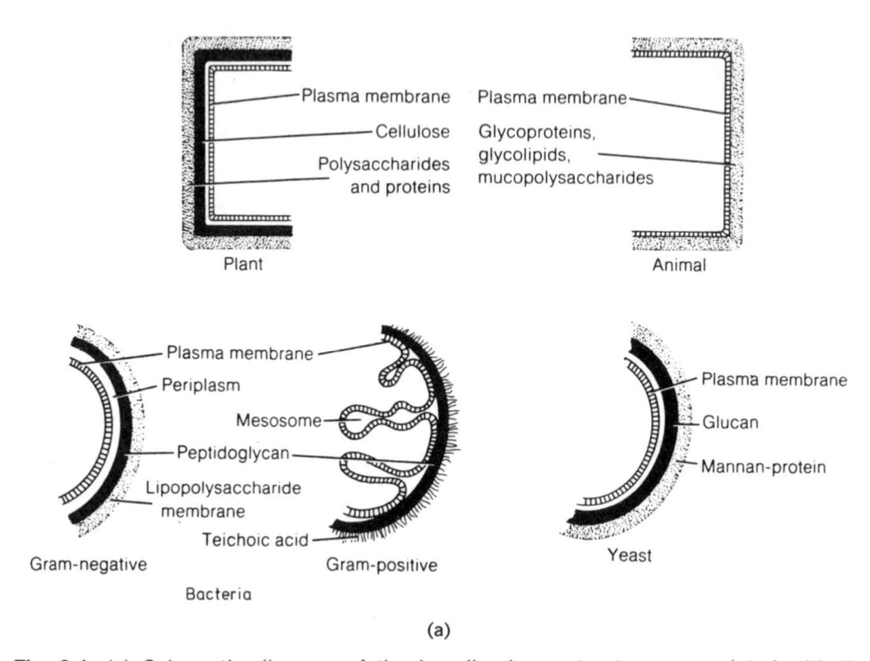

Fig. 2-1. (a) Schematic diagram of the lamellar layer structures associated with the plasma membrane in the envelope of cells from plant, animal, bacteria, and yeast. (b) Major features of the membrane bound components in a cell. The **plasma membrane** constitutes the outer boundary of a cell. Its matrix is a bilayer of the type shown in Fig. 1-3. The **Golgi** apparatus is a system of stacked, membrane-bounded, flattened sacs involved in modifying, sorting, and packaging macromolecules for secretion or for delivery to other organelles. Around the Golgi apparatus there are numerous vesicles (50 nm or larger) that are probably involved in bulk transfer of solutes, membranes, and proteins between different compartments of the cell. The **endoplasmic reticulum** (ER) network is made up of flattened sheets, sacs, and tubes of membrane that extend throughout the cytoplasm of eukaryotic cells. ER specializes in the synthesis and transport of lipids and membrane proteins. ER membrane is structurally continuous with the outer membrane of the nuclear envelope. The rough ER generally occurs as flattened sheets studded on its outer face with ribosomes engaged in protein synthesis. The smooth ER is more tubular and its major function is in lipid metabolism. **Lysosomes** contain hydrolytic enzymes involved in intracellular digestion, and **peroxisomes** contain oxidative enzymes that generate and oxidize hydrogen peroxide. **Chloroplasts** are double-membrane-bound organelles that contain photosynthetic apparatus (thylakoid and grana) of higher plants. **Mitochondria** are energy-transducing (oxidative phosphorylation) devices found in all eukaryotic cells. **Nucleus** is separated from cytoplasm by an envelope of two membranes. All the chromosomal DNA is held in the nucleus packaged into chromatic fibers by its association with an equal mass of histone proteins. (Adapted from Alberts et al., 1983.)

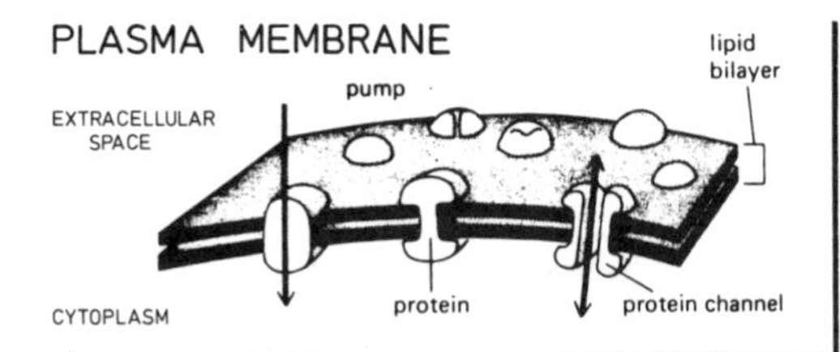
PLASMA MEMBRANE
pump
lipid bilayer
EXTRACELLULAR SPACE
CYTOPLASM
protein
protein channel

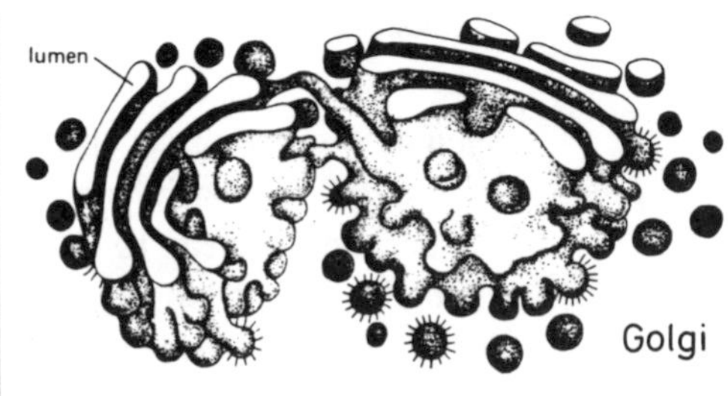
lumen
Golgi

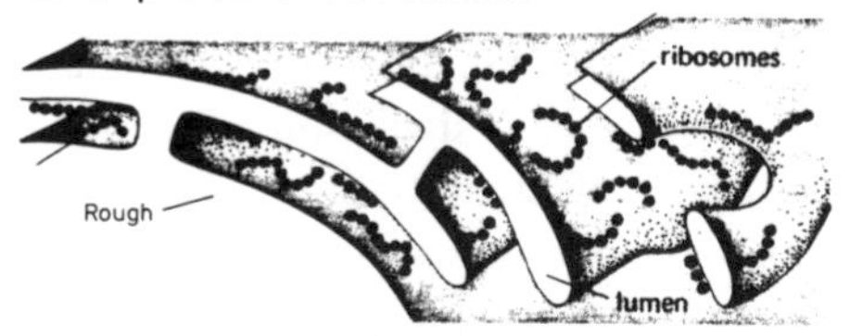
Endoplasmic reticulum
ribosomes
Rough
lumen

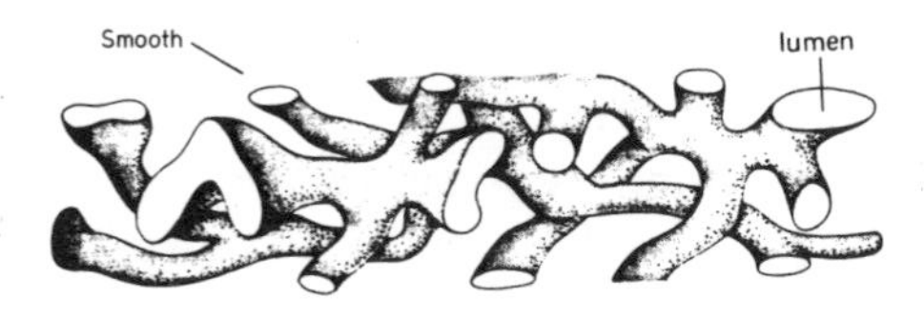
Smooth
lumen

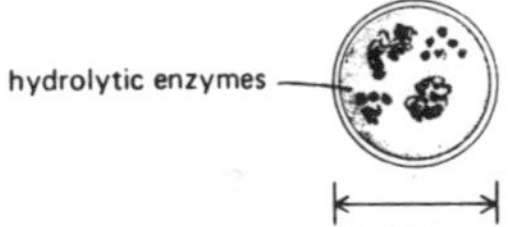
LYSOSOMES
hydrolytic enzymes
0.2–0.5 μm

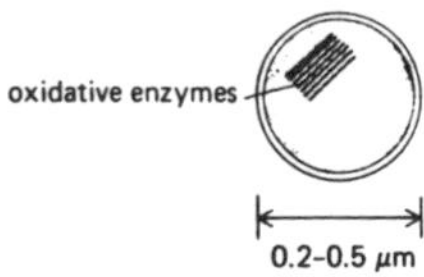
PEROXISOMES
oxidative enzymes
0.2–0.5 μm

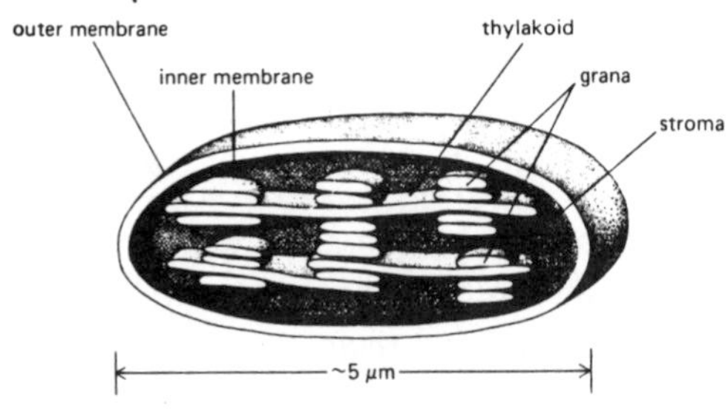
Chloroplast
outer membrane
inner membrane
thylakoid
grana
stroma
~5 μm

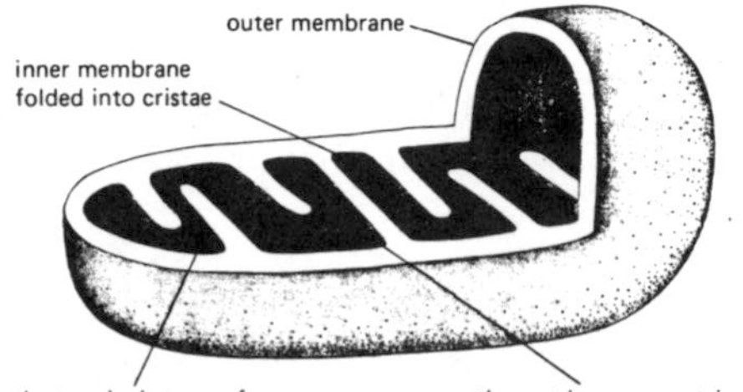
Mitochondria
outer membrane
inner membrane folded into cristae
the terminal stages of oxidation occur at the inner membrane
the matrix space contains a concentrated solution of many different enzymes

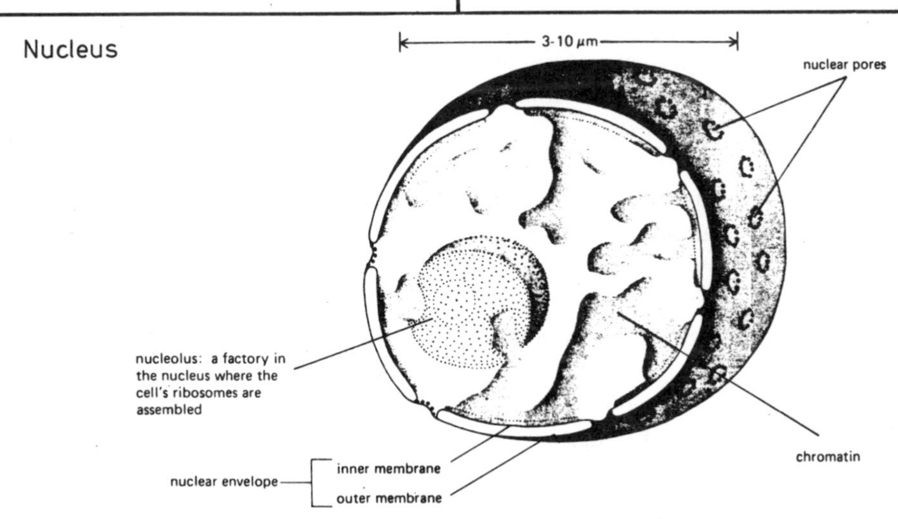
Nucleus
3-10 μm
nuclear pores
nucleolus: a factory in the nucleus where the cell's ribosomes are assembled
chromatin
nuclear envelope
inner membrane
outer membrane

(b)

such as sucrose or sorbitol are often used in the dispersing medium to adjust the osmolarity and density. Ionic strength, composition, and temperature appear to influence significantly the intermembrane exchangeability of loosely bound membrane components, the sidedness of the resealed vesicles, and also the fracture boundaries of the plasma membrane. Buffers, metal ion chelators, and multivalent ions in the homogenization medium also serve the specific functions of stabilizing pH, destabilizing the membrane, and inducing vesiculation. The final protocol is usually established by trial and error because different types of cells exhibit environmentally sensitive topographic heterogeneity, transverse asymmetry, and functional complexity that require integration of several membrane components.

3. Separation of vesiculated membranes and organelles can be achieved on the basis of one or more of the following properties: volume, density, electrical properties, and the presence of receptors for specific ligands. Density gradient centrifugation on glycerol, sucrose, Ficoll, or dextran gradients is often used as the first step to be followed by other more specific methods. As shown in Fig. 2-2, with increasing centrifugal force the nuclei, mitochondria, lysosomes, and microsomes sediment in that order.

A choice of a specific protocol for isolation of a membrane fraction depends upon the source and the nature of the contaminants. Mammalian erythrocytes (red cells) have only one membrane, and their *ghosts* can be readily prepared by hypotonic lysis followed by centrifugation. Bacterial protoplasts and spheroplasts are isolated by the removal of the cell wall (generally by enzymic degradation), and the plasma membranes are then isolated by subjecting the spheroplasts to hypotonic lysis. In more complex tissues, such as intestinal mucosa, it is not unusual to follow a six- or seven-step procedure such as the one shown in Fig. 2-3. The homogenized tissues are centrifuged at low speeds to sediment the nuclei and debris of connective tissues. The supernatant is then sedimented at a high g-force to separate the *soluble* and *released* proteins. The resultant pellet can be fractionated on a density gradient. Thus, brain homogenate yields, in the order of increasing density, fractions rich in myelin, synaptosomes, mitochondria, and nuclei. Further fractionation can be used to resolve these fractions into relatively homogeneous individual components. Similar procedures have been used for fractionation of liver, muscle, brain, fat pad, and epithelial tissues, as well as plant and bacterial cells (Fleischer and Packer, 1974; Maddy, 1976).

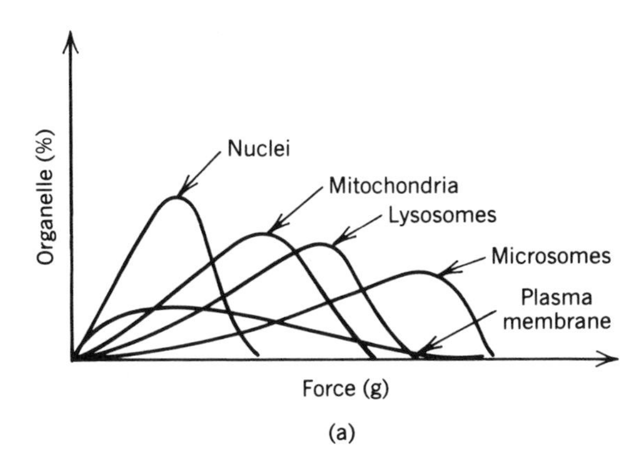

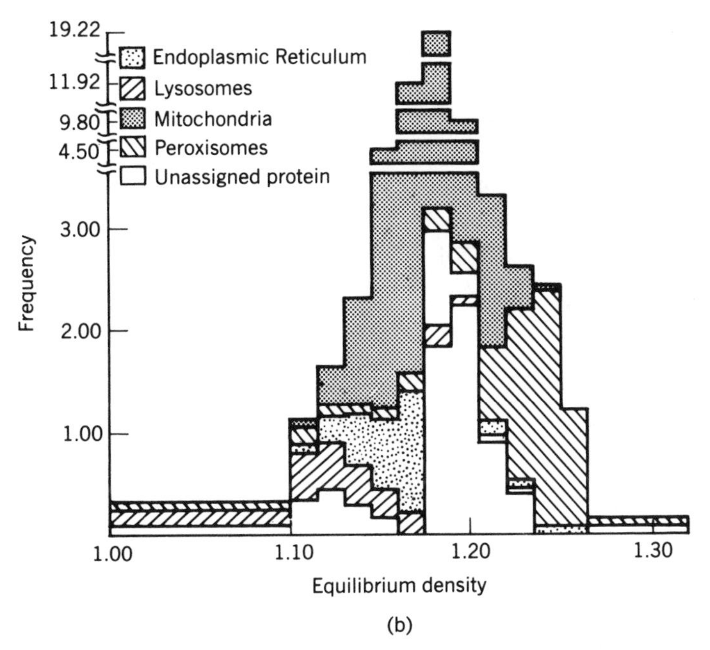

Fig. 2-2. (a) Relative distribution of organelles during differential centrifugation. The lighter organelles require more force, and differential centrifugation provides an indication of their relative densities. (b) Equilibrium density distribution of some rat liver organelles on sucrose density gradient. A series of density steps were collected and identified with marker enzymes. The ordinate represents the percent of the total protein in the histogram contributed by the components depicted. The unassigned protein may represent Golgi, plasma membrane, or other particles. The low density of lysosome was induced by injecting Triton WR-1339 to the animals prior to sacrifice. (From Leighton et al., *J. Cell Biol.* 37, 482, 1968.)

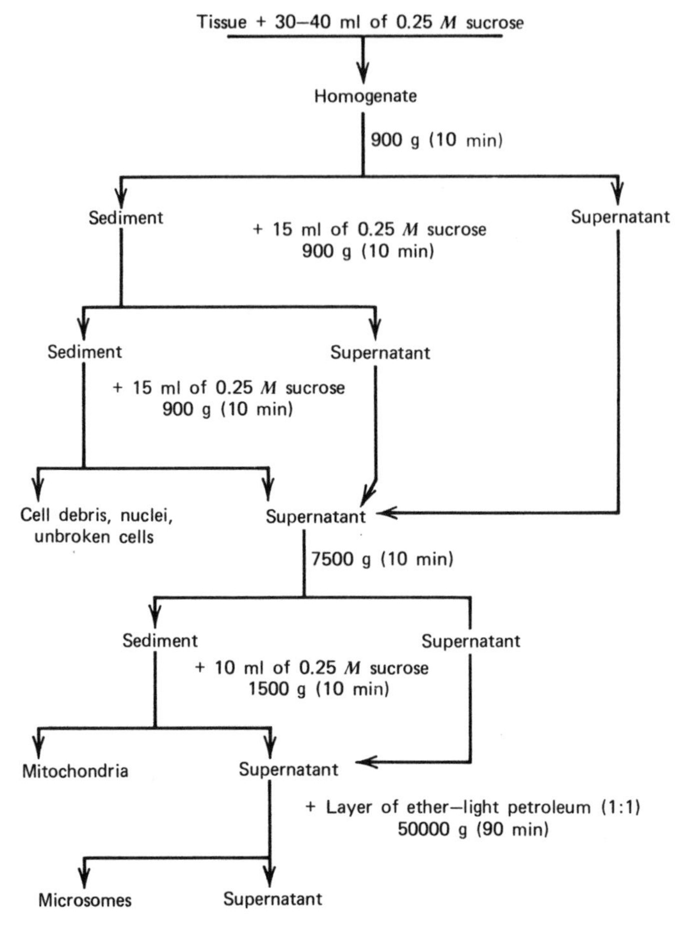

Fig. 2-3. A general outline of the protocol commonly used to fractionate tissue to obtain its membrane components. The pellet from 7500 or 50,000 *g* is generally used for density gradient centrifugation.

The fractions obtained by density-gradient centrifugation of homogenized cells are not homogeneous. These mixed populations arise from the presence of membranes of other cell types; heterogeneity in the size of the released vesicles; a difference in the fracture plane of the fragments; entrapment of other vesicles and cytosol; inversion of membranes during vesiculation to give right-side-out and inside-out vesicles; a complete or partial removal or exchange of some of the membrane components; and fusion of vesiculated membrane fragments. Contributions from these factors should not be underestimated, as illustrated by demonstrable differences in the preparation of as simple a membrane as erythrocyte ghosts (Kirk, 1968; Steck et al., 1970; Billah et al., 1976).

CHARACTERIZATION OF ISOLATED MEMBRANES

Obviously it is necessary to identify and quantitate each class of membrane that is obtained by a fractionation procedure in order to determine the topography (*inside-out, right-side-out*), and to establish the nature and the amount of the contaminants. Generally two types of techniques are used for characterization of isolated membranes: (i) electron microscopy and histochemical characterization provide morphological information about the possible origin of the membrane fragments; (ii) intrinsic markers like enzymes, receptors, and antigens are used to identify biochemical functions in a given membrane. Such functions are correlated ultimately with the source of the membrane. Thus, rigorous fractionation requires not only the enrichment of a specific marker or markers, but also a depletion of markers characteristic of the contaminants.

Morphology of Membrane Fragments

Whole cells examined by electron microscopy exhibit characteristic membrane-bound structures (Fig. 2-1b). Once cells are homogenized, much of the gross morphology is lost and the membrane fragments form vesicles that become more difficult to specifically identify. Organelles like nuclei, mitochondria, and lysosomes retain their gross morphology and can be identified by comparison with their morphology in intact cells. Rough endoplasmic reticulum forms vesicles, but can be recognized by associated ribosomes. Vesiculated Golgi membranes, smooth endoplasmic reticulum, and plasma membrane fragments are more difficult to distinguish from each other. Smooth endoplasmic reticulum forms featureless vesicles, while plasma membranes may possess characteristic features such as desmosomes, microvilli, or gap junctions. Features such as density of staining, thickness, asymmetry, attached structures, presence of microvilli, and the propensity for specific electron stains have been successfully employed by experienced investigators to identify the origin of a particular membrane preparation.

Histochemical methods of identifying membranes utilize staining of particular constituents. Thus, DNA can be demonstrated in the nucleus by use of Feulgen stain; acid phosphatase can be histochemically shown to be limited to lysosomes; and most carbohydrate-bearing constituents can be demonstrated to reside on the outer surface of the plasma membrane.

Techniques for identification of membrane fractions on the basis of their morphological or histochemical features are not always unambiguous; they cannot be easily quantitated, and as such these criteria can be applied only for those membranes for which adequate information already exists.

Markers for Membranes

Each subcellular organelle has a characteristic set of functions, which are often due to specific components. Many of the components responsible for these functions are localized in the membrane of an organelle. The presence of such characteristic enzymes (Table 2-2), binding and recognition sites, and chemical functional groups provides a means of distinguishing membrane fractions. Transport functions of vesicles can also be used for characterizing a membrane preparation and also the transporter (e.g., see Stevens et al., 1984). Ideally, for a constituent to serve as a marker, it must be confined only to one class of organelles and be present in every organelle of that class. The specificity of a potential marker can be judged on the basis of its group, site, or topographic accessibility. Group-specific markers (Table 2-3) interact with a single functional group or recognize a segment of a macromolecule. Site-specific markers are ligands directed toward specific functional groups on a membrane. Specific topographical markers provide information about membrane domains (hydrophobicity, mobility, polarity, asymmetry) and their relationships (separation, orientation, distribution, degree of penetration).

Among the various markers, enzymes are particularly popular; they are specific and are relatively easier to quantitate, even in the presence of other protein components. In principle, even closely related enzyme activities can be distinguished by the use of a difference in pH-rate profiles, or in the substrate and inhibition specificities. Morphological localization of some enzymes is also possible by histochemical methods. The marker enzyme levels should be considered both as relative specific activities and as absolute activities, since many potenital nonmembrane contaminants are often present in membrane preparations. The use of soluble enzymes as markers is restricted because it requires isolation of intact organelles. Determination of specific activities of membrane enzymes is particularly tricky. As discussed in subsequent chapters, the kinetic behavior of many of such enzymes is appreciably modulated by changes in the phase state of the bilayer. Moreover, some of the enzymes may be exchanged between different types of membranes, or the enzyme may not be uniformly distributed in different

TABLE 2-2 Some Enzymic Markers for Membranes and Organelles (based on the rat liver data)

Organelle/Fragment	Density (g/ml)	Size (μM)[a]	%P[d]	Marker Enzyme	#E[e]	Comments
Nuclei	1.32	5–10	15	NAD pyrophosphorylase DNA polymerase		DNA content for whole nuclei
Mitochondria (inner)	1.2	0.5–2	25	Cytochrome oxidase Succinate dehydrogenase	70	
Mitochondria (outer)				Monoamine oxidase Kynurenine hydroxylase		
Microsomes	1.15	3–20				RNA content for rough ER is not uniformly distributed
Endoplasmic reticulum	1.2	—	20	Glucose-6-phosphatase[b]	75	
Smooth				Cytochrome b_5 reductase Glucose-6-phosphatase Uridine diphosphatase		
Rough				Various esterases		
Sarcoplasmic reticulum	1.15	—	10	Ca-ATPase		
Brush border	1.15	—	5	Sucrase Enterokinase		
Plasma membrane	1.15	—	2	5′-Nucleotidase[b] Alkaline phosphatase[b,c] Adenylate cyclase Na,K-ATPase Aminopeptidase Hormone and viral receptors Phosphodiesterase	24	Low recovery; cholesterol content is relatively high (>20%) in plasma membrane
Synaptosomes	1.17	0.01	1	Acetylcholine receptor		
Lysosomes	1.25	0.5–0.8	2	Acid phosphatase β-Glucoronidase Aryl sulfatase Acid DNase	14	Must be disrupted before assay
Golgi bodies	1.14	1	2	Sugar Transferases Thiamine pyrophosphatase		Low recovery
Peroxisomes		0.5–0.8		Catalase, D-amino acid oxidase Urate oxidase		
Chloroplast		0.5–2		Ribulose-diP carboxykinase	>15	Chlorophyll as a marker
Cytosol		—	30	Lactate dehydrogenase	>30	

[a] For the longest axis.

[b] Occurs only in certain tissues.

[c] Can be located in other membranes.

[d] Percent of total protein present in the cell.

[e] Number of enzymes.

TABLE 2-3 Covalent Markers for Functional Groups

Reagent	Size (Å)	Remarks
For amino group		
2-Acetamido-4′-isothiocyanostilbene 2,2′-disulfonate (SITS)	>20 × 12	Permeable; also reacts with histidine and guanidine
Pyridoxal phosphate + $NaBT_3$	12 × 6	Impermeable
Isothionylacetamide	4 × 4	Impermeable; reacts with lipids and proteins in RBC
Methyl 4-azidobenzoimidate		Photosensitive bifunctional reagent; reacts with several other groups
1,5-Difluoro-2,4-dinitrobenzene	10 × 9	Cross-links groups 5 Å apart
2,4,6-Trinitrobenzene sulfonate (TNBS)	10 × 9	Impermeable
4,4′-Difluoro-3′,3′-dinitrophenylsulfone (FEDS)	15 × 10	Cross-links 9 Å apart
N-(3-Fluoro-4,6-dinitrophenylcysteamine (FDPC)	22 × 10	Cross-links 18 Å apart
For thiol group		
N-(3-Mercuri-5-methoxypropyl)poly-DL-alanylamide	>50 × 50	Impermeable
PCMB-aminoethyldextran	>50 × 50	Binds selectively to Na,K-ATPase in RBC
For tyrosyl residue		
Iodine + lactoperoxidase	>50 × 50	Impermeable, generally useful
For glutamyl residue		
Primary amine + transglutaminase		Impermeable, several types of probes can be attached
For alkyl groups		
Azidonaphathalene		Photoactive, lipid soluble
Azidophospholipids		Photoactive, cross-links vesicles
For serine and threonine		
ATP + phosphoprotein kinase		Quite a few membrane proteins are phosphorylated
For sugars		
$NaBT_3$ + galactose oxidase		Only terminal sugars are modified
CMP-NANA + sialyltransferase		Terminal sialyl residues are modified

regions of the membrane. Finally it should be emphasized that isolation of a membrane preparation containing its characteristic markers is no assurance that the isolated preparation is identical in every aspect to the intact membrane. Marker enzymes present in the various membrane preparations and organelles are listed in Table 2-2.

Cell membranes also contain very specific receptors and antigens. These can serve as high-affinity binding sites. Thus, hormones bind to their receptors, lectins to glycoconjugates, and antibodies to appropriate antigens. If the apparent dissociation constants for these complexes are low enough, then under appropriate conditions radiolabeled ligand can be bound before homogenization. Recovery of the membrane-bound label can be followed quantitatively during subsequent stages of fractionation and purification. Under most conditions a variety of binding assays can be used to determine dissociation constants of >0.01 m*M* to <0.01 n*M*. Similarly, ligands bound to an immobilized support can be used for purification of certain proteins. For example, affinity chromatography has been used to purify receptors for insulin, acetylcholine, toxins, and epinephrine, and also for retention of membrane fragments and whole cells on immobilized affinity ligands.

Labeling with covalent reagents for marking proteins can be very specific under certain conditions, as in affinity labels based on natural ligands and substrates. Ideally, a covalent marker or probe should be impermeable, react under physiological conditions, form stable covalent linkage with only the target component, cause a minimal perturbation of other membrane components, and be detectable in small amounts. Typically, covalent labeling involves tagging the membrane of intact cells or intact organelles with radioactive, fluorescent, or paramagnetic agents. Some of these covalent probes are summarized in Table 2-3. Most of the labels by their very nature modify their target. Sometimes such perturbations are minimal, and under optimal conditions only a functional state of the membrane is *frozen* in time. Impermeable probes localize only on the outer surface, while others cross membranes and localize on all possible membranes in a cell. The characteristics of probes vary according to their proposed use, such as by identification of the substrate binding sites, quantitation of the functional groups, or establishment of the arrangement of functional components. Thus, labels give information about the nature, number, distribution, and mobility of the binding sites. Many examples are given in subsequent chapters. Some of the more imaginative applications of covalent probes involve use of fluorescently labeled phospholipids to follow their transfer to different organelles by optical microscopy (Pagano and Sleight, 1985).

COMPONENTS OF BIOMEMBRANES

Isolated membranes are white, flocculent solids of specific gravity 1.05–1.35 g/ml. Thus, depending upon the fractionation procedure, it is possible to separate plasma membrane, microsomal membranes, smooth and rough ER, basolateral and brush-border membranes from appropriate tissues. Some membranes contain chromophores which impart their color; e.g., inner mitochondrial membranes are pale green from cytochromes, and membranes of retinal rod outer segment and *Halobacterium* are purple due to carotenoids bound to proteins. The density of isolated membranes is directly proportional to their protein content, and the protein content of membranes appears to increase with the number of metabolic functions carried out by the membrane. Thus the density of isolated membranes increases in the order (Table 2-2): myelin < plasma membrane < endoplasmic reticulum < mitochondria. Isolated nuclei are heavier because trapped DNA increases their density even though their membranes are lighter.

Separation into Components

Since membrane components are held together by noncovalent interactions, isolated membranes are readily disrupted by treatment with organic solvents. Typically, membranes treated with chloroform + methanol mixture (3 : 2 or 2 : 1 v/v) partition their lipids into a heavier organic phase, whereas most of the protein and nucleic acids remain in the aqueous phase. Carbohydrates, which are always covalently attached or conjugated to lipids or proteins, distribute into appropriate phases. Highly nonpolar proteolipids, which may contain covalently attached acyl chains, also partition into the heavier organic phase. However, denatured proteolipids do not dissolve in organic solvents when the solvent used for their initial extraction is removed by evaporation. Quantitative separation of lipids from the wet membranes requires extensive extraction–evaporation cycles. Separation of lipids according to their class is often accomplished with column chromatography on silica gel or carboxymethyl cellulose. Separation of individual lipid species according to their chain composition is possible only under a very limited set of conditions. Use of thin-layer chromatography plates impregnated with silver nitrate has been successful in a few cases. However, use of the capillary columns for gas chromatography coupled with mass spectrometry (Myher and Kuksis, 1982) and high-pressure liquid chromatography (HPLC) (Robins and Patton, 1986) has become routine now. Derivatization of phospholipids for gas chromatography is done by silylation of diacylglycerols obtained by treatment of phospholipids with phospholipase C. Although cumbersome, the assay for phosphate in completely oxidized phospholipids is one of the most common methods for determining total phospholipids (Mrsny et al., 1986). Special mass detectors for quantitation of phospholipids in effluents from HPLC offer great promise (Christie, 1985).

Phospholipids and Other Components

The composition of membranes isolated from certain sources is summarized in Table 2-4. Membranes contain lipids and proteins as well as their glyco-

TABLE 2-4 Composition of Membrane Preparations

Source	Average Density (g/ml)	Dry Weight (%)		Lipid Composition (lipid percent)									
		Lipid	Protein	Choles-terol	PC[a]	SM[b]	PE[c]	PI[d]	PS[e]	PG[f]	DPG[g]	PA[h]	Glyco-lipids
Rat liver													
Plasma	1.15	30–50	50–70	20	64		17	11		2			
Endoplasmic reticulum (rough)	1.20	15–30	60–80	6	55	3	16	8	3	—	—	—	
Endoplasmic reticulum (smooth)	1.15	60	40	10	55	12	21	6.7			1.9		
Mitochondria (inner)	1.19	20–25	70–80	<3	45	2.5	25	6	1	2	18	0.7	
Mitochondria (outer)	1.12	30–40	60–70	<5	50	5	23	13	2	2.5	3.5	1.3	
Nuclear	1.25	15–40	60–80	10	55	3	20	7	3	—	—	1	
Golgi	1.14	60	40	7.5	40	10	15	6	3.5	—	—	—	
Lysosomes	1.20	20–25	70–80	14	25	24	13	7	7	—	5	—	
Rat brain													
Myelin	1.06	60–70	20–30	22	11	6	14		7				
Synaptosome	1.17	50	50	20	24	3.5	20	2	8			1	21
Rat erythrocyte	1.2	40	60	24	31	8.5	15	2.2	7			<0.1	
Rat rod outer segment (ROS)	1.11	50	40	<3	41		37	2	13				3
Mycoplasma		20–30	70	0	—								
E. coli		20–30	70	0	—		80			15	5		
B. subtilis		20–30	70	0	0		30			12			
Chloroplast		35–50	50–65	0	4			1.5		6			55
Sindbis virus				0	26	18	35		20				

[a] Phosphatidylcholine.
[b] Sphingomyelin.
[c] Phosphatidyl ethanolamine.
[d] Phosphatidyl inositol.
[e] Phosphatidyl serine.
[f] Phosphatidyl glycerol.
[g] Diphosphatidyl glycerol.
[h] Phosphatidic acid.

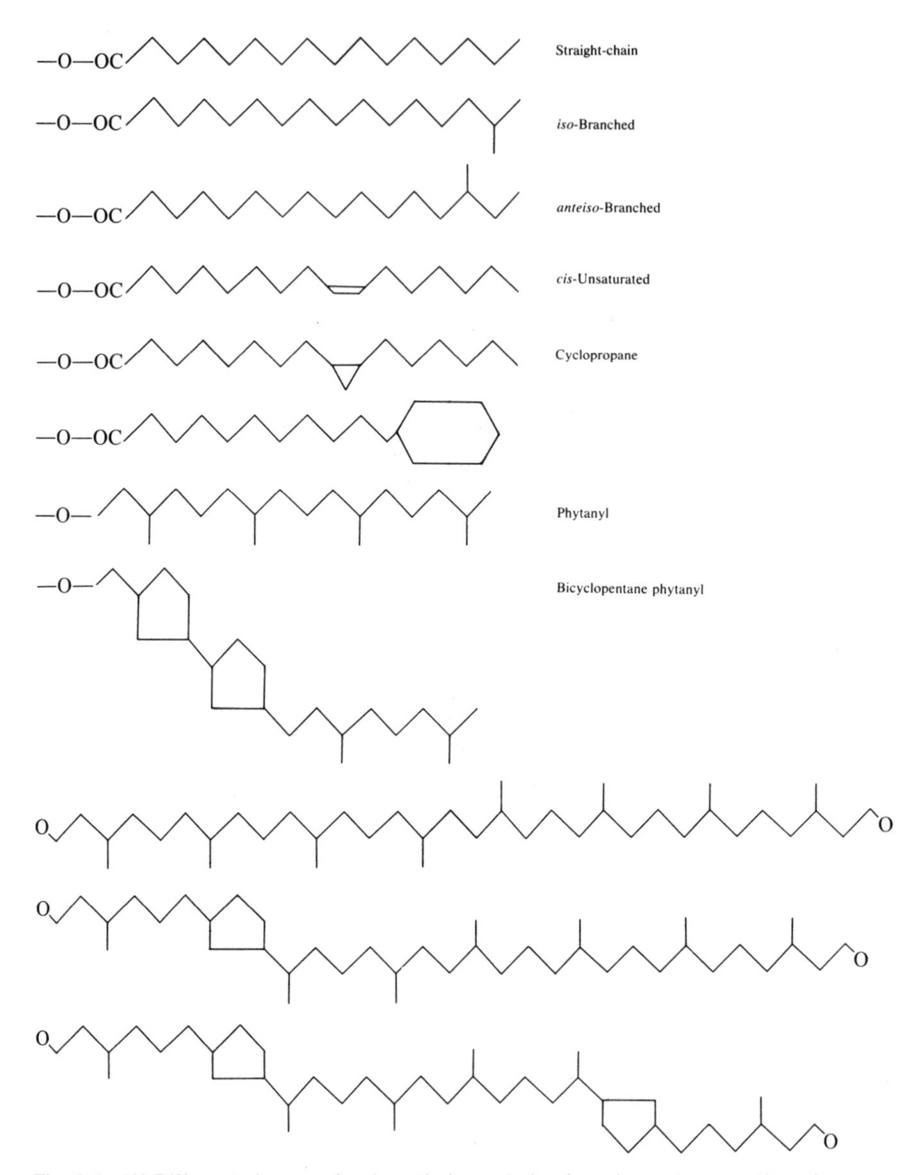

Fig. 2-4. (A) Different classes of polymethylene chains found as esters or ethers in membranes. Structures of *n*-fatty acids are shown in the adjoining table. The bipolar lipids shown in the middle are found in phospho- and glycoglycero-lipids of *Archaebacteria*. Hydroxy- and amino- (sphingosine) fatty acids shown in the bottom panel are found in many classes of lipids. (B) Common names and systematic names of straight-chain and derivative fatty acids that are found as membrane components.

Names of Straight Chain Fatty Acids

Common Name	Systematic Designation	Abbreviations
Saturated		
Lauric	Dodecanoic	12 : 0
Myristic	Tetradecanoic	14 : 0
Palmitic	Hexadecanoic	16 : 0
Stearic	Octadecanoic	18 : 0
Arachidic	Eicosanoic	20 : 0
Behenic	Docosanoic	22 : 0
Monoenoic		
Palmitoleic	9-Hexadecenoic	16 : 1Δ9
Oleic	9-Octadecenoic	18 : 1Δ9 or 18 : 1 (n − 9)
Vaccenic	11-Octadecenoic	18 : 1Δ11
Gadoleic	9-Eicosenoic	20 : 1Δ9
Petroselinic	6-Octadecenoic	18 : 1Δ6
Cetoleic	11-Docosenoic	22 : 1Δ11
Erucic	13-Docosenoic	22 : 1Δ13
Nervonic	15-Tetracosenoic	24 : 1Δ15
Elaidic	9-*trans*-Octadecenoic	18 : 1Δ9t
Dienoic		
Linoleic	9,12-Octadecadienoic	18 : 2Δ9,12 or 18 : 2 (n − 6)
Trienoic		
α-Linolenic	9,12,15-Octadecatrienoic	18 : 3Δ9,12,15 or 18 : 3 (n − 3)
γ-Linolenic	6,9,12-Octadecatrienoic	18 : 3Δ6,9,12 or 18 : 3 (n − 6)
Dihomo-γ-linolenic	8,11,14-Eicosatrienoic	20 : 3Δ8,11,14 or 20 : 3 (n − 6)
α-Eleostearic	9-*cis*,11-*trans*,13-*trans*-Octadecatrienoic	18 : 3Δ9*c*,11*t*,13*t*
Mead acid	5,8,11-Eicosatrienoic	20 : 3Δ5,8,11 or 20 : 3 (n − 9)
Tetraenoic		
Arachidonic	5,8,11,14-Eicosatetraenoic	20 : 4Δ5,8,11,14 or 20 : 4 (n − 6)
Parinaric	9,11,13,15-Octadecatetraenoic	18 : 4Δ9,11,13,15
Adrenic	7,10,13,16-Docosatetraenoic	22 : 4Δ7,10,13,16 or 22 : 4 (n − 6)
Pentaenoic		
Timnodonic	5,8,11,14,17-Eicosapentaenoic	20 : 5Δ5,8,11,14,17 or 20 : 5 (n − 3)
Docosapent-aenoic	4,7,10,13,16-Docosapentaenoic	22 : 5Δ4,7,10,13,16 or 22 : 5 (n − 6)
Clupanodonic	7,10,13,16,19-Docosapentaenoic	22 : 5Δ7,10,13,16,19 or 22 : 5 (n − 3)
Hexaenoic		
Cervonic	4,7,10,13,16,19-Docosahexaenoic	22 : 6Δ4,7,10,13,16,19 or 22 : 6 (n − 3)

Fig. 2-4. *(Continued)*

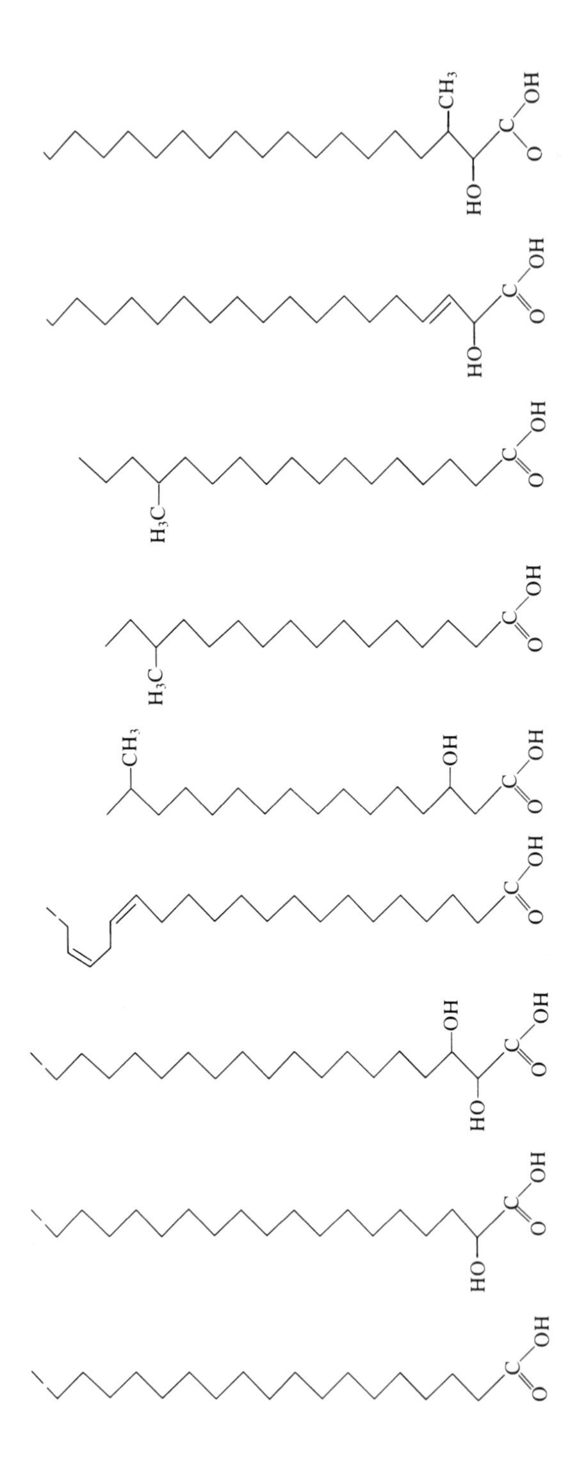

CH3
OH
HO
C
O
H3C

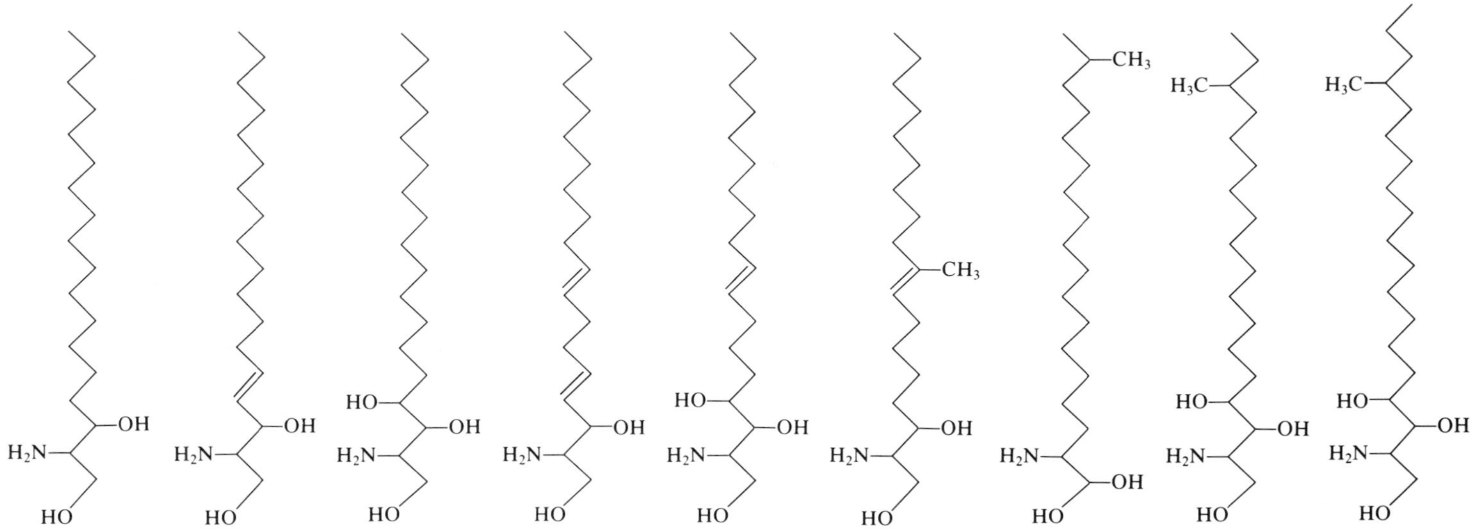

H_2N OH HO

H_2N OH HO

HO H_2N OH HO

H_2N OH HO

HO H_2N OH HO

CH_3 H_2N OH HO

CH_3 H_2N OH HO

H_3C HO H_2N OH HO

H_3C HO H_2N OH HO

Fig. 2-4. *(Continued)*

conjugates. The proportion of proteins in a membrane or organelle parallels their density as well as the level of the metabolic activity. Lipids are defined operationally as actual or potential derivatives of fatty acids and their metabolites. As a first approximation, lipids may be considered amphipathic molecules that contain nonpolar and polar regions. Typically, most of the lipid components present in an organism are found predominantly in membranes. However, some phospholipids are also found in serum lipoproteins and lung surfactants. Deposits of lipids that form only under certain pathological conditions are often rich in certain lipids. The possible number of structural and conformational isomers of lipids in membranes is very large, however, only a few classes of lipids predominate in membranes of a given class of organisms. Molecular characteristics of the various classes of lipids are described below.

The nonpolar portion of lipids in biomembranes is mostly made up of polymethylene chains (Fig. 2-4a, b). Free fatty acids rarely constitute more than 1% of the membrane; however, they are intermediates in the metabolism of eicosanoids leading to the synthesis of prostaglandins, thromboxanes, and leukotrienes (see Fig. 2-9). Similarly, acyl coenzyme A is an intermediate in the transacylation of phospholipids. Therefore, under certain conditions the concentrations of these metabolic intermediates in a membrane fluctuates considerably. It is also interesting to note that many medium-chain (8–12 methylenes) fatty acids exhibit antibiotic and antibacterial activity.

Covalently attached fatty acids form the bulk of membrane lipids as acyl or alkyl derivatives. The polar head groups to which these chains are attached belong to a few distinct classes (see below). It should be emphasized that the classes of lipids with the same polar group but different fatty acids are truly different molecules, with different metabolic and physical properties, i.e., patterns of biosynthesis, transport, degradation, and distribution. In most eukaryotes the fatty acids are the straight-chain type with an even number of carbon atoms and zero to six double bonds (Fig. 2-4). Most of the fatty acids present in membrane lipids are synthesized *in vivo*; however, some of the polyunsaturated fatty acids are required as dietary supplements in humans. Bacterial lipids also contain iso-, anteiso-, cyclopropane, ω-cyclohexyl acyl chains, as well as polymethylene and isoprenoid alkyl chains. Fatty acids with hydroxyl group in position 2 or 3 from the polar end have been found in bacterial membranes. The fatty acid profiles for membranes change considerably with a change in the age, metabolic state, diet, and other growth conditions. Similarly, the fatty acid composition of membranes under constant growth conditions is remarkably consistent and diagnostic. In fact, the identification of most aerobic and anaerobic bacteria can be accomplished with >95% certainty simply on the basis of fatty acid profiles obtained from the bacteria grown under standardized conditions (Moss, 1981; also done by Microbial ID, Newark, Del.).

Structures of the polymethylene chains found as major components in membranes of eucaryotes and prokaryotes are shown in Fig. 2-4. Most prevalent are the acyl derivatives of *n*-polymethylene chains with an even number of carbon atoms, typically 14–20, with a polar (carboxylic, hydroxylic, or amino) group on one end. Unsaturated acyl chains in membrane lipids contain double bonds in the 9-, 9,12-, 6,9,12-, and 5,8,11,14- positions, and invariably they are in the *cis* configuration. The double bond in the 2-position of plasmalogens is also *cis*. Unusual acyl chains are found in membranes that encounter unusual environments. Polyisoprenoids with polar groups on one or both ends are found in membranes of *Archaebacteria* which do not contain alkanoic acids (see below). Ether-linked bipolar lipids containing isoprenoid chains with methyl or cyclopentyl groups as branches apparently provide stability to bilayers containing such lipids. Lipids containing ω-cyclohexane fatty acids are found in *Bacillus acidocalderius*.

Sterols constitute the second most important class of nonpolar moiety playing a unique role in the architecture of the apolar region of certain types of biomembranes. The generalized structure of sterols found in membranes is shown in Fig. 2-5, including a variety of sterols that have been found in marine organisms (Carlson et al., 1980). These lipids have a planar steroid nucleus due to *trans* ring junctions, 3β-, 17β, and $20R$-configurations, and a long aliphatic chain with 6–10 carbon atoms. The structural changes that occur in sterols from different sources are usually in the side-chain or in the substitution of methyl groups on the β-face of the planar steroid nucleus. Importance of conservation of these structural features will become apparent in the subsequent chapters: Both sterols and fatty acids retain structural features that may be important for their incorporation into bilayers and for their interaction with other membrane components.

Membranes of prokaryotes do not contain sterols, although membranes of fungi and wall-less prokaryotes like *Methanococcus* and *Mycoplasma* can incorporate sterols if added to the growth medium. Some species of *Mycoplasma* also require sterols for growth. Eukaryotes contain a variety of sterols. Cholesterol is the sterol most commonly found in mammalian membranes. Characteristic sterols have been identified in other classes of organisms. For example, sitosterol is found in plants, diplopterol and tetrahymenol in *Tetrahymena* species, and dinosterol in marine organisms. Sterols like fucosterol, clondrillasterol, and porferasterol, along with ergosterol and cholesterol, are found in protists. Most sterols present in membranes are free, however, glycolipids derived from hopane have been isolated from

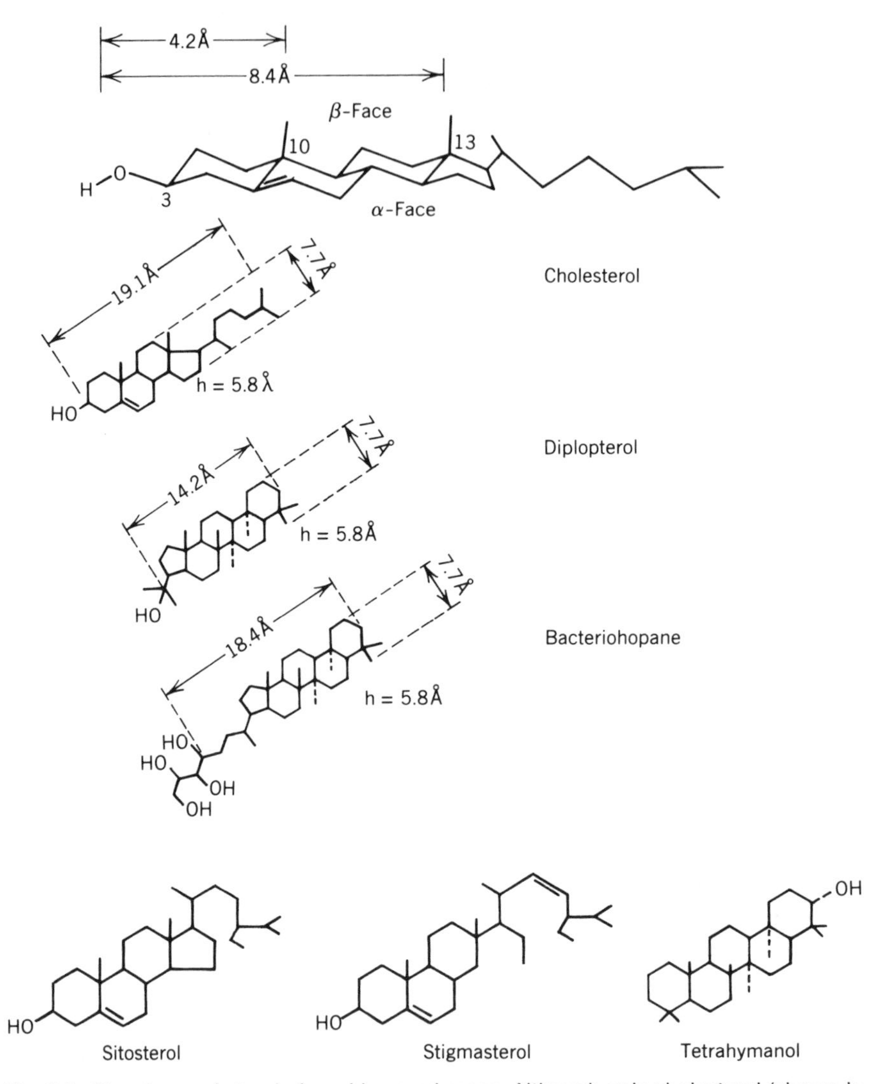

Fig. 2-5. Structures of sterols found in membranes. Although only cholesterol (shown in two different conventions) is found in animal plasma membranes, other sterols are found in membranes from plants (stigmasterol, sitosterol) and other organisms (tetrahymenol from *Tetrahymena*.

Bacillus acidocaldarius (Kannenberg et al., 1985). Acylated sterols present in serum lipoproteins are formed by enzymes present in membranes. Since acylated sterols are not amphipathic, their solubility in membranes would be very limited (less than 5%), as is the case for triglycerides (Hamilton, et al., 1983).

All lipids present in membranes also contain a polar region. In sterols it is an hydroxyl group on one end; however, as shown in Figs. 2-6, 2-7, and 2-8 and as discussed below, a variety of polar groups provide a molecular basis for the broad classification of membrane lipids.

Glycerophospholipids or 1,2-Diacyl-*sn*-glycerophospholipids are the most abundant phospholipids in most prokaryotic and eukaryotic membranes. The corresponding 2,3-diacyl derivative does not occur naturally except in Archaebacteria. As shown in Fig. 2-6, the second substituent on phosphate is usually an alcohol such as choline (in phosphatidyl choline, PC), ethanolamine (in phosphatidyl ethanolamine, PE), serine (in phosphatidyl serine, PS), glycerol (in phosphatidyl glycerol, PG), glycerol phosphate, threonine, and inositol (in phosphatidyl inositol, PI). PCs are a major class of lipids in animal membranes and they are absent in most (but not all) prokaryotic membranes. On the other hand, PE predominates in gram-negative bacteria and PG in mycoplasma. In humans, PC is also a critical component of lung surfactants, serum lipoproteins, and bile. Phosphatidic acid (PA) is only a minor component of membranes; however, it is produced by the action of phospholipase D (an enzyme found only in plant tissues) on substituted phospholipids. PA is a key intermediate in biosynthesis of phospholipids. In the PI cycle it is produced by phosphorylation of 1,2-diacyglycerols (cf. Chapter 14). Typically, the acyl chain in the *sn*-2 (*sn* refers to stereospecifically numbered) position of most glycerophospholipids is unsaturated, whereas the acyl chain in *sn*-1-position is saturated in most phospholipids. Hydrolysis of the phosphodiester bond to yield diacylglycerol is catalyzed by phospholipase C, whereas the hydrolysis of the ester linkage in the position *sn*-1 and the position *sn*-2 is catalyzed by phospholipase A1 and A2, respectively.

Phosphonolipids are found as major components in *Tetrahymena* (about 23%) and ciliary membranes, and in trace amounts in certain tissues (brain, heart) in humans. They occur extensively in molluscs, coelenterates, and protozoa. As shown in Fig. 2-6, they contain a phosphorus-to-carbon bond which can influence hydration and charge associated with the bilayer interface. Phosphonolipids in *Tetrahymena* also

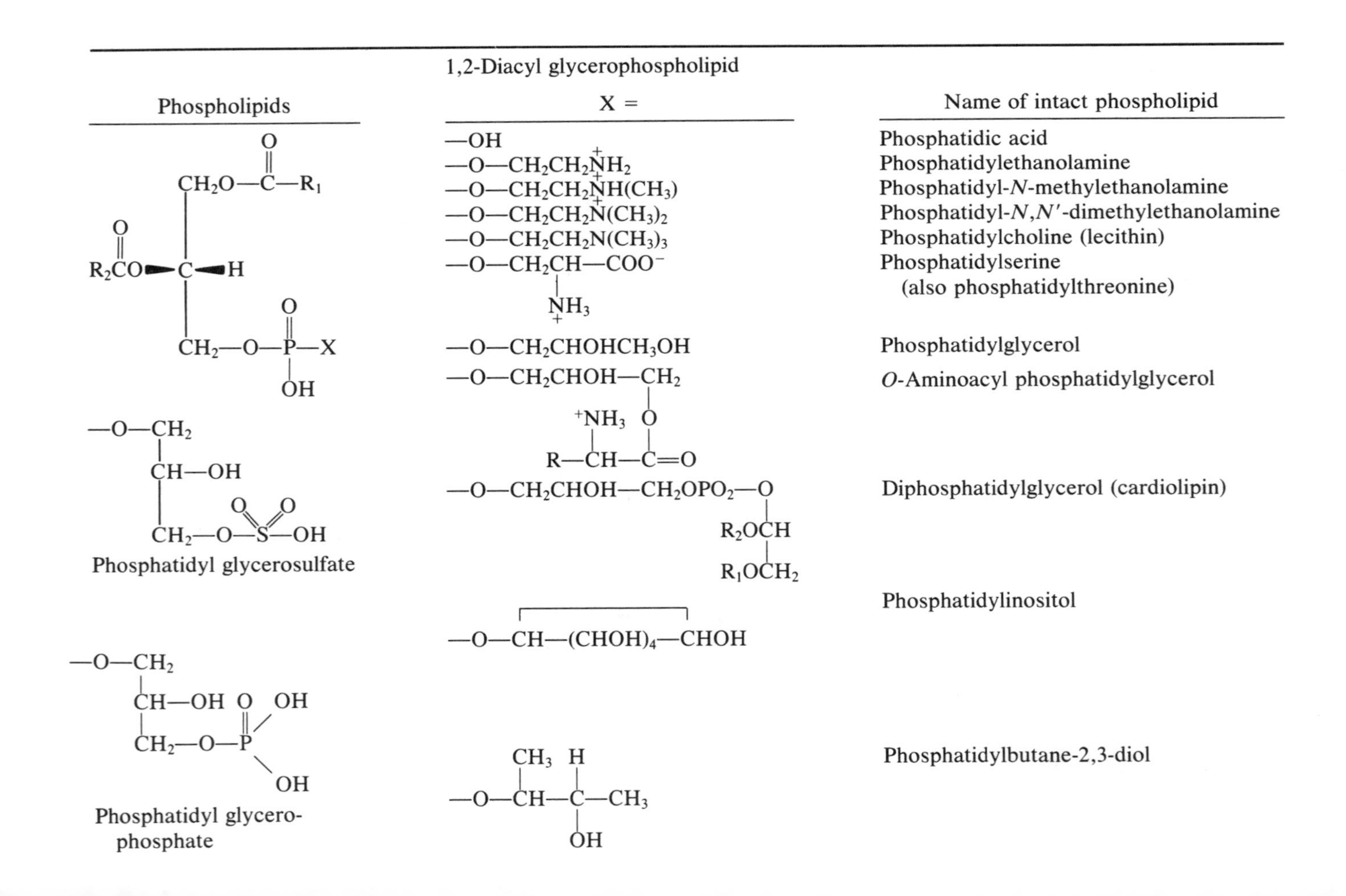

Phospholipids	1,2-Diacyl glycerophospholipid X =	Name of intact phospholipid
	—OH	Phosphatidic acid
	$—O—CH_2CH_2\overset{+}{N}H_2$	Phosphatidylethanolamine
	$—O—CH_2CH_2\overset{+}{N}H(CH_3)$	Phosphatidyl-*N*-methylethanolamine
	$—O—CH_2CH_2\overset{+}{N}(CH_3)_2$	Phosphatidyl-*N*,*N*′-dimethylethanolamine
	$—O—CH_2CH_2N(CH_3)_3$	Phosphatidylcholine (lecithin)
	$—O—CH_2CH(\overset{+}{N}H_3)—COO^-$	Phosphatidylserine (also phosphatidylthreonine)
	$—O—CH_2CHOHCH_3OH$	Phosphatidylglycerol
	$—O—CH_2CHOH—CH_2—O—C(=O)—CH(^+NH_3)—R$	*O*-Aminoacyl phosphatidylglycerol
Phosphatidyl glycerosulfate	$—O—CH_2CHOH—CH_2OPO_2—O—CH(R_2O)—CH_2OR_1$	Diphosphatidylglycerol (cardiolipin)
	$—O—CH—(CHOH)_4—CHOH$ (ring)	Phosphatidylinositol
Phosphatidyl glycerophosphate	$—O—CH(CH_3)—C(H)(OH)—CH_3$	Phosphatidylbutane-2,3-diol

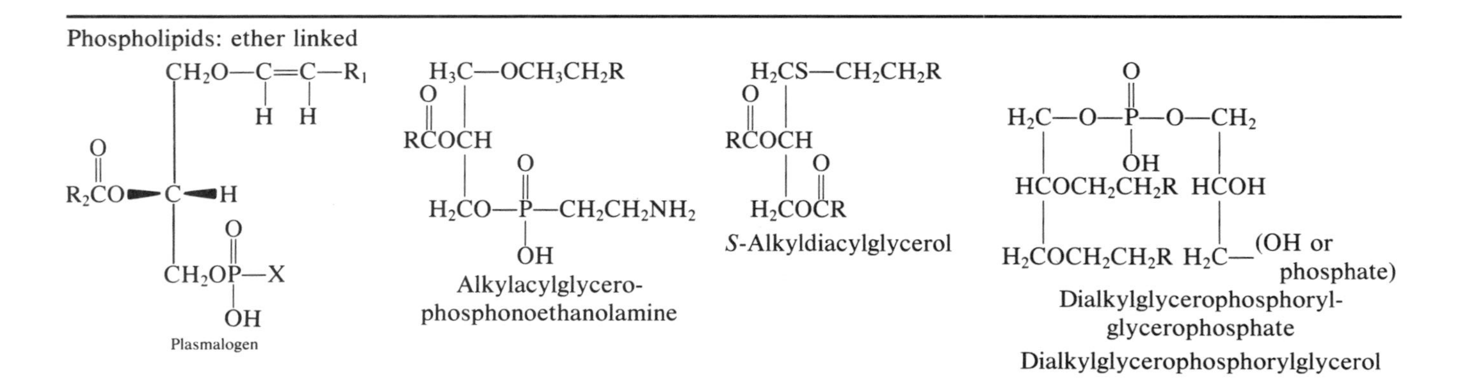

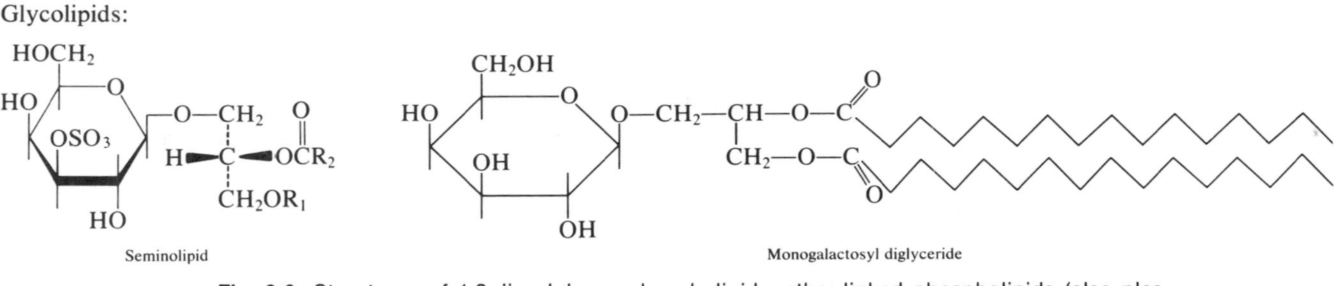

Fig. 2-6. Structures of 1,2-diacylglycerophospholipids, ether-linked phospholipids (also plasmalogen), glycoglycerolipids, and minor lipids. See Figs. 2-8 and 8-2 for the structures of other glycolipids.

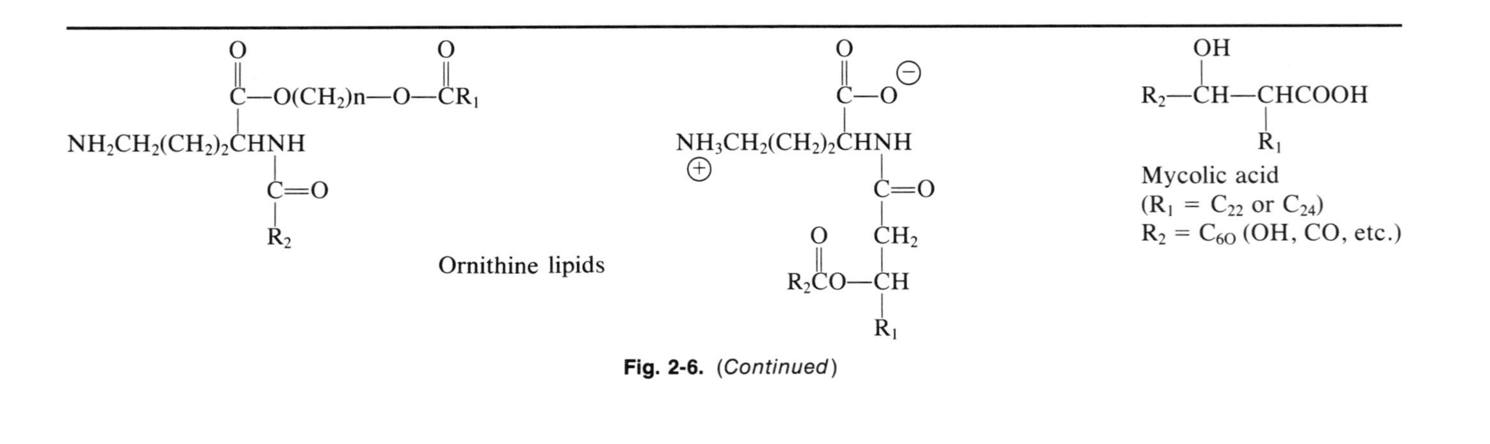

Fig. 2-6. *(Continued)*

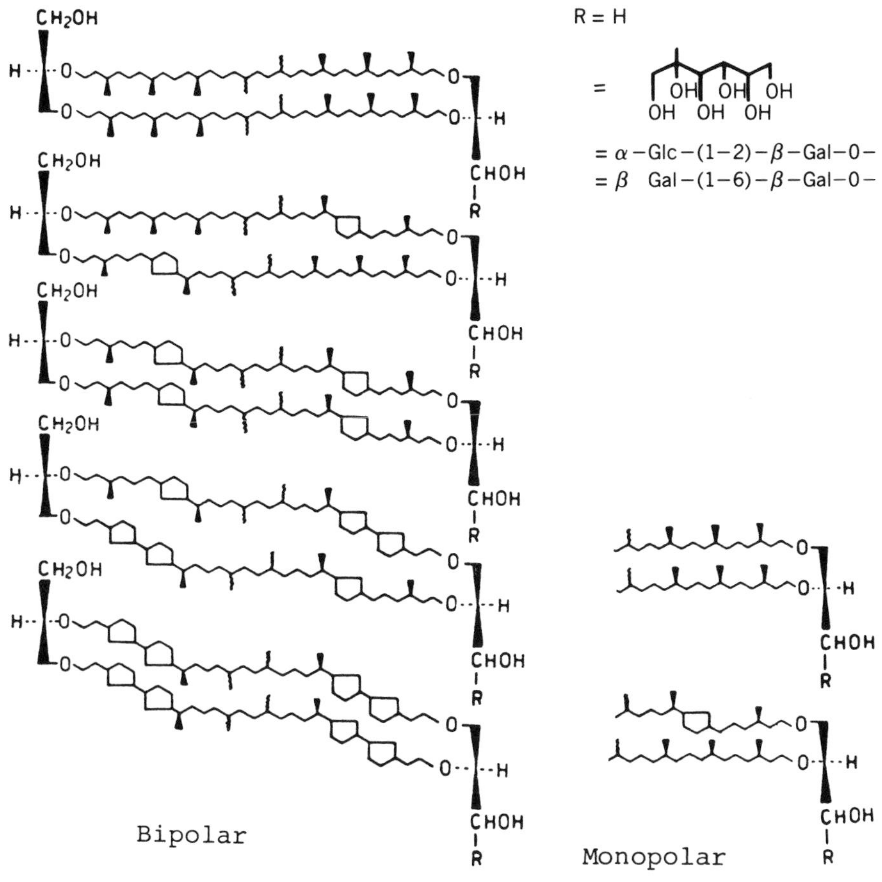

Fig. 2-7. Structures of monopolar and dipolar lipids with isoprenyl ether chains. These are major components of membranes from *Archaebacteria*. Note that the stereochemistry of the glycerol backbone is *sn*-2,3-dialkyl- rather than 1,2-dialkyl.

contain an ether linkage in the *sn*-1 position. Thus, these lipids are remarkably insensitive to chemical and enzymatic hydrolysis.

Plasmalogen or 1-*O*-alk-1′-enyl,2-*O*-acyl phosphoglycerides (Fig. 2-6) are minor but metabolically important components of mammalian membranes. The aldehyde in position *sn*-1 is usually saturated and the acyl group in position 2 has double bonds. Other phosphodiester substituents are typically choline in muscle, and ethanolamine or serine in brain tissues.

Glycolipids are glycosides of 1,2-diacylglycerols (Fig. 2-6) that are widely distributed in bacteria and plants, but extremely rare in animals. As they form more than 50% of the total polar lipids in plants, they are

Structure	Name
$CH_3(CH_2)_{14}—CH(OH)—CH(NH_2)—CH_2(OH)$	Sphinganine (Dihydrosphingosine)
$CH_3(CH_2)_{13}—CH(OH)—CH(OH)—CH(NH_2)—CH_2(OH)$	4-D-Hydroxysphinganine (Phytosphingosine)
$CH_3(CH_2)_{14}—CH{=}CH—CH(OH)—CH(NH_2)—CH_2(OH)$	Eicosasphingenine (D-*erythro*-2-Amino-4-*trans*-eicosene-1,3-diol)

$$CH_3—N(CH_3)_2—CH_2—CH_2—O—P(=O)(O^-)—O—CH_2—CH(NH—C(=O)—R)—CH(OH)—CH{=}CH—R'$$

Sphingomyelin

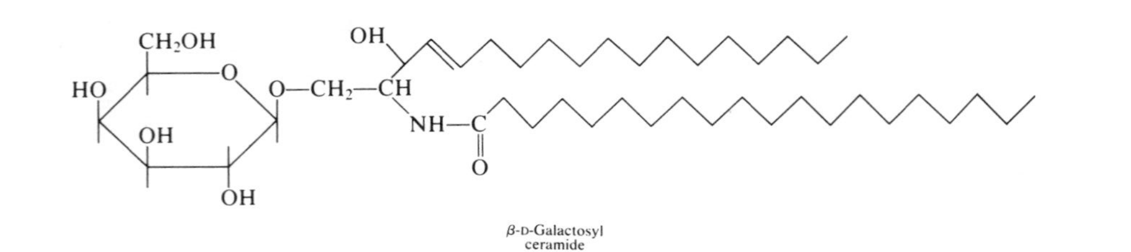

β-D-Galactosyl ceramide

Sphingosine: $CH_3 \cdot (CH_2)_{12} \cdot CH : CH \cdot CH(OH) \cdot CH(NH_2) \cdot CH_2OH$
Ceramide: $CH_3 \cdot (CH_2)_{12} \cdot CH : CH \cdot CH(OH) \cdot CH(NH) \cdot CH_3OH$
$CH_3 \cdot (CH_2)_{16\text{–}22} \cdot C : O$
Glucocerebroside: ceramide-glucose
Ceramide lactoside: ceramide-glucose-galactose
Globoside: ceramide-glucose-galactose-galactose-*N*-acetylgalactosamine
Hematoside: ceramide-glucose-galactose-*N*-acetylneuraminic acid
Ganglioside: ceramide-glucose-galactose-*N*-acetylgalactosamine-galactose
N-acetylneuraminic acid
Sulfatide: ceramide-galactose-3-sulfate
Sphingomyelin: ceramide-phosphorylcholine
Ceramide trihexoside: ceramide-glucose-galactose-galactose
Galactocerebroside: ceramide-galactose
Tay-Sachs ganglioside: ceramide-glucose-galactose-*N*-acetylgalactosamine
N-acetylneuraminic acid

Fig. 2-8. Structures of sphingolipids derived from ceramide. See Fig. 8-2 for the structures of more complex gangliolipids. Sphingosine bases of different structures are shown in Fig. 2-4A.

probably the most abundant lipids in the biosphere. Sulfoquinovoside diglyceride is a minor (7%) component in plant membranes. Seminolipid, 1-*O*-palmitoyl-2-*O*-palmityl- 3-β-D-(3′-sulfo) galactopyranosyl-*sn*-glycerol (Fig. 2-6) is abundant in the mammalian testis and in spermatozoa. Monogalactosyl diacylglycerol (MGDG) is a major (about 50%) component of chloroplast thylakoids and oxygen-evolving photosynthetic organisms. The major lipid of the radiation-resistant bacterium *Deinocossus radiodurans* is 2′-*O*-(1,2-diacyl *sn*-glycerol-3-phospho-3′-*O*-(α-galactosyl) *N*-D-glycerol heptadecylamine (Anderson and Hanson, 1985).

Diphosphatidyl glycerol (DPG) or cardiolipins are tetraacyl phospholipids (Fig. 2-6) that are the major components of mitochondrial and bacterial membranes. This attests to their common evolutionary origin.

Sphingolipids are derived from spingosine obtained by amidination of ceramide (Fig. 2-8). In sphingomyelin (SM), sphingosine is esterified to cholinephosphate (Fig. 2-6). SM is usually found in mammalian plasma membranes and is a major component of nerve membranes. Hydrolysis of the amide bond in SM is catalyzed by sphingomyelinase. Sphingolipids derived from dihydro-, phyto-, and eicosasphinganine (Fig. 2-4a) are found in plant membranes.

Gangliosides and glycospingolipids are glycoconjugates derived from conjugation of ceramide with carbohydrate residues (Fig. 2-8). Ceramide is a generic term for *N*-acyl (fatty acid) derivatives of sphinganine. The length of the acyl chain may vary. For example, the most common ceramides in the central nervous system are based on fatty acids containing 24 carbon atoms like lignoceric, cerebronic (2-hydroxy-), and neproscic (*cis*-Δ^{15}) acids. These glycolipids are classified according to the nature of the carbohydrate backbone. For example, ceramide is conjugated to galactosides in cerebrosides, to amino sugars in globosides, and to sialic acids at terminal positions in gangliosides. Detailed structures of glycolipids are given in Chapter 8. These complex lipids are minor but important components of animal membranes, where they serve recognition and antigenic functions.

Minor lipid components in membranes have also been reported. Some of these are metabolic intermediates, such as arachidonate, prostaglandins, thromboxanes, leukotrienes, and lipoxins from eicosanoid pathways (Fig. 2-9). PA and diacylglycerols are intermediates in the biosynthesis of phospholipids (Fig. 2-10) and they are also metabolites in polyphosphoinositol cycle (Chapter 14). Membranes of organelles, like chromaffin granules, contain significant amounts (up to 17%) of ly-

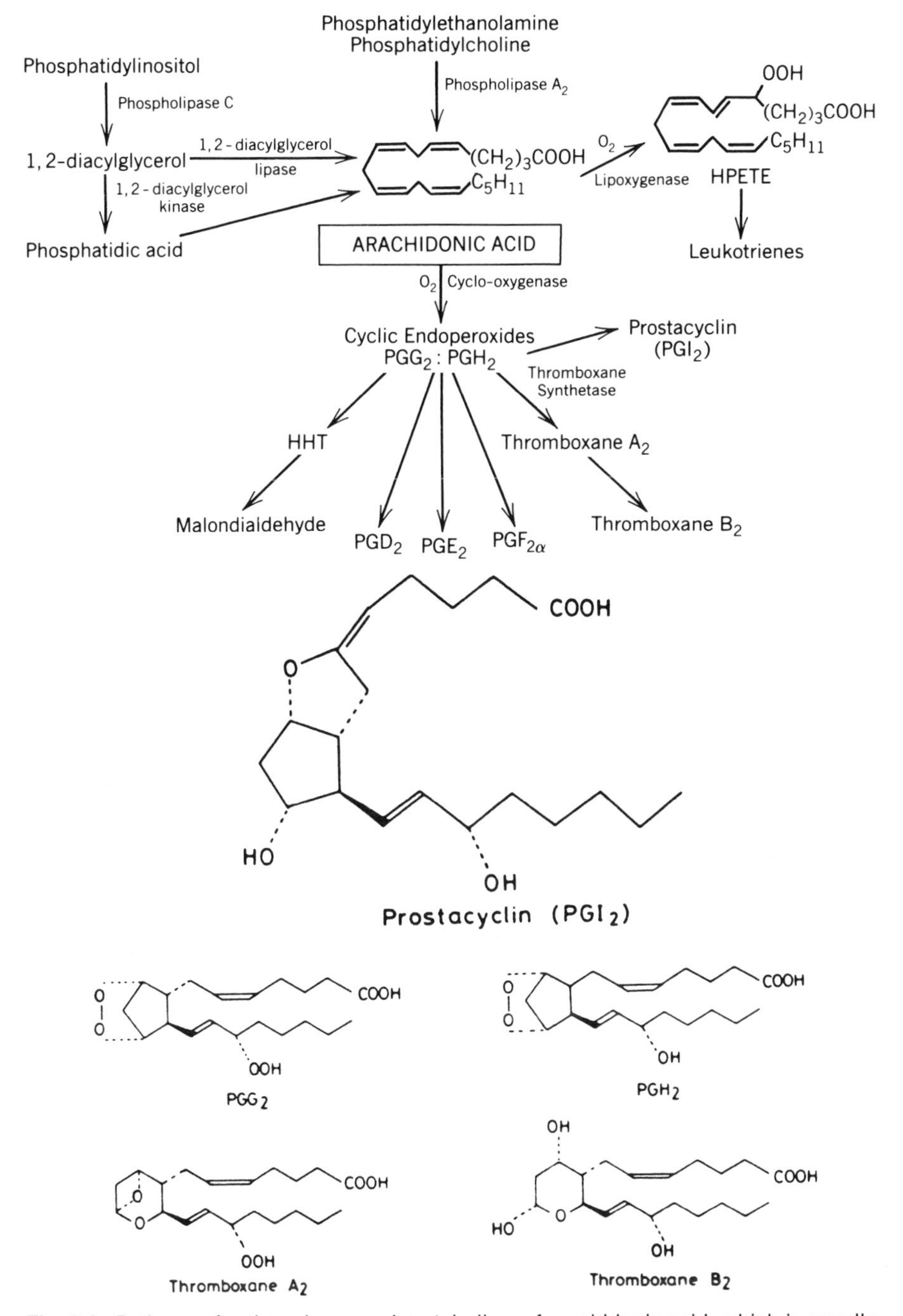

Fig. 2-9. Pathways for the release and metabolism of arachidonic acid, which is usually present as an *sn*-2- ester in phospholipids from higher animals. It is released by the action of phospholipase A2 on PC, PE, or PA, however, a possible role of diglyceride lipase has also been suggested. Normally, the concentration of free arachidonate in cells is negligible, therefore its release from phospholipids is the rate-limiting step in the synthesis of prostaglandins, thromboxanes, prostacyclin, and leukotrienes.

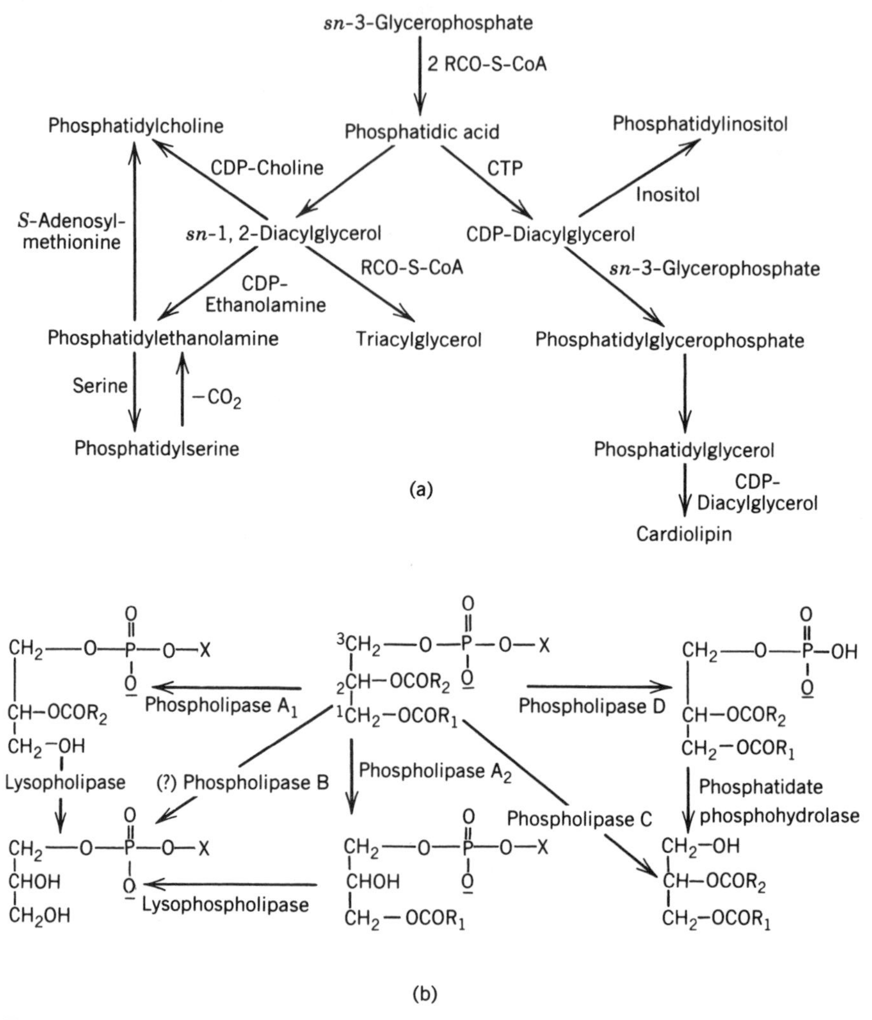

Fig. 2-10. (a) Pathways for the synthesis of major phospholipids. (b) The position specificity for the site of action of phospholipases on phospholipids. (c) Hydrolytic enzymes that degrade different classes of lipids from membranes.

sophospholipids and fatty acids; diol lipids based on ethylene glycol have been found in a variety of animals, plants, and microorganisms (Vaver et al., 1977); metabolically dormant cysts of *Azotobacter vinelandii* have been shown to contain 5-*n*-alkylresorcinols and 6-alkylpurones (Reusch and Sadoff, 1983). Platelet activating factor (PAF; 1-alkyl,2-acetyl-PC) has been shown to induce aggregation of platelets and neutrophils at concentrations <10 n*M* and it is a potent vasode-

Lipid Class	Enzyme
Phospholipids	α-Glycerophosphate dehydrogenase
	Diglyceride kinase
	Phospholipase A_1
	Phospholipase A_2
	Phospholipase C
	Phospholipase D
	Phosphatide acylhydrolase
Digalactose diglyceride	α-Galactosidase
Plasmalogen	Plasmalogenase
Cerebroside	Cerebroside galactosidase
Ceramide	Ceramidase
Sphingolipids	Neuraminidase
	β-N-Acetylhexosaminidase
	β-Galactosidase
	β-Glucosidase
	Sphingomyelinase
	Sialidase

(c)

pressor (Hanahan, 1986). 2-Methyl analogs of PAF have also been found to inhibit the growth of tumor cell lines. These findings add yet another dimension to the physiological role of minor ether phospholipids (Mangold and Paltauf, 1983). Di- and triacyl bis-phosphatidic acids (Fig. 2-6) have been found in lysosomes (Longmuir, 1985), where they are synthesized from PG and DPG. Their precise function is not known. Lipid A (cf. Chapter 8) from gram-negative bacteria show endotoxicity and related functions. Cord factor may have a similar function. There is also a large variety of nonextractable lipids that are removed from tissues and cells only by hydrolysis. For example, gram-negative bacteria have acyl chains in lipopolysaccharides; gram-positives have lipoteichoic acids; *Micrococcus lysodeikticus* have acylated mannan; *Mycobacteria*, *Corynebacteria*, *Nocardia*, and *Actinomycetes* have wax D (Goldfine, 1983). Post-translational acylation of proteins in eukaryotic cells has also been observed in many cases (see Chapter 7).

Unusual membrane lipids (Fig. 2-4, 2-7) occur as major components in membranes of *Archaebacteria* (methanogens, halophiles, thermophiles) which encounter and thrive under unusual conditions of pH, temperature, high salt concentration, and absence of oxygen. Most organisms, particularly *poikilotherms*, alter their acyl chain composition, and much less often their polar head groups, in response to the changing environment. On the other hand, membranes of *Sulfolobus sulfataricus*, which grows at 85°C at pH 2, is rich in polyisoprenoids containing 0–4 cyclopentane rings. Most of the lipids of *Archaebacteria* are based on the ether linkage obtained by condensation of glycerol or more complex polyols (nonitol) with two polyisoprenoid alcohols with 20, 25, or 40 carbon atoms (Fig. 2-7). Polyisoprenoids with 40 carbon chains are found attached to polar head groups on both ends, i.e., they are bipolar. Although most *Archaebacteria* contain only polymethylene or only polyisoprenoid chains, the proportion of mono- and bipolar lipids depends upon the growth conditions. Also the proportion of cyclopentane rings increases with increasing temperature (DeRosa et al., 1986). Halobacteria, which grow at high salt (above 2 *M*) concentrations, contain polyisoprenoid (C_{20}) chains linked via an ether bond to glycerol backbone with sugar residues as polar substituents. Psychrophils, which grow at lower temperatures, contain shorter and more unsaturated chains. On the other hand, methanogenic bacteria and *Thermoplasma acidophilum*, which optimally grow at higher temperatures (50–85°C), contain a bilayer of tetraether glucosylglycerylphosphoryl derivatives with two polyisoprenoid chains with 40 carbon atoms (Blocher et al., 1985). Membranes of *Thermomicrobium roseum*, which grow at 75°C, contain about 85% stearic acid and the head group consists of alkyl-1,3-propanediol (Pond et al., 1986). The functional significance of the structural variation in these lipids can be appreciated in terms of their stability (ether linkage) and bipolarity, lower melting points due to branching, and the charged groups that could regulate the surface charge and hydration profiles of the bilayer.

Specific functions of lipids in biomembranes are implied by their characteristic patterns of distribution. Many of these functions may be appreciated in terms of the properties of the membrane in specific environments, and not on the basis of the structure or the reactivity of the components *per se*. The phylogenic and taxonomic correlations are found in distribution of specific phospholipids in organelles. This pattern changes with age, diet, growth environment, and the pathological conditions. Also most organisms do not indiscriminantly accept and incorporate phospholipids from the growth medium. The presence or a lack of a specific class of lipids may also be viewed to have functional significance. For example, a lack of PC in most bacteria may be related to their interaction with hosts. Bacterial sulfatides of mycobacteria aid in the intracellular growth of the pathogenic strains by interfering with the formation of phagolysosomes.

The diverse classes of lipids found as membrane components have several features in common. They are amphipathic, i.e., the polar and apolar regions

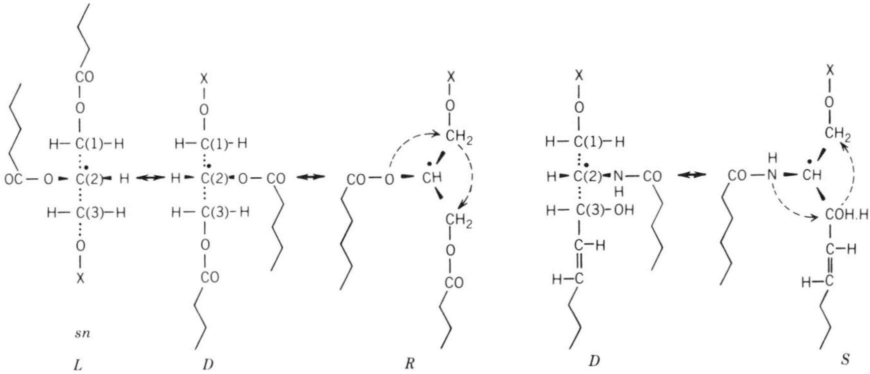

sn

L *D* *R* *D* *S*

Fig. 2-11. Stereochemistry at *sn*-2 (*sn* for stereospecifically numbered) of glycero- and sphingo-phospholipids. Different convensions (R/S; D/L; *sn*) have been used to describe the asymmetry or chirality at this position. Most naturally occurring glycero-lipids are 1,2-diacyl-*sn*-, however the lipids from *Archaebacteria* are 2,3-dialkyl-*sn*-.

of the molecule are separate (Fig. 2-11). Except for sterols, all the major classes of lipids in membranes have three features (Fig. 2-12) in common: (i) the hydrophobic region contains two polymethylene chains with one or more double bonds with branching toward the methyl end; (ii) the backbone is made up of glycerol, propanediol, glycol, and carbohydrates like nonitol, ceramide, or ornithine; (iii) a hydrophilic group incorporating phosphate,

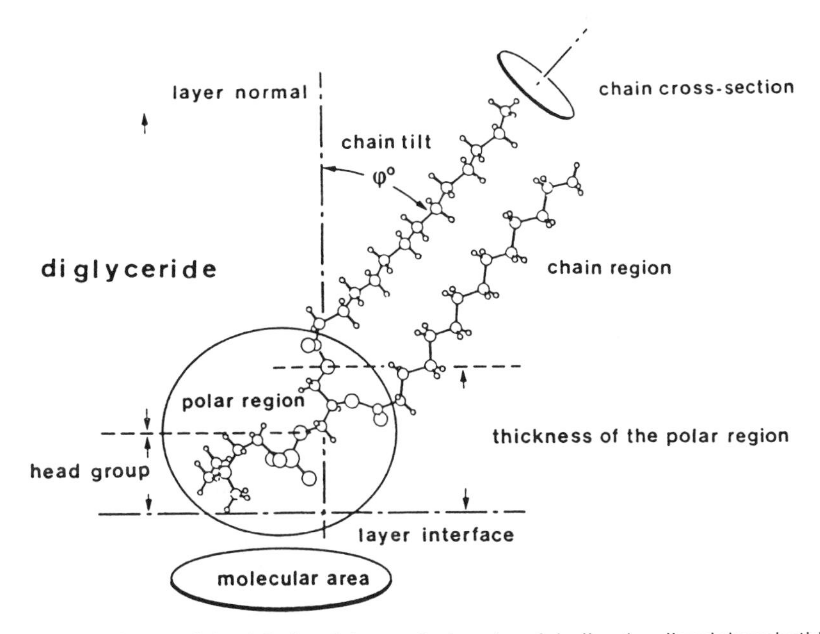

Fig. 2-12. Anatomy of the 1,2-diacylglycero-3-phosphorylcholine (or diacylphosphatidylcholine, PC) molecule to define its structural regions. (From Hauser, 1985.)

amino acids, sugar, and amines. The variability in the sterol content and the acyl chain composition, is designed to augment environmental effects and to restore appropriate membrane properties. This is in accord with the observed variability in the lipid composition of cells under a variety of growth conditions, and also by the phospholipid substitutions in intact cells. Thus, the survival time, clearance time, shape change, and viability of cells are appreciably altered even when only 0.5–1% of the total phospholipid present in their membranes is substituted by different molecular species (Op den Kamp et al., 1985).

CONFORMATION OF PHOSPHOLIPIDS

Phospholipids are amphipathic molecules in which structurally different regions can be distinguished. The polar regions comprise the glycerophosphocholine moiety and the two carboxyl ester groups. The apolar chains extend away from the plane of the head group. Detailed information on molecular conformation and packing of a variety of synthetic phospholipids in single crystals has been obtained by X-ray analysis (Table 2-5) and also in solution by spectroscopic methods. The results are reasonably consistent (Hauser et al., 1981). Phospholipids pack preferentially in bilayer form in the crystalline anhydrous state as well as in aqueous dispersions (Chapter 3). As shown in Fig. 2-13, the hydrocarbon chains of phospholipids in crystals are stacked

TABLE 2-5 X-Ray Crystal Structure Data on Phospholipids

Phospholipid[a] (Ref.)	Features
Dilauroyl PE (Elder et al., 1977)	PN dipole is layer parallel to the surface), ON (2.6-2.7A), area 36–40 Å^2
Dilauroyl *N,N*-dimethyl PE (Pascher and Sundell, 1986)	PN dipole is layer perpendicular to the surface area 45 Å^2, O-N 4Å. Stacked bilayers, 33° tilt, orthorhombic
Dimyristoyl PC (Pearson and Pascher, 1979)	PN dipole is layer parallel to the surface O-N (4 Å), area 51 Å^2, P is hydrated
Dimyristoyl PA, Dilauroyl PA (Harlos et al., 1984)	Glycerol is parallel to the interface
L-2-Deoxy-phosphatidyl choline (Hauser et al., 1980)	O...N (2.6–2.7 Å), glycerol backbone is perpendicular; interdigitated chains
Cerebroside (Sundell, 1977)	Layer perpendicular glycerol
Dilauroyl-Glycerol (Pascher et al., 1981)	Layer parallel glycerol
1-octadecyl 2-methyl-PC (Pascher et al., 1986)	Stacked bilayers with interdigitated and tilted chains. PC dipols are layer parallel, S = 74 Å^2

[a] PE, phosphatidylethanolamine; PC, phosphatidylcholine; PA, phosphatidic acid.

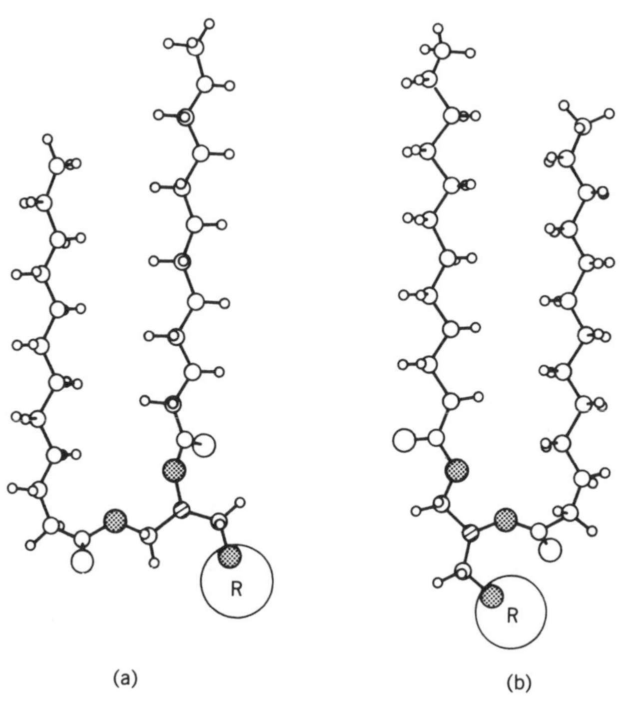

Fig. 2-13. Preferred molecular conformations of dilauroyl phosphatidic acid (a) and dimyristoyl phosphatidylcholine (b) showing the difference in glycerol orientation and chain stacking. In zwitterionic phospholipids, there is an approximate 90° bend at the phosphodiester linkage, such that the orientation of the base substituent is parallel to the bilayer plane (not shown here, however see Fig. 2-14). Note that despite the different orientation of the glycerol backbone (hatched atoms) the molecular conformation of the glycerol-dicarboxylester group is practically identical. R = head group substituent. See Table 2-5 for appropriate references.

parallel to each other, and are oriented perpendicular to the plane of the bilayer. Typically, the head group is oriented parallel to the plane of the bilayer, and the chains are not equivalent. In such a packing, the projected area of the head group and that of the chains at the interface must match so as to achieve a cylindrical shape that is necessary for packing in a planar bilayer (see Chapter 3).

The molecular conformations of diacyl- as well as monoacyl-PC are strikingly similar (Fig. 2-13). The glycerol group is oriented approximately perpendicular (except in PA, where it is parallel) to the bilayer plane forming a continuous zig-zag with the *sn*-1-fatty acyl chain. The initial part of the *sn*-2-chain is at a right angle to the glycerol group but makes a 90° bend at C_2 to become aligned parallel to the *sn*-1 chain. This conformation displaces the methyl end of the *sn*-2 chain by three methylene groups. The various conformations of the glycerol backbone are the result of rotational motions and axial translations of the hydrocarbon chain. Nuclear magnetic resonance (NMR) data suggest that the layer parallel and layer perpendicular conformers of glycerol exist in a 1 : 2 ratio in monomeric solutions, however, the layer perpendicular conformation predominates in an aggregated state.

Alignment of the zwitterionic head groups is phospholipids is determined by electrostatic (for PC) and H-bonding interactions (for PE) between the N-H and P=O moieties. A transition from layer parallel ($DLPEMe_2$) to layer perpendicular (DMPC or DLPE) orientation of the glycerol backbone can be achieved by rotating the head group about the C_1-C_2 glycerol bond or by an inversion of all α-torsion angles of the head group chain. In mono- and dilauroyl-PE, N is within 0.26–0.27 nm of P=O, and these bonds appear to have H-bonding character. The quaternary N to O distance in DMPC is about 0.4 nm, and phosphate is hydrated. This implies that electrostatic interactions are important determinants of the head group conformation. In crystals of dimyristoyl-PC, two water molecules located between the PC molecules form a hydrogen-bonded phosphate-water-phosphate ribbon. Two other water molecules are H-bonded to one lipid phosphate and to one other water molecule. This H-bonded network of the head group and water molecules provides stabilization of the interface.

In PCs restrictions in the head group conformation are imposed by the tendency of the zwitterionic head groups to optimize their intra- and intermolecular interactions, as well as by steric factors that determine conformations about the P-O ester bonds. The phosphodiester group has a double-*gauche* conformation that gives rise to a 90° kink at the phosphate group. As a result, the head group lies parallel to the surface. However the relative orientation of diacylglycerol moiety is different in PA and PE (Fig. 2-13). Thus, the inclination of the P-N dipole toward the layer plane ranges from 7 to 27°.

The quaternary nitrogen is inclined toward the hydrophobic part in dilauroyl PE. Two conformations of the phosphocholine moiety are seen in dimyristoyl PC, which arise from different torsion angles about the C_1-C_2 bond of glycerol. This implies that there are no rotational restrictions about this bond.

Hydrocarbon chains in the condensed crystalline state pack with a cross-sectional area of 0.185 to 0.21 nm^2. The exact value depends on the chain length and the temperature. The molecular cross section of DLPE and DMPC is about 0.388 nm^2, which is very close to the sum of the area of two acyl chains in orthorhombic packing. This suggests that the chains are almost perpendicular to the layer plane. In less ordered states the molecular areas of hexagonally packed chains (pair) expands to 0.41 nm^2. In compounds with relatively large head groups and only one acyl chain like, lyso-PC (about 0.52 nm^2) the molecules pack in the bilayer with extremely tilted hydrocarbon chains (about 36°) to establish a close-packing contact. These observations suggest that the packing behavior of phospholipids is determined by the cross-sectional area of their acyl chains (see Chapter 3).

In the disordered fluid state of bilayers, the phospholipid molecules retain an average orientation perpendicular to the layer surface. Rapid rotational transitions occur along the long axis of the polymethylene chain (Fig. 2-14). Due to coupled *trans-gauche-trans* isomerizations, the chains remain extended and essentially parallel to each other. The average cross-sectional area of such conformationally disordered chains is about 0.25 nm^2, which

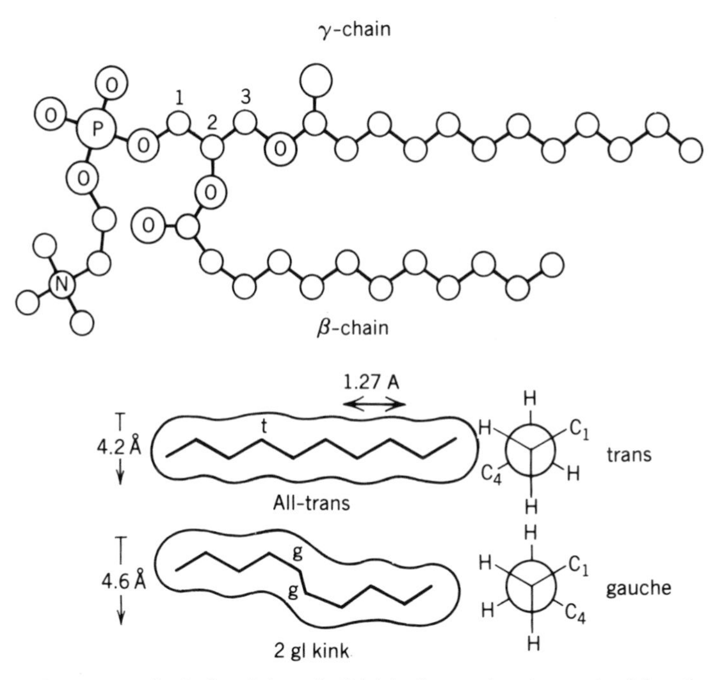

Fig. 2-14. Structure of 1,2-diacylphosphatidylcholine molecule emphasizing the relative orientation and conformation of the head group and of the acyl chain. The lower part of the figure illustrates an alkyl chain in an all-*trans* and part-*gauche* conformations. The Newman projections for the *trans* and *gauche* conformations are also shown. The dihedral angle (C_1-C_4) is 180° for the *trans* conformation and 120° or 240° for the *gauche* conformation. The 2*gl* kink is formed by rotating about one C—C bond by an angle of 120° and then rotating about either of the two next nearest neighboring bonds by −120°.

corresponds to the molecular area of 0.5 nm^2 for diacyl phospholipids. The effective chain length is reduced by about 8%. To accomodate the increased molecular area the head group is also expanded by hydration.

SYNTHESIS OF PHOSPHOLIPIDS

A combination of synthetic and enzymatic methods has been used for total or partial synthesis of phospholipids (Stepanov and Shvets, 1979; Eibl, 1980; Eibl and Woolley, 1986) and glycolipids (Schmidt, 1986). Naturally occurring fatty acids have been used extensively for acylation reactions. However, fatty acids with unusual chain length (Mena and Djerassi, 1985), branching (Nuhn et al., 1986), double bonds (Barton and Gunstone, 1975), spin label (Watts et al., 1978), fluorescent label (Somerharju et al., 1985), and isotopic labels (DasGupta et al., 1982) have been used for the synthesis of phospholipids.

Glycerol-*sn*-3-phosphocholine (GPC) prepared by deacylation of egg PC by methanolic KOH or tetra-*t*-butylammonium hydroxide (Chadha, 1970) is commonly used for synthesis of most other phospholipids. Typically, anhydrides of fatty acids are used for esterification of GPC in the presence of 4-(dimethylamino)pyridine in chloroform or tetrahydrofuran in anhydrous inert atmosphere at room temperature while the mixture is sonicated (Gupta et al., 1977). 1,2-Diacyl-PCs have been used for the synthesis of other lipids by the use of enzymatic methods.

Coupling of phosphatidic acid with choline (as tetraphenylborate salt) can be accomplished employing 2,4,6-triisopropylbenzene sulfonyl chloride as the condensing agent (Harbison and Griffin, 1984). Transphosphatidylation of PC with appropriate primary alcohols by phospholipase D is used for the synthesis of PE, PG, PS, and other analogs (Comfurius and Zwaal, 1977; Mena and Djerassi, 1985). Similarly, phospholipase A2 (crude snake venoms) is used for deacylation at position *sn*-2. The resulting lysophospholipid can be reacylated to obtain the appropriate asymmetrical diacylphospholipid. Under the acylation conditions described above transmigration of the acyl chains is $<3\%$.

Complete synthesis of a variety of phospholipids and their structural analogs has been achieved (Eibl, 1980). DPG, bisphosphatidic acid, phosphatidylcholesterol, PC, and several other phosphatidyl alkanols have been synthesized by the reaction of appropriate alcohols with diacylglyceryl cyclic endiol phosphates (Ramirez and Marecek, 1985), or with diacylglycerylphosphorodichloridate (Eibl, 1980). Thiophospholipids chiral at phosphate (Jiang et al., 1984) have also been synthesized.

TABLE 2-6 Lipid Storage Diseases and Enzymic Defects

Disease	Enzyme Defect	Lipid Stored	Biopsy Source	Comments
GM_1 gangliosidosis	Ganglioside-β-galactosidase	GM_1 ganglioside	Rectum Appendix	Acute infantile onset Mental retardation; short survival
GM_2 gangliosidoses				
Tay Sach's disease	Hexosaminidase A	GM_2 ganglioside	Rectum	Mental retardation and blindness
Sandhoff's disease	Hexosaminidase A + B	GM_2 ganglioside	Appendix	Similar but rapidly progressive
Batten's disease	?	Ceroid-lipofuscin	Rectum Appendix	Short survival Therapy: restrict Vitamin A
Farber's lipogranulomatosis	Acid ceramidase	Ceramide	Skin	Fatal during second year of life
Fabry's disease	Ceramide trihexosidase and α-galactosidase	Ceramide trihexoside	Skin Kidney	Purple skin rash; renal failure; burning pains in extremities
Refsun's syndrome	Phytanic acid hydrolase	Phytanic acid	Liver	Therapy: restrict chlorophyll
Wolman's disease	Cholesterol esterase	Cholesterol esters and triglycerides	Liver Lymph node	Fatal during first year
Cholesterol ester storage disease	Cholesterol esterase	Cholesterol esters and triglycerides	Liver Lymph node	More benign
Tangier disease	Impaired synthesis of apoprotein A	Cholesterol esters	Tonsils	Follows a fairly benign course
Niemann–Pick's disease	Sphingomyelinase	Sphingomyelin and cholesterol	Bone marrow Liver; rectum	Heptatosplenomegaly and mental retardation
Gaucher's disease	Glucocerebrosidase	Glucocerebroside	Bone marrow Liver	Hepatosplenomegaly; mental retardation in infantile form only
Krabbe's leukodystrophy	Galactocerebroside-β-galactosidase	Galactocerebroside	Brain	Multinucleate "globoid" cells around vessels: mental retardation; fatal by 2 years
Metachromatic leukodystrophy	Arylsulfatase A	Sulphatide	Urine deposit Sural nerve	Loss of myelin; mental retardation; fatal during first decade
Fucosidosis	α-Fucosidase	H-isoantigen		Cerebral degeneration
Mucolipidosis II (I-cell disease)	Acid neuraminidase			
Ceramide lipidosis	β-Galactosidase	Cer-lactoside	Liver Spleen	Brain damage

Degradation of Phospholipids

Catabolism of phospholipids is extremely complex because phospholipid molecules in membranes undergo exchange between various compartments, as well as a variety of acyl- or base-exchange reactions occur (see Fig. 2-10). Lipid degradation and acyl-transfer reactions occur not only in lysosomes, but also by phospholipases associated with plasma membrane, Golgi, and microsomes. There are many interesting features of lipid metabolism. For example, many genetic defects associated with degradation of sphingolipids in lysosomes have been identified (Table 2-6). However, defects attributed to lysosomal phospholipase deficiency or those associated with metabolism of glycerophospholipids have not been reported, probably because such defects are lethal. Phospholipids are not only the structural components of membrane but they are also source of certain fatty acids for further metabolism (for example, see Fig. 2-9), for transfer of acyl chains by lecithin–cholesterol acyl transferase, and exchange of base from the head group as by choline transferase, methyl transferase, and PS-decarboxylase.

The general biochemical principles underlying metabolism of phospholipids and sterols are outside the scope of this book. See, however, Vance and Vance, 1985.

3 Self-Association of Phospholipids

It takes a great deal of Christianity to wipe out uncivilized Eastern instincts such as falling in love at first sight.

RUDYARD KIPLING
Plain Tales from the Hills

The attractive interaction between nonpolar molecules in water is referred to as the hydrophobic effect. It is the major driving force not only for the self-association of amphipaths but also for folding and aggregation of proteins, subunit interactions leading to the formation of oligomeric proteins and viral particles, stacking of bases in polynucleotides, and the peculiar thermodynamic properties of liquids and solutions. Thus, the consequences of the hydrophobic effects in the organization of molecules into larger aggregates are far reaching (Kauzman, 1959; Tanford, 1980; Small, 1985; Israelachvilli, 1985; Cevc and Marsh, 1987).

Hydrophobic effect arises from the molecular and physical characteristics of water (Table 3-1). Hydrophobic substances cannot interact favorably with water via ionic, polar, or hydrogen-bonding interactions (for an introductory text, see Noggle, 1985). Amphipathic molecules aggregate in water so as to maximize appropriate interactions, i.e., the interactions that maximize orientational and configurational entropy. Interaction of acyl chains with water as well as with the polar groups is, for example, minimized in the aggregates (Figs. 3-1 and 3-2). Two factors are at work.

1. The net free energy change for the removal of a methylene group from water to apolar environment depends upon the nature of the apolar environment, but it is always negative and about 600–850 cal/mole. The gain in the free energy arising from the hydrophobic effect increases linearly with the chain length. These values suggest that the

TABLE 3-1 Properties of Liquid Water at 20°C

Property	Value	Units
Viscosity (η)	1.00	Centipoise
Self-diffusion coefficient (D)	2.1	cm^2/sec
Molecular dipole moment	1.84	Debye
Dielectric constant (ε)	80.1	
Dielectric relaxation time	9.5	psec
Lifetime of single H_3O^+ ion	~1	psec
O—H bond length	0.957	Å
H—O—H bond angle	104.52	Degrees
Average nearest neighbor (O—O distance)	2.85	Å
Concentration of pure liquid	55.34	M
Volume per molecule	30.0	$Å^3$

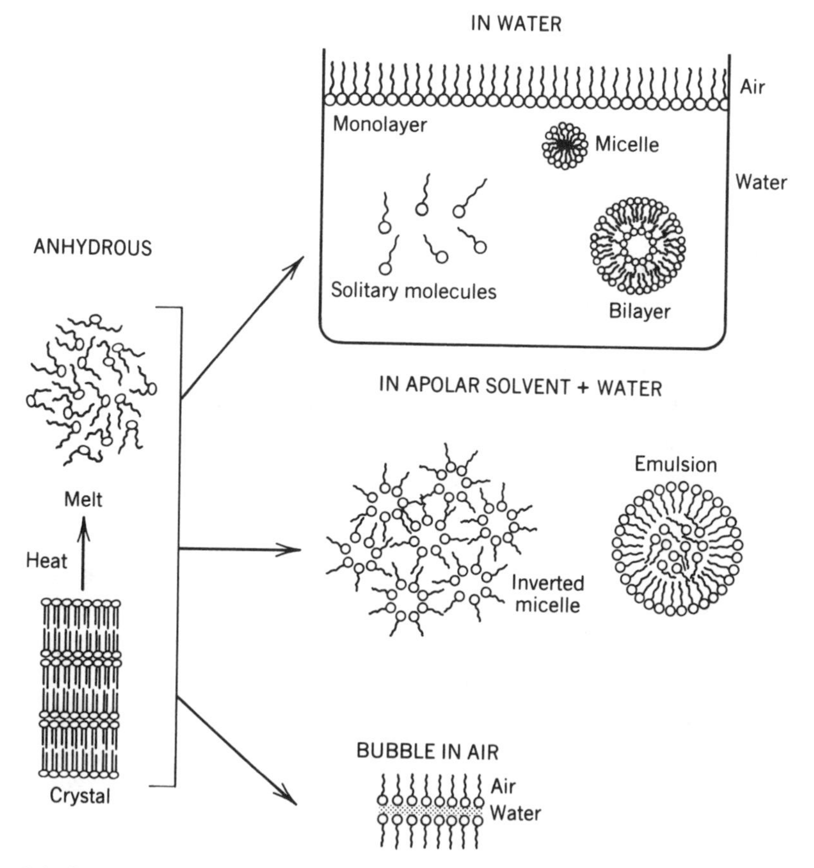

Fig. 3-1. Organized structures in dispersions of lipids in water. In the absence of any solvent, when solid lipids are heated the crystal lattice is disrupted and a neat liquid phase is formed. In the aqueous phase, solitary monomeric molecules may coexist with monolayer, micelle, or bilayer phases. In nonpolar solvents containing water, inverted micelles or emulsions are formed. The lipid–water film separating air (as in a soap bubble) ultimately thins down to an inverted bilayer that separates two compartments of air.

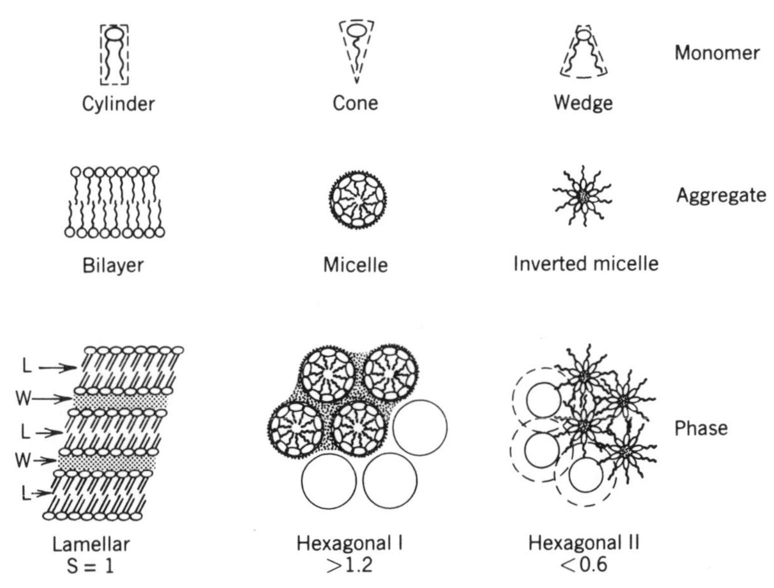

Fig. 3-2. To a first approximation the "generalized" shape (defined as S = volume/(area of the polar head group × length) of the phospholipid molecule determines the shape of the aggregate and the long-range order (or the phase characteristics) in its aqueous dispersions. Cylindrical molecules organize into bilayers, cone-shaped molecules into micelles, and the wedge-shaped molecules into inverted micelles and hexagonal rods.

head-group repulsion, branching, extent of chain–chain interactions, and interfacial area influence the configurational entropy of methylene groups.

2. The electrostatic energy change for the removal of a charged group from water to an apolar environment is given by the Born equation. The difference in electrostatic free energy required to transfer ions of radius r Å and charge q from vacuum ($\epsilon_v = 1$) to a solvent of dielectric constant ϵ_s is

$$\Delta F = -\frac{q^2}{2r}\,[(1/\epsilon_v) - (1/\epsilon_s)] = -\frac{332}{2r}\,[1 - (1/\epsilon_s)] \text{ kcal/mole}$$

For most polar solvents $\epsilon_s >> 1$ ($\epsilon_s = 80$ for water). Note that ΔF will be the same for the transfer of ions of the same radius. For the transfer of a point charge from the medium of dielectric constant 80 (water) to 2 (hydrocarbon), the overall change in the free energy is about 15 kcal/mole or 9.10^{-11} ergs. ($q = 5 \times 10^{-10}$ esu, $kT = 4.1 \times 10^{-14}$ ergs, and $NkT = 0.6$ kcal/mole at 25°C.)

Thus, the overall gain in free energy for the transfer of a phospholipid molecule from water or from a hydrocarbon medium to a polar–nonpolar interface is well over −15 kcals/mole.

The phenomenological basis of hydrophobic effect leading to the formation of aggregates of amphipaths is usually sought in the observation that the hydrocarbons are only slightly soluble in water, and the polar solutes are not soluble in hydrocarbons (Langmuir, 1917). This differential solubility results from an unfavorable (negative) standard entropy of solution, presumably due to structuring of the water by the solute. The nature of the molecular interactions giving rise to the hydrophobic effect is beginning to emerge in the case of simpler solutions (for a theoretical basis, see Pratt, 1985). Model calculations suggest that the first hydration shell around smaller hydrocarbons dissolved in water retains the structure of labile clatharate hydrates without losing any hydrogen bonds, yet permitting solvent exchange into the clatharate structure on a picosecond scale. Water next to large hydrophobic surfaces loses about one hydrogen bond per molecule, and the total energetic contribution can add up to a substantial change in the orientational and configurational entropy. The van der Waals type of attractive dispersion interactions between hydrophobic solutes will also contribute to the formation of aggregates. However, the extent of such contributions remains to be established in most cases. Since few H bonds are actually broken during hydrophobic interactions, the main contribution to the free energies comes from packing, configurational, and orientational factors. The entropic interactions are significantly longer range than any typical bonding interactions, but they decay exponentially with distance. Also the attractive interactions due to hydrophobic effect would become stronger at higher temperature, as the contribution of the $T\Delta S$ term would make the overall free energy more negative.

POLYMORPHISM OF LIPID–WATER AGGREGATES

Considerations of geometrical factors, molecular organization and motion, and conformational transitions in an aggregate provide a basis for understanding the shape and size of the aggregates that are formed by dispersing amphipaths in aqueous solutions (Israelachvilli et al., 1980; Israelachvilli, 1985). Amphipathic molecules aggregate in the aqueous phase to form a variety of phases, including monolayers, micelles, hexagonal, cubic, and lamellar phases (Fig. 3-2). Which phase predominates in a given system depends on the structure and conformation of the amphipath, the composition and water content of the mixture, and environmental factors such as ionic strength, pH, and temperature (Luzatti and Tardieu, 1974). As shown in Fig. 3-1, the geometrical organization of aggregates differs considerably. This difference in the structural order gives rise to unique macroscopic phase and molecular properties such as conductivity, osmotic pressure, mobility, birefringence, X-ray diffraction characteristics arising from the structural order and orientation, and the spectral characteristics arising from the degrees of freedom that give rise to segmental and molecular motions. Thus, the physical basis of these differences is to be found not only in the overall shape and size of the aggregate but also in the orientational and motional parameters of the component molecules.

The generalized shape of an amphipathic molecule determines the macroscopic organization (Fig. 3-2) in aqueous dispersions (Israelachvilli et al., 1980). Thus, not only the primary structure and conformation but also the state of ionization and hydration contribute to the overall shape and size of a molecule in an organized phase. The packing parameter in an aggregate can be defined as:

$$S = \frac{v}{a \cdot l}$$

where v is the hydrophobic volume of the amphipath, a is the area per head group at the interface, and l is the length of the hydrocarbon chain. For molecules in a monolayer or a bilayer organization $S = 1$. A smaller head group area ($S > 1$) leads to hexagonal II phase, whereas a larger head group area ($S < 1$) would promote micellar organization in H_I phase. Characterization of aggregates and phases of amphipaths requires the use of many techniques; each provides a unique window of advantages and pitfalls. No attempt has been made in this book to introduce the use and relative merits of the various techniques. Some reviews that may be useful for beginners are given in Table 3-2, and for an advanced discussion it is probably worthwhile to follow through with specific articles cited in the text and to cultivate an appreciation of intrinsic limitations of these methods.

Two of the most successful methods for characterizing lipid–water dispersions are X-ray and neutron diffraction. Phases with long-range order exhibit characteristic small-angle X-ray diffraction patterns ($S = 2 \sin \Theta/\lambda < 10\ A^{-1}$). The lamellar phases, for example, exhibit long spacings of the first and higher orders in the ratio 1 : 1/2 : 1/3. The micellar phase is characterized by much simpler diffraction pattern. The small-angle reflections ($s > 5\ A^{-1}$) also provide information about the symmetry and dimensions of the lattice, including molecular parameters such as area per chain. Quantitative interpretation of this data requires consideration of symmetry,

TABLE 3-2 Physical Techniques Used for the Study of Lipid–Water Dispersions and Membranes

Technique	Comments
Fourier transform IR	CH_2 bending or scissoring mode (1470 cm^{-1}); CH_2 rocking (1150–700); C=O stretching (*sn*-1 1742, *sn*-2 1725); P=O stretching (1250 antisymmetric; 1085 symmetric); P—O stretching (900–800). (Casel and Mantsch, 1984).
Resonance Raman spectroscopy	Trans- and gauche-conformers give characteristic peaks. (Wong, 1984).
Nuclear magnetic resonance	
1H	Line position and intensity; shift reagents to discern signal from the outer monolayer of vesicles; not very useful for biomembranes or large vesicles.
2H	Chemical shift anisotropy, quadrupole splitting, spin-lattice relaxation provide information about chain order of specifically labeled residues. Quadrupolar splittings for C-D in different positions depends on the local environment as well as on the surface charge. Results can be interpreted to provide information about orientation as well as the amplitude of the motion (Davis, 1983; Smith and Oldfield, 1984; Meier et al., 1986).
13C	Line width and chemical shift provide information about relaxation time and order parameters on natural abundant or specifically labeled residues (Griffin, 1981).
14N	Quadrupole coupling constant is sensitive to solute-induced perturbation of the head-group region of bilayers, preferably in oriented samples (Rothgeb and Oldfield, 1981).
19F	Labeling of lipids and protein has provided information about order parameter and motional freedom. Labeled solutes can be used to detect perturbation of bilayer organization (McDonald et al., 1985).
31P	Chemical shift anisotropy arising from orientation of phosphate tensor. Commonly used for distinguishing bilayer, hexagonal II, and isotropic phases because a difference in the anisotropy of rotation of the phospholipid molecule in these phases; the line shapes and line widths are also different (Smith and Ekiel, 1984; Tilcock et al., 1986).
Ions	^{23}Na, ^{45}Ca, ^{59}Cd have been used for binding to the head groups.
Shift reagents used with NMR	For asymmetry of bilayers. (Barsukov et al., 1980)
Spin-labeled probes	Organization and motional parameters. (Devaux and Seigneuret, 1985)

TABLE 3-2 (*Continued*)

Technique	Comments
Differential scanning calorimetry	Temperature, enthalpy, and cooperativity of transition have been used to obtain phase diagrams. Phase separation is also indicated in appropriate cases.
Fluorescent lipids	Pyrene, anthracene, NBD-, rhodamine labeled phospholipids have been used to follow the movement and mixing of lipids in the bilayers (Yguerabide and Foster, 1981; Somerharju et al., 1985; Bergelson et al., 1985; Devaux and Seigenuret, 1985). These probes have also been used for the measurement of anisotropy, excimer formation (pyrene), phase separation (prodan, octadecylrhodamine), asymmetry, fusion, membrane potentials.

dimensions of lattice, chemical composition, and partial specific volumes. Relative intensity of the reflections and the electron density have also been used to obtain additional detailed information. (Blaurock, 1982; Boldt et al., 1979; Knoll et al., 1985).

The X-ray diffraction parameters as well as other physicochemical parameters have been used to reconstruct phase diagrams of type shown in Fig. 3-3. From such data, it is usually possible to establish lyotropic and thermotropic mesomorphism between the various phases that exist in a given temperature and concentration range (Chapman et al., 1967; Luzatti and Tardieu, 1974). Thus, with increasing water content the crystal lattice of DPPC is broken to form lamellar dispersions. It can be calculated from the data of type shown in Fig. 3-3 that about 12 water molecules per dipalmitoyl phosphatidylcholine molecule are required to disrupt the crystal lattice; however, the exact number of water molecules required for complete hydration of the head group depends upon the nature of the phospholipid as well upon the conformational state of the acyl chains. As shown in Fig. 3-3, hydration of DPPC is accompanied by an increase in the interlamellar distance, and a downward shift in the temperature for melting of the acyl chains. In excess water, the interlamellar distance and the chain melting temperature for zwitterionic phospholipids remains relatively independent of the water content, and the excess water freezes at 0°C. For anionic phospholipids the interlamellar distance increases even after the head group is completely hydrated.

The information about molecular motions in an organized phase can be

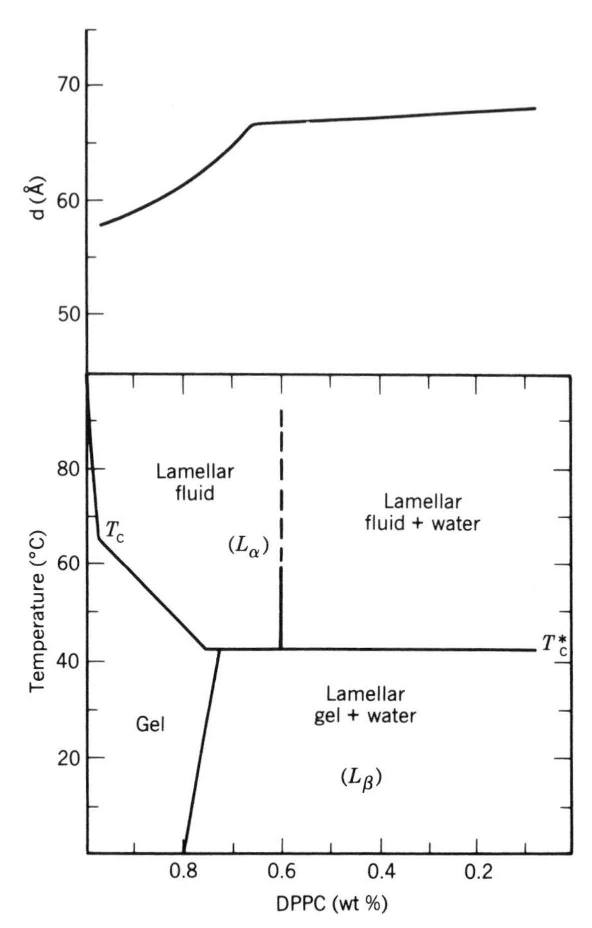

Fig. 3-3. A phase diagram to illustrate dependence of the interlamellar separation (top) and the acyl chain melting temperature (bottom) on the mole fraction of dipalmitoyl PC in water. Detailed description of the lamellar fluid (L_α) and gel (L_β) state bilayers that coexist with excess water is given in Chapter 5. The interlamellar distance (top) increases due to hydration or swelling of the lamellae that are also present in crystal lattice.

TABLE 3-3 Properties of Some Aggregates in Lipid–Water Dispersions

Property	Micelle/HI[a]	HII[b]	Lamellar[c]
Shape (S)	<1/3 (spherical)	>1	1
Residence time	0.1 msec	>1 sec	>1000 min
Freeze-fracture		Cylinders	Plane of fracture
^{31}P-NMR	Isotropic	anisotropic	anisotropic
Long spacing		1 : 1/3 : 1/2 : 1/7	1 : 1/2 : 1/3
Lipids	Lyso-PL, PI gangliosides	PE, DAMG	PC, SM, CL PA, PS, PG

[a] PI, Phosphatidylinositol.

[b] PE, Phosphatidylethanolamine; DAMG, diacylmonogalactoside.

[c] PC, Phosphatidylcholine; SM, sphingomyelin; CL, cardiolipin; PA, phosphatidic acid; PS, phosphatidylserine; PG, phosphatidylglycerol.

obtained with spectroscopic methods. Recent progress in multiple dynamic nuclear magnetic resonance (NMR) (Meier et al., 1986) is particularly promising because it offers the possibility of measuring motions with correlation times in the range of milli- to nanoseconds.

The phases in the lipid–water system can be characterized from the phase diagrams generated on the basis of the organization or motional parameters. Properties of some of the major organizational states of phospholipids (Table 3-3) in the phases formed in water are discussed below.

MONOLAYERS

Most amphipaths with a long polymethylene chain form a thin film at the air-water interface. Such an arrangement avoids energetically unfavorable interactions between acyl chains and polar solvents like water, and it also avoids unbalanced cohesive forces between the water molecules at the interface. As there is considerable difference between the polarity of the two media, amphipathic molecules localize such that the polar groups face the water and the acyl chains extend into the air (Fig. 3-1). An interface between water and a nonpolar solvent like benzene also offers an ideal environment for the formation of monolayers where the acyl chains are in the organic medium. Such an arrangement avoids energetically unfavorable interactions between acyl chains and polar solvents like water (Taylor et al., 1976; Yue et al., 1976). We have not considered monolayers at oil-water interface, although such monolayers provide very useful information about the organized interface without complications from interchain cohesion.

Formation of monolayers at the interface of media of differing polarity is an equilibrium phenomenon. If the hydrophobic effect is dominant, as is the case with most of the naturally occurring phospholipids, amphipathic molecules tend to segregate at the air–water interface to form *insoluble* monolayers, that is, the equilibrium lies essentially completely in the favor of the monolayer. The concentration of amphipaths in the interface can be altered within reasonable limits by changing the amount of the amphipath or by varying the area of the interface. Depending upon the number of molecules in an interface, the surface tension or interfacial pressure changes. Surface pressure (π) is defined as the difference between the surface tension of pure water and the surface tension of the surface covered with the monolayer. Thus, surface pressure corresponds to the reduction in water activity in the interface due to the presence of the amphipathic molecules. This is analogous to the origin of the osmotic pressure of a solution as a function of the solvent activity. Viewed another way, pressure arises from the momentum transfer during collisions between the lipid molecules, as well as from the steric, electrostatic, and hydration forces.

In one of the most commonly used methods (Wilhelmley plate) to measure interfacial pressure, the force (in dynes/cm or milliNewton/meter, mN/m) necessary to remove a disc of platinum from the interface is determined. The pressure versus area (π-A) curves for a given amount of an amphipath in the interface are obtained by varying the surface area of a monolayer at the air–water interface by means of a movable barrier. Idealized π-A curves are shown in Fig. 3-4. At any given point along this curve the area per molecule is the total area divided by the number of amphipathic molecules in the system. Therefore, from π-A curves of pure amphipaths it is possible to derive useful information about the orientation, conformation, and intermolecular interactions between the molecules at the interface. Several interesting features of an idealized π-A curve of the type shown in Fig. 3-4 may be noted (Albrecht et al., 1978).

1. At very low pressure (region I) the area per molecule is very high and therefore the molecules are in a random motion and probably lie flat on the surface. In this (gas phase) region, changing the surface area alters the surface pressure only slightly (Gershfeld, 1976; Gershfeld and Tajima, 1977).
2. Beyond a certain point, the pressure begins to change when the surface area is compressed and $\pi A = RT$ relationship is obeyed. In this liquid-expanded state or fluid (region II) molecules are coherently oriented with their long axis pependicular to the interface, however the area per molecule is large enough so that the molecules have considerable degree of rotational and segmental motion.
3. On further compression the segmental motion of acyl chains ceases, and a liquid condensed state or gel (region IV) is formed. In this state the chains are vertical, hexagonally close-packed, and have rotational freedom without any significant segmental motion.

As shown in Fig. 3-4, the transition between the regions II and IV is quite sensitive to temperature. This is further illustrated in Fig. 3-5. For example, the abrupt transition from the liquid-expanded (fluid) to liquid-condensed (gel) state is not observed above a certain critical temperature. This temperature usually corresponds to the transition temperature for bilayer in gel to the fluid phase (to be discussed in Chapter 5). It may however be emphasized that all these features of π-A curve do not appear in monolayers of all phospholipids. As also shown in Fig. 3-5c, the π-A relationships are dis-

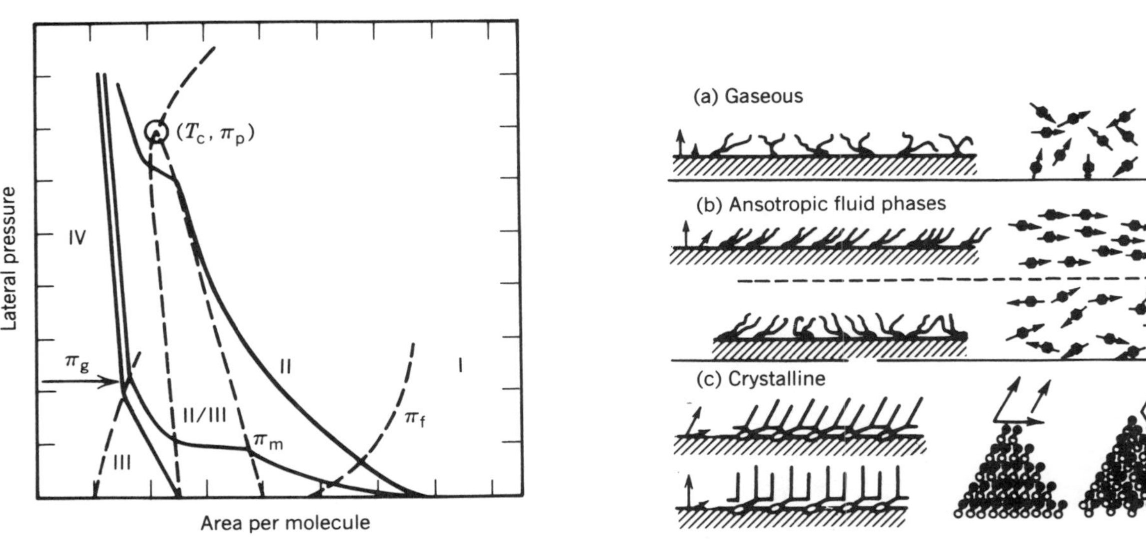

Fig. 3-4. (Left) Idealized isotherms of pressure (π) versus area (A) relationship of phospholipid monolayers showing different phases and coexistence regions. The full lines are representative isotherms while the broken lines indicate phase transitions or boundaries of coexistence regions. π_f and π_g represent fluid–fluid and gel–gel transitions. π_c, T_c represents a triple critical point. Regions I (flat) and II (phospholipid molecules are perpendicular to the aqueous interface) represent two fluid phases with different orientations of phospholipid molecules; III represents a gel phase; IV represents a completely compressed solid state. The area enclosed by the dotted line represents the coexistence region for the perpendicular fluid and gel phases. (Right) Probable arrangement of molecules in the gaseous, fluid, and gel or crystalline states of phospholipid molecules in a monolayer is shown with the cross-sectional view or the top view. The arrows indicate the centers of mass of the hydrocarbon chains.

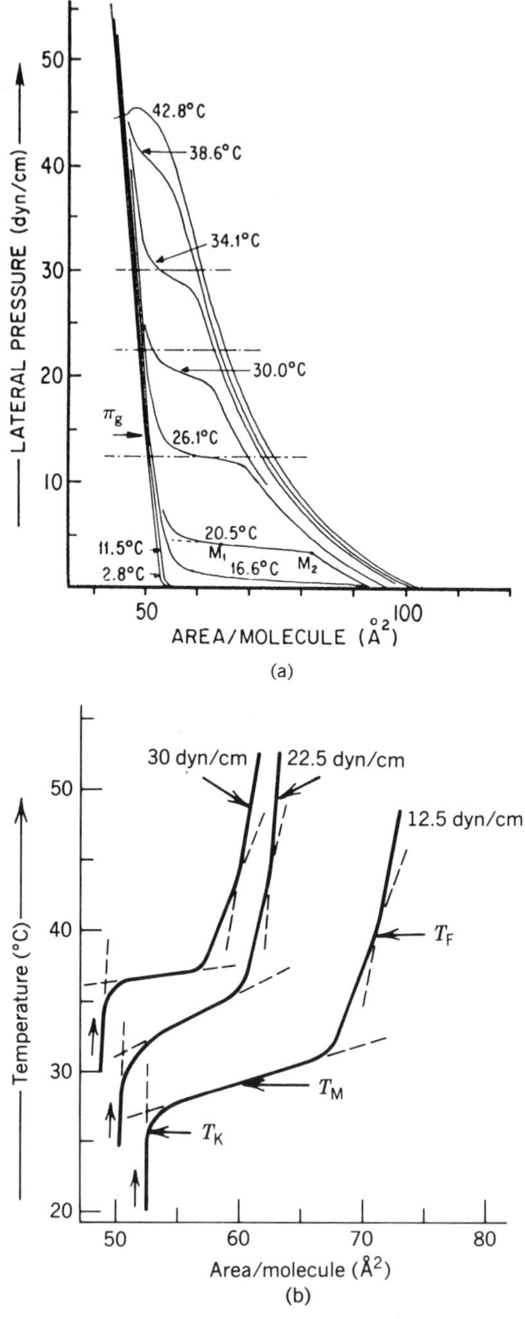

Fig. 3-5. (A) Pressure-area isotherms or monolayers of dipalmitoyl PC on water at the temperatures indicated. (From Albrecht et al., 1978.) (B) The temperature-area isobars for DPPC monolayer on water at pressure indicated (Albrecht et al., 1978). (C) The pressure-area isotherms for (a) 22 : 0, (b) 16 : 0, and (c) egg PC at 22°C. Egg PC has on the average 1.7 double bonds per *sn*-2 acyl chain. Egg PC remains fluid and 22 : 0 PC remains gel, while 16 : 0 PC undergoes a phase change when compressed.

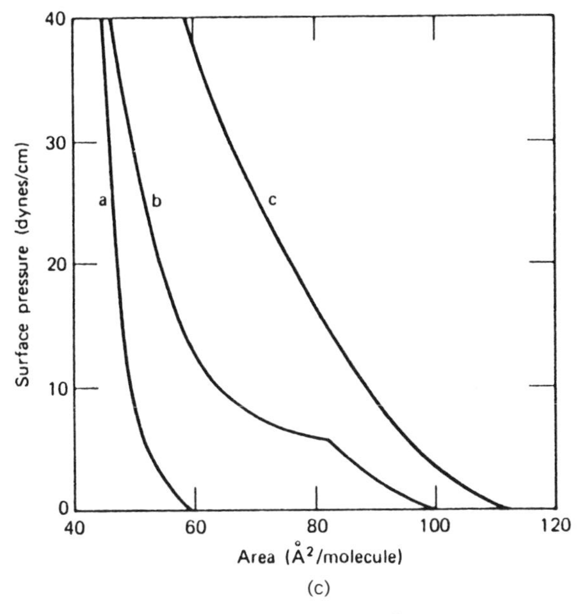

Fig. 3-5c. *(Continued)*

tinctly different for lipids with different chain lengths and degrees of unsaturation (Hui et al., 1975; Albrecht et al., 1978; Hawco et al., 1981; Fisher and Sackman, 1984).

The π-A curve in the condensed state (region IV) is steep. That is, the area per molecule does not change appreciably even when the pressure is increased. Beyond a certain pressure the monolayer collapses or folds. The steep region of the π-A curve in the condensed phase can be extrapolated to zero pressure to obtain the molecular area of an amphipath. The values summarized in Table 3-4 show that a variety of derivatives containing saturated polymethylene chains but different polar head groups have the same cross-sectional area per chain, i.e., about 20 Å^2. The molecular area of cholesterol also suggests a similar orientation. The area of acyl chains at a given surface pressure increases with the number of double bonds, decreases slightly with the increasing length of the acyl chain, and increases due to repulsion of the charged groups (Fig. 3-6). The areas of isomeric phospholipids containing dissimilar acyl chains in the two positions are also identical. All these characteristics would be expected if the amphipathic molecules in the insoluble monolayers are oriented such that their long axes are perpendicular to the plane of the bilayer. This is also consistent with the permeability of monolayers as reflected in the retarded rate of evaporation of

TABLE 3-4 Molecular Areas of Some Amphipaths in Monolayers

Lipid	Limiting Area (A^2) at 20–25°C
Stearic acid (18 : 0)	20.6
Oleic acid (18 : 1, *cis*)	32
Cholesterol	39
Dipalmitoyl PC[a]	44.5
Dioleoyl PC	72
Egg PC	62
Sphingomyelin	42
Phosphatidyl serine	50–100
Egg PE	42
Ganglioside	140 (Head)
	31 (chain)

[a] PC, Phosphatidylcholine; PE, phosphatidylethanolamine.

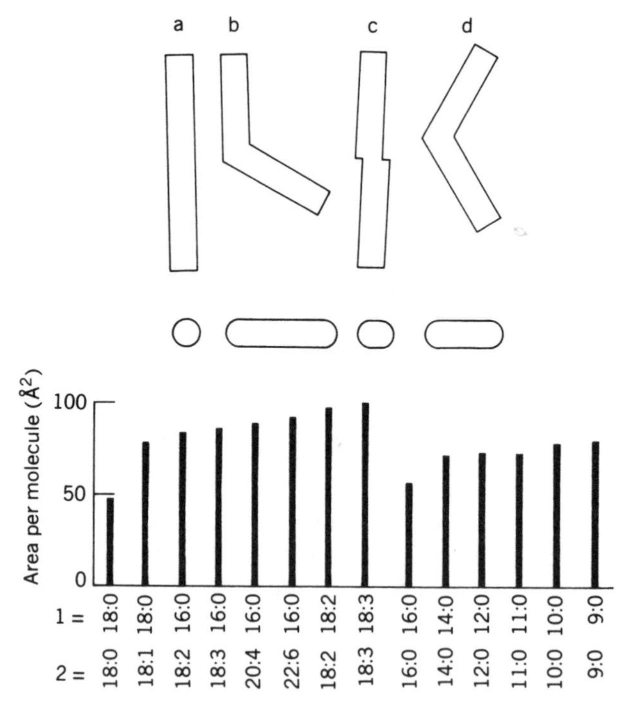

Fig. 3-6. (Top) Projected areas of hydrocarbon chains. (a) A hydrocarbon chain in an all-*trans* conformation assumes a cylindrical shape; (b) a *gauche* rotation along one C-C bond leads to a 60° angle between the upper and lower portions of the chain; (c) a chain with one *trans* double bond and tilted to minimize the projected area; (d) a chain with one *cis* double bond and tilted such that the bottom of the chain is directly below the top. (Bottom) Area per molecule at 12 dyne/cm (22°C) for synthetic diacylphosphatidylcholines. The chain lengths in the positions 1 and 2 are given at the bottom of each bar. (From R. Demel, *Methods Enzymol.* 32, 539-544, 1974.)

water. The rate of evaporation of water, for example, decreases with the increasing length of the polymethylene chain.

The collapse pressure (the pressure at which the monolayer folds) of monolayers depends largely on the structure of the amphipath. Phospholipids containing less than eight methylene residues in each chain do not form stable monolayers, presumably because their hydrophobicity is low, and therefore exchange between the monolayer and the aqueous phase is appreciable. For monolayers of PC above a critical temperature, the collapse rate is so high that they can not be compressed beyond a certain maximum surface pressure. At a very fast compression rate, it is possible to *overcompress* a monolayer to produce surface pressures approaching about 55 dyne/cm. The highly compressed liquid condensed phase is a metastable phase that slowly relaxes to about 25 dyne/cm. The collapse pressure and the rate of collapse change appreciably when other amphipathic agents are also present in the monolayer (see Keough, 1985).

Localization of the polar groups at the aqueous interface of the monolayer give rise to a surface potential[1] that reflects the strength and orientation of surface dipoles attenuated by the redistribution of ions in the aqueous phase. From such measurements it can be shown that the p*K*a values at the monolayer interface are quite different from the values in the bulk aqueous phase. There is an apparent shift of 0.6 units toward neutrality, i.e., an increase in p*K*a for acids and a decrease for bases. This effect is apparently due to a decrease in the surface pH compared with the bulk pH, which will reduce the apparent ionization of the acid and therefore increase the apparent p*K*a. Such effects could also arise from a change in the local dielectric in the interface.

Specific interactions between two or more components present in a monolayer are generally indicated by departure from the additivity of their respective areas (Fig. 3-7). For example, the presence of divalent ions reduces the area of anionic phospholipids, whereas ionization increases the molecular

[1] When a metal plate coated with a weak α-emitter, such as $_{241}$Am or $_{210}$Po, is placed just over the surface of an aqueous solution, it renders the air conductive, and the electrical potential of the plate is identical to that of a point immediately outside the aqueous phase. The potential difference between the plate and an immersed reference (calomel) electrode yields the contact potential of the interface. The surface potential of the monolayer is defined as the difference in the surface potential in the absence and in the presence of a monolayer at the interface. Typical experimental values are of the order of a few hundred millivolts. The surface potentials (ΔV) of monolayers yield information about molecular surface dipole moments, $\mu = \pi\Delta V/4n$ where n is the number of molecules per cm^2 at surface pressure π. Large fluctuations in surface potentials arise due to surface heterogeneity, phase transition, and chemical reactions, as reflected in time-dependent changes, in the orientation angle (θ) for a dipole ($\mu = \mu_1 \cos\theta$), and in the state of ionization of the surface group.

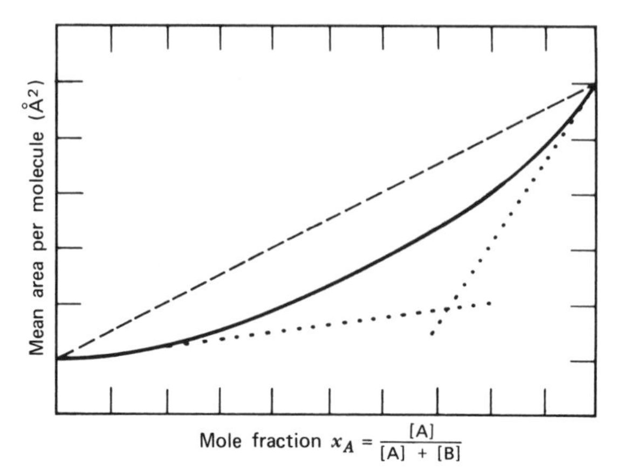

Fig. 3-7. Average area per molecule in mixed lipid (A + B) monolayers as a function of composition. The dashed line represents the variation that would correspond to a simple additivity of the molecular area. The dotted lines represent the extrapolated contributions from A and B. The point of intersection corresponds to the mole proportion of A in the complex.

area. Cholesterol and phospholipids together occupy a smaller area, and from the discontinuities or nonidealities in mole fraction plots it appears that the composition of the complex is approximately 1 : 1. Immiscibility and demixing of the two components in a monolayer can also be demonstrated from measurement of the collapse pressure. Monolayers containing two separate phases show a constant collapse pressure that is independent of the mole fraction of the components. This collapse pressure will be the lower of the two for the two separate components. On the other hand, monolayers of two miscible components will have a variable collapse pressure depending upon the composition. Nonideal mixing of the components is also reflected in many properties of monolayers including the kinetics of action of lipolytic enzymes.

In contrast to the situation in bulk systems, the phase relations in lipid monolayers are not directly accessible by spectroscopic methods. However, the phase rule for monolayers can be described as follows. For the systems in which the independent intensive variables (F) that can be varied independently without altering the phase relations and the number of components (C) are known, the phase rule is $F = C - P^s - P^b + 2$, where P^s and P^b are coexisting surface and bulk phases (Defay et al., 1966).

The molecular area at a given pressure increases with increasing temperature. As shown in the schematatic phase diagrams in Figs. 3-4 and 3-5, several transitions are observed. The transition π_m is associated with a large change in the molecular area and it corresponds to a transition from expanded (fluid) to condensed (gel) state. Above a certain temperature [about 42–46°C for dipalmitoyl PC (DPPC)] the monolayer does not enter the liquid condensed phase, even at pressures approaching the collapse pressure. This critical temperature decreases with increased unsaturation, branching, and shorter chain length. Based on such observations it appears that the upper limit for the critical temperature corresponds to the gel-to-fluid transition in bilayers (see Chapter 5). To a first approximation, a monolayer at about 33 dyne/cm^2 may be considered to be in the same state of packing as a bilayer. This is because at this pressure, the molecular areas are similar, and the monolayer at the air–water interface is in equilibrium with the outer monolayer of a layer of adsorbed vesicles (Schindler, 1979). However it may be stressed that the properties of a monolayer do not necessarily always correspond to the properties of half-a-bilayer (Lis et al., 1982) because the internal energies are appreciably different in two systems. For other complexities see Gershfeld and Tajima (1977), Albrecht et al. (1978), Fischer and Sackman (1984), and Loesche and Mohwald (1985).

Lateral distribution of phospholipid molecules in the solid and fluid patches or domains of well-defined geometry have been visualized by doping monolayers with fluorescent probes like 4-nitrobenz-2-oxa-1,3-diazol-4-yl-PE (NBD-PE) (Peters and Beck, 1983) or carbocyanines (Loesche and Mohwald, 1985). Use of epifluorescence optical microscopy has demonstrated the formation of compressed (solid)-phase domains from the expanded (fluid)-phase domain on compression of monolayers (Peters and Beck, 1983). As shown in Fig. 3-8, lateral (translational) mobility of probes in a monolayer shows significant changes at a characteristic pressure, and a phase separation is also indicated. The size, shape, and handedness of the domains formed by phase separation depends not only on the temperature and the presence of other components, but also on the stereochemical configuration of the glycerol backbone (Weiss and McConnell, 1984).

Monolayers represent a novel lamellar phase of matter, and therefore they are an excellent tool to study organization, orientation, and motion of phospholipid and protein molecules at the polar–apolar interface. Monolayers at the air–water interface have been used as models for a variety of interfacial processes such as reactions at the interface, enzymatic catalysis at the interface, and exchange of phospholipid molecules between a monolayer and vesicle or cell membranes. Recently there has been considerable interest in the study of bilayers formed by sequential transfer of two monolayers (usually at a constant surface pressure, near 30 mN/m, to emulate the packing or lateral pressure of bilayers) on an alkylated solid support such as oxidized silicon, glass, quartz, mica, or tin sulfide plates (e.g., see Fisher and Sackman, 1984). The electron diffraction patterns obtained from multi-

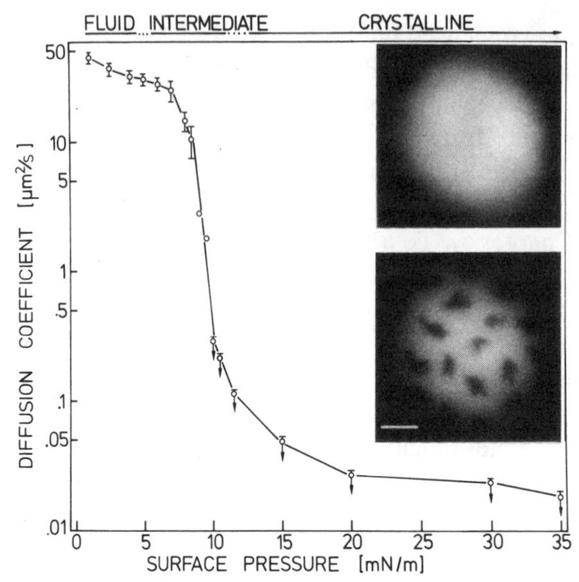

Fig. 3-8. The fluid–gel transition in DPPC monolayers monitored by the rate of lateral diffusion of NBD-PE (at 22°C). The micrographs shown as inserts are those seen in the fluid (beginning) or the intermediate region where both the gel and the fluid domains coexist. (From Peters and Beck, 1983.)

ple stacked monolayers suggests that in the liquid condensed films the molecules are in a crystalline two-dimensional hexagonal lattice. These films have also been useful for the study of proteins incorporated into monolayers under defined pressure.

Lung Surfactant

The surface forces regulating the mechanics and stability of lung alveoli are modulated by the presence of amphipathic substances at the air–water interface. Alveoli are small, air-filled spherical units interconnected by tubules. Not only do these structures not collapse due to surface tension but they also can undergo appreciable expansion and contraction. The pressure–volume and pressure–area relationships in alveoli are maintained by surfactants. The tendency to collapse is low at low lung volume, and this reduces the pressure required to keep the lungs open. At normal air pressure the surfactant reduces the work required for breathing, increases lung compliance, prevents lung edema, and facilitates opening of fluid-filled alveoli. Lungs of immature infants have abnormal lung surfactant and there is a strong tendency for the airways to collapse. This situation can be corrected by applying a suspension of pulmonary surfactant to lungs.

Lung surfactant can be isolated by washing lung with saline. The surfactant from human lungs consists of 9% protein and 91% lipid made up of 40% dipalmitoyl phosphatidylcholine (PC), 30% unsaturated PC, 8% phosphatidylglycerol, 2% phosphatidylethanolamine, and about 8% neutral lipid, which is mostly cholesterol (Keough, 1985). The monolayer of lung surfactant at the air–water interface exhibits a gel- (liquid condensed) to-fluid (liquid expanded) transition that depends upon the pressure. For dipalmitoyl PC monolayers, the critical temperature (above which the transition from expanded to condensed state is not observed in the pressure area curves as shown in Fig. 3-5a) is approximately the same as the T_m for DPPC bilayers. This dependence on temperature is the basis for the function of lung surfactants at the normal human body temperature. For most lung volumes the surface tension is below 20 mN/m in the absence of the surfactant. At 37°C a monolayer of DPPC can achieve such surface tension, but a monolayer of 1-palmitoyl-2-oleyl phosphatidylcholine could not. The liquid condensed-phase state of the monolayer, and not the structure or the composition of acyl chains, is important for the stability of lungs. While saturated lipids are required to make monolayers of very low surface pressure, unsaturated lipids are required for insertion of lipids into the monolayer. Similarly, the role of other components in the lung surfactant could be associated with more subtle functions, such as the rapid rate of dispersal and the phase separation during compression. This is because the monolayer of the lung surfactant is rapidly converted from a relatively fluid to a relatively condensed phase. The surfactant of human lungs at 37°C is poised to do so.

MICELLES

Amphipaths with a relatively large head group ($S < 1/2$) form aggregated structures called micelles (*micella* in Latin means ''small bits'') in which the polar groups are on the surface and the chains are segregated from the aqueous phase (Fig. 3-1). In organic phases containing small amounts of water, the energetically favorable organization is inverted; in such inverted micelles the polar groups are oriented toward the center and enclose a small doplet of water (Fig. 3-2). Depending on the structure of the monomer and the environment, the size of micelles ranges from a sphere of about 5 nm (for dihexanoyl- and diheptanoyl-PC) to oblate cylinders of several hundred nanometers in length for dioctanoyl-PC. Micelles of lysolecithin are also cylindrical. Spherical micelles of charged amphipaths like dodecylsulfate typically contain about 50–80 monomers (aggregation number), which are in equilibrium with the free monomers in the aqueous phase.

The exchange of monomers between micelles is rapid. The half-time for formation and breakdown of micelles is about 1 to 100 msec (Okubo et al., 1979), and the residence time of monomer amphipaths in a micelle is of the order of about 1 msec. This emphasizes not only the dynamic nature of micelles, but also the fact that on the average, the acyl chain of the amphipathic molecule spends a fair amount of time in contact with water, which can be viewed at two levels: the dynamics of monomer–aggregate equilibrium as implied in the concept of critical micelle concentration, and the molecular dynamics of packing of chains in micelles.

Critical micelle concentration (cmc) is the concentration of monomer amphipaths that remains constant in the aqueous phase in equilibrium with micelles (Fig. 3-9). Further addition of amphipath results in the formation of more aggregate. The cmc is related to the interfacial free energy per unit area, γ:

$$\mathrm{cmc} = \exp\left[\frac{-4\pi r^2 \gamma}{kT}\right]$$

where r is the effective cross-sectional radius of the molecule. The value of γ is typically 50 mN/m for these systems. The size of spherical micelles is relatively insensitive to lipid concentration above the cmc, and the micelles

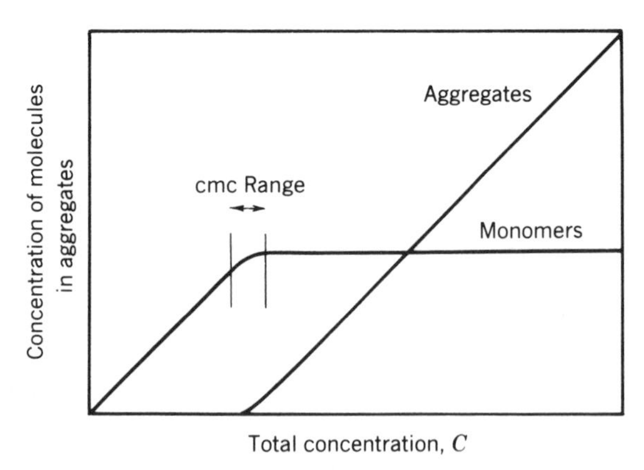

Fig. 3-9. Idealized relationship between monomer and aggregate (micelle) concentration as a function of the total amphipath concentration. Abrupt changes in many molecular and colligative (bulk) properties of the solution are seen in the narrow concentration range called critical micelle concentration (cmc).

TABLE 3-5 Properties of Micelles Derived from Some Commonly Used Detergents

Detergent	Monomeric MW	Critical Micelle Concentration (*M*)	Aggregation Number	$\bar{v}_D$ (cm^3/g)
Octyl-β-D-glucoside	292	2.5×10^{-2}	27	—
Dodecyl-maltoside	528	1.7×10^{-4}	140	0.81
Dodecyldimethyl-*N*-amineoxide (DDAO)	229	2.2×10^{-3}	75	1.122
Lauramido-*N,N*-dimethyl-3-*n*-propylamineoxide (LAPAO)	302	3.3×10^{-3}	—	1.067
Dodecyl-*N*-betaine (zwittergent 3–12)	336	8×10^{-2}	87	—
Tetradecyl-*N*-betaine (zwittergent 3–14)	350	6×10^{-3}	130	—
Myristoylphosphoglycerocholine	486	9×10^{-5}	—	0.97
Palmitoylphosphoglycerocholine	500	1×10^{-5}	—	0.976
3-[(3-cholamidopropyl)-dimethylammonio]-1-propanesulfonate (CHAPS)	615	5×10^{-3}	—	—
Deoxycholic acid	393	3×10^{-3}	22	0.778
Cholic acid	409	1×10^{-2}	4	0.771
Taurodeoxycholic acid	500	1.3×10^{-3}	20	0.75
Glycocholic acid	466	—	6	0.77
Sodium dodecyl sulfate (SDS) in 50 m*M* NaCl		8×10^{-3}	62	0.87
Dodecyl ammonium · Cl^-		15×10^{-3}	55	
Ganglioside GM_1		10^{-9}	150	
PEG-dodecanol		1×10^{-4}	130	
Polyoxyethylene glycol detergents				
C_8E_6	394	1×10^{-2}	32	0.963
$C_{10}E_6$	422	9×10^{-4}	73	—
$C_{12}E_6$	450	8.2×10^{-5}	105[e]	0.989
$C_{12}E_8$	538	8.7×10^{-5}	120	0.973
$C_{12\&14}E_{9.5}$(Lubrol PX)	620	—	100	0.958
$C_{12}E_{12}$	710	9×10^{-5}	80	—
$C_{12}E_{23}$(Brij 35)	1200	$9\ 10^{-5}$	40	—
$C_{16\&18}E_{17}$(Lubrol WX)	1000	—	90	0.929
tert-p-$C_8ØE_{9.5}$(Triton X-100)	1625	3×10^{-4}	140	0.908
tert-p-$C_8ØR_{7\text{-}8}$(Triton X-114)	540	2×10^{-4}	—	0.869
$C_9ØE_{10}$(Triton N-101)	670	1×10^{-4}	100	0.922
C_{12} sorbitan E_{20}(Tween 20)	1240	6×10^{-5}	—	0.869
$C_{18:1}$ sorbitan E_{20}(Tween 80)	1320	1.2×10^{-5}	60	0.896

are fairly monodisperse. The standard deviation for the aggregation number is approximately equal to the square root of the aggregation number. Rod-like or cylindrical micelles are usually large, polydisperse, and change their properties with the lipid concentration.

The surface and bulk properties of the aqueous solution of amphipaths exhibit an abrupt change at cmc. Such a change in the surface tension, viscosity, conductivity, solubility of foreign solutes, fluorescence spectrum of lipid soluble dyes, scattering, line shape in NMR, refractive index, and sedimentation coefficient demonstrate that aggregation of monomeric amphipathic molecules into micelles begins above cmc, and micelles disappear rapidly upon dilution to concentrations below cmc. Under certain conditions there is indication that microaggregates of amphipathic molecules probably exist below cmc. It should be appreciated that the monomer-micelle aggregation is a multistate equilibrium:

$$L + L \rightarrow L_2 + L \cdot \cdot \rightarrow \cdot \cdot L_n \text{ (micelle)}$$

for which there is no uniquely defined cmc. If, however, $n > 50$ the equilibrium behavior approaches phase separation as predicted by Gibb's phase rule. This approach also accounts for the fact that cmc represents a narrow but finite concentration range.

As summarized in Table 3-6 a large variety of synthetic and biological amphipaths (like lysophospholipids, short-chain diacylphospholipids, gangliosides, phosphatidylinositols) form micelles when dispersed in water. Structures of several amphipaths used in the study of membranes are also given in Fig. 3-10. The monomer–aggregate equilibrium process for the formation of micelles is largely driven by the hydrophobic effect. Thus, cmc decreases with the increasing acyl chain length, and the incremental free energy is about −600 to −800 cal/mole for a methylene group. This is somewhat smaller than the incremental free energy for the transfer of alkanes and

TABLE 3-6 Phase Preferences of Membrane Lipid Classes

Classes	Lamellar	H_H	Micellar
Zwitterionic phospholipids			
Phosphatidylcholine	+	−	−
Sphingomyelin	+	−	−
Phosphatidylethanolamine	+	+	−
Negatively charged phospholipids			
Phosphatidylserine	+	+	−
Phosphatidylglycerol	+	+	−
Phosphatidylinositol	+	−	−
Phosphatidic acid	+	+	−
Cardiolipin	+	+	−
Glycolipids			
Monoglucosyldiglyceride	−	+	−
Monogalactosyldiglyceride	−	+	−
Diglucosyldiglyceride	+	−	−
Digalactosyldiglyceride	+	−	−
Cerebroside	+	−	−
Cerebroside sulfate	+	−	−
Gangliosides	−	−	+
Lysophospholipids	+	−	+

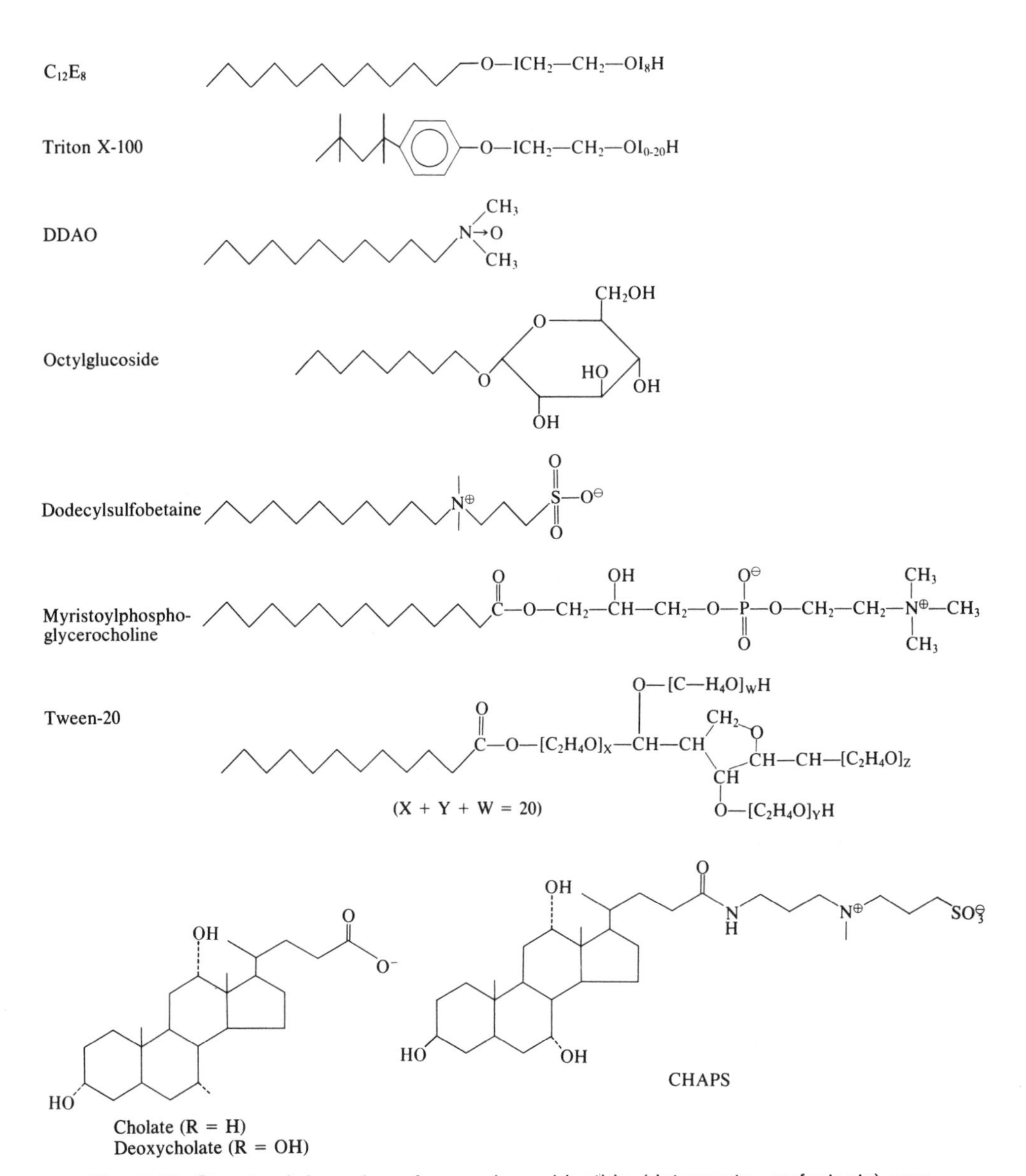

Fig. 3-10. Structural formulas of several amphipathic (detergents, surfactants) compounds that form micelles when dispersed in water.

alkanols from water to bilayers. This could be partly due to ordering of the chains in the micelle and partly due to a more polar averaged environment in micelles. The large negative free energies of micellization, which becomes increasingly negative as the chain length increases, have a small positive enthalpy term, which diminishes slightly with the increasing chain length, and a large positive entropy term, which becomes more positive with increasing chain length.

Both the aggregation number and cmc are sensitive to the structure and conformation of the amphipath, and the nature of the counterions. Typically, cmc decreases with increasing chain length, with shielding of the surface charge, and with decreasing temperature. Salts, also, change the cmc of zwitterionic phospholipids. The effect of temperature on these systems is rather complex. The variation of cmc and the solubility of the amphipath with temperature shows that at a critical temperature (Krafft temperature) solubility is equal to cmc. Above the Krafft temperature, solubility increases rapidly, because hydration of the crystal lattice is facilitated by thermal disordering of the acyl chains. As expected, the Kraft temperature decreases for shorter chain lengths as well as with increasing polarity of the head group.

It may be emphasized that formation of micelles in aqueous dispersions of amphipaths is but one possibility to minimize the overall free energy. Often at higher mole fractions of amphipaths, micelles are in lyotropic and thermotropic equilibrium with other organized phases. Transition of micelles to other polymorphs like hexagonal I, bilayer, or cubic phases has been observed. Lysophospholipids at low water content form long cylindrical tubes with the head groups at the surface and the chains in the interior (hexagonal I phase). Thermotropic transition of micelles into other organized phases also occurs under certain conditions. For example, lysophosphatidylcholine and its analogs form micelles when dissolved in aqueous solutions by heating. However, on keeping these micellar solutions below a certain critical temperature, an interdigitated lamellar phase is slowly formed (see below).

The ability of an amphipath to form micelles is appreciably altered in the presence of other amphipaths (Lichtenberg et al., 1983). For example, lysophospholipids form stable lamellar structure in the presence of fatty acid (Jain et al., 1980; Allegrini et al., 1983) or cholesterol. Similarly, salts of fatty acids (soaps) codispersed with un-ionized fatty acids or alkanols form lamellar structures (Hargreaves and Deamer, 1978).

The organization of amphipaths in a micelle is still being debated, although several models have been proposed. The "oil-droplet in a polar coat" (Fig. 3-11) model (Hartley, 1936), which is based on the Langmuir principle, has reasonably accommodated a considerable range of experimen-

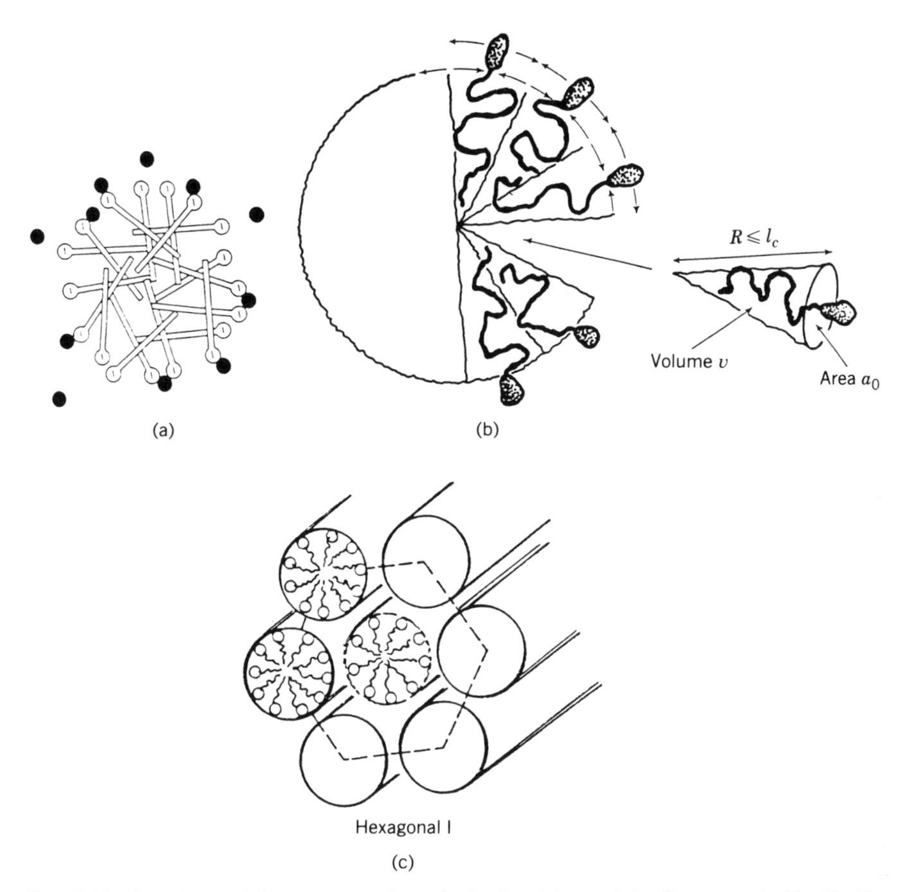

Fig. 3-11. A cartoon of the cross section of micelles (a) as originally suggested by Hartley (1936); (b) the current consensus articulated by Israelachvilli (1985). Repulsive forces between the head-groups and the attractive hydrophobic forces between acyl chains determine the optimum shape of the cone. The cross-sectional area of the head group, the chain volume v and the chain length l determines the generalized shape parameter $S = v/a \cdot l$, which is <1 for micelles. (c) The phase with hexagonal I symmetry consisting of long cylinderical micelles separated by the aqueous phase is formed with certain amphipaths.

tal findings on the properties and spectroscopic behavior of micelles, such as size, cmc, thermodynamics of micelle formation, ability of micelles to dissolve nonpolar and amphipathic solutes, effect of counterions and salts, and other related aspects of colloid behavior (Wennerstrom and Lindman, 1980; Romstead, 1984; Mittal and Bothorel, 1985). While the packing of the acyl chains of cone-shaped molecules does tend to maximize hydrophobic effect, such as packing into spherical aggregates raises difficulties due to steric crowding of the acyl chains in the center of a micelle. Similarly, the subtle effects in the behavior of micelles also arise from head-group repulsion, hydration, and conformational disorder of the acyl chains. The steric requirements for chain packing have been difficult to accommodate into any consistent model. In recent years, however, the oil-droplet model has been elaborated to suggest that the acyl chains in micelles have jogs and kinks such that all segments of chains spend an appreciable proportion of time near the surface (Cantor and Dill, 1984; Gruen, 1985); that the surface of a micelle has fluctuations of 1–2 Å, which is about one C-C bond length (Pratt et al., 1985); and that the interior of micelles is loose, porous, and highly disorganized (Menger and Doll, 1984). At the very least these models emphasize a need for a better description of the constraints underlying chain conformation and chain packing that is consistent with hydrophobic effect, chain dynamics, and the geometrical constraints of the aggregate (for some model calculations, see Ben-Shaul and Gelbart, 1985).

Micellar Phase in Living Systems

Absorption of lipids in the gastrointestinal tract is mediated by solubilization as micelles of bile salts, which also act as an interface for the action of a variety of lipases and phospholipases. The aggregation number of micelles of bile salts is typically 4–30. The amphipathic character of bile salts is primarily due to the presence of polar groups on one face of the sterol nucleus, therefore, these amphipaths could stabilize discs of bilayers in which edges are *sealed* with cholate molecules. Other effects of amphipaths that form micelles include dispersal of water-insoluble substrates for kinetic studies, solubilization of water-insoluble solutes and membranes, denaturation of soluble proteins, reconstitution of bilayers, activation and inactivation of soluble as well as membrane-bound enzymes, and stabilization of the membrane-active conformation of proteins and peptides (Lichtenberg et al., 1983; see also Martinek et al., 1986 for a discussion of reverse micelles). Incorporation of proteins into micelles appears to be a reasonably specific process in the sense that the underlying interaction is dominated by ionic or by hydrophobic interactions, and that usually the ratio is detergent to protein and the size of the resulting micelles is constant.

Hexagonal (H_{II}) Phase

In the H_{II} phase the lipid molecules are organized into long tubes (5–12 nm diameter, Fig. 3-12). These tubes are arranged with hexagonal symmetry, which gives rise to a characteristic X-ray diffraction pattern and ^{2}H-NMR line shape). The polar head groups surrounding the aqueous channel are

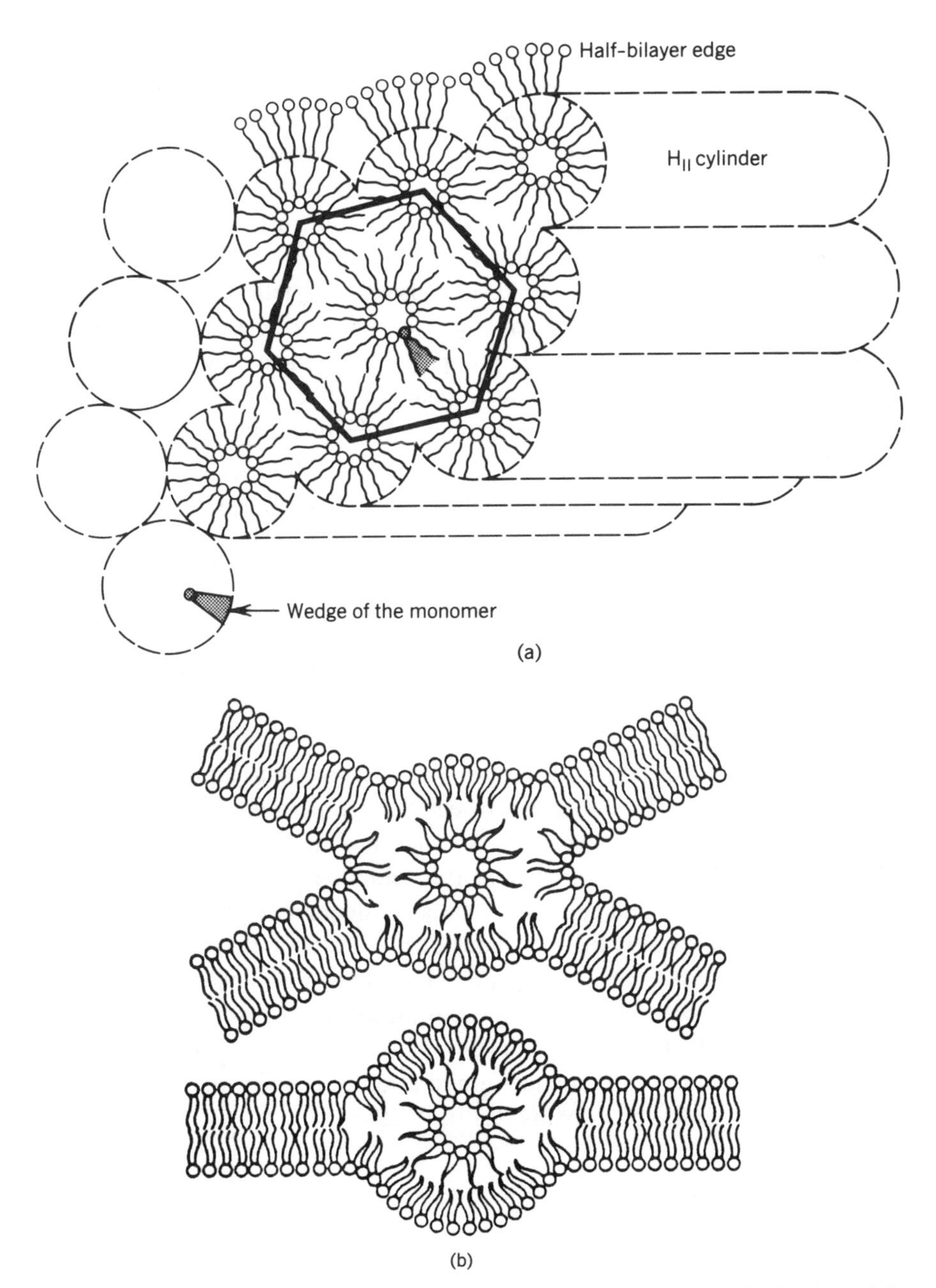

Fig. 3-12. (a) A cross-sectional view of a bundle of water-filled tubes in hexagonal II symmetry. Note that the individual molecules in a tube are wedge-shaped ($s > 1$), that the interaction between cylinders is hydrophobic, and that the bundle of cylinders is separated from excess water by a bilayer. (b) In a nonpolar solvent some lipids can form inverted micelles. Two possible modes of integration of inverted micelles into bilayers are shown. (From Cullis et al., 1985.)

present within these tubes (Cullis et al., 1985; Caffrey, 1985; Grunner, 1985a). Individual molecules with a tube have rapid averaging rotation along two axes but not the third. Compared with other phases formed in a lipid–water system, this polymorph is characterized by hydrophobic surface and low water content. In excess buffer the H_{II} tubes separate from the excess aqueous phase, and the phase boundary along the surface of the packet of tubes is formed by a lipid monolayer.

Several naturally occurring lipids like phosphatidylethanolamine (PE) and glucosyldiglycerides form hexagonal phases when dispersed alone in water; however, deprotonation of $-NH_3$ group in PE stabilizes the bilayer phase. Similarly, the bilayer to hexagonal phase change occurs on heating or isothermally by protonation or by adding solutes (like calcium ions, gramicidin, dolichols) to the lamellar dispersions of unsaturated PE anionic phospholipids like phosphatidylserine (PS), phosphatidylglycerol (PG), phosphatidic acid (PA), and diphosphatidylglycerol (DPG) (Killian and deKruijff, 1986). This behavior can be correlated to the generalized shape of these molecules. The wedge-shaped molecules ($s > 1$) with a relatively small head group tend to form the H_{II} phase. Unsaturated PE and the calcium salt of unsaturated cardiolipins have been shown to prefer the H_{II} phase at ambient temperatures, and such phases have been found to coexist as separate phases along with the lamellar phase.

The H_{II}-to-bilayer thermotropic transition involves an interbilayer fusion event, which occurs in less than 1 msec. The temperature ($T_{b\text{-}h}$) for bilayer-to-H_{II} transition is usually above (sometimes it is coincident with) the gel-to-fluid transition, and the enthalpy of this endothermic transition is usually less than 2 kcal/mole and apparently does not depend upon the chain length. $T_{b\text{-}h}$ decreases at lower water content (Caffrey, 1985), in the presence of salts (Seddon et al., 1984) with increasing chain length or unsaturation. Ether analogs have a lower $T_{b\text{-}h}$ transition than the ester analogs of phospholipids (Boggs, 1984). The bilayer-to-H_{II} phase transition is rapid, cooperative, and occurs without significant change in hydration or a major reorganization or exposure of hydrocarbon chains (Caffrey, 1985).

Lipidic Particles

Freeze-fracture study of some biological membranes revealed the presence of lipidic particles (6–13 nm diameter) that are not removed by the presence of protease but can be destroyed by phospholipase. It appears that these particles are cubic phases in which both the hydrocarbon and the water phases are continuous (Rilfors et al., 1986). Lipid particles appear to be labile structures that can readily exchange lipid molecules from an adjacent bilayer. The functional significance of such particles is not known but they have been implicated in fusion, transport and transbilayer movement, protein insertion during translation, and secretion (Cullis et al., 1985).

Cubic Phase

In a mixture of bilayer and the H_{II}-type of organized structures, sometimes cubic symmetry (Fig. 3-13) is also found (Fontell, 1982). It appears that some of the lipid particles have the architecture of the cubic phase. Such phases in dispersions of phospholipids give an isotropic ^{31}P-NMR signal, similar to that obtained with micelles, lipidic particles, or sonicated vesicles. Probably the best-characterized cubic phases arise from aqueous dispersions of monooleoylglycerol, consisting of bilayer walls that form complex lattice channels. Dialkylglycerol tetraether lipids of *Archaebacteria* and *Sulfolobus sulfataricus* form cubic phase under physiological conditions, i.e., 85°C, pH 3.0 (Gulik et al., 1985). Transitions between different phases of this lipid are slow and considerable metastability is observed.

Emulsions

A mixture of amphipaths can sometimes form emulsions in which a bulk "oil" phase is in contact with a bulk water phase. The size of emulsion particles depends upon the ratios of the oil and water phases. It is not uncommon that emulsions and aerosols also contain varying proportions of other phases described in this chapter.

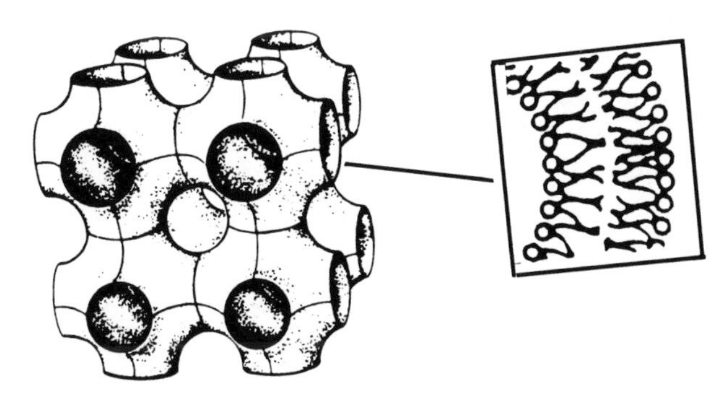

Fig. 3-13. Schematic representation of the phase with cubic symmetry. The inner and outer compartments are water filled and connected to each other. The organization of the interface is as a bilayer.

LAMELLAR PHASE

Most naturally occurring phospholipids (alone or as naturally occuring mixtures) form a lamellar phase (Fig. 3-14) when dispersed in excess water. Presence of the lamellar phase can be demonstrated or inferred by a wide range of techniques ranging from polarization and electron microscopy, X-ray and neutron diffraction, to the use of appropriate probes. A typical electron-density profile calculated from the X-ray diffraction data is shown in Fig. 3-15. It is dominated by a strong electron density contrast between the high-density polar head groups and the low-density hydrocarbon core. The principal feature of the profile is that the high-density peaks are separated by about 5.2 nm for dipalmitoyl PC (DPPC), i.e., the separation between the middle of phosphate and the glycerol backbone on the two interfaces (4.1 nm) plus the distance to the outer edge of the PC head group. The minimum in electron density corresponds to the position of the terminal methyl group. From the wide-angle reflections, it is possible to obtain the separation between the acyl chains. For DPPC dispersions it is found to be 0.42 nm (sharp) in the gel phase and 0.46 nm (diffuse) in the fluid phase. In the gel phase a diffuse reflection at 0.41 nm has been interpreted to suggest the chain tilt of about 23° toward the plane of the bilayer.

Interdigitated Lamellar Phase

Transition from the lamellar to the micelle phase has been observed for some diacylphospholipids (Ranck, 1983), lysophospholipids and its analogs (Jain et al., 1985b), platelet activating factor (Huang and Mason, 1986), as well as for mixed-chain PC and sphingomyelin (SM), whose two acyl chains have considerably different lengths (Hui et al., 1984; McIntosh et al., 1984; Huang and Mason, 1986). The transition on heating is rapid and occurs sharply at a characteristic temperature. Both the enthalpy and temperature for transition increase with increasing chain length. Such systems show considerable hysteresis on cooling. The monomer concentration in equilibrium with interdigitated bilayers is very low, and the equilibrium is not dynamic.

Lamellar Bilayer Organization of Biological Membranes

Polymorphism of aqueous dispersions of phospholipids raises several interesting questions about the organization of biological membranes. The polymorphs described in this chapter are thermodynamic phases only as an approximation. Thus, the question of their stability has limited significance because energetically other phases differ only slightly. Although aqueous

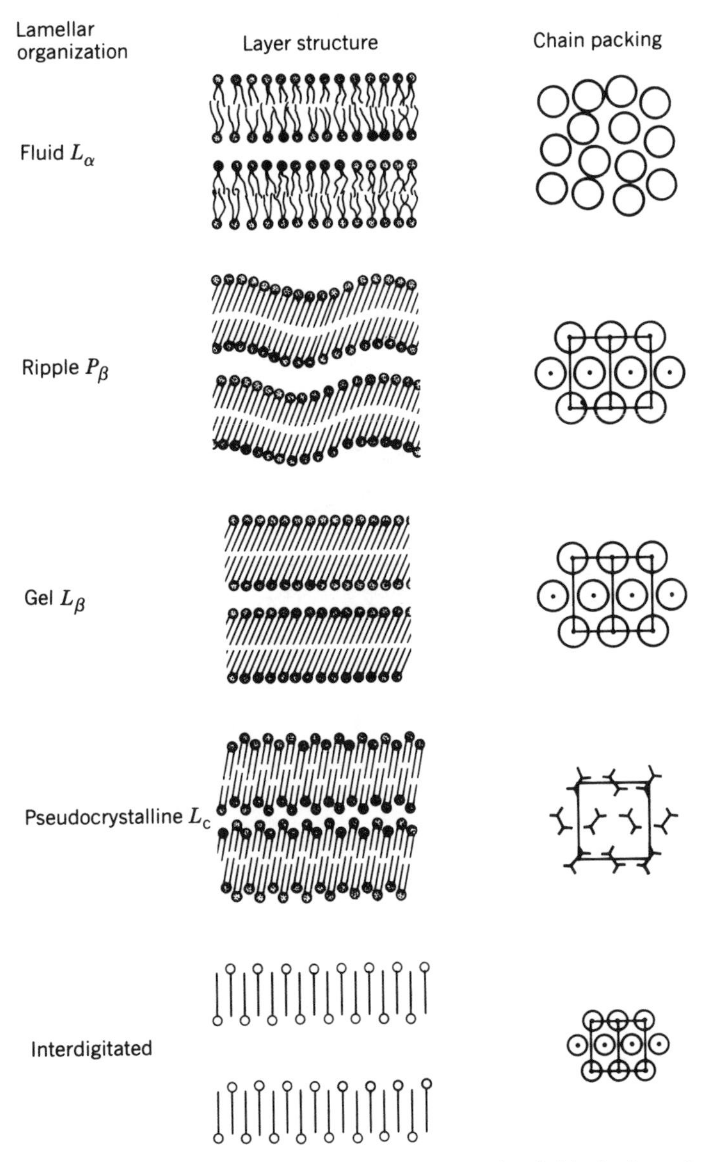

Fig. 3-14. Organization of the lamellar bilayer phase in the fluid, ripple, gel, pseudocrystalline, and interdigitated states (top to bottom). The top view is shown in the last column.

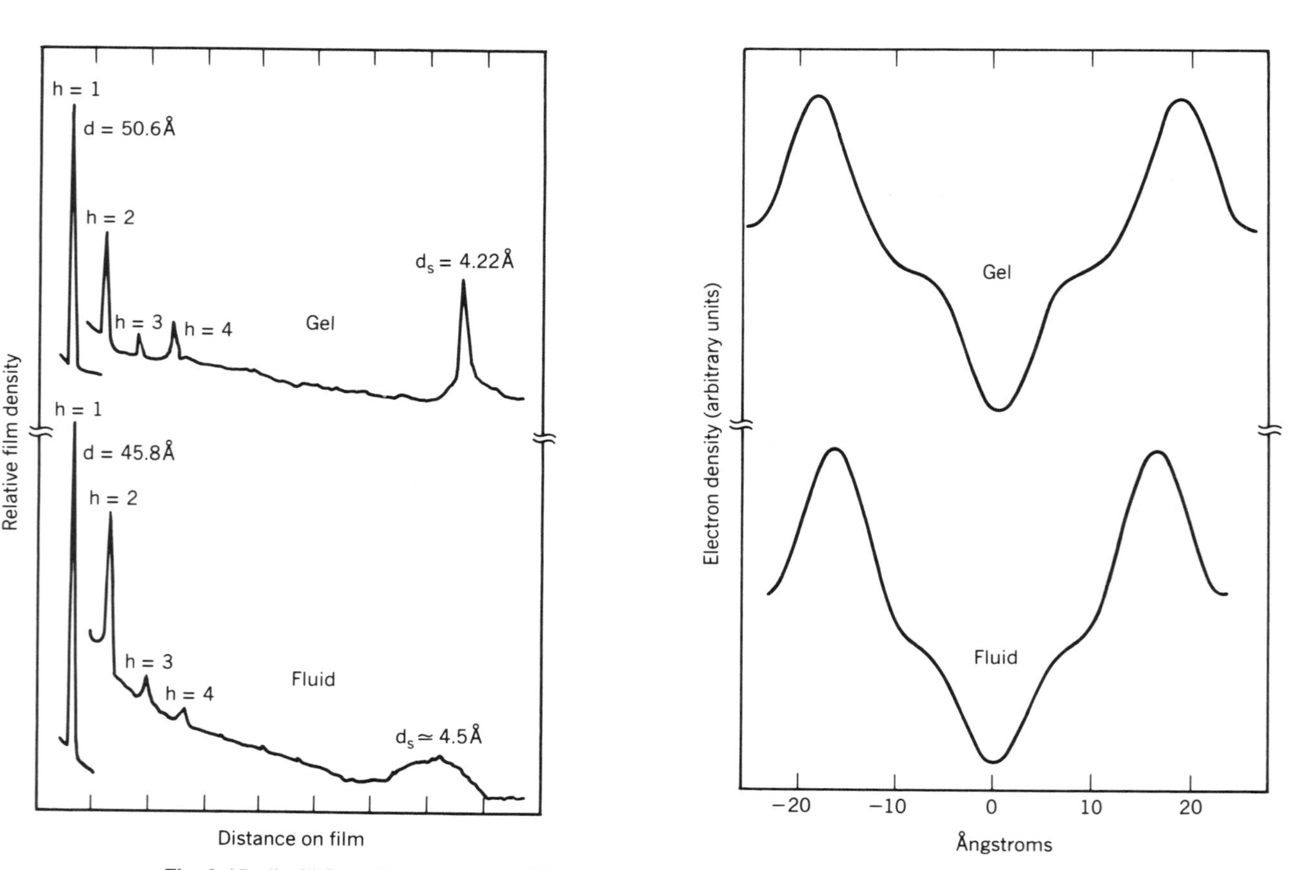

Fig. 3-15. (Left) Densitomer traces of X-ray diffraction profiles of hydrated dilauroylphosphatidyl ethanolamine in the gel and the fluid phases. For both phases, four orders (h = 1–4) of a lamellar repeat period are observed. (Right) Electron density profiles of DLPE in the gel and the liquid crystalline phase. (Courtesy of Dr. T. McIntosh.)

dispersions of many pure naturally occurring phospholipids exhibit tendency to form the H_{II} phase, the propensity of naturally occurring mixtures of phospholipids to form lamellar bilayer structures is overwhelming (Table 3-6). On the basis of properties of extracted phospholipids, it appears that the bilayer matrix could form a significant proportion of many biomembranes. The balance of the matrix may be interrupted with proteins. It is also possible that a small proportion of the membrane matrix may be interrupted by nonlamellar phases within the overall constraints of the bilayer organization. The metastability of the various conformational states of phospholipid molecules also adds yet another level of complexity to the overall question of the organization of biomembranes. The composition of biomembranes also suggests that a disordered conformation of acyl chains predominates and conformational fluctuations corresponding to the conformation of phospholipids

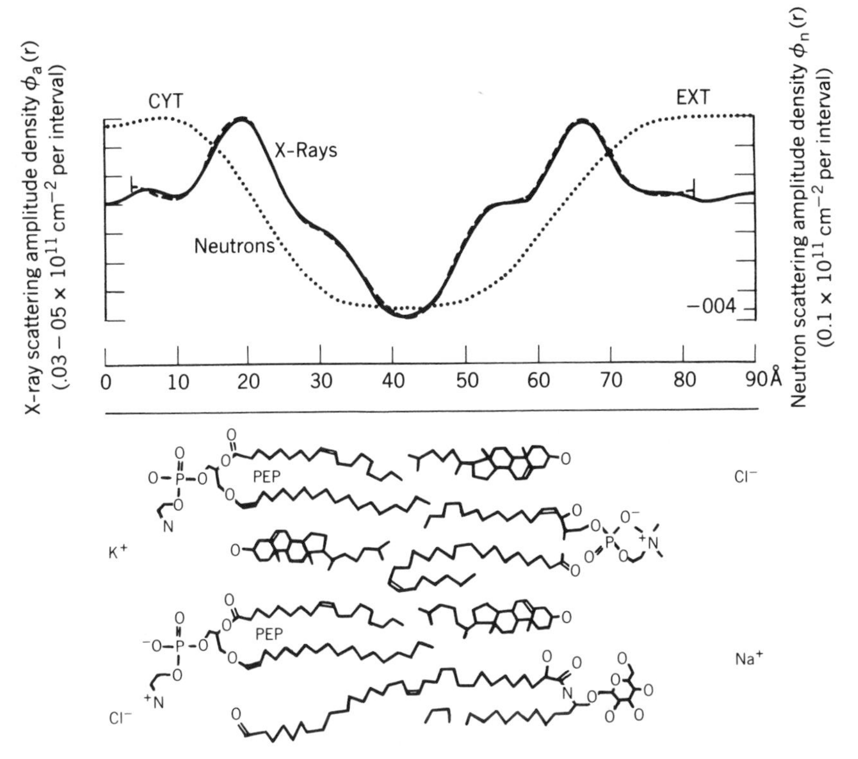

Fig. 3-16. Schematic representation of the bilayer structure in myelin based on the electron-density profile (top) from rabbit sciatic (solid line) and rabbit optic nerve (dashed line) and the neutron diffraction profile from sciatic nerve (dotted line); (bottom) interpretation of the profile in terms of the arrangement of the components. For the morphological and functional significance of myelin see Fig. 12-2. (From Kirschner and Casper, 1979.)

in the various polymorphs described here could occur. These observations imply that a long-range bilayer organization with topological continuity could persist in spite of a lack of a repeat unit and in spite of the presence of local disorder.

X-ray diffraction profiles from a variety of natural membranes have shown the presence of lamellar structures. For example, as shown in Fig. 3-16, the diffraction profile of myelin can be accounted for essentially in terms of the presence of the bilayer organization of its lipid components (Kirschner and Casper, 1979). Asymmetry of the profile is due to asymmetric distribution of lipids between the two halves of the bilayer. The step between the polar head groups is due to the presence of cholesterol, and an asymmetry suggests that there is about twice as much cholesterol in the extracellular half of the bilayer as in the cytoplasmic half. Other lipids like PE plasmalogen are also located mainly on the cytoplasmic surface, as shown by binding of mercuric chloride. On immersion in water the lamellar repeat distance increases from about 17 to 27 nm, largely due to swelling of the extracellular space.

Electron diffraction profiles of pellets of isolated membranes have been obtained for sarcoplasmic reticulum, erythrocyte membranes, rod outer-segment membranes, thylakoid membranes, and others. These results are generally consistent with the suggestion that the matrix of these membranes is due to a lipid bilayer and they are generally asymmetric. Subtle differences, probably due to the presence of proteins, are also indicated.

HYDRATION OF THE INTERFACE

The state of water molecules at the interface of bilayers has been a matter of considerable debate. Several lines of investigation appear to suggest that a certain degree of order is experienced by a few molecules of water at the interface:

1. A variety of experiments suggest that a few (between 5 and 20) molecules of water per molecule of phospholipid in the aqueous dispersions do not have the properties of liquid water, e.g., they are incapable of dissolving nonelectrolytes or undergoing ice–water transition.
2. Diffusion of water in the interlamellar aqueous phase increases with increasing water content up to 24% water. Immediately beyond this, the diffusion constant falls by a factor of 4 and remains constant for further increase in the water content (Lange and Gary-Bobo, 1974). Similar behavior is seen for the diffusion of other solutes.

3. The capacitance of a bilayer is found to be smaller than the thickness of the bilayer calculated on the basis of the acyl chain length. One of the possible explanations for this discrepancy is that some water molecules penetrate to the depth of the third or the fourth methylene residue along the acyl chain in the bilayer.

To account for the interlamellar distance of close approach in dispersions of phospholipids, Rand and Parsegian (1984) have suggested that a layer of water molecules at the interface exhibits a continuous monotonic interbilayer repulsive pressure P.

$$P = P_s \exp\left[\frac{d\omega}{\lambda}\right]$$

where $d\omega$ is interbilayer separation and P_s is a coefficient of exponential decay that reflects the degree of perturbation of water at the interface. By using osmotically stressed lamellar dispersions to remove the interlamellar layer of water, it has been shown that the bilayer thickness and the area per molecule remain constant (within 4%) over a wide range of pressure, and the decay constant, λ is found to be 0.14 nm for DPPC in the gel phase and 0.17 nm for egg PC (McIntosh and Simon, 1986). From these measurements it has been calculated that 8.5 water molecules per DPPC (and 10.8 per molecule of egg PC) are removed when the thickness of the bilayer is reduced from 5.9 to 5.2 nm under the osmotic pressure. Full implications of this repulsive hydration force are not understood yet (however see Gruen, 1985; McIntosh and Simon, 1986). It may be emphasized that the overall magnitude of this energy does not change appreciably on fusion of bilayers. Also, the water of hydration is readily exchangeable on the submicrosecond scale, therefore it appears unlikely that a hydration layer could offer a kinetic barrier.

4 Properties of Bilayers

Its chief merit is its simplicity—simplicity so pure, so profound, in a word, so simple, that no word will fitly describe it.

LEWIS CARROLL

Swelling of naturally occurring phospholipids in excess water gives rise to lamellar phases that coexist with excess water. The organization of phospholipid molecules in lamellar aqueous dispersions gives rise to a variety of structures that have provided insights into the properties of the bilayer. Characteristics and properties of some of the structures containing bilayers are summarized in this chapter.

LIPOSOMES

Edges of a bilayer in lamellar dispersions seal readily to form enclosed structures so that the acyl chains are not exposed to the aqueous phase (Fig. 4-1). Due to a low interfacial pressure of the bilayer, such structures are generally spherical and their diameters range from 20 nm to well over 5000 nm, depending upon the method of preparation as well as the structure and composition of phospholipids (Szoka and Papahadjopoulos, 1980; Deamer and Uster, 1983). Swelling and shrinking in response to osmotic shock, isotopic efflux, and trapping of markers show that liposomes are topologically continuous closed structures. Because of their ability to compartmentalize the aqueous phase, liposomes have been used as models for cells to characterize the properties of bilayers, as microcapsules, and as matrix for reconstitution of membrane functions from purified membrane components. Thus, the properties of liposomes relate to the molecular properties of phos-

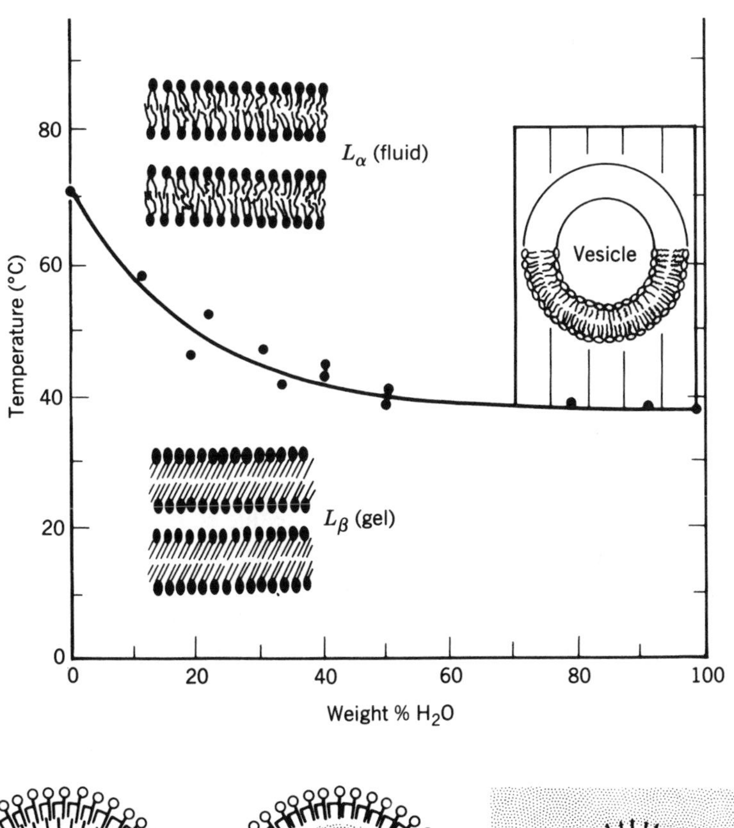

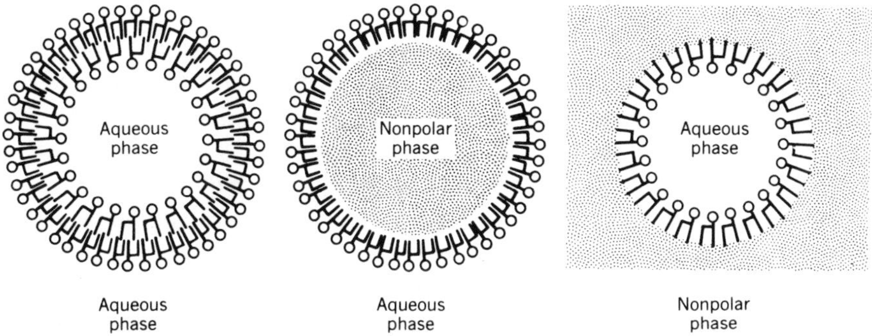

Fig. 4-1. (Top) A phase diagram of dimyristoyl phosphatidylserine (ammonium salt) in water. The solid line represents the transition from the gel to fully hydrated fluid bilayer. The hatched area represents the region where a spontaneous transition from multilamellar to unilamellar vesicles occurs above the chain melting temperature. It is the transition from a one-phase system (fully swollen, smectic, lamellar phase) to a two-phase system (unilamellar vesicles dispersed in excess water). It may be noted that in the dispersion of PC (Fig. 3-3), excess water coexists with multi- or unilamellar liposomes both below and above the chain melting temperature. (Bottom) The bilayer vesicles (left) should be distinguished from other enclosed structures that are formed in stable emulsions of phospholipids in water and immiscible organic solvents. In excess water, droplets of the nonpolar phase are separated from excess water by a monolayer of lipid (middle). The opposite orientation is seen in the excess nonpolar phase (right). The size of these droplets enclosed with a monolayer depends upon the ratio of the bulk phases and the amount of phospholipid.

pholipids, as well as to the organizational and phase properties of the bilayer.

Methods for preparation of liposomes take advantage of the ability of water to penetrate the crystal lattice of lipids. As the dispersions swell, in most cases bilayer-enclosed structures are formed and excess water forms a separate phase. The balance of elastic and hydrophobic forces appears to determine the tendency of the bilayer to vesiculate (Lasic, 1982). Most naturally occurring diacyl phospholipids alone or as mixtures will form liposomes when water content reaches >30% by weight. The critical micelle concentration of most phospholipids is less than 0.1 n*M*. Bilayer-enclosed structures are also formed from dispersions of single-chain amphipaths like acid soaps, a mixture of ionized and protonated fatty acids (Hargreaves and Deamer, 1978). Indeed, propensity of amphipathic compounds, like alcohols and fatty acids, to form bilayer-enclosed structures is so pronounced that such compounds could have served as prebiotic precursors of cell membranes (Deamer, 1986). An equimolar mixture of lysophospholipid and fatty acid (Jain and DeHaas, 1981) and of lysophospholipid and cholesterol (Ramasammy and Brockerhoff, 1982) also form lamellar dispersions. It is not established whether interdigitated bilayers in lamellar dispersions of lysophospholipids exist as liposomes. Synthetic amphipaths like dioctadecyldimethylammonium chloride and dihexadecylphosphate (Fendler, 1982) and several other types of dialkyl amphiphiles (Murakami et al., 1985) also form liposomes. The primary structural requirement for the formation of the lamellar phase appears to be that the molecule be cylindrical either as a monomer or a functional dimer or oligomer. Additional stability is achieved if the segment of the chain near the head group is rigid and the terminal region is flexible. It appears that most amphipaths with straight alkyl chains longer than C_{11} and above their Krafft point can form bilayers if they approach a cylindrical shape either as a monomer or as a functional dimer.

Liposomes resulting from aqueous dispersion of most phospholipids have diameters ranging from 20 to 5000 nm and contain multiple bilayers forming concentric shells (Fig. 4-2). The surface area of vesicles per mole of phospholipid depends somewhat on the conformation of the acyl chains. However, in the fluid state the total area of vesicles obtained from 1 gram of dioleoylphosphatidylcholine is about 300 m^2. Unless there is a net repulsive force between bilayers, the amount of aqueous phase between lamellae is very small. On sonication multilamellar vesicles (MLV) disperse readily into small unilamellar vesicles (SLV) of 20–50 nm diameter (cf. Table 4-1). The smaller vesicles approach a lower limit of size where packing constraints prevent any further shrinkage. The smallest vesicles of egg PC (22 nm diameter) have 1900 and 1100 molecules in the outer and the inner monolayers,

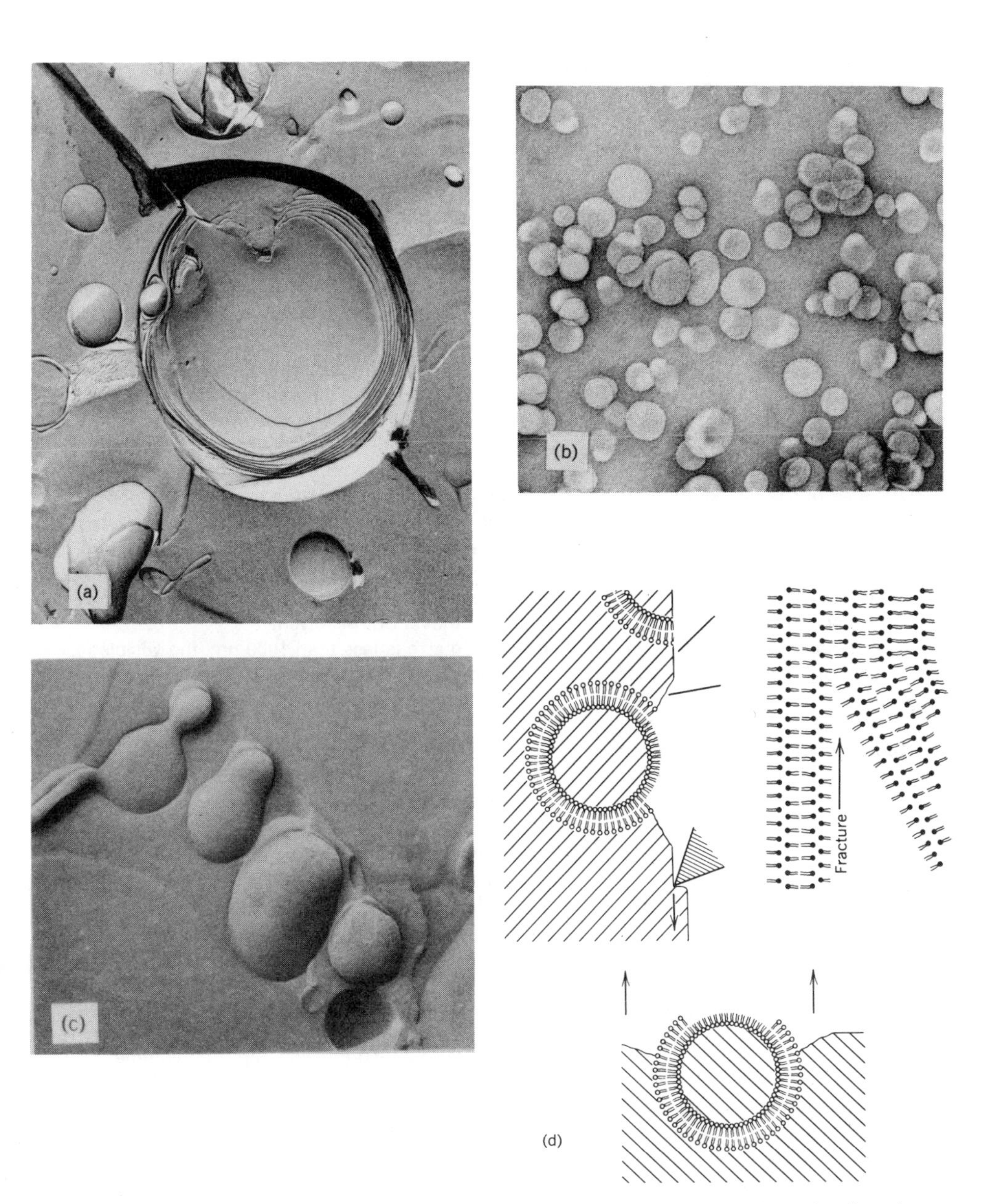

Fig. 4-2. Electron micrographs of (a) multilamellar vesicles (MLV), (b) sonicated unilamellar vesicles (SUV), and (c) large unilamellar vesicles (LUV). Micrographs of MLV and LUV are obtained by freeze-fracture as shown in (d). When the sample is frozen the fracture plane passes through the middle of the bilayer. In c a pair of vesicles is captured just after fusion. The images in b are obtained by negative staining with uranyl nitrate on grids coated with Piolofom FN50 and then made hydrophilic by coating with carbon. (c is from Bearer et al., *Biochim. Biophys. Acta* 693, 93, 1982; a and b are through the courtesy of P. Kaufman.)

TABLE 4-1 Methods for Preparation of Liposomes or Vesicles (see Szoka and Papahadjopoulos, 1980, for earlier references)

Liposomes or Vesicles	Comments
MLV (multilamellar vesicles with captured volume 1–4 liters/mole)	Spontaneous solvation of dry neutral or zwitterionic lipids. Time and rate of hydration regulate the trapped volume (2–10 liters/mole). Polydisperse (diameter 0.2 μm to 50 μm) Stable plurilammelar vesicles (Gruner et al., 1985).
LUV (large unilamellar vesicles)	Size (trapped volume up to 50 liters/mole; diameter 100 nm to $>10 \mu$m) depends on the method of preparation.
Solvent injection	Phospholipids in ether or ethanol are injected in aqueous phase at temperatures near the boiling point of the solvent. The size of the resulting vesicles depends upon the rate of evaporation and concentration of the injected solution.
Reverse-phase evaporation (Duzguenes et al., 1983)	Reverse micelles in a mixture of water and organic solvent are evaporated. Size and dispersity depend upon the nature and the proportion of the solvent.
Hydration-dehydration cycle (Shew and Deamer, 1985)	Fusion is induced in the presence of the solute by slow removal of the solvent
Extrusion of larger liposomes through polycarbonate filters of definite pore size provides vesicles of smaller size	If the pore size is less than 1000 nm, the vesicles are unilamellar (Hope et al., 1985)
Slow dispersal on a support (Dimitrov et al., 1984)	
Fusion	PC vesicles fuse slowly when kept at 4°C for several days; the rate of fusion is erratic and it is considerably enhanced by small amount of the products of degradation. Fusion of anionic vesicles is promoted by Ca, the resulting preparations are unilamellar only under limited range of conditions.
SUV (small unilamellar vesicles)	Formed spontaneously by dispersion of lipids when net surface charge density exceeds 2 $\mu C/cm^2$ or $>10\%$ charged additive. The vesicles are of diameter 20–50 nm with captured volume 0.15–1 liter/mole.
Codispersion with short-chain phosphatidylcholine (Gabriel and Roberts, 1984), lysophospholipids (Jain et al., unpublished), amphipaths like chlorpromazine (Hauser, 1985)	Phospholipids dissolved in detergent are diluted and excess detergent removed by dialysis, adsorption of Bio-beads SM-2 or gel-filtration. Lysophospholipids, if used as detergents, can be removed by acyltransferase.
Detergent removal	

respectively, and the trapped internal volume is about 0.13 L/mole phospholipid. The volume to area is significantly higher in large unilamellar vesicles (LUV) of 50–150 nm diameter (Fig. 4-3), which are prepared by fusion of SUV by fusogenic agents or by freezing and thawing, or by extrusion through polycarbonate filters or French press, or by spontaneous evaporation of the organic solvent from reverse micelles of phospholipids, or by removal of detergent. Vesicle fusion by freezing and thawing at very high (>1 *M*) electrolyte concentration have been used to make *giant* unilamellar liposomes of a diameter greater than 10,000 nm (Oku and MacDonald, 1983a). These methods of making LUV leave trace amounts of additives that cannot be removed even after extensive dialysis or gel filtration. The pres-

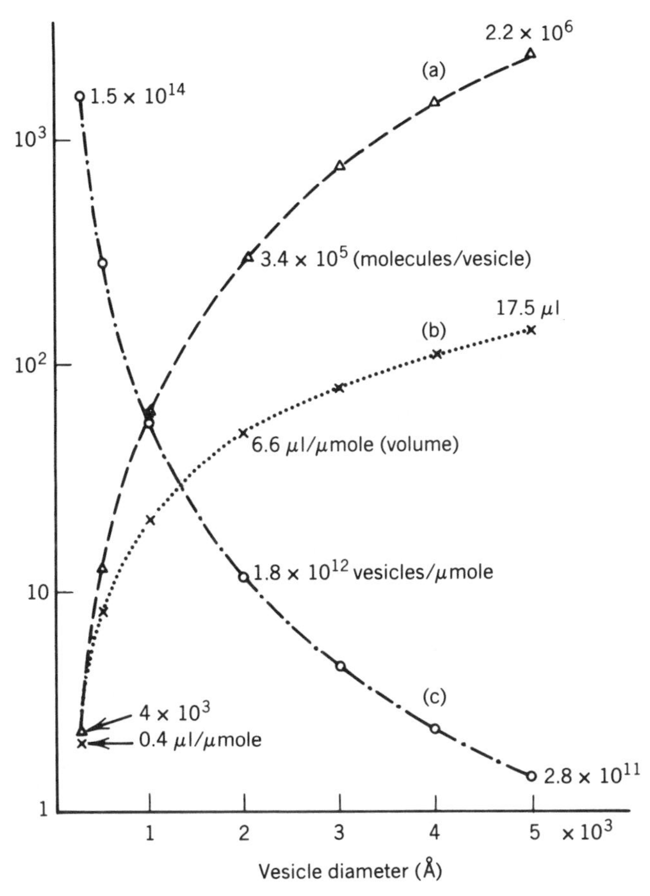

Fig. 4-3. Relationship between vesicle diameter and the number of egg phospholipid molecules per vesicle (a), the trapped volume per liter/mole (b), and the number of vesicles per mole (c). Note that the ordinate is in log units.

ence of trace impurities significantly modulates leakage of ions, aggregation and fusion of vesicles, and incorporation of proteins into vesicles. The size and dispersity of the resulting LUVs can vary over a very wide range, depending on the mixture of lipids and on the methods for their dispersal. The properties of liposomes that depend upon the size include:

1. Captured volume, i.e., liters of the aqueous phase captured per mole of lipid.
2. Encapsulation efficiency is the faction of the total volume sequestered by bilayers.
3. Fraction of the total lipid molecules in the outer monolayer of the liposomes.

Captured volumes and encapsulation efficiency for several types of liposomes are summarized in Fig. 4-3. The faction of total lipid in the outer monolayer of handshaken MLV is typically 0.07, for LUV it is 0.5, and for SUV it increases with decreasing diameter approaching 0.66 when the diameter is about 22 n*M*. Obviously this ratio depends upon the structure and composition of the lipid, and the distribution of lipid in the two monolayer halves may not necessarily be the same for different conformational states of a phospholipid molecule. It appears that vesicles prepared below and above the phase transition temperature of the lipid also have different proportions of lipid in the inner and the outer monolayers, which could account for unusual hysteresis and annealing behavior.

Methods to characterize vesicles include gel filtration (Schurtenberger and Hauser, 1984), quasi-elastic light scattering, and electron microscopy, all of which give information about the particle size. Trapping and leakage of probes like tempocholine quenched with ascorbate, calcein quenched with cobalt ions, or self-quenched calcein or carboxyfluorescein provide quantitative information about trapped volume (Watts et al., 1978). Intervesicle transfer of monomer lipids, and increased aggregation and fusion of vesicles provide a measure of the instability of the bilayer in vesicles. While many of these properties do not depend significantly upon the history of the vesicle, other properties such as anomalous leakage of ions at the phase-transition temperature or the susceptibility of liposomes to phospholipase A2 (pig pancreas) appear to be very sensitive to annealing, i.e., the rate of cooling of the vesicles through the phase transition. Thus, the organization of phospholipid molecules in the bilayer and the barrier properties of liposomes are optimized by slow cooling of the liposomes from a high temperature to lower temperatures.

Properties of DMPC vesicles have been characterized as a function of temperature (Watts et al., 1978). Apparently the size of the vesicles remains constant as a function of temperature with the molecular weight corresponding to 2850 lipid molecules. However, on heating from 15 to 30°C (T_m 21°C), the outer radius of vesicles increases from 9.7 to 12.8 nm; the internal volume increases from 0.1 to 0.6 ml/gram lipid; thickness of the bilayer decreases from 5.1 to 3 nm; the cross-sectional area per molecule increases from 0.45 nm^2 to 0.75 nm^2.

Transbilayer Asymmetry in Vesicles

Large vesicles of mixed lipids have symmetrical composition at the two interfaces. However, because of a significant difference in the packing at the two interfaces in small vesicles, some asymmetry is expected in SUV containing two or more phospholipids. For example, in cosonicated phosphatidylserine (PS) + phosphatidylinositol (PI) vesicles, PI is found mostly in the inner monolayer (Barsukov et al., 1980). However, if PI vesicles are mixed with phosphatidylcholine (PC) vesicles, spontaneous exchange results in formation of highly asymmetric vesicles in which PI is localized in the outer monolayer. In these vesicles peroxidation of lipid promotes transbilayer movement which ultimately leads to opposite asymmetry resembling that of cosonicated vesicles.

Vesicles of Polymerized Lipids

Single-walled vesicles of high mechanical stability are very desirable for theoretical and practical reasons. For example, stability of membranes of *Archaebacteria* is probably due to the bipolar lipids (cf. Chapter 2). Bilayers supported on strong polymer capsules (*corked capsule membranes*) that respond to changes in the external environment have also been prepared (Okahata, 1986). Covalent cross-linking of phospholipids in vesicles, either laterally or near the terminal methyl ends, also gives rise to very stable vesicles. Thus, vesicles of dioctadecadienoyl ammonium bromide can be polymerized into large vesicles of over 50-micron diameter in the gel as well as the fluid phase (Gaub et al., 1985). Vesicles with a heterogeneous outer shell containing monomeric and polymerized lipids combine mechanical stability with flexibility which is very desirable for certain applications. The viscoelastic properties of such vesicles are expected to resemble those of erythrocytes which are stabilized by the spectrin–actin network on the cytoplasmic surface. Besides mechanical stability, vesicles of polymerized lipids exhibit appropriately modified properties: Somewhat lower permeability to water and polar solutes, ability to incorporate nonpolar solutes, gel-to-fluid phase transition, lateral mobility of monomeric lipids, and phase separation.

Planar Bilayer Lipid Membrane

Formation of a bilayer across a hole on a hydrophobic septum of Teflon, Kel-F, or polypropylene or on the tip of a glass pipette (Fig. 4-2) offers an attractive system for the measurement of electrical properties of bilayers (Jain, 1972; Tien, 1985). Appropriate electrodes can be inserted into the two aqueous compartments whose contents can also be readily altered. In this configuration a bilayer lipid membrane (BLM) can be made spontaneously either by applying a solution of lipids, or by apposing two monolayers from the two sides of the septum (Montal and Mueller, 1972), or by spreading monolayers from liposomes (Schindler, 1980), or by forming a bilayer on the tip of a pipette. The use of lipid-impregnated filters for reconstitution has the advantage that such membranes are not as unstable as the unsupported bilayers. Besides addition of proteins to the aqueous phase, proteins can also be incorporated into BLM by fusing vesicles containing proteins (Jain et al., 1972; Miller 1986). Attachment of membrane fragments to BLM has also yielded useful information about transport processes (Fendler et al., 1985; Latorre et al., 1985). The properties of BLM (Table 4-2) are consistent with the bilayer organization. In spite of their poor stability and difficulties in controlling their composition, BLM is one of the best models available for the study of the ion transport properties of channels and carriers for ions (Chapter 9). Planar *monolayer* membranes separating two aqueous compartments have also been prepared from bipolar isoprenoids, and exhibit some unusual properties.

As shown in Fig. 4-4 lipid bilayers can be formed on the tip of a pipette

TABLE 4-2 Comparison Between Natural and Artificial Membrane Properties

Property	Natural Membranes	Black Lipid Membrane
Hydrocarbon thickness (Å)	20–40	25–70
Capacitance (μF/cm^2)	0.8–1.5	0.3–0.9
Specific conductance (S)	10^{-2}–10^{-5}	10^{-3}–10^{-9}
Dielectric breakdown (mV)	100–400	100–500
Refraction index	1.6	1.3–1.6
Water permeability (μm/sec)	0.25–400	8–50
Reflectance	—	0.0001
Interfacial tension (dyne/cm)	0.2	1
Conductance	10 mS	0.1 nS
Ion selectivity	for K	None
Dielectric constant	5	2
Thickness	4 nm	4 nm

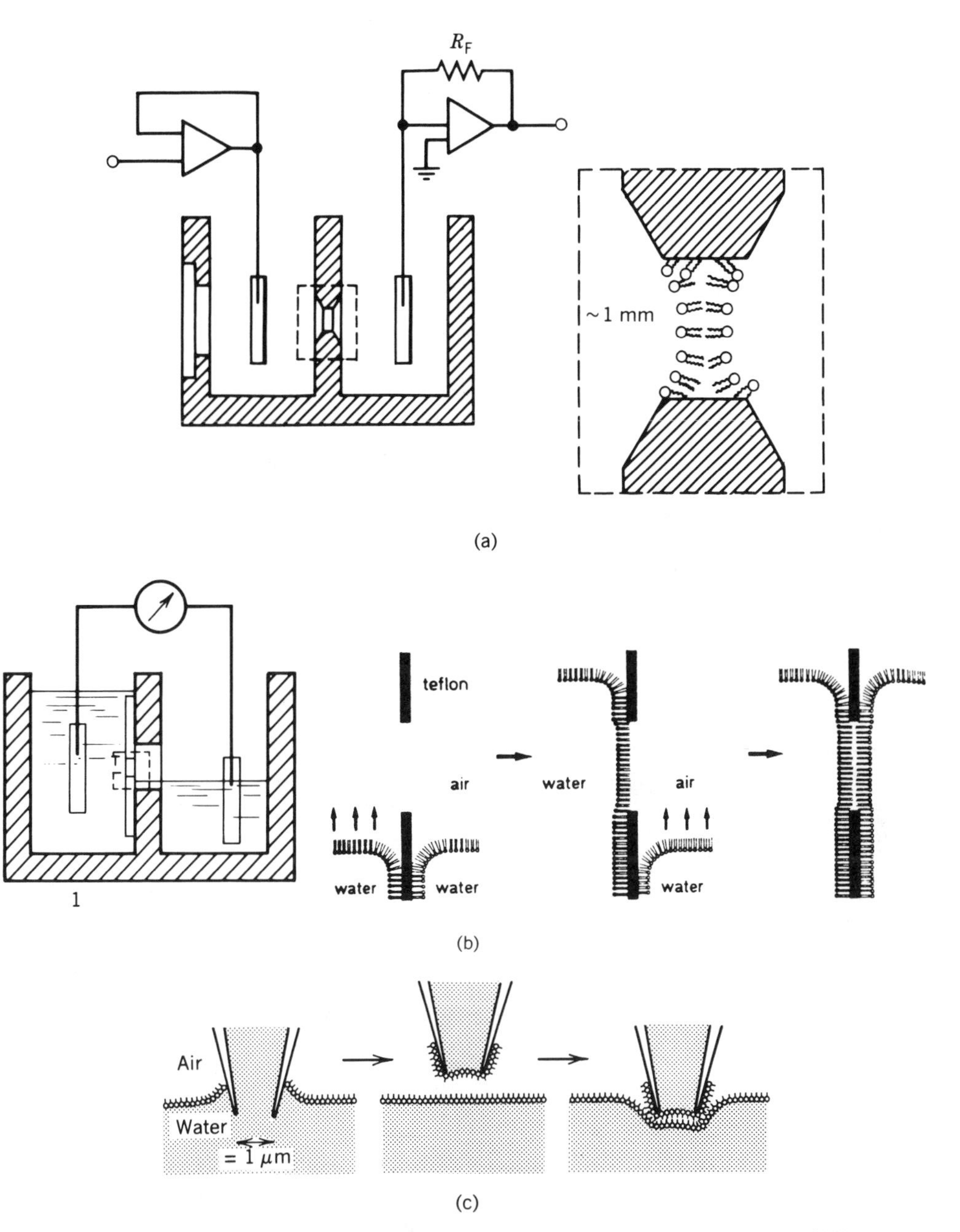

Fig. 4-4. Schematic representation of the experimental set-up for the formation of bilayer lipid membrane (BLM) by painting a film of lipid on a hole on Teflon cup (a), by apposing two monolayers on a hydrophobic partition by raising the level of the aqueous phase on which a monolayer has been formed (b), by apposing two monolayers on the tip of a micropipette by first removing the tip from the aqueous phase on which a monolayer is present (c), and then reinserting the tip through the monolayer. (Courtesy of P. Lauger.)

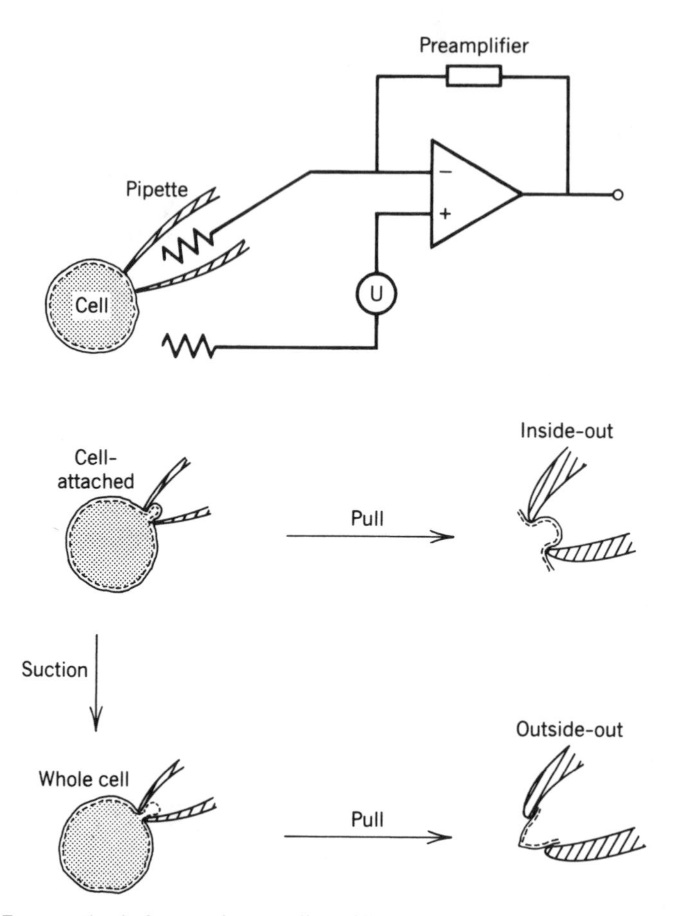

Fig. 4-5. Four methods for patch-recording. All methods start with a clean fire-polished pipette pressed against an intact cell to form a gigaohm seal between the pipette and the membrane it touches. Conductance pathways can be recorded in this "cell-attached" mode. Other modes of recording are attained by pulling (inside-out patch), suction (whole cell), or suction followed by pull (outside-out patch). A technique for perfusion of the contents of a cell has been described (Jauch and Lauger, 1986). (Adopted from Neher and Struhmer, 1985.)

with a small area, typically 1 μm^2. Electrically isolated patches of biological membranes can also be sealed on the polished tip of a pipette in many different forms (Fig. 4-5). Such preparations are very useful not only to study the conductance behavior of a few channels that may be present in the patch but also to study lateral mobility and distribution of channels in a living cell. Secretory events that bring about a change in the area of the membrane also exhibit fluctuations in electrical capacitance.

DISPERSIONS OF IONIC PHOSPHOLIPIDS

The swelling of charged phospholipids exhibits a pronounced dependence on pH and salt concentrations. In the presence of monovalent cations and at high pH, anionic phospholipids show continuous swelling and ultimately small unilamellar vesicles are formed (Hauser, 1984). These effects are due to interlamellar electrostatic repulsion, and, therefore, lowering pH, increasing the salt concentration, or the presence of multivalent cations leads to limited swelling and formation of multilamellar vesicles.

Phosphatidylserine

Phosphatidylserine (PS) is a major anionic lipid component of mammalian plasma membranes like myelin, erythrocytes, platelets, and rod outer-segment disk membranes where it is preferentially localized on the cytoplasmic side. Deprotonation of the ammonium ($pKa > 8$) and the carboxylic ($pKa < 7$) groups can occur under physiological conditions, however at pH 7–8 both of these groups are ionized. The surface charge density of PS vesicles ranges from 23 to 36 μC per cm^2, depending upon the nature of the fatty acyl chains.

The properties of bilayers of PS are altered appreciably in the presence of ions. In the presence of monovalent cations the transition temperature and enthalpy of the melting of acyl chains are somewhat higher compared to the corresponding PC. The effect of divalent ions is complex. Equilibrium binding studies suggest that for the reaction $2PS + Ca \rightleftharpoons Ca(PS)_2$, the binding constant is determined only by the concentration of free calcium ions (Feigenson, 1986). Formation of the $Ca(PS)_2$ complex is accompanied by extensive fusion of vesicles followed by formation of multilamellar aggregates in which there is extensive dehydration of the interlamellar space and solidification of acyl chains (Hauser and Shipley, 1984). The transition temperature of $Ca(PS)_2$ dispersions is very high. These observations suggest that interlamellar bridging by calcium occurs. Dehydration does not occur with monovalent cations, however the interlamellar repeat distance decreases [61 for Å dimyristoyl PS (DMPS) in 0.5 *M* NaCl] due to shielding of ionic groups. Li causes progressive crystallization of the acyl chains and an appreciable reduction in the interlamellar distance. Two forms of packing of acyl chains have been observed in the dispersions of PS. Each methylene group contributes 2.2 Å to the thickness of the bilayer in form II and 2.1 Å in form I. The head group contributes 6 Å in both cases. The $Ca(PS)_2$ complex prefers the tilted form II. The primary site of binding of Li and of multivalent cations appears to be the phosphate group, which is also dehydrated along with the cation.

In mixed bilayers containing PS with a zwitterionic phospholipid like PC, divalent ions induce isothermal phase separation (Stewart et al., 1979). This implies that lateral Ca-bridging occurs between PS molecules.

Phosphatidic Acid

The presence of two ionizable groups on phosphatidic acid (PA) (p*K*a 3 and 7) gives rise to several distinguishing features to its aqueous dispersions. For example, varying pH induces a systematic change in the phase properties (Eibl, 1983). Full protonation or the presence of divalent ions leads to crystallization of chains. These changes are considerably moderated by the substitution of a proton by methyl group.

Cardiolipin

Both glycerol moieties of diphosphatidylglycerol (DPG) adopt identical conformations and are oriented perpendicular to the plane of the bilayer (Allegrini et al., 1984).

Lateral Compressibility

The isothermal area compressibility of fluid bilayers has been measured by the pipette-aspiration technique (Kwok and Evans, 1981). The compressibility modulus is about 0.13 N/m, which is in excellent agreement with the estimates of the values of internal surface pressure obtained from the monolayer data. The isothermal thickness compressibility is estimated to be $5 \cdot 10^7$ N/m^2.

ELECTRICAL DOUBLE LAYER AT THE INTERFACE

The total membrane electrical profile of a phospholipid bilayer has two major components (Fig. 4-6). The **transmembrane potential** is the net electric potential difference between the bulk aqueous phases. It arises from net charge separations across the bilayer due to selective permeability for diffusion of ions (to be discussed further in Chapter 9). The **surface potential** is the electrostatic potential at the bilayer interface relative to that in the corresponding bulk solution, and this arises from a net charge on the membrane surface. For a review of other factors that contribute to the total membrane potential see Honig et al. (1986).

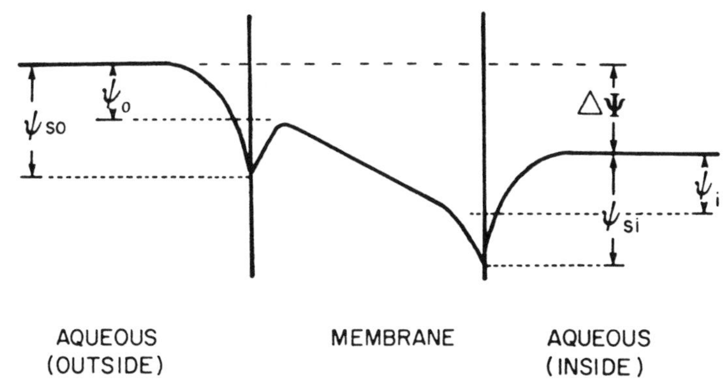

Fig. 4-6. Schematic representation of total membrane electric potential profile as the sum of the transmembrane potential ($\Delta\Psi$), surface potentials (ψ_s) and internal (e.g., dipole or adsorption) potentials (ψ). Suffixes i and o refer to the potentials on the inner and the outer interfaces. An intramembrane potential is defined as the net potential drop across the membrane. In the absence of surface and internal potentials, the intramembrane potential is the same as the transmembrane potential; in other cases intramembrane potential is obtained as a difference between the surface or the adsorption potentials at the two interfaces.

The interface between the hydrophobic region of the bilayer and the polar aqueous phase exhibits an uneven distribution of ions. The interface of PC is zwitterionic between pH 3 and 12, whereas the p*K*a of the NH_2 group in PE is about 8.5. On the other hand dispersions of PS, PG, PA, DPG, PI, globosides, and galacerebrosides have a net negative charge at pH 5–8. A charged bilayer in an aqueous solution will attract ions of the opposite charge (counterion) and repel ions of the same charge. This region of unequal positive and negative ion concentrations in the vicinity of the interface is commonly referred to as the electrical double layer. The charge profile at the interface is discrete and probably periodic, and it is difficult to treat this situation quantitatively. Therefore, it is often assumed that away from the interface the charge is smeared out and decays exponentially with the distance away from the interface.

Several types of interactions determine the behavior of counterions in the vicinity of a charged surface. A steady drop in the electrical potential in the aqueous phase away from the charged surface arises from a dynamic competition of the thermal motion with the attractive forces for the counterions. The potential drop away from the charged interface experienced by a counterion can be described by either the Donnon equilibrium potential or the diffuse double layer at the interface. However, Ohshima and Ohki (1985a) have shown that these possibilities are not fundamentally different. The underlying model consists of a planar charged membrane that is in equilibrium with the electrolyte solution. The interface consists of a layer of thickness d, which contains fixed charges and is permeable to counterions. In the Donnan potential, d is very large, and in the diffuse double layer $d = 0$.

The theories for electrical potential at or near interfaces make certain assumptions, which are schematically illustrated in Fig. 4-7. If ions in direct contact with the fixed charge at the interface are completely or partially

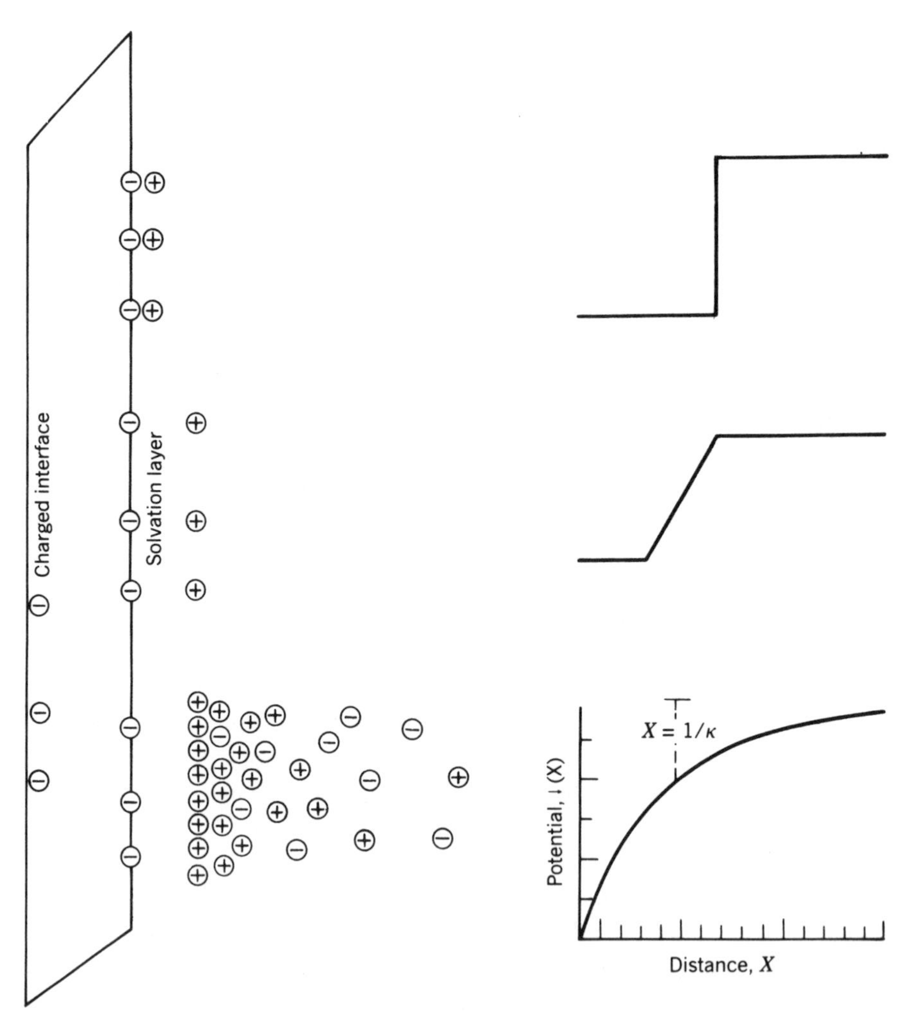

Fig. 4-7. Distribution of ions (left) and the potential drop (right) near a negatively charged surface. (Top) Discontinuous potential drop for interaction between desolvated ions; (middle) continuous linear potential drop approximated for interaction of solvated ions; and (bottom) surface potential drop due to diffusion of solvated counterions as predicted by Gouy–Chapman theory.

desolvated, binding of counterions would be very strong, thus establishing local electroneutrality. Under these conditions the potential drop will be essentially discontinuous and abrupt. If the electrostatic attraction of the counterions is not strong enough to cause desolvation, the counterions will be effectively stopped at a distance equal to the radius of the hydration sheath. However, counterions could readily diffuse away from the interface. Under these conditions electroneutrality is maintained because the total number of counterions near the interface remains constant. Also in a mixture of counterions, only the ions with relatively stronger affinity would approach the charged interface. Thus, the binding of a counterion is shielded by other counterions, and the extent of shielding will depend upon the charge densities of the interacting species. This situation is approximated as Helmholtz double layer potential drop given by:

$$V = \frac{4\pi r\sigma}{\varepsilon}$$

where σ is the charge density (=surface charge/area), r is the separation between the fixed ions and the counterions at the interface, and ε is the relative dielectric constant of the medium. The potential drop experienced by the counterion approaching the interface is assumed to vary linearly with the distance. Several other assumptions are implicit in this parallel plate condenser model of the interfacial electrical double layer. For example, thermal motion would make the layer of counterions more diffuse, and thus the potential drop would be a more complex function as developed below.

The Gouy–Chapman theory attempts to describe the diffuse double layer through nonlinearized Poisson-Boltzman law for the distribution of counterions in a two-dimensional space moving away from a planar surface containing point charges. With such assumptions, the potential drop in a diffuse double layer at a distance x from the interface is given by:

$$V_{(x)} = \frac{\sigma}{\varepsilon\kappa} \exp\left[-\kappa (x - a)\right]$$

Here κ determines how rapidly the decay of the potential occurs. Debye length, $1/\kappa$, represents the distance from the interface where the potential has dropped to $1/e$ of its value at the interface. Thus, Debye length would decrease as ionic strength increases. At a distance less than a, there is no diffusion of ions. Under these conditions the Helmholtz double-layer approximation holds, as can be seen by substituting $x = 0$. Note that the use of

$1/\kappa$ in the Guy–Chapman theory produces the potential at the interface rather than at the Debye length from the interface.

Several other treatments and more rigorous solutions of the diffuse double-layer theory have appeared (Toko and Yamafuji, 1980, and references therein; Carnie and McLaughlin, 1983). Some of the more promising treatments of the diffuse double layer take a combination of the following factors into considerations (Starzak, 1984):

1. The Graham equation for the diffuse double layer relates the surface charge to surface potential, ψ, as:

$$\sigma^2 = 2\varepsilon RTC\left[\exp\frac{-zF\psi}{RT} - 1\right]$$

   where C is the bulk concentration of ions of valence z, and F is the Faraday constant.

2. The Boltzmann equation relates the concentrations at the interface (C_o) to the bulk concentrations as

$$C_0 = C\exp\left[\frac{-zF\psi}{RT}\right]$$

3. The Langmuir adsorption isotherm relates the concentration of bound ions to the concentration of ions at the interface to the concentration of the fixed charges on the interface as:

$$K = \frac{[CL]}{[L][C_0]}$$

A numerical solution of these three equations is possible.

A negative charge at the interface produces an electric field that attracts counterions. The surface potential falls off exponentially to $1/e$ value at Debye length, which is between 1 and 10 nm when the monovalent salt concentration is 0.1 and 0.001 M. The magnitude of the electrostatic potential for a bilayer with about 25% of the lipids (area 70 Å^2 as anionic lipids and the concentration of salt about 0.1 M, the potential at the interface would be −60 mV. In other words for a surface potential of −60 mV, the surface concentration of monovalent anions is depleted 10-fold, while that for monovalent cations is enhanced 10-fold:

$$C(\text{surface}) = C(\text{bulk}) \cdot \exp(-zF\Psi/RT)$$

Such simple calculations exmphasize the fact that the surface potentials do not contribute to the transmembrane potentials measured by placing the electrodes in the bulk aqueous phase. However, the interfacial potentials do influence the ionic concentrations in the vicinity of the interface and therefore the p*K*a of the ionizable groups:

$$\mathrm{p}K(\text{surface}) = \mathrm{p}K(\text{bulk}) - (\Psi F/2{\cdot}3RT)$$

The surface potential along with the transmembrane potential could, however, contribute to the orientation of dipoles in the membrane, and thus influence the effective activation energy for many of the membrane processes, including transport, current–voltage relationships, and gating of channels. Some of these are discussed in subsequent chapters.

The consequences and general predictions arising from electrical double-layer theory have been used successfully to account for a variety of effects, including shielding or screening effect, polarization (Hainesworth and Hladky, 1987), electrocapillarity, electrophoresis, electroosmosis, and streaming potential (Starzak, 1984). On a macroscopic level, binding of ions to an interface induces perturbation of the interfacial charge properties, such as the electrophoretic mobility of vesicles (Fig. 4-8) and electro-osmosis (Balasubraminan and McLaughlin, 1982) of vesicles. From such measurements it is possible to obtain ζ potential at about 2 Å from the surface. This distance probably corresponds to the plane of the hydrodynamic shear. The surface potential is perturbed by screening by high electrolyte concentration, by changing pH, by ionic amphiphiles, by multivalent ions that bind to the interface, by changing the mole fraction of charged phospholipids, and by chemical modification of charged groups in the interface.

Zwitterionic lipids like DMPC do not have a net charge, however, specific binding of anions induces significant negative ζ potential in the decreasing order (Tatulian, 1983): trinitrophenol > perchlorate > iodide > thiocyanate > bromide > nitrate > chloride = sulfate. A sharp increase in the ζ potential is detected at the phase transition from gel to fluid phase due to a change in the number of the binding sites per unit area rather than a change in the binding constant. The effect of anions on the bilayer of zwitterionic phospholipids is also substantiated by the fact that the interlamellar distance and the thickness of the dipalmitoyl PC bilayer increases in the presence of 1 *M* KBr, whereas KCNS appears to induce interdigitation of the acyl chains (Cunningham and Lis, 1986). The anions of the Hofmeister series are known to influence organization of water (Collins and Washabaugh, 1985), and their effect on the bilayer could occur through the hydration layer at the interface.

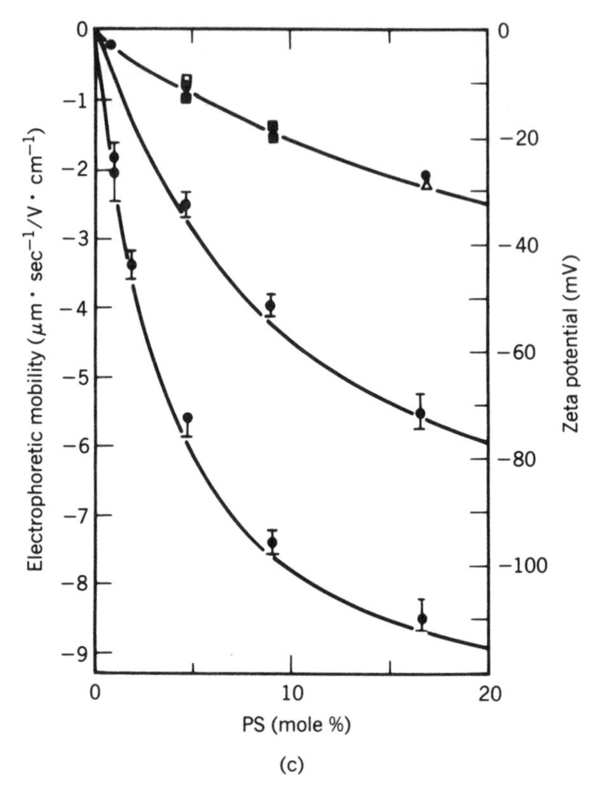

Fig. 4-8. Electrophoretic mobility and ζ potential of multilamellar vesicles formed from a mixture of PC with PS. The curves illustrate the predictions of Gouy–Chapman–Stern theory if the intrinsic association constant of sodium with PS is 1 M^{-1}. The three curves are (from top) for 0.1, 0.01, and 0.001 *M* NaCl buffered to pH 7.5 at 25°C with 10, 1, and 0.1 m*M* MOPS, respectively. Such a good fit is seen only when there is no desolvation of the interface or the ion.

In addition to screening the fixed charges, the ions in solution can bind to and neutralize the fixed charges, thus actually decreasing the surface charge density. This has several interesting consequences. For example, not only the concentration but also the nature of the ions determine the surface potential and surface ion concentration; binding of ions would lower the surface charge density. Multivalent ions are much more potent in screening surface potential.

Strong evidence for binding of calcium to PS phosphate (but not to carboxylate anion) with dehydration of the phosphodiester group has been obtained (Dluhy et al., 1983). Similarly, dehydration and phase-dependent dissociation of calcium has been noted for DPPG (Boggs et al., 1986). Formation of $Ca(PS)_2$ complex lowers the surface charge density and therefore the electrophoretic mobility and the ζ potential of vesicles. Similarly, intermolecular bridging of PS molecules by calcium ions can increase the surface tension at the air–water interface. From such measurements it is possible to obtain interfacial association constants. Unfortunately, these association constants are dependent upon the underlying models, and therefore their values can not be compared directly. The association constants for divalent ions (1 : 1 complex) are about 8–15/mole ($=K_2$), and 0.1–1 per mole ($= K_1$) for monovalent ions (Hauser et al., 1977; Lau et al., 1981; McLaughlin et al., 1981). The association constant for $Ca(PS)_2$ complex is about 30–50/mole (Ohshima and Ohki, 1985b). These values are considerably lower than the apparent association constants implicated from direct binding studies (Feigenson, 1986). Apparently, $Ca(PS)_2$ complex is formed when the concentration of calcium exceeds a critical value (about 1 μM) which depends upon the nature of the acyl chains. These results have been interpreted by assuming that free calcium along with PS and $Ca(PS)_2$ coexist as separate phases. This situation is analogous to the decomposition equilibrium of $CaCO_3$ (s) to CaO (s) + CO_2 (g), where the equilibrium constant is solely determined by the partial pressure of CO_2. The full significance of these observations remains to be elaborated.

Usefulness of electrical double-layer theory has been somewhat limited because the mathematical relationships are complex, and many of the underlying assumptions are probably not valid. However, some of these assumptions fortuitously cancel each other out. The empirical evidence also indicates that assumptions about some other parameters simply do not affect the accuracy with which the theory describes the electrostatic and ionic environment near a phospholipid bilayer. Many of the experimental complications arise from competition between protons, monovalent and divalent cations, change in p*K*a of the ionizable groups in the interface, desolvation of the groups in the interface during binding, effect of bound ions on the hydrogen-bonding characteristics of the groups in the interface, and effect of inter- and intralamellar bridging by counterions on the organization of acyl chains. It is also probably relevant to consider the nature of interfacial equilibrium if it involves partial desolvation. Under such conditions the counterion could jump from one site to another without leaving the interface, and therefore the apparent dissociation constant would be smaller.

TRANSBILAYER MOVEMENT OF PHOSPHOLIPIDS

The rate of transbilayer movement of phospholipids in liposomes is immeasurably slow, with a lower limit for the half-time as >50 days. This is to be

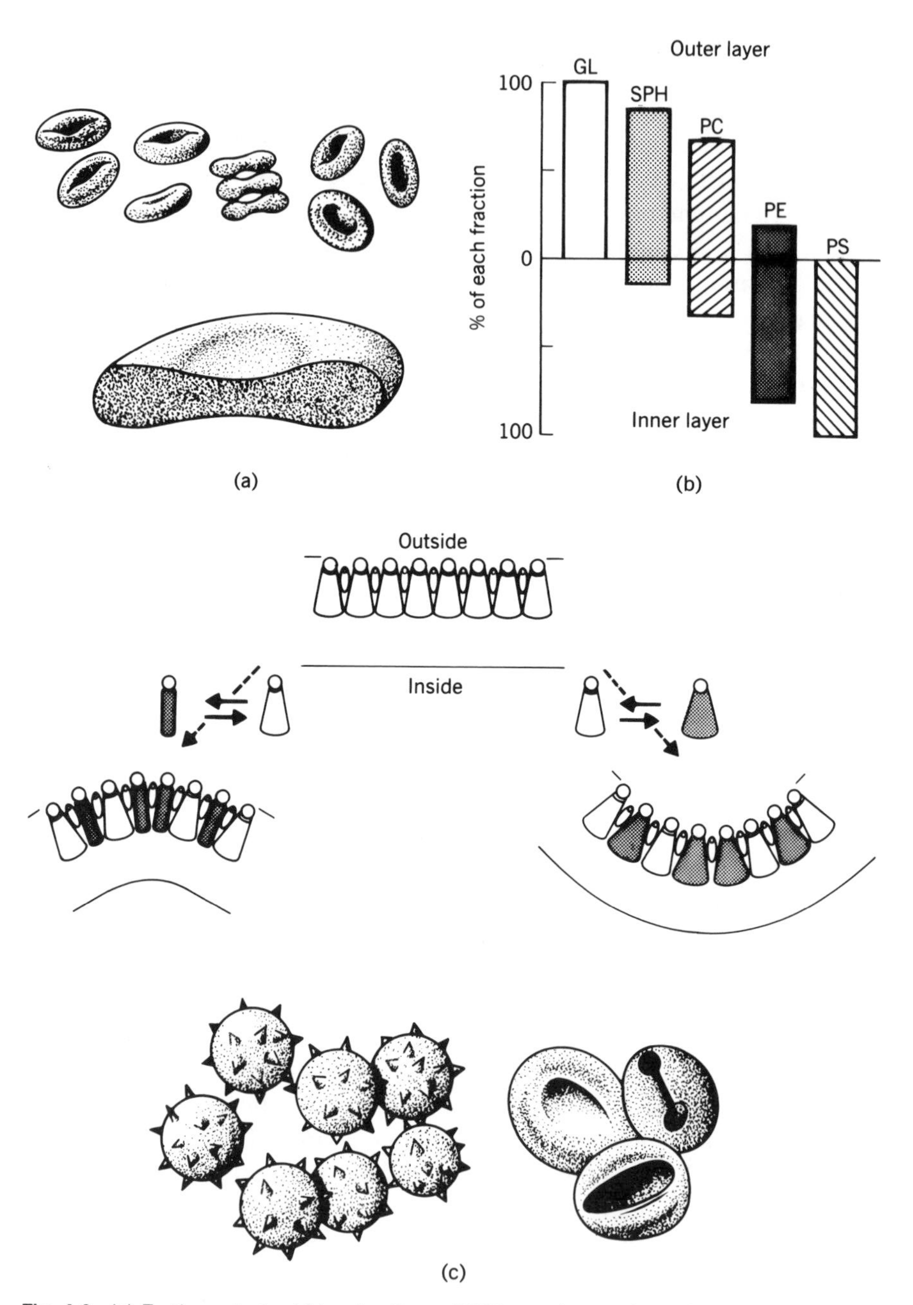

Fig. 4-9. (a) Erythrocyte (red blood cells, or RBC) membranes have been studied extensively because they are relatively simple and abundant. There are 5×10^9 erythrocytes in 1 ml of blood. Erythrocytes are small cells without nucleus or internal membranes, and they are stuffed full of oxygen-binding protein hemoglobin. Their normal shape is biconcave, but it is very sensitive to perturbations by lipophilic solutes and phospholipid composi-

TABLE 4-3 Half-Times for Transbilayer Movement across Erythrocyte Membranes

Lipid	Half-Time
PC[a]	3–30 hr
Lyso-PC[b]	4–10 hr
Cholesterol	<1 min
Diacylglycerol	<1 min
Fatty acid	<1 min

[a] PC, Phosphatidylcholine; Middlekoop et al. (1986)
[b] Bergmann et al. (1984).

expected because more than 15 kcal/mole is required for the transfer of a polar group through the hydrophobic interior of the bilayer. The rate of spontaneous transbilayer transfer of lipids depends upon their structure (Table 4-3), i.e., chain length, temperature, pH, and cholesterol content. The rate of transbilayer movement of phospholipids in biological membranes is found to be considerably faster, i.e., from 20 hours in virus and erythrocyte membranes to a few minutes in *Bacillus megaterium* membranes capable of conducting *de novo* phospholipid biosynthesis. ATP-dependent transbilayer distribution of phospholipids has also been noted in erythrocytes (Seigneuret and Devaux, 1984). The very fast rates of transbilayer movement of phospholipids have been attributed to a hypothetical enzyme christened as *flippase*. However, other factors could also accelerate the rate of transbilayer movement. Inclusion of membrane proteins that desolvate the protein-bilayer microinterface, small amounts of detergents, and lipid-soluble solutes have been shown to lower the half-time for transmembrane movement of phospholipids in liposomes to a few hours. Similarly, conditions that promote formation of nonbilayer phases also accelerate transbilayer movement.

One of the consequences of slow transbilayer movement is that the phospholipid composition of most biological membranes is asymmetrical (Fig. 4-9 and Table 4-4) as reviewed by Op den Kamp (1979). Possible functions of membrane asymmetry include selective activation of proteins and asymmetric vesiculation of membranes for endo- and exocytosis. The physiological role of transbilayer movement of phospholipids is in biogenesis of mem-

tion. (b) Asymmetric disposition of polar lipids in erythrocyte membranes. (c) Shape changes in erythrocytes can also be induced by replacing the acyl chains. Native PC commonly contains palmitoyl, oleoyl, palmitoleoyl, and linoleoyl species. If they are replaced by more saturated (left) or unsaturated (right) species, the curvature of the membrane changes. This causes a major change in their overall shape and the viscoelastic properties.

TABLE 4-4 Asymmetric Distribution of Lipids in Membranes

Species	% Lipid Detected in Outer Leaflet[a]	
Erythrocyte		
Human	PC	76
	PE	20
	SM	82
	PE	33
	PS	0
	Cholesterol	66
Rat	Cholesterol	69
Mouse fibroblast plasma membrane	Cholesterol	20
Chick embryo fibroblast plasma membrane	PE	35
	PS	16
Chick embryo myoblast plasma membrane	PE	62
	PS	45
Rat liver microsomes	PE	33
Beef heart mitochondria inner membrane	PC	72
	PE	30
	CL	25
Rat liver mitochondria inner membrane	PE	30–40

[a] PC, Phosphatidylcholine; PE, phosphatidylethanolamine; SM, sphingomyelin; PS, phosphatidylserine; CL, cardiolipin.

branes. However, slow transbilayer movement of hydrophobic solutes and drugs could be responsible for the blood–brain barrier function (Jain et al., 1985b). Also the rate of transbilayer movement of lipids is increased in the presence of certain solutes and during the dielectric breakdown of the bilayer.

LATERAL OR TRANSLATIONAL DIFFUSION

Diffusion of membrane components in the plane of the bilayer has been extensively studied; (see Clegg and Vaz, 1985 for a critical review of the various models). A simple hydrodynamic model (Fig. 4-10) is based on the assumption that the lipid bilayer is a fluid of viscosity η. Thus, the lateral diffusion coefficient (D) and rotational diffusion coefficient (R) of a protein of radius (a) spanning a membrane of width h is (Saffman, 1976)

$$D = \frac{kT}{4\pi\eta h}\left(-\gamma + \log\frac{\eta h}{\eta_w a}\right)$$

$$\text{and } \frac{D}{R} = a^2\left\{\ln\left(\frac{\eta h}{\eta_w a}\right) - \gamma\right\}$$

η_w is the viscosity of the surrounding aqueous phase. γ is Euler's constant $= 0.5772$. These equations are approximations that are valid when η_w (about 0.01 poise) is much less than η for membrane (about 0.5 poise; see Chapter 5), and they appear to be valid in many cases (Axelrod, 1983). It may be noted that for two-dimensional diffusion, the value of D does not change significantly with the molecular weight. However, a critical test of this and the free volume theory of lateral diffusion is on the basis of the values of lateral diffusion coefficients for small solutes (Vaz et al., 1984) whose size is comparable to the size of the phospholipid molecule. The available evidence seems to favor the free volume model.

From a physical point of view, lateral diffusion is the net result of a series of collisions (Fig. 4-10). Under these conditions the mean free path is much smaller than the diameter of the particles. The average of the square of the displacement, x during time t is related to lateral translational diffusion coefficient D in two dimensions as

$$x^2 = 4Dt$$

For a cell with a radius of 5000 nm, the square of the circumference is 10^9 nm^2. If $D = 10^{-8}$ cm^2/sec, it will take about 8 min for a protein to diffuse around a cell.

There is considerable experimental evidence suggesting that the components of membrane, both lipids and proteins, can move past each other in the plane of the membrane. The lateral diffusion coefficient (D) has been calculated to be in the range of 10^{-7} to 10^{-9} cm^2/sec for phospholipids and probes (e.g., pyrene) for bilayers in the fluid phase, and the values are 100- to 1000-fold smaller in the gel phase. The activation energy is about 4–6 kcal in the fluid phase and about 10 kcal/mole in the gel phase.

The values of D for lateral diffusion of phospholipids in bilayers are about 1000-fold smaller than the corresponding values for free diffusion of solutes of comparable size in the aqueous medium. For proteins D is about 10^{-10} cm^2/sec in the absence of extramembrane constraints such as cytoskel-

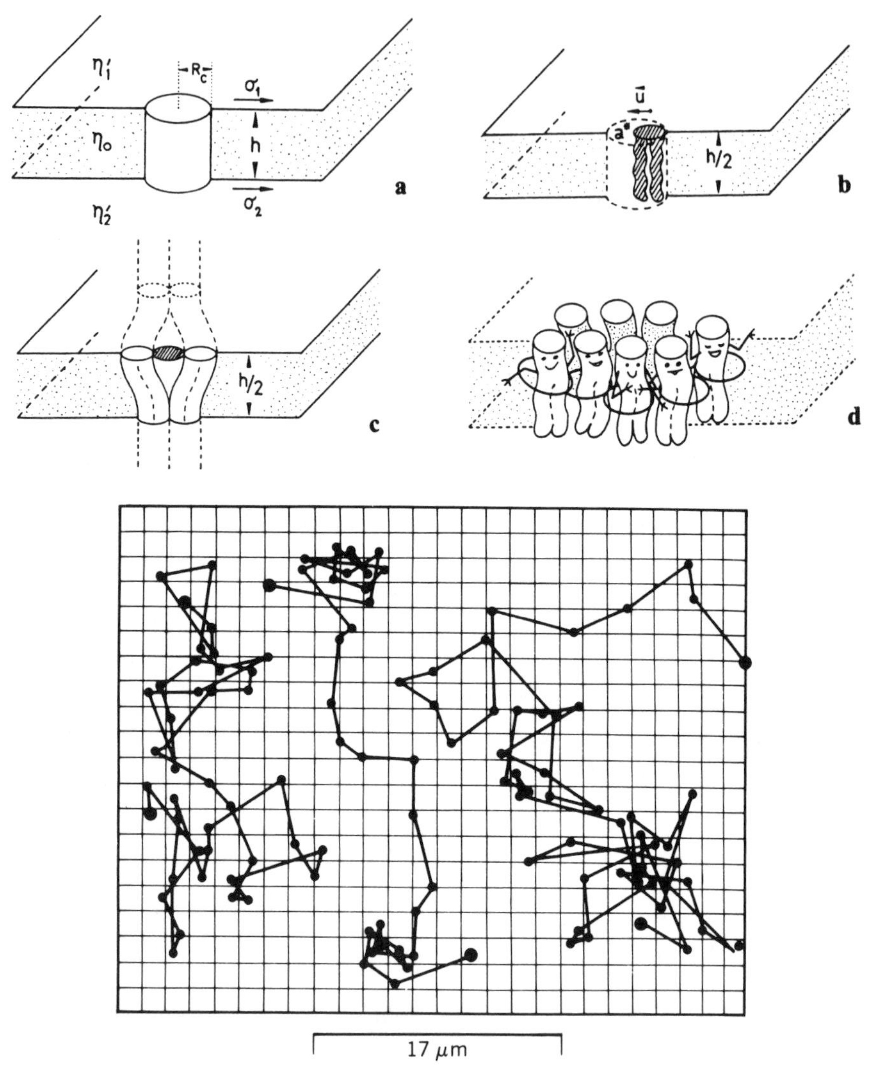

Fig. 4-10. (Top). Schematic representations of the main models for lateral diffusion. h is the thickness of the bilayer in all cases. **(a)** *Hydrodynamic model.* The hydrodynamic description for a diffusing cylinder (radius a) spanning the bilayer. The surface stress (force/area) at the aqueous-membrane interface, are caused by the fluid velocity gradients in the aqueous phase. The membrane *fluid* is not allowed any velocity gradients perpendicular to the sheet, so the σ's produce a *body* force (force/volume) within the membrane sheet which is tangential to the membrane. The energy dissipation within the sheet is modelled by hypothesizing a three-dimensional viscosity which is uniform and constant within the sheet (see text). **(b)** *Free-volume model.* The free volume of a molecule (in this case a lipid with two chains) is considered to be the volume in which the molecule can move unhindered by the adjacent molecules. The corresponding free area of the two-dimensional membrane is indicated. We have pictured the average *cage* as existing within a continuum fluid and given it a cylindrical shape (circular area). The free volume (free

eton interactions (see Chapter 7 for further discussion on this point). A lateral diffusion coefficient of $5 \cdot 10^{-9}$ cm²/sec corresponds to a root-mean-square molecular displacement, x (where $D = x^2/4t$) of 300 nm/sec. Therefore, the average time between collisions of two molecules separated by a distance d is $\tau = d^2/4D$. Similarly the relaxation time for randomization (τ) of a point solute inserted in a vesicle of radius r, is given by $r^2/2D$. Thus, randomization of a newly synthesized lipid in a bacteria of radius 500 nm would occur in about 1 sec, whereas randomization of a newly synthesized protein may take about 100 sec. However, there is evidence showing that not all proteins in membranes are randomly distributed.

Of the various methods for the measurement of lateral diffusion coefficient, fluorescence photobleach recovery (FPR) is widely applicable. Using a brief, intense pulse of laser light of appropriate cross section and wavelength, an initial concentration gradient is created by irreversible destruction of fluorophores. The rate of fluorescence recovery due to the lateral diffusion of unbleached fluorophores in the observation region can be readily measured because the rate of intervesicle hopping of the probes covalently attached to the lipid or transmembrane membrane protein is relatively slow. Lateral diffusion coefficients have been measured for a variety of solutes. The data suggest that D increases with increasing temperature, and decreases somewhat with the cholesterol content of the bilayer. Similarly, cross-linking of the membrane component not only retards diffusion and induces patching, but the diffusion of the species that are not cross-linked is also retarded. For small solutes like pyrene and pyrene-labeled phospholipids, D is roughly proportional to the square root of the molecular weight, however, the lateral mobility of macromolecules (proteins) shows only a weak dependence on the size. Lateral diffusion coefficients for polymerized

---

area) around any molecule fluctuates and if sufficient space is available, i.e., $a_f > a^*$, the molecule may translocate to a neighbouring site (the site is not depicted). The figure is obviously an idealization. **(c)** *Flexible rod model.* The solid lines depict the separating chains for the monolayer part of the bilayer, and the dashed lines are the appropriate extensions for the infinitely long polymer model. The dark circle represents either the penetrant for the polymer model or the polar head group of a neighbouring lipid for the lipid bilayer model. The configuration shown is an activated state where the maximum separation is sufficient to allow the entrance of a neighbouring head group (or penetrant). **(d)** Perhaps a more realistic representation of the thermal motions of the lipid molecules in a membrane. From Clegg and Vaz, 1985. (Bottom) An example of trajectories of diffusing particles. This random path is followed by three mastic particles undergoing Brownian motion (diffusion) as seen under a microscope. The positions of the particles were measured every 30 sec and joined in straight lines to make the drawing. To appreciate the full complexity of the movement, one must imagine replacing each of the line segments with a trajectory as complex as one of those drawn, and the replacing each of those segments with an equally complex trajectory, and so on.

patches of lipids are also slightly smaller (Gaub et al., 1984). This is consistent with the modified Stokes–Einstein theory which predicts that D is proportional to log a, where a is the radius of the diffusing part of the molecule that penetrates in bilayer.

For the calculation of D by most (but not all) methods, diffusion is assumed to occur in a homogeneous two-dimensional matrix. There is considerable experimental evidence suggesting that such an assumption is not valid. In most bilayers, the fluorescence recovery or decay curve exhibits two or more rate constants, and often the recovery is far from total. Such observations suggest that two or more parallel pathways exist for lateral diffusion. Comparative measurements of D in erythrocytes with the excimer probe and photobleaching technique show that the random walk over a distance of about 10 nm is characterized by $D = 10^{-7}$ cm^2/sec, while for long-distance transport, D is about 10^{-9}/sec (Kapitza and Sackman, 1980).

The functional consequences of lateral diffusion could arise from lateral interactions of membrane proteins with other proteins as well as with lipophilic ligands, e.g., hormones, substrates, cofactors, and lipids (Axelrod, 1983; Berg and von Hippel, 1985).

EXCHANGE OF PHOSPHOLIPIDS BETWEEN BILAYERS

Redistribution of membrane components from the site of their synthesis or uptake to the site of ultimate location involves several unique processes that are not normally encountered by other soluble metabolites. Transfer of membrane lipids between organs, cells, and organelles occurs by a variety of pathways involving transmembrane movement, intervesicle transfer, spontaneous exchange, fusion, and protein-mediated transfer (Dawidowicz, 1987). In this section we discuss the properties of phospholipid exchange proteins that are found as cytoplasmic components. It may be pointed out that several lipid transfer proteins present in plasma are (Tall, 1986) not discussed here.

Spontaneous Exchange

Unmediated net transfer and exchange of phospholipids, glycolipids, and sphingolipids between bilayers is a slow process with half-times in excess of 50 h. This is because the monomer concentration of lipids (the transition state for this process) in the aqueous phase is very low. Relatively rapid rates of spontaneous transfer of phospholipids could arise from collisionally dependent transfer processes such as fusion. On the other hand, the rate of intervesicle transfer of short-chain phospholipids is considerably more rapid, and it occurs through soluble monomers (Nichols, 1985). The off-rate constant is about 0.89/min for 4-nitrobenz-2-oxa-1,3-diazole derivative of dihexanoyl PC, C_6-NBD-PC, whereas for the C_{12} analog it is about 0.0046/ min. The rate limiting step for the intervesicle transfer of solutes appears to be the escape of the lipid monomers from the bilayer into water. The rate of transfer increases in the order expected on the basis of their solubility in the aqueous phase (Pownall et al., 1983). From the value of the equilibrium constant it is also possible to calculate the forward or association rate constant. For pyrene nonanoic acid, it is at least 100-fold slower than that expected for a diffusion-controlled process (Doody et al., 1980) As expected, the values of the thermodynamic parameters also depend upon the chain length, which suggests that the free energy of transfer from water to bilayer is largely enthalpic. Based on such considerations the rates of spontaneous exchange of phospholipids between vesicles is very slow, less than 0.0001/sec (McLean and Phillips, 1984). This is expected on the basis of their monomer concentration in the aqueous phase, which is less than 1 n*M*.

Half-time for the exchange of cholesterol is relatively rapid (about 1 hr) among a variety of aggregates containing cholesterol, such as erythrocyte membrane, liposomes, lipoproteins, and viruses. The spontaneous exchange of cholesterol has been used to modulate cholesterol content of plasma membranes. As expected the rate of exchange is independent of metabolic energy, concentration of the donor and acceptor particles, ionic strength, and surface charge. However, the rate of exchange does depend upon temperature, composition, phase state, and the presence of lipid-soluble solutes. These observations are consistent with the suggestion that the exchange of cholesterol occurs via free cholesterol in the aqueous phase and the desorption of cholesterol is the rate-limiting step. Critical micelle concentration of cholesterol in aqueous phase is 25–40 n*M* at 25°C. This would indicate that the spontaneous transfer of cholesterol between aggregated phases could occur via micellar or aggregated cholesterol, however, direct evidence is not available.

Phospholipid Transfer Proteins

Exchange and net transfer of phospholipids between membranes is accelerated by a variety of transfer proteins (TP) that are specific for the head group of phospholipids. Thus, TP have been isolated for a large variety of lipids including PC, PI, cerebrosides, cholesterol, and cholesterol esters (Table 4-5). Similarly, TP have been isolated from tissues ranging from human plasma, homogenates of intestinal mucosa, brain, and liver, to yeast and

TABLE 4-5 Properties of Some Phospholipid Exchange Proteins[a]

Source	Lipid Specificity	MW	pI
Beef heart	PC >> SM	25,900	5.5
		21,000	
Beef brain	PI >> PC	I 29,000	5.2
		II 30,000	5.5
Beef liver	PC	20,000	5.8
Rat liver	PC	16,000	8.4
Rat liver	PI >> PC	>25,000	5.1
		>25,000	5.3
Rat liver	PE > PC > PI > SM	13,000	—
Potato tuber	PI > PC > PE	22,000	—

[a] PI, phosphatidylinositol; PC, phosphatidylcholine; SM, sphingomyelin; PE, phosphatidylethanolamine.

maize seedlings. TP for different phospholipids, but isolated from the same organism, do not show any immunological cross-reactivity. Thus, for example, antisera raised against two PI exchange proteins from beef brain are cross-reactive, yet they fail to react with the PC-TP from beef liver

Most of the TP are water soluble and acidic, with molecular weights of 15,000–30,000. They are head group-specific and noncovalently bind one phospholipid molecule. The exchange catalyzed by most TP is very specific, however nonspecific transfer proteins have also been isolated from beef liver. They catalyze insertion as well as extraction of one phospholipid molecule from interfaces of plasma membranes, membranes of organelles, bilayers, monolayers, and serum lipoproteins. In bilayers only the lipid molecules in the outer monolayer are exchanged. If the phospholipid molecule for which TP is specific is not present in the acceptor interface, then most TP can mediate net transfer of the phospholipid.

TP enhance the off-rate of the phospholipid from the donor. Exchange of phospholipids mediated by TP is characterized by the following: it is time and temperature dependent; it is bidirectional; intact phospholipids are exchanged; and it does not require energy or any other component or cofactor. The TP-mediated exchange is promoted in the fluid phase and by the presence of lipophilic impurities. The rate of exchange increases with the concentration of TP, and reaches a maximum value at high protein concentration. The turnover number of most exchange proteins is about 3000 per min between plasma membrane and liposomes. TP do not seem to perturb the bilayer organization, however, the exchange is stimulated by low concentration of bilayer-perturbing agents like deoxycholate. The rate of exchange also depends upon the phase properties of the donor membrane. Exchange occurs rapidly only when both donor and acceptor vesicles are in the fluid phase. Thus, membranes containing unsaturated lipids are better donors than those containing saturated lipids. PC-TP from bovine liver, for example, cannot donate PC to vesicles in the gel phase. Similarly, the rate of exchange is inhibited by >25 mole% cholesterol in donor as well as in acceptor bilayers. Although removal of phospholipid monomers from donor vesicles is expected to be the rate-limiting step in the intervesicle exchange of phospholipids, a direct proof for this is lacking. Also, it is not certain if the exchange protein actually delivers the phospholipid to the acceptor membrane. It is likely that the phospholipid extracted from the donor vesicle is released in the medium, from which it can diffuse to the donor and acceptor vesicles in proportion to respective surface area and respective rates of uptake.

TP have been useful for measurement of the rate of transbilayer movement and for modulating the composition and asymmetry of membranes. Despite their presence in the cytoplasm of a variety of tissues and a possible role in biogenesis of membranes, the physiological role of TP remains to be settled.

Pathways for Intracellular Translocation of Lipids

Most of the enzymes responsible for lipid biosynthesis in animal cells reside on rough and smooth endoplasmic reticulum. Yet lipids are found in all the membrane systems of the cell, each with its characteristic composition. This raises the question of intracellular translocation, sorting, and targeting of molecular components destined for the various compartments. Methods for resolution of these questions are just beginning to emerge (Mills et al., 1984; Pagano and Sleight, 1985). Indications are that both the uptake and distribution of lipids occurs not only through insertion of monomers but also via an intracellular vesicular transport mechanism, membrane recycling, and endocytosis (cf. Chapter 15). These processes apparently show considerable specificity for different classes of lipids as well as for the target organelles.

TRANSVERSE DIFFUSION OF SOLUTES ACROSS BILAYERS

The bilayer of liposomes is typically surrounded by a variety of solutes, including water, salts, and protons. The bilayer of liposomes is more permeable to water and therefore subject to osmotic stress, i.e., a change in volume of liposomes occurs when a gradient of impermeable solutes is present. In small vesicles such changes are restricted by packing density and surface tension. In fact, the smallest vesicles are relatively insensitive to osmotic changes.

The osmotic pressure ($\pi = mRT$, where m is the molality of the solution) that develops at equilibrium across a semipermeable membrane is the result of an asymmetry in the chemical potential of water in the two compartments. On a molecular level this difference in the chemical potential would manifest itself as a difference in the probability of diffusion of the solvent from one interface to the other, and at equilibrium this is compensated by dilution of or by an increase in the hydrostatic pressure on the side of the more concentrated solution. Swelling and shrinking of MLV or LUV in response to a change in the osmolarity can be seen readily by monitoring a change in the turbidity.

Diffusion (Fig. 4-10) is the prime mover of solutes across membranes by a solubility diffusion mechanism based on partitioning of solutes in the bilayer. The driving force for the net movement of solutes by diffusion is the concentration gradient, and also the coulombic field force for ionic solutes. It is a statistical force describing the increase of randomness due to an increasing entropy of dilution without a change in the thermodynamic internal energy. Permeation across bilayers may also be viewed as a rate process that tends to equalize the activity of a solute at the two interfaces. Because of the thermal motion, solute molecules in the aqueous phase collide with the interface. The equilibrium concentration of solutes in the bilayer is rapidly established as determined by their membrane–water partition coefficients, which depend largely on the hydrophobicity of the solute. However, the rate of transbilayer movement of solutes depends on additional factors that require modification of solute–water interactions, such as disruption of hydrogen bonds.

The concept of membrane as a barrier implies that at least some substances can pass across it and that the flux of these species is causally related to forces. For the permeation by solubility diffusion mechanism, the driving force is the concentration gradient of the solute (dc/dx) over the thickness of the bilayer dx. Thus, the flux per unit area per unit time (Fick's first law):

$$J = -D\,(dc/dx)$$

The negative sign indicates that the net diffusion occurs in the direction of decreasing concentration. The proportionality factor D, diffusion coefficient, has dimensions of L^2/T and is usually expressed in cm^2/s. The flux has dimensions of M/TL^2 (flow of mass per unit time per unit area).

This equation can not be used readily because the thickness of the diffusion barrier as well as the concentration of the solute at the two interfaces is not known. If it is assumed that the concentration of the solute at each interface is related to its bilayer–water partition coefficient K; the concentration difference between the two compartments is C. Then

$$J = -D \cdot K\,(\Delta C/\Delta x) = -P\Delta C$$

By defining permeability coefficient as $DK/\Delta x$, the flux across membrane (dC/dt) can be expressed in terms of the concentration difference across the membrane. The expression for permeability is thus reduced to a simple first-order exponential relationship if it is assumed that initially the solute is present only on one side of the bilayer:

$$\frac{C_t}{C_o} = -\ \exp(-Pt)$$

Where C_o is the initial concentration of the solute and C_t is the concentration at time t. Thus, P is the first-order diffusion constant by which uniformity of concentration across the membrane is achieved. These equations show a clear relationship between the permeability and partition coefficient inherent in the concept of bilayers. Thus, permeability is a rate process that contains contributions from both an equilibrium (partitioning) and a nonequilibrium (transverse diffusion) step.

The assumptions of instantaneous partitioning of a solute is not always valid. A treatment based on Eyring's absolute reaction rate theory has been developed (Zwolinsky et al., 1949). The flow of molecules through the membrane can be viewed as a series of successive molecular jumps of length λ from one energy minimum to the other (Fig. 4-11). The rate constants for adsorption and desorption on the interface are k_{sm} and k_{ms}, and k_{pm} is the rate constant for diffusion in the membrane. By assuming equal flux across a series of barriers, the permeability coefficient, P is given by:

$$\frac{1}{P} = \frac{2}{k_{sm}\lambda} + \frac{mk_{ms}}{k_{sm}k_m\lambda}$$

If the rate constant for adsorption $k_{sm}\lambda = k_s$, the membrane thickness as $\Delta x = m\lambda$, the diffusion coefficient $D = k_m/\lambda$, and the partition coefficient $K = k_{sm}/k_{ms}$, then

$$\frac{1}{P} = \frac{2}{k_s} + \frac{\Delta x}{KD}$$

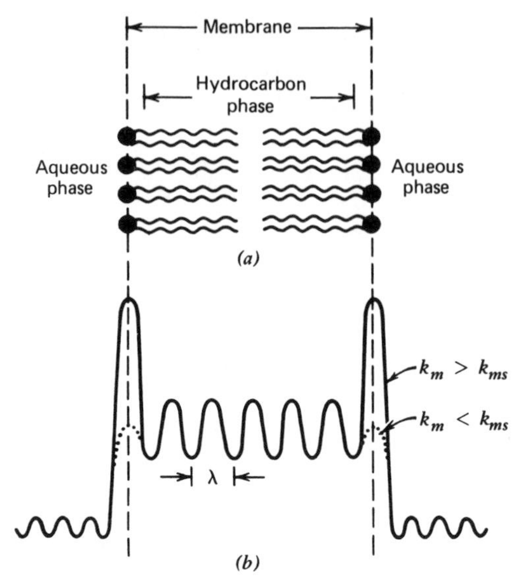

Fig. 4-11. A representation of the interfaces involved in transfer of a solute from one interface to the other. Possible potential energy profiles for a solute molecule diffusing through the membrane are also shown. (From Trauble, 1971).

If diffusion across the membrane is rate-limiting, i.e., $2/k_s << \Delta x/DK$, then $P = DK/\Delta x$ as derived from the continuum model. If adsorption is rate-limiting $2/k_m >> \Delta x/DK$, then P is equal to the rate of adsorption on the membrane. The apparent proportionality between partition and permeability indicates that the same forces that determine the rate of adsorption to the membrane also determine partitioning into the membrane.

Since the permeabilities of the interface (P_i) and the acyl chain region (P_c) are in series, total permeability (P) is

$$\frac{1}{P} = \frac{1}{P_i} + \frac{1}{P_c}$$

Experimentally, P is obtained by measuring the rate of permeation as a function of the concentration gradient of a solute. This can be achieved by a variety of methods: (i) by monitoring exchange of radiolabeled solutes in the absence of a concentration gradient; (ii) by monitoring osmolarity of a solute relative to an impermeable solute; (iii) from the exchange times obtained by nuclear magnetic resonance (NMR) (Morariu et al., 1981; Stilbs et al., 1984); (iv) by fluorescence microphotolysis (Peters, 1986); (v) or by directly measuring the rate of appearance or disappearance of the solute with an appropriate spectroscopic probe such as Quinn-2 for calcium ions or 6-methoxy-*N*-(3-sulfopropyl)quinolinium for chloride ions (Illsley and Verkman, 1987). Ion permeability can also be obtained from conductance or potential measurements, or monitored by ion-selective electrodes or dyes. A method of general applicability that has been used for binding as well as permeability measurements is the *rapid filtration* technique. The basic procedure for the rapid filtration technique consists of the following steps (see Kaback, 1983; Turner, 1983). Vesicles are combined with an incubation medium containing radioactively labeled ligands and the other constituents as required. After an appropriate time a stop solution is added and the vesicles are collected on a filter, which is subsequently washed and counted for radioactivity. Efflux studies can be carried out in the same way using vesicles preloaded with labeled substrate.

Movement of Water across Membranes

Permeability coefficients for water and a variety of solutes across bilayers are summarized in Table 4-6. The net permeability coefficient of water

TABLE 4-6 Water Permeation in Various Cells

Cell	Osmotic Permeation ($\mu m \cdot s^{-1}$)	Temp. (°C)	Activation Energy (kJ/mol)	Isotopic Permeation ($\mu m \cdot s^{1}$)	Temp. (°C)	Activation Energy (kJ/mol)	Cell Radius (μm)
Frog egg							
Ovarian	41 ± 3	22	46	$\rightarrow\infty$	22	—	750
Body cavity	2.4 ± 0.6	22		2.9 ± 0.5	22	59	800
Pike egg				0.18	9		1400
Salmon egg				0.01	5.5		3000
Mouse ova							
Unfertilized	9.9	20	61				75
Fertilized	9.6	20	54				
Lymphocytes							
Human	10.3 ± 0.4	25	59				4
Erythrocytes							
Human	136	22	14 ± 2	32	22	25 ± 1	4
	144 ± 40	25(?)		140	25		
	200	23					
	240	25(?)					
Dog	190	22	16 ± 2	55 ± 7	22	21 ± 1	3.5
				57	20		
Novikoff hepatoma	8.2 ± 0.3	20	44	9.7 ± 1.0	20		12
Chinese hamster lung fibroblast V-79-S 171-W1	17.6 ± 0.5	24	96 ± 13 (above 21°C) 29 ± 7 (below 21°C)				7
Black lipid membrane	2–50	25	42–63	1–11	25	63 ± 4	—

across bilayers is about 0.004 cm/sec versus 0.015 for self-diffusion of water. Comparable values of P are observed for exchange and net-flux if precautions are taken to minimize the effects of unstirred layer at the interface. One of the simplest demonstrations of permeability of water across bilayers comes from the swelling or shrinking of osmotically shocked vesicles. Such changes can be readily quantified by monitoring the change in the turbidity or scattering. This is because under the driving force of osmotic pressure (= mRT, where m is the molality of the solute, R = 0.08205 lit · atm/deg · mole) water moves to the more concentrated solution.

Both P and the activation energy for P show considerable dependence on the lipid composition and phase transition (Carruthers and Melchior, 1983; Engelbert and Lawaczeck, 1985). Cholesterol at 5 mole% increases permeability in the gel phase, and under most other conditions cholesterol lowers permeability. The rate of swelling for PC is in the order: egg PC > dimyristoyl PC (DMPC) = distearoyl PC (DSPC) > dipalmitoyl PC (DPPC), where the overall difference is about 50-fold (Carruthers and Melchior, 1983). Similarly, proteoliposomes containing transmembrane proteins with or without channels (glycophorin, band 3, gramicidin) enhance permeability coefficient for water by 30- to 100-fold.

The bilayer organization of cellular membranes entails the problem of osmotic stress that arises from osmotic influx of water driven by enclosed metabolites and macromolecules. Plants and most bacteria are protected from the swelling by their rigid wall. Most other organisms equalize the osmotic gradient by modulating ion fluxes. Evaluation of the movement of water in cells across their membranes has been complicated because several factors contribute:

1. Flow of water across unstirred layers (Barry and Diamond, 1984) gives lower values for P_e than those obtained under net osmotic flux (P_o).
2. Presence of aqueous pores or channels also gives $P_o > P_e$. Using Poiseuille's law for volume flow, pore diameters of 1–30 Å have been calculated. Most of these values are probably wrong. However, P_o/P_e of 4.5 for human erythrocytes, which corresponds a pore diameter of 4.5 Å is ascribed to the presence of protein channels, probably formed by band 3 protein (see Chapter 10). The activation energy for flow of water through such channels is about 4–7 kcal/mole versus 10 to 15 kcals/mole for the diffusion through acyl chains of bilayer. These channels are also blocked by thiol reagents (Solomon et al., 1983).
3. Diffusion of water in cell cytoplasm (D = 500 μm^2 per sec) is slower than the self-diffusion of water. Therefore, bulk diffusion becomes a

TABLE 4-7 Permeability and Related Parameters for Human Red Cell

Permeant	Basal *P* (cm sec^{-1})	Volume *V* (cm^3 mol^{-1})	Partition Coefficients between Solvent and Water			
			K-Hexadecane	K-Oil	K-Octanol	K-Lipid
Water	1.2×10^{-3}	10.6	4.2×10^{-5}	1.3×10^{-3}	0.041	—
Methanol	3.7×10^{-3}	21.7	3.8×10^{-3}	9.5×10^{-3}	0.18	0.21
Ethanol	2.1×10^{-3}	31.9	5.7×10^{-3}	3.6×10^{-2}	0.48	0.44
Urea	7.7×10^{-7}	32.6	2.8×10^{-7}	1.5×10^{-4}	0.0022	0.23
Ethanediol	2.9×10^{-5}	36.5	1.7×10^{-5}	4.9×10^{-4}	0.012	0.12
Thiourea	1.1×10^{-6}	39.5	—	1.2×10^{-3}	0.072	—
n-Propanol	6.5×10^{-3}	42.2	3.3×10^{-2}	1.4×10^{-1}	2.2	1.3
Glycerol	1.6×10^{-7}	51.4	2.0×10^{-6}	7.0×10^{-5}	0.0028	0.050
Erythritol	6.7×10^{-9}	66.2	—	3.0×10^{-5}	0.0012	0.026
n-Hexanol	8.7×10^{-3}	72.9	1.3	7.6	110	—

rate-limiting step in large cells. The Arrhenius plot for diffusion of water in cytoplasm also exhibits an anomalous increase at 15°C. This is probably due to a structural transition between two different structural forms of water (Mild and Lovtrup, 1985).

4. Water fluxes regulate tissue and cell volume, which in turn regulates the concentration (therefore the rate of transport) of a variety of other ions and solutes (Grinstein, 1948). Maintenance of iso-osmolarity is essential for the survival of cells that do not have mechanically resistant envelopes. In such eukaryotic cells, osmotic regulation is mediated by antidiurectic hormones, like vasopressin, which change the permeability of toad urinary bladder by a factor of 80. These changes are apparently due to a change in the permeation pathways at cell junctions.
5. Basal nonspecific permeability for water and a variety of solutes across erythrocyte membrane have been measured (Table 4-7). As discussed by Lieb and Stein (1986) the permeability coefficients are related to partition coefficients in a solvent that has the apolarity of *n*-hexadecane. The transverse diffusion behavior of solutes is apparently consistent with a model in which the solute molecule jumps from one hole to the other. The size of such a hole is about 8 cm^3/mole, which could account for the steep selectivity on the basis of size.

5 Order and Dynamics in Bilayers

"The question is" said Alice, "whether you can make words mean so many different things."
"The question is which is to be master, that's all." Humpty Dumpty continued in a scornful tone, "when I use a word, it means just what I choose it to mean . . . neither more nor less."
"Contrariwise," said Tweedledee, "if it was so, it might be; and if it were so, it would be, but as it isn't. That's logic."
"Tut, tut, child," said the Duchess. "Everything has got a moral if only you can find it."

Taken somewhat out of context from *Through the Looking Glass*

The observations summarized in the preceding chapters suggest that a wide variety of amphipaths are found in biological membranes, and these amphipaths form lamellar bilayers when dispersed in water. Although a very diverse group of amphipaths can form topologically continuous bilayers, most living organisms can tolerate only a minor change in their phospholipid and acyl chain composition, and the composition of biomembrane is retailored when the growth environment of the organism changes. For example, replacement of about 30% of the native phosphatidylcholine (PC) by dipalmitoyl PC (DPPC), distearoyl PC (DSPC), di-18 : 3-PC, or 16,20 : 4-PC results in an increased osmotic fragility and lysis of erythrocytes (Kuypers et al., 1984). Significant perturbations in the organization of phospholipid molecules in bilayers that alter viability of cells are known to occur under a variety of conditions, such as by changing temperature, by mixing other lipids, and by introducing solutes. Such compositional and conformational changes occur largely within the organizational constraints of a bilayer. The ability of the acyl chains to accommodate perturbations probably accounts for the presence of a large number of phospholipid species in most biological membranes. As elaborated in this chapter, biomembranes apparently maintain their phospholipid composition not only to maintain characteristic macroscopic phase properties but also to maintain microscopically well-defined organization.

Orientation of a phospholipid molecule in a bilayer implies strong anisotropic constraints for its orientation, conformation, and motion. For example, rotation along the long axis of phospholipid molecules is much more facilitated than it is along other axes. Traditionally two parameters have been invoked to characterize the general state of disorder in bilayers: (i) The orientation and conformation of acyl chains and (ii) the motion and flexibility of a particular region of the chain that determines the dynamic properties like rotation, lateral diffusion, and permeation of solutes. Thus, acyl chains give rise to not only the hydrophobic effect that accounts for the thermodynamic stability of the organization of bilayers, but the conformation and dynamics of acyl chains also plays an important role in determining the properties of bilayers.

The conformational states of a polymethylene chain arise primarily from rotations about the C—C bond, although the configuration and position of a double bond also contributes significantly. The lowest potential energy conformation of a polymethylene chain is an all-*trans* conformation. As shown in Fig. 2-14, the diheral angle between the alternate C—C (say C_1—C_2 and C_3—C_4 bonds in butane) bonds is 180°. In this fully stretched conformation, a polymethylene chain would be cylindrical with a theoretical cross-sectional (Van der Waals contact) area of 0.18 nm^2 and the length of 0.127 nm per methylene residue. In the all-*trans* state, saturated acyl chains are longest, thinnest, and most closely packed. In a bilayer, all-*trans* chains pack in a pseudo-orthombic, monoclinic, or hexagonal lattice, each with its characteristic interchain separation ranging from 0.386 to 0.42 nm.

There is always a rotation about C—C bond (Fig. 2-14). The two higher energy minima corresponding to the dihedral angles of 120 ($g+$) and 240 ($g-$) between the alternate bonds give rise to *gauche* (g) conformers. The free energy of such *gauche* conformers is about 0.5 to 0.8 kcal/mole higher than that of the *trans* (t) conformers because of the greater steric hindrance for rotation (Flory, 1969). The activation energy barrier for t to g conformational change is about 3.5 kcal/mole in bulk alkanes, and it is only 2 kcal/mole for C—C bonds adjacent to a double bond. Typically, the C—C bond oscillation frequency is about $5 \cdot 10^{12}$/sec and the jump frequency is about 10^{10}/sec. The statistical-mechanical description of an acyl chain undergoing rotational isomerism in bilayers is reviewed by Cevc and Marsh (1987).

In a close-packed bilayer a single g conformer in an acyl chain causes a large change in the length, width, and packing ability of the neighboring acyl chains (see Fig. 3-6). This is partly overcome by correlating the conformation of the neighboring acyl chains which gives rise to the cooperative change in the organization of the acyl chains. Manifestations of such changes are described below. However, combinations of t and g conformers also give rise to kinks, jogs and fold in polymethylene chains. A $g+ \cdot t \cdot g-$ kink, for example, increases the cross-sectional area to about 0.22 nm^2 while the length of the chain is effectively reduced by one methylene unit length, about 0.127 nm (Trauble, 1971).

Order Parameters

In a hexadecyl chain about 4–6 *g* conformers are predicted at room temperature. However, in bilayers in fluid phase the steric hindrance to rotation about C—C bond varies throughout the length of the acyl chain. Rotation of a C—C bond out of an all-*trans* chain will cause the rest of the acyl chain to rotate along the locus of a cone (Fig. 3-6). This could lead to energetically unfavorable interactions in a close-packed lattice of acyl chains. The maximally allowable amplitude of oscillation away from the *trans* conformation would therefore increase markedly toward the methyl end of the acyl chains. An increase in the disorder experienced by methylene residues along the acyl chain can be measured by a variety of techniques. The molecular order and dynamics have been measured as order parameters. A decrease in the time and ensemble-averaged order parameter experienced by the C—D bond along the acyl chain length is shown in Fig. 5-1. By definition for the environment of *trans* conformers $S = 1$ in a crystalline state, and $S = 0$ when methylene residues have no conformational preference as in an isotropic liquid. Note the constancy of the deuterium order parameter for the first 10 methylene residues, and a major change in the order parameter occurs beyond C_{10} in membranes of *Acholeplasma laidlawii*, as well as for bilayers of a variety of synthetic phospholipids. While this could be solely attributed to the disorder created by the double bond in oleoyl chains, at least a significant decrease in the order parameter is also seen in bilayers of saturated phospholipids. The change in the order parameters suggests that the methylene residues near the polar region are quite efficiently packed, whereas the middle of the bilayer may be as disorganized as liquid hydrocarbons.

The order parameter is an averaged parameter that is related to the behavior of a label like deuterium, nitroxide, or fluorescent probes. To report faithfully about its environment the probe should not perturb the bilayer organization, should be cylindrical as to accommodate between the acyl chains, and the intrinsic life-time of the probe should be in the range of the motion to be monitored. In principle, under appropriate conditions the order parameter provides insights into the organization, orientation, and motional characteristics of the ensemble but not on the individual chains. It may be emphasized that the order and the motion of chains are not necessarily correlated, as it is often implied in a simplistic interpretation of order param-

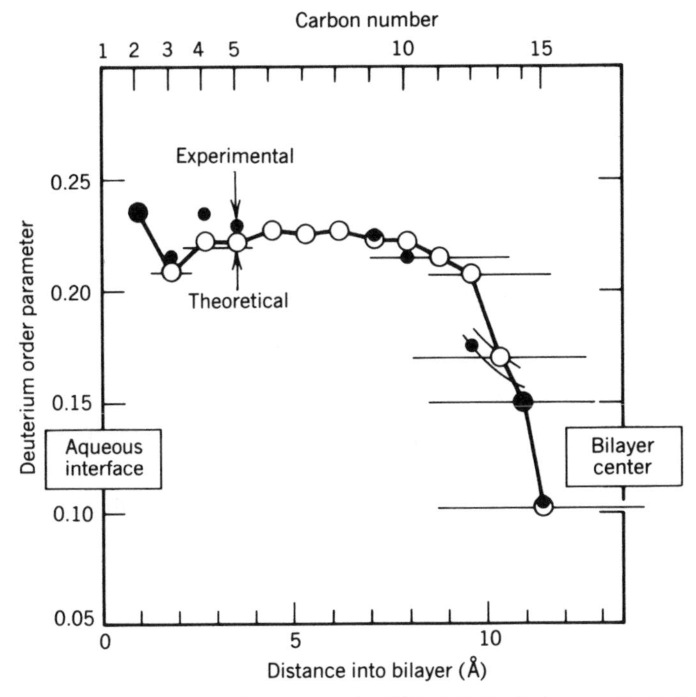

Fig. 5-1. Measured deuterium order parameter (filled circles) along a dipalmitoyl phosphatidylcholine chain in bilayers. Theoretical values are given as open circles. Note that the mean positions of C atoms (top axes and open circles) are not linear with the distance in the bilayer. The horizontal bars indicate standard deviation of the spatial motion of each carbon atom about its mean position. (From Seelig and Seelig, 1980.)

eters. Attempts to resolve and interpret the underlying motional characteristics from the anisotropy and asymmetry of orientation of the functional groups are in their early stages (Smith, 1985; Mulder et al., 1986), although a significant progress has been made (Meier et al., 1986).

Viscosity and Fluidity

Disorder in the organization of bilayers has also been expressed in terms of "fluidity" or "microviscosity," as calculated from the fluorescence anisotropy of a probe like diphenylhexatriene (Shinitzky and Barenholz, 1978). On the basis of such measurements it has been suggested that the microviscosity of biological membranes is similar to that of olive oil, i.e., between 1 and 3 cp as it changes with the metabolic state of the cells and with the environmental conditions. It appears that over long periods of time organisms tend biosynthetically to change the composition of the membrane phospholipids so as to maintain viscosity in a narrow range (Stubbs and Smith, 1984). There are several problems in interpretation of such data. For example, diphenylhexatriene is readily distributed in the plasma membrane as well as in membranes of organelles. Any change in the relative contribution to the anisotropy from these compartments could manifest a change in the overall microviscosity. As vague as the concept of homeoviscous adaptation may appear, it has generated considerable literature suggesting that such changes are seen in response to temperature, pH, ionic strength, diet, and long-term drug uptake.

It may be emphasized that the concept of fluidity and microviscosity is based on rather untenable assumptions about the nature of rotation of probes. For example, the rotation of diphenylhexatriene is not isotropic (based on the decay of fluorescence anisotropy). The equation used for calculation of microviscosity is based on three assumptions (Y. Levine, personal communication): first, the rotating molecule is much larger than the molecules of the surrounding medium; second, the effect of the structure of the medium is negligible; and third, the origin of the viscosity is due to the solvent drag by the rotating molecule so that there is no relative motion between the solute and the solvent.

THERMOTROPIC CHANGES

Correlating the conformation of neighboring acyl chains can give rise to a cooperative change in the organization or phase behavior of the acyl chains in a bilayer. A phase transition is an abrupt change in the macroscopic properties of a system as a result of a very small change in one or more of the parameters of the system, e.g., temperature, degree of ionization, pH, additives, and pressure. Intrinsic changes in the organization of bilayers as a function of temperature have been detected by a variety of techniques. One of the simplest demonstrations of a cooperative thermotropic change in the conformation of the acyl chains in bilayers is by differential scanning calorimetry (DSC). In DSC scans the difference in the heat flow between the sample and a reference is plotted as a function of temperature. At the phase transition, more heat has to be applied or withdrawn from the sample, which shows up as a peak in the DSC trace (Chapman et al., 1973). A similar curve obtained with the static adiabatic calorimeter is shown in Fig. 5-2. For dispersions of DPPC three transitions are observed with most techniques: sub-, pre-, and main transitions. The organization of molecules in the four states of a bilayer (crystalline, gel, ripple, and fluid) separated by these three transi-

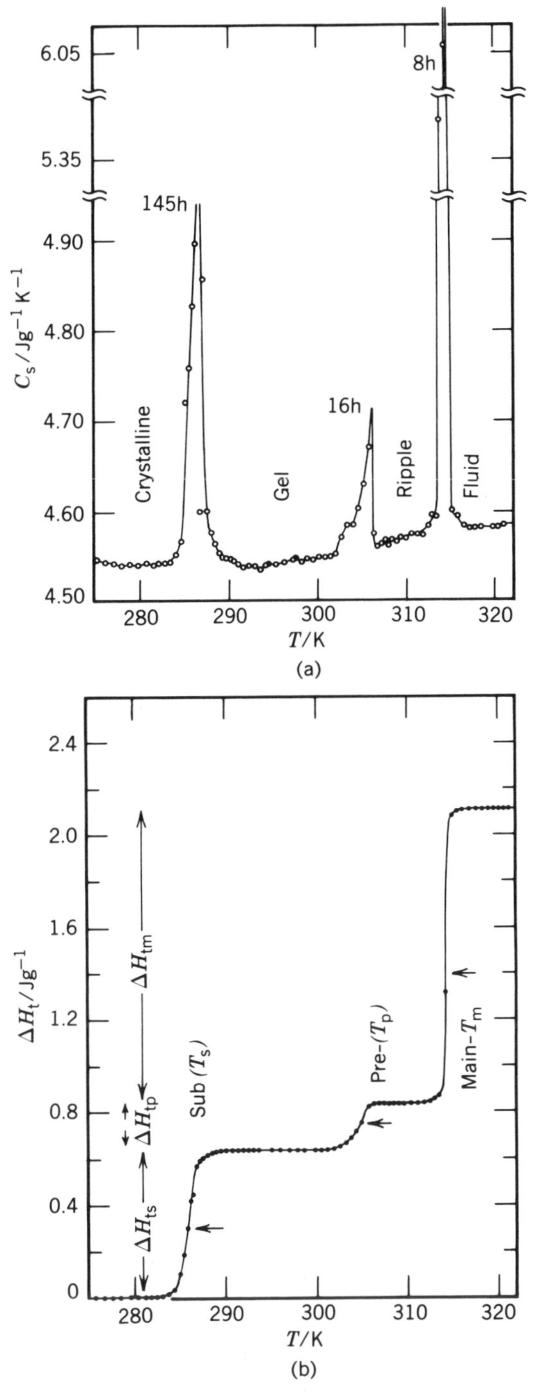

Fig. 5-2. (A) Heat capacity versus temperature curve for 1,2-dipalmitoyl *sn*-3-phosphatidylcholine in 97.6% water. (B) Excess enthalpy versus temperature for the same system. The curve in a was obtained by static adiabatic calorimeter, and the time periods at the transition points express the time required for the completion of the transition. See Fig. 3-14 for the organization of chains. (From Kodama et al., 1985.)

tions is shown in Fig. 3-14. A typical transition profile obtained by calorimetry contains following information:

1. Phase transition is a continuous process, i.e., it occurs over an appreciable range of temperature. The transition temperature is defined as the temperature at the beginning of the transition. Usually, the midpoint of a transition can be more accurately measured.
2. The area of the transition peak gives the enthalpy, ΔH. For a reversible two-state first-order transition (which is probably not a reasonable assumption for transitions in bilayers) the free energy change $\Delta G = 0$ at T_m, and $\Delta S = \Delta H / T_m$.
3. The width of the phase-transition profile and the rate of change of the excess enthalpy (σ, i.e., the height of the transition peak) at T_m is related to the cooperativity (σ = square root of $\Delta H / \Delta H_{vh}$ or the size of the cooperative unit ($N = 1/\sigma$). The Van't Hoff enthalpy of the phase change (only for a first-order discontinuous two-step process) is given by the relationship:

$$\Delta H_{vh} = \frac{\sigma R T_m^2}{\Delta H_{cal}}$$

The thermotropic phase changes in bilayers are accompanied by not only a change in the enthalpy but also the thickness (Watts et al., 1981; Seddon et al., 1984), volume (Nagle and Wilkinson, 1982) and the various orientational and motional parameters as detected by the spectroscopic techniques. The assumptions required for quantitative interpretation of a phase-transition profile are still being debated (Lee, 1977; Berde et al., 1980; Nagle, 1980; Jain, 1983; Pink, 1984). The questions that are yet to be settled relate to the meaning of the bilayer as a thermodynamic phase, the significance of coexistence of phases within a bilayer, metastability of phases, the thermodynamic order of the transition, and the molecular significance of the cooperative unit. For the purpose of the present discussion, we have adopted a qualitative interpretation of phase-transition parameters on a descriptive basis.

The phase-transition profiles of biomembranes are very complex as shown in Fig. 5-3. Such transitions have been studied extensively in membranes of bacterial auxotrophs that grow off externally supplied fatty acids. They often contain multiple peaks which are broad and unresolved. Some of the transitions at higher temperatures are not seen on repeat scans or in the dispersions prepared from extracted phospholipids, which suggests that such transitions are probably due to irreversible denaturation of proteins.

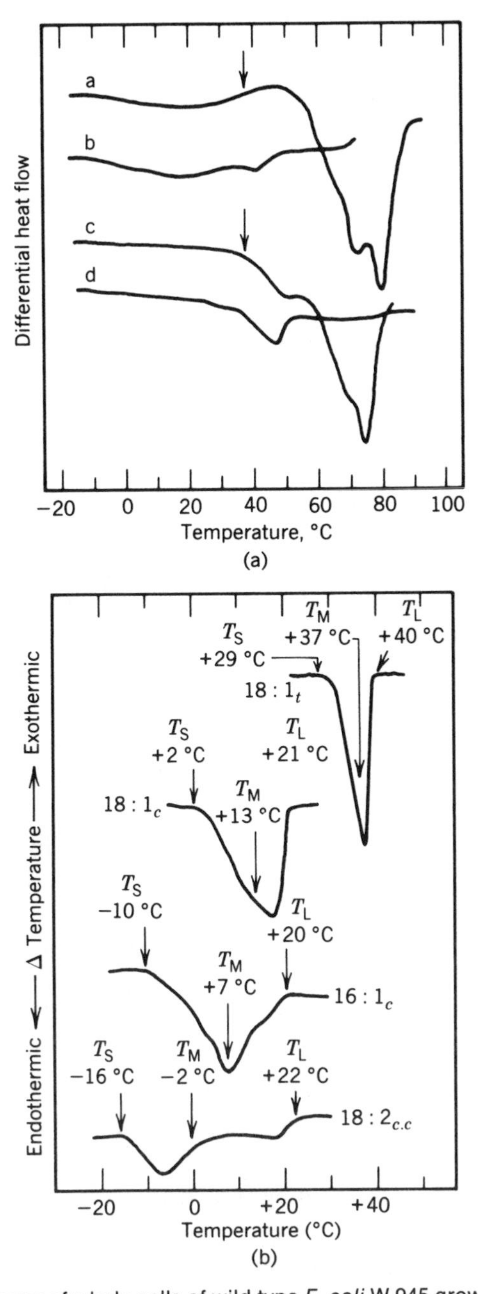

Fig. 5-3. (A) DSC scans of whole cells of wild type *E. coli* W 945 grown at 37°C: (from top) through protein denaturation, and after protein denaturation. The membrane transition extends from −10 to 40°C. The higher-temperature outer-membrane transition appears in b after heating. Some of the same cells grown in the same medium with an inhibitor of unsaturated fatty acid synthesis appears as c and d. (From Melchior, 1982.) (B) Temperature-based thermograms of isolated lipids from membranes of *E. coli* K1060 grown in the presence of glycerol plus various unsaturated fatty acids (from top): elaidic (18 : 1t), oleic (18 : 1c), palmitoleic (16 : 1c), linoleic (18 : 2c). The lower and upper boundaries of the lipid-phase transitions are indicated by T_s and T_L, respectively; the temperature at which the transition is half-complete, as measured by the area under the endothermic peaks, is designated T_m. Courtesy of R. McElhaney.

Lipid-dependent transitions have been shown to occur in biomembranes not only by physical techniques but also through the interpretation of the breaks and discontinuities in the Arrhenius plots for functions of membrane proteins (Chapter 7). While such correlations are strongly suggestive, a precise interpretation of cooperative thermotropic phase transitions and phase separations in biomembranes still remains an elusive goal.

Phosphatidylcholine Dispersions

The most commonly studied transition in the aqueous dispersions of phospholipids is the gel-to-fluid transition. However, as shown in Fig. 5-2, the overall process is considerably more complex. For example, with the aqueous dispersions of DPPC, at least (for others see Casal et al., 1983) three transitions are observed:

1. Subtransition at about 20°C is broad with ethalpy of about 4 kcal/mole. This transition is observed only when the samples are annealed for several hours at <10°C, and is not seen during the cooling scan until cooled below 4°C. It is believed that in this transition from subgel to gel phase the rotational motion of PC molecules increases, so does the specific volume by 0.009 ml/g. This transition is prominent only in dispersions of stereochemically pure phospholipids.
2. Pretransition occurs at about 37°C with enthalpy of about 1.5 kcal/mole. It is not usually seen on cooling, and appears on reheating only when DPPC dispersion is kept below 32°C for more than 15 min. The origin of the changes leading to the pretransition is not known. Some of the possibilities include a change in the conformation of the glycerol backbone, tilt of the acyl chains, and the loss of the interdigitation of acyl chains near the terminal methyl groups. Specific volume increases by 0.035 ml/g during the pretransition.
3. Main chain-melting transition or gel-to-fluid transition occurs at 41.6°C (T_m) with ethalpy of 8.9 kcal/mol. This transition is observed in the cooling as well as the heating runs without any significant hysteresis, and the transition characteristics are essentially identical on repeated reheating scans. The population of *gauche* conformers in an acyl chain increases significantly on chain melting. The specific volume increases by 0.035 ml/g or 3.5% during the main transition. This small change in volume is consistent with the conclusion that the overall change in the state of the acyl chains is small and cooperative. Thus, the main transition occurs with a disruption of hydrophobic and van der Waals interactions.

4. Pressure and temperature have opposite effects on *trans* to *gauche* change, chain mobility, and interchain interactions. The effect of change of pressure (ΔP), on the change in (ΔT) T_m and the volume change (ΔV) is given by:

$$\frac{\Delta P}{\Delta T} = \frac{\Delta H}{T_m(\Delta V)}$$

Therefore, pressure is expected to augment the effect of increasing temperature. Indeed, T_m and T_p increase linearly with pressure: dT_m/dp = 20.8C/Kbar, and $dT_pnp/$ = 16.2C/Kbar for DPPC (Wong and Mantsch, 1985). The volume change with increasing pressure is not constant.

Several features of organization and dynamics of molecules in a bilayer emerge from the study of the thermotropic transition properties of synthetic phospholipids. The four phases delineated by the three transitions in DPPC dispersions as described above exhibit increasing degree of disorder, although the bilayer organization is retained. Only within the transition range do the two phases coexist. The pre- and subtransitions in phospholipid bilayers show considerable metastability and hysteresis. Other features of the thermotropic transition between the various organizational states of a bilayer may be noted.

1. The endothermic transitions in the aqueous dispersions of phospholipids are observed only when the mole fraction of water in the dispersion exceeds a certain ratio. Typically, complete hydration of phospholipid molecules is required for the formation of lamellar bilayer structures. This is because adequate solvation of the head-group region is required to disrupt the crystal lattice, and these water molecules are so tightly bound that they do not undergo usual ice–water transition at 0°C. The amount of water tightly bound to phospholipid head groups in the bilayer phase depends upon the charge and the structure of the polar head group. Between 7 and 30 water molecules are required for hydration, depending upon the organization and the structure of the head group. In the presence of excess water the transition characteristics do not depend upon the mole fraction of water in the dispersions.
2. A sequential transition between different organizational states of a bilayer are not obligatory. For example, dispersions of lysophosphatidylcholine form an interdigitated bilayer that melts directly to the mi-

cellar phase, and this transition is not lyotropic (Jain et al., 1984). Similarly, dispersions of diacyl cyclopentanophospholipids show a transition from subgel to liquid crystalline bilayer (Jain et al., 1984a), whereas, the dispersions of bisphosphatidic acid undergo a transition directly from the gel bilayer to the hexagonal phase (Grunner and Jain, 1985).

3. The thermotropic phase-transition parameters exhibit a strong dependence upon the structure of the head group and of the acyl chains (Table 5-1). As shown in Fig. 5-4, the transition temperature in vesicles (Barton and Gunstone, 1975) as well as *Acholeplasma laidlawii* membranes decreases when the position of the double bond shifts toward the center of the acyl chain. All the factors that give rise to more disorder (shorter length, more double bonds, or branching toward the

TABLE 5-1 Recent Studies on the Phase Properties of Phospholipids (for earlier data see Silvius, 1982; Keough and Davis, 1984)

Phospholipid	Comments
Phosphatidylcholines (PC)	Ether and ester PC compared (Ruocco et al., 1985a,b); length and position of chains (Stumpel et al., 1983); branched acyl chain (Nuhn et al., 1986; Lewis and McElhaney, 1985a,b); 9- or 10-BrDPPC does not show sharp gel-to-fluid transition (Lytz et al., 1984); 1,2-dipentadecylmethylidine phospholipids (Blume and Eibl, 1981); omega-cyclohexane fatty acid-PC (Kannenberg et al., 1984; Lewis and McElhaney, 1985); mixed acid (Coolbar et al., 1983; Lewis et al., 1984); thiophospholipids (Jiang et al., 1984); fluorinated chains (MacDonald et al., 1985).
Phosphatidyl-X	Several polar and apolar substituents (Jain et al., 1986c)
Phosphatidylethanolamines (PE)	Gel–fluid–H_{II} transition of acyl and alkyl PE (Seddon et al., 1983); NMR study (Blume et al., 1982a,b); cyclopropane fatty acid (Perly et al., 1985)
Phosphatidic acid (PA)	DSC (Blume, 1983)
bis-Phosphatidic acid	DSC studies (Rainier et al., 1979)
Glucolipids	DSc and monolayer studies (Hinz et al., 1985)
Sphingomyelins (SM)	(Barenholtz, 1983)
Cardiolipin (DPG)	DSC on salts (Rainier et al., 1979). ^{31}P-*NMR* and X-ray on with one to five chains (Powell and Marsh, 1985). Natural DPG shows bilayer-to-H_{II} transition with Ca or >1 *M* NaCl.
Cerebrosides	DSc and X-ray data (Ruocco et al., 1981).
Galactolipids	Form H_{II} phase
Di- and tetraether lipid including phytanyl derivatives (Blocher et al., 1985).	

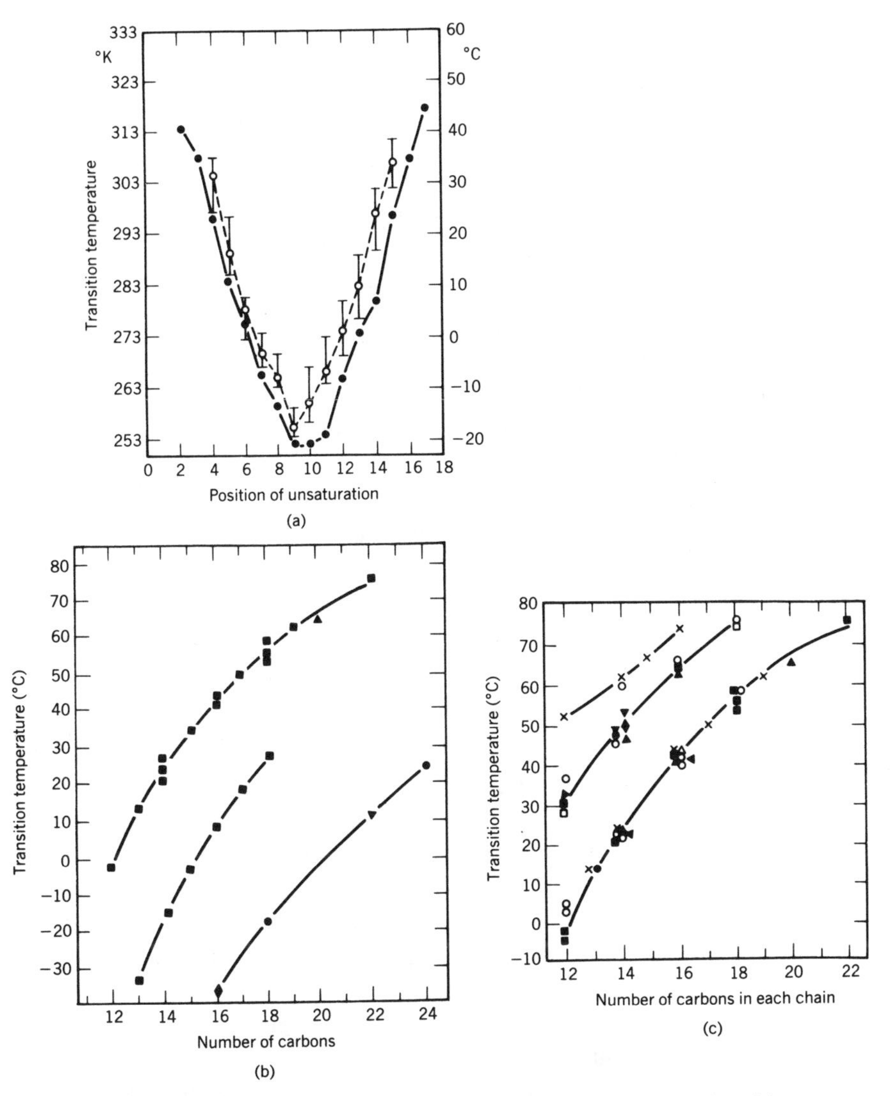

Fig. 5-4. (a) Gel-to-fluid T_m as a function of the position of the double bond in aqueous dispersions of *Acholeplasma laidlawii* polar lipids (open circles) and aqueous dispersions of 1,2-dioctadecanoyl phosphatidylcholines (closed circles). (From MacDonald et al., 1985.) (b) T_m as a function of chain length: (from top) saturated PC; anteisobranched PC; *trans*-monoenoic PC; *cis*-monoenoic PC. (c) As a function of the head group; (from top) DPG · Na, PS · Na, PE, PA · Na, PC and PG · Na. (Adopted from Keough and Davis, 1984.)

center of the chain, *cis*-double bond) tend to favor more disorganized phases (Barton and Gunstone, 1975; Silvius, 1982). Thus increasing the chain length increases T_m (about 14°C per $CH_2—CH_2$) and ΔH (about 2 kcal/mole per $CH_2—CH_2$), and the even–odd alternation (that is observed in the melting of hydrocarbon crystals, see Small, 1985) is not observed for the gel-to-fluid transitions. In fact, such alternation would be expected in the subtransitions because the chains are in pseudocrystalline array in the subgel phase. The enthalpy change per methylene residue during the melting of hydrocarbons is much higher than that is observed during the main transition in bilayers (1.2 kcal/mole versus 0.6 kcal/mole), and it is approximately the same as the sum of the enthalpy of transition from subgel to the fluid phase. This suggests that the gross conformational change during the transition from the crystalline subgel to the fluid bilayer is the same as during the transition from solid to the fluid melts. Indeed, the organization of acyl chains in the subgel bilayer of the aqueous dispersions of DPPC is pseudo-orthorhombic, very similar to some crystal forms of hydrocarbons.

4. The overall change in the enthalpy of chain melting in bilayer has contributions of about 3–4 kcal/mole from *trans–gauche* conformational change interchain from dipolar, hydrogen-bonding and hydration interactions (Nagle, 1980).
5. In mixed-chain phospholipids, the position of the more disordered chain in the *sn*-2-position lowers the transition temperature; this is because the two chains in diacyl phospholipids are conformationally nonequivalent and therefore give rise to packing faults in the terminal methyl region (Davis et al., 1981; Stumpel et al., 1983). The *sn*-1 and *sn*-3 stereoisomers of most glycerophospholipids do not appear to alter appreciably the main transition temperature, however a difference has been noted in the pre- and subtransition temperatures (Eklund et al., 1984). The ether analogs have higher (2–4°C) T_m and higher ΔH, presumably because the number of methylene residues is effectively higher, and their organization is also somewhat different (Ruocco et al., 1985a,b).

Phosphatidylethanolamine Dispersions

The thermotropic phase-transition behavior of the aqueous dispersions of phosphatidylethanolamine (PE) is very complex (Finegold et al., 1985). They exhibit not only several metastable phases but also fluid bilayer to H_{II}

phase transition (Seddon et al., 1984; Perly et al., 1985). T_m for PE is about 20°C higher than that or PC, although the enthalpies of transition are comparable. Sub- and pretransitions have not been observed. It appears that the ability of the substituents in the head group to form intermolecular hydrogen bonds (—NH. .O=P—) increases the T_m of PE dispersions (Boggs, 1984). Lower hydration, lower permeability, and reduced chemical reactivity of the amino group toward trinitrobenzene sulfonylchloride are also consistent with this suggestion. It may be pointed out that the conformation of ethanolamine and choline residues in the crystals of the respective phospholipids is the same, i.e., parallel to the plane of the bilayer (cf. Chapter 2, Table 2-5).

The gel–fluid and bilayer-to-H_{II} transition characteristics of PE exhibit significant differences (Seddon et al., 1984): T_{b-h} decreases with acyl chain length and with salt concentration; and it increases with increasing pH and water content.

Dispersions of Other Phospholipids

As summarized in Fig. 5-4, the gel-to-fluid transition temperature of a variety of phospholipids exhibits a systematic variation with the chain length, unsaturation, and structure of the head group. In all cases examined so far, T_m changes not only with the chain length but also with the state of ionization and hydrogen bonding. As expected charge repulsion lowers T_m about 15°C per net charge, and T_m increases by about 20°C per H bond in the head-group region. The hydrogen bonding and ion binding effects are much more pronounced in phosphatidic acid (PA). Thus, an isothermal chain melting can be induced in bilayers of such phospholipids by changing pH. A 10–15° increase in the transition temperature at 0.5 M monovalent cation concentration is predicted from the electrostatic interactions implicit in the diffuse double layer. However, multivalent cations like calcium appear to be desolvated on binding (Rainier, 1979). Evidence for intermolecular interactions in modulating the phase properties of bilayers of PE, phosphatidylserine (PS), PA, sphingomyelin (SM), glycolipids, and other synthetic lipids has been reviewed (Boggs, 1984).

Thickness and Molecular Areas

In the gel state of the bilayer acyl chains are tilted relative to the plane of the membrane, and in the fluid state chains lie perpendicular to the surface. A fluid phase bilayer is about 15% thinner than in the gel state (Melchior, 1982; Seddon et al., 1984; Watts et al., 1981). An increase in the acyl chain length is accompanied by an increase in the thickness of the fluid bilayer that is only 40% of the value expected for the fully extended chain, and the area per molecule also increases with the chain length (Cornell and Separovic, 1983). These observations imply that the methyl ends of the chains are folded. In highly asymmetric PC in the gel state, there is evidence for interdigitation of the acyl chains of the two monolayer halves (Hui et al., 1984).

Kinetics of Gel-to-Fluid Transition

Since the gel–fluid transition in a bilayer is accompanied by an increase in specific volume, a pressure drop would trigger an isothermal transition. This has been used to study the kinetics of the phase change. For PA three relaxation times are observed (Elamrani and Blume, 1984). Although these rate constants are independent of the vesicle concentration, they exhibit a maximum at the transition. For dilauroyl phosphatidic acid, the relaxation times at T_m are about 13, 36, and 1000 msec. These relaxation times are found to be higher for the ether analog, and higher in the presence of 1 *M* NaCl. This is also consistent with the limiting values (halftime <1 sec) obtained with X-ray diffraction studies (Caffrey, 1985). The origin of these processes is not known.

The Dynamic State of Molecules in Bilayers

Nuclear magnetic resonance (NMR) technique has been used extensively to obtain information about the motion of phospholipid molecules in bilayers. The following information is obtained readily from these techniques: segmental motion of chains and the orientation of the polar group by deuterium NMR and conformational change in glycerol backbone and the orientation of carbonyl groups from ^{13}C-NMR. These changes are often associated with gel-to-fluid transition. General conclusions based on the study of bilayers of pure phospholipids are:

1. Different phospholipids in the fluid bilayer exhibit a similar behavior: the segmental motion of the acyl chain decreases abruptly beyond C_9 (Fig. 5-1); head groups like choline have greater conformational freedom than is available in the acyl chain region. The effective rotational correlation times for the first half of the chain is about 0·1 nsec, and it decreases to 0·01 nsec at the end of the chain. These changes are mainly due to a change in the order profile (Brown and Williams, 1985), whereas the intrinsic rotational correlation times probably remain constant (about 0·01 nsec) along the length of the chain.
2. The rate of appearance of disorder increases sharply near the main phase transition.

3. It is more difficult to obtain NMR data in the gel phase. Results on DPPE suggest that the motion of the head group is slower by 3–4 orders of magnitude (Blume et al., 1982).

BILAYERS OF MIXTURES OF LIPIDS

The gel-to-fluid transition characteristics of bilayers are appreciably modified in the presence of proteins (McElhaney, 1986) and other lipids. The temperature versus composition diagrams provide some useful information if it is assumed that the bilayers in the gel and the fluid states represent two distinct phases. From the phase diagrams it is possible to ascertain ideal miscibility, immiscibility, and metastability of the phases. The theoretical basis of such phase diagrams has been discussed (De Fay et al., 1966; Lee, 1977; Jacobs et al., 1977; Cevc and Marsh, 1987). In dispersions of phospholipids in excess water the Gibb's phase rule ($F = C - P + i$) reduces to $F = C - P + 1$ at constant pressure. Here F is the number of experimental variables (also called degrees of freedom or variance) than can be independently varied; P is the number of phases; C is the number of chemical components; and i is the number of intensive variables such as temperature, pressure, electric field, and salt concentration. This equation is a consequence of the condition for the equilibrium in a heterogeneous system, and at equilibrium there is no net difference in the partial molar free energy of a chemical species between any of the phases. For melting of ice $i = 1$ and $C = 1$, so that $P = 2$ and $F = 0$, that is melting of ice, when solid and liquid water coexist, is an isothermal event. For binary lipid mixtures $C = 2$ in the presence of excess water. Therefore, the phase rule reduces to $F = 3 - P$, i.e., a maximum of three phases can coexist at the eutectic point at a fixed temperature and composition. In a one-phase region both temperature and composition can be varied, whereas in the two-phase region either temperature or the composition can be varied (see also pp. 58 and 87).

An idealized temperature versus composition phase diagram for binary aqueous dispersions of phospholipids exhibiting complete miscibility in the solid and fluid phases is shown in Fig. 5-5. It has three regions. Below the lower curve both components exist in an ideally mixed gel phase, and above the upper convex curve both the components exist in an ideally mixed fluid phase. The composition of the fluid and the gel phase is determined only by the relative mole ratio of the components and not by the temperature. At temperatures $>T_2$ and $<T_4$, which also correspond to the range of DSC transition profile, and the mole fractions <1 (described by the region between the two curves), the gel and fluid phases coexist in equilibrium. In this

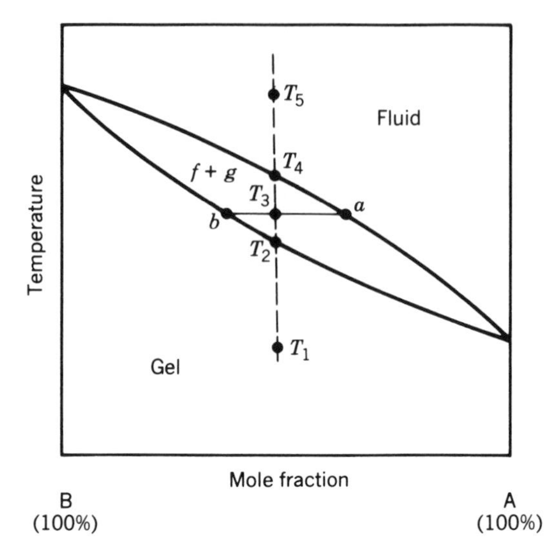

Fig. 5-5. Phase diagram for thermotropic transition of ideally mixed binary codispersions. The regions of the phase diagram containing gel (g) and fluid (f) and (g + f) bilayers are shown.

region there is only one degree of freedom, therefore the composition of the two coexisting phases varies both with the temperature and with the mole fractions of the two lipids. In DSC profiles such an ideal mixing is exhibited by symmetrical transition profiles.

The composition and the relative amount of gel and fluid phases under a given set of conditions can be obtained from a phase diagram of type shown in Fig. 5-5. At all temperatures above the fluidus curve the system consists of an ideally mixed fluid phase. On cooling from a point T_5 (where both solutes are in a single fluid phase), the solid phase separates only at temperature T_4 where the composition of the solid phase is g, the intercept on the lower (solidus) curve. Thus, the gel phase is enriched in the higher melting component, but it is not pure higher melting component. On further cooling from T_4 to T_2, the composition of the gel phase changes along the lower line, and the gel phase becomes richer in the lower melting component. The composition of the coexisting fluid phase would be given by the intercept on the upper (fluidus) curve. The ratio of the gel and fluid phases is given by the ratio of the horizontal lines b to a (lever rule).

In the temperature range enveloped by the solidus and fluidus curves, the fluid and gel phases coexist. This can only happen by a preferential removal of A from the fluid phase and mixing with the gel phase that is previously formed. Below T_2 both the solutes exist in the gel phase, and the last of the bilayer that was in the fluid phase had the composition corresponding to the intercept on the fluidus curve, which is enriched in A but is not pure A. Such ideal mixing is possible only when both the components adopt identical structures, so that one can be substituted by the other in the liquid as well as the gel phase. This condition is apparently satisfied in binary mixtures of phospholipids that differ by less than four methylene residues, e.g., DPPC + DMPC or DPPC + DSPC, but not DMPC + DSPC.

Immiscibility in the solid phase is, however, observed in binary codispersions of DMPC + DSPC (Fig. 5-6) and two transitions are often observed in

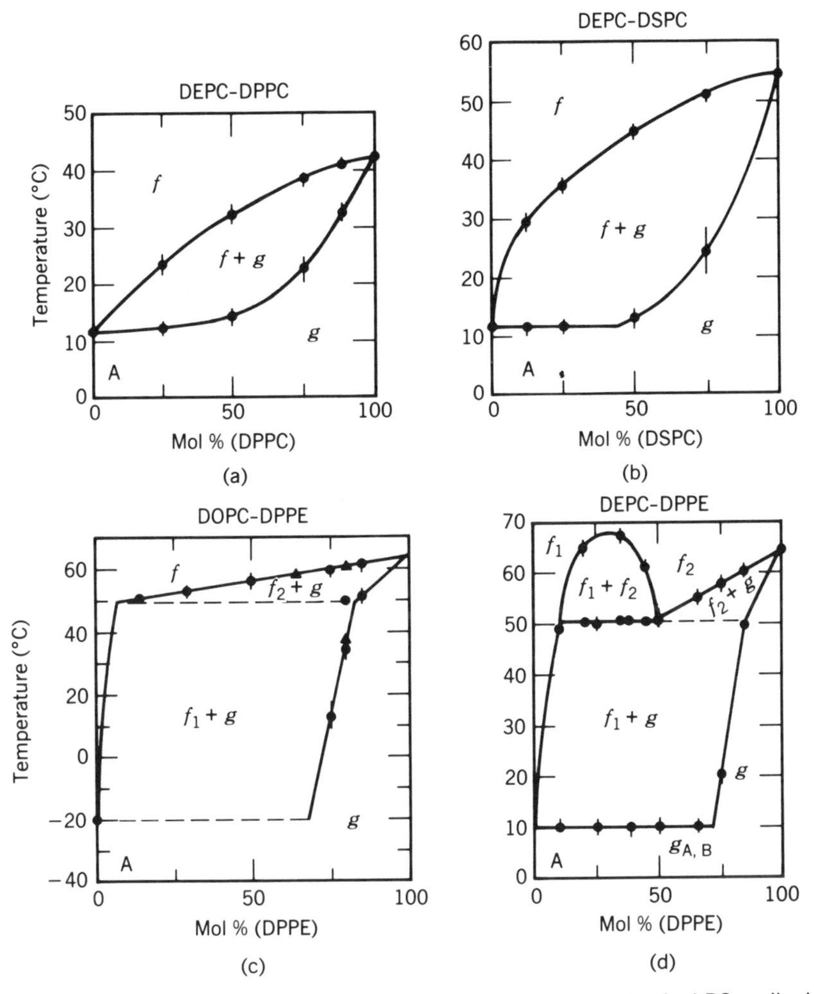

Fig. 5-6. Phase diagrams for aqueous binary dispersions of (A) dielaedoyl PC + dipalmitoyl PC; (B) dielaedoyl PC + distearoyl PC; (C) dioleoyl PC + dipalmitoyl PE; and (D) dielaedoyl PC + dipalmitoyl PE. The data are derived from measurement of the Tempo spectral parameter as a function of temperature. (From Wu and McConnell, 1975.)

their DSC profiles. Partial miscibility has been observed with a variety of other lipids like DMPC + DMPE, DMPC + DMPA, DPPC + DOPC, and anionic lipids in the presence of multivalent cations, and the phase separation is manifested in the properties of bilayers. From such studies it is clear that the mixing properties of lipids in a bilayer depend on the nature of the acyl chains as well as the polar head groups. Hydration and the hydrogen bonding tendency of the head group also play a role. Nonideal mixing in the gel phase of bilayers is commonly observed, however nonideal mixing has also been reported in the fluid phase (see Fig. 5-6).

A quantitative interpretation of phase diagrams is possible only for systems at equilibrium (Cevc and Marsh, 1987), and several other assumptions are also implicit. While determining phase diagrams, the rate of cooling and heating (van Dijk et al., 1977) or the rate of exchange across the phase boundaries (Blume et al., 1982b) can become a limiting factor. Under such conditions the rate of diffusion of molecules across gel-phase domains can be significantly slower than the NMR time scale or the scanning rate in DSC. This gives rise to anomalous quantitative measures of phases.

The nature of the bilayer as a thermodynamic phase that undergoes a first-order transition can only be taken as an approximation. In several cases separation and redistribution of lipids into different compartments (two halves of the bilayer, formation of smaller vesicles, formation of other phases) has been shown to account for apparent phase separation. Significance of the phase boundaries remains to be established, although such boundaries are implicated in certain experimental observations. Anomalous permeability for polar and ionic solutes has been interpreted in terms of the phase boundaries formed by lateral phase separation (Tse and Singer, 1984). Many other processes do not appear to be sensitive to lateral phase separation of mixtures of phospholipids, while these same processes are appreciably affected in the presence of impurities. The life-time of defects or phase boundaries could be different in the presence of impurities. Although the description of phases given by a phase diagram obtained from calorimetric data is generally but not always consistent with the information obtained by other techniques, in some cases there are thermally silent transitions (Kohn and Schullery, 1985). Similarly not many techniques are available that could distinguish presence of different subphases in the fluid phase.

Interactions in Bilayers with Two Components

Methods to characterize intermolecular interactions within binary bilayers are yet to be established. Most of the evidence for such interactions is indirect. For example at a mole fraction of 0.67 palmitic acid in DPPC probably forms a stoichiometric complex which melts from gel-to-H_{II} phase (Kohn and Schullery, 1985; Marsh and Seddon, 1982). Distinct clusters of PS and PE which can be chemically cross-linked have been detected in erythrocyte membranes (Marinetti and Crain, 1978). For circumstantial evidence suggesting seggregation of lipids in bilayers see Jain and White (1977), Jain (1983).

The order parameters for the *sn*-2 acyl chains and the ^{31}P-NMR line shape suggests, POPC, DPPE, DMPC, DMPG form lamellar dispersions. DMPG + DMPC are miscible in binary bilayers over a wide range of mole fractions. However, the order parameter for deuterium label in the head group of dimyristoyl PC is quite sensitive to the presence of DMPG (Sixl and Watts, 1983). Such results suggest that DMPG induces a change in the H-bonding and water structure. In bilayers of DPPE + DPPC, coexistence of fluid and gel phase is also indicated (Blume et al., 1982b). While these results are preliminary, considerable experimental data have accumulated on the binary dispersions containing cholesterol.

Cholesterol in Bilayers of Phospholipids

In spite of considerable publicity (so far eleven Nobel prizes have been awarded for the work on sterols), our understanding of the biological roles of cholesterol is only rudimentary. The sterol content of animal tissues exceeds, by several order of magnitudes, their bodily needs for the production of bile acids, steroid hormones, and vitamin D. Since most of the sterols in tissues is present in plasma membranes, it is believed that it is a primary modulator of the properties of bilayers. Plasma membranes of mammalian cells require cholesterol for normal functions and bacterial membranes do not. *Mycoplasma* is the only true sterol auxotroph. Cholesterol requirement in mammalian cells is fulfilled by endogenous biosynthesis and by extracellular supply. Low-density and high-density lipoproteins are the two major carriers of cholesterol in human blood, and high concentrations of cholesterol are found in plaques formed in arteries of patients suffering from atherosclerosis. Factors governing distribution of cholesterol in membranes and the plaques and homeostasis of cholesterol are beginning to emerge (Goldstein and Brown, 1983).

The cholesterol content of endoplasmic reticulum (ER) (about 12%) and other organelle membranes (less than 20%) is considerably smaller than that of the plasma membranes (about 30–50 mole%). The origin of such differences in the levels of cholesterol in the membranes of the same cell and in serium lipoproteins is surprising because the rate of intervesicle exchange as well as the rate of transbilayer transfer of cholesterol in liposomes is much faster than that for phospholipids. It may also be noted that isolated membranes of different cholesterol content apparently retain differences in their cholesterol content, even when they are incubated for several hours during isolation and fractionation procedures. However, cholesterol content of intact erythrocytes can be altered by incubation with sonicated vesicles or serum lipoproteins containing cholesterol. This implies that cholesterol may preferentially interact with certain components of plasma membrane that are not available in bilayer vesicles.

Endogenous cholesterol is synthesized in ER from squalene epoxide by enzymes that are localized in membranes, and very little is known about these enzymes. The biosynthetic intermediates with the 4,4,14-methyl groups on the β-face are not as effective as cholesterol. The main role of cholesterol in biomembranes, which is not fulfilled by these biosynthetic intermediates (Bloch, 1983), is in modulating the conformational state of acyl chains (*fluidity*), stability, permeability, and utilization of unsaturated fatty acids. Besides this, cholesterol also modifies functions of several membrane-bound enzymes, and it acts as a precursor for bile acids, steroidal hormones, and vitamin D. Cholesterol has also been shown to promote growth in *Mycoplasma capricolum*; ergosterol has a similar effect in yeast and *Tetrahymena*.

The cholesterol content of membranes is often determined after isolation; however, cholesterol content of membranes in intact cell can be determined by filipin treatment (Miller, 1984). In freeze-fracture electron micrographs of filipin-treated membranes large pits are observed in cholesterol-rich regions. Other studies suggest that even in the same membrane, lateral as well as the transbilayer distribution of cholesterol is not uniform (Yeagle, 1985). Asymmetric distribution of cholesterol found in some membranes is surprising in the light of the observation that the rate of transbilayer movement of cholesterol is rapid; half-time is less than 1 min in erythrocytes and also in vesicles. However, the rate of transbilayer movement of cholesterol may be considerably slower in other cell types if specific interaction occurs with an asymmetrically distributed membrane component.

The conformation of the four fused rings of cholesterol gives rise to a flat amphipathic structure (Fig. 2-5) with the 3-β-hydroxyl group on one end, and the angular methyl groups on the β-face, thus making the α-face planar and smooth. Cholesterol molecule is predominantly hydrophobic, and therefore only slightly soluble in water, <30 nM. Cholesterol may exist as a dimer in aqueous solution as well as in its crystal lattice. However, the shape of cholesterol is such that it is amphipathic and it is readily incorporated into phospholipid bilayers in varying mole fractions. Cholesterol incorporated into vesicles of the various phospholipids undergoes a slow intervesicle transfer or a transfer from vesicles to plasma membrane of intact cells. The rate-limiting step is the desorption of cholesterol from donor membranes, and the transfer appears to occur by transfer of cholesterol via the aqueous phase. The nature of phospholipid and the protein content of the donor membranes also regulates the rate of intervesicle transfer. The overall cholesterol content of membranes is therefore determined both by the thermodynamic factors dictated by its composition, and by the kinetic factors dictated by the metabolic flux of cholesterol.

Physical studies show that cholesterol incorporated into phospholipid bilayers as well as biomembranes is oriented perpendicular to the membrane surface. The cholesterol molecule is located such that the hydroxyl group is in the immediate vicinity of the phospholipid ester carbonyl group, although hydrogen bonding between these groups is not required for the effects of cholesterol on the phase properties of the bilayer. The rotational correlation time along the long axis is 0.1 nsec. The tail of cholesterol is in a somewhat less ordered environment than the rings. In the liquid crystalline phase, cholesterol increases the order parameter of methylene residues 2 through 11 of the acyl chains of phospholipids, and it has little or no effect on the properties of the residues 12 through the end region of the acyl chains. In the gel phase all positions in the hydrocarbon chain experience an increase in the motional order. Effect of cholesterol on the thickness of the bilayer also depends upon the acyl chain length. The thickness of bilayers containing acyl chains shorter than 16 carbons increases, and that of longer acyl chains decreases in the presence of cholesterol. The length of the cholesterol molecule is approximately equal to the length of hexadecanoyl chain in a phospholipid.

Several sterols can be incorporated into bilayers, which is consistent with the suggestion that hydrophobic interactions dominate. Subtle differences are, however, observed when relatively high mole fractions (more than 20 mole%) of sterols are present. For example, ergosta-5,7,9-trienol and cholesta-5,7,9-trienol have been used as fluorescent probes (Rogers et al., 1979; Jain et al., 1984b). These molecules are indistinguishable from cholesterol as far as their effect on permeability of water across the bilayer and on cell growth is concerned. Studies with other analogs of cholesterol suggest that the precise geometry of the ring has considerable influence on intermolecular association and lateral distribution of sterols. Hopanoids (Kannenberg et al., 1985), marine sterols (Carlson et al., 1980), and ergosterols have effects on bilayers that are similar to cholesterol. On the other hand, the ordering influence of cholesterol and ergosterol is much more pronounced than that of lanosterol (Yeagle, 1985) and epicholesterol. Deuterium NMR data suggests that the orientation of the sterol nucleus could be influenced by the orientation of the hydroxyl group (Murari et al., 1986). The ordering effect of cholesterol on lipids in fluid bilayer depends on the availability of the smooth α-face. Any modification leading to a change in the α-face leads to an ineffective sterol. Such ordering effects of sterols are seen in the permeability properties, gel–fluid transition, and the condensing effect on monolayers of phospholipids (Fig. 5-7). The origin of these effects in terms of the molecular organization is not understood. Some explanations invoke a change in the thickness or packing of phospholipids, or a change in the stability and mem-

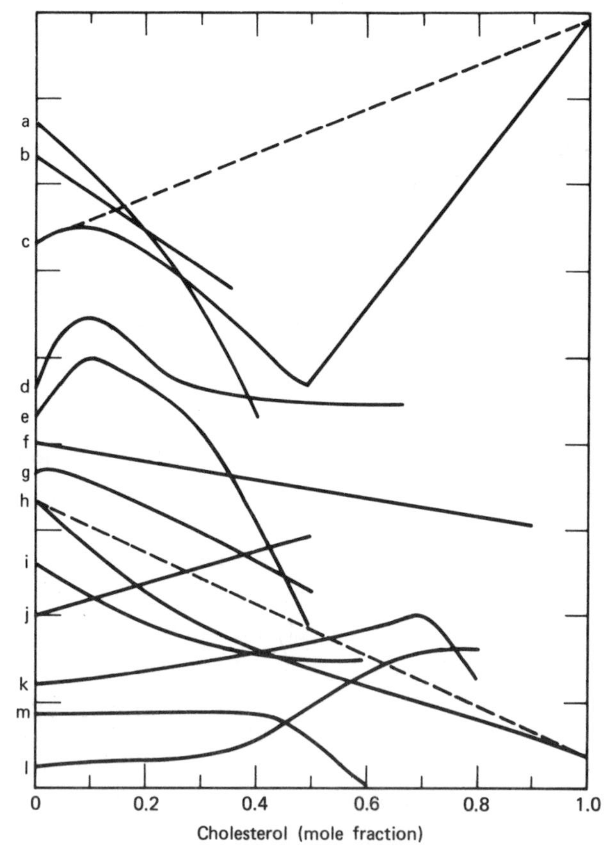

Fig. 5-7. Some membrane-associated phenomena that change as a function of the mole fraction of cholesterol in bilayers. From top: T_m for hydrated DPPC (a); enthalpy of the transition (b); heat absorbed in the ice transition (c); interlamellar distance of DPPC in excess water (d); rate of swelling (e); permeability of water (f); glucose release (g); molecular areas in monolayers with 18 : 0–18 : 4 PC (h); hyperfine splitting in multibilayers of PE (i); ratio of the low field peak height to center peak height in multibilayers of PE (j); electrical resistance of bilayer lipid membrane (BLM) (k); electrical capacitance of BLM (l); and V_m for the action of phospholipase A2 (bee venom) on egg PC (m). (From Jain, 1975.)

brane morphology leading to the formation of nonlamellar phases, or a change in the solubility of small solutes, or a change in the number and distribution of the kinks, or a change in the lateral organization of the cholesterol + phospholipids complexes in the bilayer, or a change in the head group organization (Jain, 1975; Yeagle, 1985). Many of these changes are not unrelated.

Physical studies suggest that cholesterol changes the lateral organization of phospholipids, and therefore the gel-to-fluid phase transition properties and the order parameters of bilayers (Presti, 1985). Early suggestions regarding limiting solubility of cholesterol up to 33% are not valid. In the gel bilayer of DMPC for example, up to 8 mole% cholesterol is apparently ideally mixed without altering the phase transition or the ripple pattern in freeze-fracture. Phase separation occurs between 8 and 23 mole% cholesterol, and yet another phase boundary exists between 23 and 45% cholesterol (Knoll et al., 1985). In the fluid phase complete miscibility is seen up to 44%. Beyond 45 mole% crystals of cholesterol are formed. There is also indication that cholesterol does not mix ideally with DPPC in fluid bilayers (Recktenwald and McConnell, 1981).

In monolayers it has been shown that 7 : 1 and 3 : 1 chain/cholesterol mixtures allow for a particularly dense packing (Cadenhead and Muller-Landau, 1984). The 7 : 1 (chain/sterol) ratio also corresponds to the phase boundary at 0.24 because two adjoining coexistence regions meet at 24%. This is in apparent contradiction with the phase rule, i.e., there can only be $n + 1$ phases for n components. Formation of a 7 : 1 stoichiometric complex is equivalent to the introduction of a new component so that two coexistence regions may meet at such a mixture. Thus, complex formation leading to phase separation of 7 : 1 bilayer/cholesterol (or 22 mole% cholesterol in diacylphospholipids) occurs. The exchange rate of phospholipids in such coexisting domains in the gel phase is 500/sec. Time for axial rotation of cholesterol is calculated to be 0.1 nsec, compared to about 1–10 nsec for phospholipids in bilayers.

While sterol has a condensing effect on chains, it increases the disorder in the head-group region. The condensing effect of cholesterol is seen with a variety of phospholipids, including branched-chain PC, PE, ω-cyclohexane fatty acid containing PC, and analogs of phospholipids. The ester carbonyl group on phospholipids is not necessary for the condensing effect and for the decrease in permeability that is induced by cholesterol in bilayers. However, some subtle differences have been noted (e.g., see Blume and Griffin, 1982). For example, sterols have more pronounced effect on oleoyl-stearoyl PC dispersions than on stearoyl-oleoyl-PC dispersions (Davis et al., 1986).

Lateral organization of cholesterol in bilayers is subject of active investi-

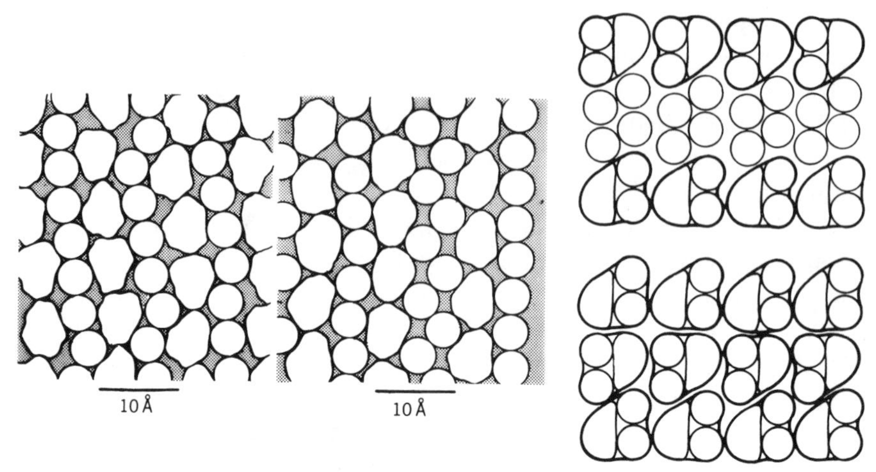

Fig. 5-8. Possible models for the lateral packing of phospholipid + cholesterol 1 : 1 (left) and 1 : 2 (right) viewed perpendicular to the plane of the lamellar phase. These arrangements emphasize long-range order of acyl chains (circles) which are closely packed near cholesterol (irregular shapes). (From Rogers et al., 1979, and Presti, 1985.)

gation. DSC data suggests that beyond 22 mole% cholesterol, the bilayer regions containing only phospholipid molecules disappear. The models adequately describe the chain/cholesterol stoichiometry of 7 : 1 (22%), 4 : 1 (31–33%) and 2 : 1 (45–50%) (Martin and Yeagle, 1978). However, a variety of experiments suggest that a long-range order leading to cooperative interactions between the acyl chain can persist even at 50 mole% cholesterol (Jain et al., 1984b, and references therein). These observations imply that in a 1 : 1 complex of cholesterol with diacylphospholipids a long-range order of acyl chains (Fig. 5-8) could exist under certain conditions. Relatively strong interaction of cholesterol with SM in bilayers has been demonstrated by using the dependence of the spectral properties of cytochrome P-450 on cholesterol content (Stevens et al., 1986). Similar specific interaction of cholesterol may be responsible for a slower rate of its efflux from cells containing certain phospholipids (Bellini et al., 1984).

6 Solutes in Bilayers

. . . but at that time I did not yet know the frightening anesthetic power of company papers, their capacity to hobble, douse, and dull every leap of intuition and every spark of talent.

PRIMO LEVI
The Periodic Table

As protective barriers, biomembranes come in contact with a wide variety of solutes ranging from water, ions, and oxygen to proteins, viruses, and bacteria. The response of membranes to these interactions is complex. Many viruses, toxins, hormones, and transmitters have specific receptors and *transporters* on some membranes and not in other. Nonspecific interactions are also important. For example, survival of a cell in a changing ionic environment depends upon osmotically driven fluxes of water; uptake of hydrophobic metabolites and drugs by cells depends upon their partitioning in membranes; membrane as a matrix enhances the catalytic activity of proteins. Specific functions associated with membranes will be discussed in subsequent chapters, and some biophysical aspects of partitioning and binding of lipophilipic solutes and the resulting perturbations in phospholipid bilayers are discussed in this chapter.

Thermal motion of solutes in aqueous phase tends to drive them into membranes. The free energy for the transfer of solutes from the aqueous phase to the bilayer phase comes from the free energy in the concentration gradient between the aqueous and the membrane phases. Entropic factors (hydrophobic effect) drive hydrophobic solutes from aqueous phase to bilayers and thus create a gradient across the interface. Interaction of solutes with specific functional groups on the surface of the bilayer (hydrogen bonding, burying charges, ionic interactions, desolvation) could provide an enthalpic contribution arising from ionic or dipolar interactions; a change in the entropy of the solute and the geometrical factors related to the shape and size could hinder their incorporation into bilayer or could lead to disruption of the bilayer. It should also be emphasized that many polar solutes like sugars and ions modulate the properties of bilayers that may be important for preparation, aggregation, storage, and stability of vesicles. For example, it appears that the membrane of nematodes is stabilized by trehalose.

The amphipathic environment of the bilayer has unique anisotropy and polarity gradients such that it can be adequately described as an isotropic bulk organic solvent. Also solutes of widely different polarity and amphipathic character can be accommodated in different regions of a bilayer. All these features relate to the questions of general interest: how much solute is incorporated into a bilayer; what are the effects of solutes on the organization of a bilayer; and what is the nature of interaction between the solute and the bilayer.

PARTITION COEFFICIENT

Incorporation by partitioning of a solute implies that to a first approximation the bilayer plays the role of a solvent. The partition coefficient of a solute is the equilibrium constant for its distribution between the two phases. Ideally, for a membrane the partition coefficient is defined as the mole fraction of the solute in the membrane divided by mole fraction of the solute in water. This unitary definition permits a meaningful comparison of derived thermodynamic parameters from the various systems. Unfortunately, very few values in literature are expressed in this fashion, partly because it is difficult to define and establish the mole fraction in a biological membrane; however, this limitation does not apply for bilayers.

Usually, the partition coefficient (P) is expressed as the ratio of equilibrium concentration of a solute in membrane (S_m) divided by its equilibrium concentration in the aqueous phase (S_w):

$$P = \frac{(S_m)}{(S_w)}$$

Although it is a convenient definition for bulk organic solvents, the assumptions involved in extrapolating this definition to the bilayer water system are probably all too obvious from the discussion in the preceding chapters. The "concentration of a solute in the bilayer" can only be approximated either by assuming that the density of the bilayer reflects all the medium that is accessible to the solute, or the concentration of the solute be expressed as moles of solute per gram of lipid. For most practical purposes partition coefficients expressed as the ratio of the moles of solute per gram of lipid divided by moles of solute per gram of aqueous phase is probably a reasonable approximation. Several methods for measurement of partition coefficients have been described (Jain and Wary, 1978; Pjura et al., 1984; Welti et al., 1984; Burke and Tritton, 1985; Jones and Lee, 1985; Luxnat and Galla, 1986; Froud et al., 1986). However, use of such data to obtain the free energy of transfer requires the assumption that there are no specific interactions between the solute and water or the bilayer phase.

BINDING OF SOLUTES

Binding of a solute to bilayer implies that only a finite number of saturable sites are available for interaction. Therefore, the equilibrium distribution of a solute between water and bilayer can be expressed as a binding constant. In its simplest form the binding equilibrium for a solute (S) between membrane (M) and the aqueous phase is:

$$\mathrm{S + M \rightleftarrows SM}$$

which can be described by a dissociation constant:

$$K_\mathrm{d} = \frac{[\text{free S}]\,[\text{free M}]}{[\text{Complex}]} = \frac{(\mathrm{S})(\mathrm{L})}{(\mathrm{MS})}$$

There are two possible methods for determining K_d from either titration of membrane by solute or by titration of solute by membrane.

1. Titration of membrane by varying concentration of the solute is feasible if the signal due to the bound solute can be readily quantitated. Under these conditions, the equilibrium concentration of the complex and the equilibrium concentration of the occupied sites is the same, say y. If the total concentration of solute is S_t and that of membrane is M_t,

$$K_d = \frac{(S_t - y)(M_t' - y)}{y}$$

or

$$\frac{M_t}{y} = \frac{1}{S_t}\left\{M_t + \frac{MK_d}{M_t - y}\right\}$$

If $y << M_t$

$$\frac{M_t}{y} = \frac{[M_t + K_d]}{S_t}$$

Also if K_d is large compared to S_t

$$\frac{M_t}{y} = \frac{K_d}{S_t}$$

Thus the plot of $1/y$ versus $1/S_t$ has slope of K_d.

2. Under most conditions it is more convenient to titrate a fixed amount of solute with varying membrane concentration. To derive a generally applicable expression, it is necessary to assume that the solute S binds to a hypothetical sites on the bilayer. If M_t is the total concentration of membrane site, and if there are n lipid molecules per site, the total lipid concentration $L_t = n \cdot M_t$. If the equilibrium concentration of the complex and the equilibrium concentration of the bound substrate are assumed to be x, then

$$K_d = \frac{(S_t - x)\left(\frac{L_t}{n} - x\right)}{x} = \frac{(S_t - x)(L_t - nx)}{nx}$$

or

$$(S_t - x) = \frac{nxK_d}{(L_t - nx)}$$

or

$$\frac{S_t}{x} = 1 + \frac{nK_d}{(L_t - nx)}$$

This is a quadratic equation in x with three unknowns n, K_d, and x. It can be solved by nonlinear regressions analysis using numerical iterative procedures. Only one physically meaningful root is obtained; however, it is necessary to assure that the three variable are independent, that is the covariance between n, K_d, and x is not significant.

In some cases simplifying assumptions can be made. If the fraction of the bound solute is very small, that is, $S_t/x >> 1$, the equation given above degenerates into:

$$\frac{S_t}{x} = \frac{nK_d}{L_t - nx}$$

In this case a plot of $1/L_t$ versus x/S_t is linear. Under these conditions often $x << L_t$, therefore

$$\frac{S_t}{x} = \frac{nK_d}{L_t}$$

The plot of $1/L_t$ versus x/S_t has the slope nK_d, which is the apparent dissociation constant. These conditions are satisfied if x/S_t is independent of L_t.

Under some conditions it is possible to vary the concentration of S by keeping the lipid concentration constant. Thus,

$$\frac{M_t}{M_s} = \frac{M_t}{S_t} + \frac{K_d \cdot M_t}{S_t\,(M_t - M_s)}$$

This can be simplified, if $M_t/M_s >> 1$ to:

$$\frac{M_t}{M_s} = \frac{M_t + K_d}{S_t}$$

$$\frac{M_t}{M_s} = \frac{K_d}{S_t}$$

According to this expression the double reciprocal plot M_s/M_t versus $1/S_t$ would give a slope of K_d.

Experimentally, K_d can be determined from a titration curve for S with increasing concentrations of M. For the example illustrated in Fig. 6-1, M is in the form of sonicated vesicles expressed as total lipid concentration Lt, and S is a tryptophan derivative whose fluorescence emission intensity increases on incorporation into a lipid bilayer. Since the increase in the fluorescence intensity (dI) is related to the fraction of the total solute in the bilayer, the fraction of the total solute incorporated in the bilayer will be:

$$\frac{I_m}{dI} = \frac{S_t}{x}$$

This expression can be used for obtaining information from the binding isotherm of type shown in Fig. 6-1. In such cases however, it is useful to be aware of the underlying assumptions.

1. It is difficult to express membrane concentration. This can be done by assuming that n lipid molecules on a bilayer make a solute binding site.

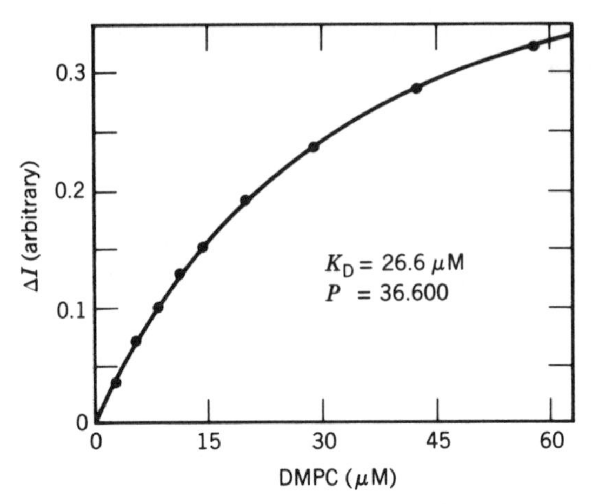

Fig. 6-1. The change in the fluorescence intensity of octyltryptophan at 333 nm versus DMPC concentration added as sonicated vesicles. The theoretical curve is for partitioning of the solute into vesicles where all the compartments are in equilibrium. (From Jain et al., 1985a.)

It is implicit that all such sites are independent. By nonlinear regression analysis the binding isotherm can be fitted to obtain n, K_d, and I_m (Dufourcq and Faucon, 1978; Araujo et al., 1979; Jain et al., 1982; Devaux and Seigneuret, 1985). As shown in Fig 6-1, for a small solute like octyl tryptophan the fit is excellent over the whole concentration range. However, before evaluating a general applicability of this treatment, significance of the binding parameters and other assumptions implicit in the above derivation should be considered.

2. When the membrane concentration is expressed in terms of the lipid concentration, it is assumed that all the lipid molecules are equally accessible for binding except those that constitute the binding site for the solute already bound to the membrane. In a closed structure like a vesicle, this condition would be satisfied only if the rate of transbilayer movement of the solute is rapid enough so that the solute is equilibrated with both monolayer halves of the bilayer. For octyltryptophan this occurs in less than 2 min, however, even for small hydrophobic dipeptides it can take several minutes. The rate of transbilayer movement is not a simple function of hydrophobicity; for example the half-time for the transbilayer movement of monoacylphospholipids is about 8 hr and more than 100 hr for diacylphospholipids. Obviously, for amphipathic molecules transfer of polar group across the apolar region is rate limiting, and the "concentration" of phospholipids with polar

head group in the middle of the chains is very small and therefore rate limiting.

3. It is also assumed that binding of a solute molecule does not effectively influence the binding of other solute molecules, i.e., binding sites are independent and the solute molecules do not associate in the bilayer. It is very difficult to test this assumption, however, the fact that the binding isotherm could be fitted to a single binding expression suggests that this is probably the case. However, this may not be a general situation, especially if solute molecules exhibit any tendency to form, say, an H bond as with alkanols (Rowe, 1985). For fluorescent solutes such solute–solute interaction could lead to self-quenching and exiplex formation. This would lead to a nonlinearity in the measured fluorescence intensity as a function of the bound solute concentration. Complications arising from self-association and self-quenching can be resolved by appropriate use of anisotropy and lifetime measurements (Burke and Tritton, 1984).
4. It is also implied that the environment of the bound solute in the bilayer is the same as it is in the absence of the solute. This assumption is valid for very small solutes that could be accommodated in kinks. However, it is not a valid assumption even for octyltryptophan; the activation energy for its binding is same in the gel as well as the fluid bilayer. This suggests that a solute creates its own microenvironment in the gel bilayer and this microenvironment is disordered (Jain et al., 1985a).
5. To interpret properly the thermodynamic significance of the binding data, it is necessary to assume that the solute is dispersed as a monomer in the aqueous phase. This becomes a serious problem with hydrophobic solutes that tend to aggregate or form microcrystals in aqueous phase. Under such conditions not only do the kinetic artefacts become unmanageable, but also a true equilibrium is never established with monomeric solutes in the aqueous phase.

Binding of a variety of solutes to phospholipid bilayers has been examined (Simon et al., 1977; Jain and Wary, 1978; Jain et al., 1985a; Pjura et al., 1984; Burke and Tritton, 1984). In a series of anthracyclins K_{app} ($= n \cdot K_d$) changes largely because there is a change in the value of n. While n does not change appreciably in the gel and fluid bilayers, K_d is always higher in the gel bilayers. Thermodynamic parameters suggest that binding can be either enthalpy driven or entropy driven, depending upon the nature of the solute (Burke and Tritton, 1984; Jain et al., 1985a).

Partitioning of Solutes in Bilayer

Nonspecific incorporation of solutes into bilayers is a very general process that underlies a wide range of membrane phenomena ranging from permeability and transport to drug-induced perturbation of protein functions and anesthesia.

Polar solutes like water have a small but finite partition coefficient. The concentration of water in bulk water is high (55 *M*), therefore a partition coefficient of about 0.002 (estimated lower limit value) would mean that the concentration of water in bilayers would exceed 0.1 *M*. This is not unreasonable and it is consistent with the permeability coefficient of about 10^{-4} cm/sec.

Bilayer–water partition coefficients have been measured for very few solutes. These values were measured at single solute concentration and it is assumed that the solute concentration in the membrane is small enough that the bilayer organization is not perturbed. Since the partition coefficient is an equilibrium constant, it is characterized by thermodynamic state functions like free energy (ΔF), enthalpy (ΔH), and entropy (ΔS) by the following relationship.

$$\Delta F = \Delta H - T\,\Delta S = -RT \ln P$$

$$= -1.34\,RT \ln \mathrm{P} \qquad \text{(at 25°C for values of } \Delta F \text{ in Kcal/mole)}$$

A cavity model has been proposed for phenomenological interpretation of partition coefficients of solutes between nonpolar and aqueous phases (Pierroti, 1976). The overall process has at least two terms: energy required to create a cavity of suitable size in the solvent and the energy resulting from solute–solvent interaction. According to this model, a small enthalpy of transfer observed for small solutes occurs because the unfavorable change in the enthalpy required to move a cavity from the nonpolar phase to the water phase is offset by a favorable change in enthalpy resulting from the solute–solvent interaction in both phases. The large favorable enthalpy of association is observed for the longer-chain phospholipids because the two enthalpy terms have the same sign.

The thermodynamic parameters for the partitioning of hexane into bilayers of different composition in micelles and in nonpolar solvents have been compared (Simon et al., 1977). Data suggest that entropically *n*-hexane prefers bulk solvents, whereas enthalpically it prefers the bilayer. The bilayer–water partition coefficients for chlorpromazine exhibits a strong dependence on the composition, cholesterol content, and chain length (Luxnat and Galla, 1986). The partition coefficient in dimyristoyl PC (DMPC) bilayers is

3200 in the fluid and 200 in the gel bilayers, and a value of 800 is observed in the rippled P_{β} phase (Muller et al., 1986).

The free energy for transfer of a solute can be considered as a sum of the incremental free energies for the various groups $\delta\Delta F(X)$ as summarized in Table 6-1. The incremental free energies for the various substituents and the ring structures are found to be fairly constant for transfer from water to a bulk organic solvent, and that is found to be only approximately true for membranes. The incremental free energies thus obtained are approximations based on the assumption that the microenvironment of a substituent is more or less independent (except for H-bonding, steric, inductive, and related effects) of its position relative to other substituents on the same molecule.

Several interesting features emerge from the partition coefficient data on small electrolytes (Diamond and Katz, 1974). There is a systematic correlation between the partition coefficients of electrolytes in bilayer and in bulk organic solvents like *n*-octanol. On this basis the polarity of the environment

TABLE 6-1 Incremental Free Energies ($\delta\ \Delta F$, in cal/mole) for Transfer from Water

Substituent	Olive Oil	*n*-Octanol	Nitella	Dimyristoyl Phosphatidylcholine Bilayer
—OH	2800	1580	3600	800–950
—O—	1400	1350	800	
—C(=O)—	2200	1650	2500	650
—C(=O)—OH	2800	925–1600	—	
—C(=O)—O—R	1400	~600	1400	1100
—C(=O)—NH_2	4800	2325	6200	
—NH—C(=O)—NH_2	5300		7300	
—C≡N		1150		
—NH_2		1580		
—CH_2—	−660	−675	−610	−450 to −800
Branching		270		370
C=C		400		
H bond (intramolecular)		880		

for these solutes in dimyristoyl PC liposomes (selectivity coefficient s = 0.87) is between isoamyl alcohol (s = 0.86) and n-octanol (s = 1.0). Solvents like olive oil (s = 1.57) and benzene (s = 2.17) are considerably less polar than membranes. These conclusions obviously apply for the average solvent properties of bilayer. The following points may however be noted about the peculiarities of bulk solvents compared with bilayers:

1. The absolute values of partition coefficients in DMPC bilayers are approximately five-fold lower compared with those in n-octanol. Anomalies in P measured in bulk solvents like n-octanol arise from the hydrogen-bonding tendency of n-octanol, the ability of octanol to dissolve substantial amounts of water, and possible dipolar interactions (Wolfenden et al., 1981).
2. Bilayer discriminates the branched solutes somewhat better than the bulk solvents.
3. The incremental free energies for the transfer of several functional groups from water to bilayer are enthalpy driven except those for —CH2— group, which is entropy driven. The incremental free energies are also found to be somewhat dependent on the position of the functional group.
4. The temperature dependence of partition coefficients of small solutes exhibit a discontinuity at the T_m for DMPC. This suggests that ordering of chains "squeezes out" solute molecules. Both the enthalpy and entropy increase by a factor of about 10 at the phase-transition temperature. This is also indicative of an ordering effect of chains.
5. A comparison with partition coefficient data for erythrocyte membranes (Seeman, 1972) suggest that the erythrocyte membrane is somewhat more disordered and hydrophobic like n-octanol, and the negative charge on the erythrocyte membranes modulates partitioning of the charged solutes.
6. The equilibrium partitioning behavior of small nonelectrolytes at low concentrations in bilayers is approximately the same as the behavior of these solutes in bulk organic solvent.

The relationship between the thermodynamic constants for P of a series of solutes can be expressed as Barkley–Butler plots. ΔH and ΔS for the transfer of a series of solutes from aqueous phase to bilayer show a linear relationship as shown in Fig. 6-2. Full implications of this enthalpy–entropy compensation relationship are not clear yet, however the slope and intercept of this plot can be reasonably interpreted in terms of the relative solute–

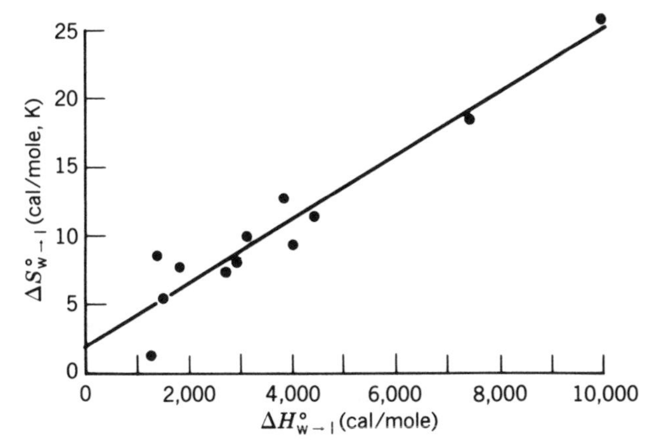

Fig. 6-2. The entropy of partitioning of solutes between dimyristoyl PC vesicles and water plotted against the enthalpy of partition. Each point represents one solute, and the straight line gives the fit for $\Delta S = -15.6 + 0.00226\ \Delta H$. (From Katz et al., *J. Membr. Biol.* 17, 113, 1974.)

solvent interactions in water and bilayer (Diamond and Katz, 1974). The slope $b = 0.00226/K$ (above 25°C for DMPC liposomes) is probably related to a loss of mobility of the solute incorporated into the bilayer. The value of intercept a depends upon the choice of the concentration unit. Use of mole fraction units permits comparison between different systems. The values of intercept are very sensitive for the phase state of the bilayer; $a = -15.6$ and -1.3 cal/mol, K respectively, for the DMPC in fluid or gel bilayers. The corresponding values for the bulk solvents are about 22 cal/mol, K. These values can also be interpreted in terms of a lower degree of disorder induced by a solute in the bilayer.

EFFECTS OF SOLUTES ON BILAYERS

As the size of a solute becomes larger than the size of the free volume between the kinks, several interesting but complex effects in the partitioning behavior as well as on the organization of the bilayer become apparent (Table 6-2). These effects are also expected at higher concentrations of smaller solutes. This is not surprising because direct but noncovalent interactions between the phospholipid molecules in a bilayer would be readily modified by incorporation or binding of a solute. The resulting effects on the organizational and motional changes have not been integrated into a cohesive model. However, such considerations are useful in evaluating effects of

TABLE 6-2 Possible Effects of Solutes on the Properties of Bilayers

Shape change
Permeability change
Phase change
Dissipation and induction of phase separation
Influence on the local surface charge density
Desolvation and solvation of the interface
Dielectric breakdown
Interdigitation of acyl chains
Solubilization of lipids, leakage, osmotic swelling
Transbilayer movement
Fusion: activation or inhibition
On the catalytic properties of membrane proteins

detergents, drugs, probes, and anesthetics on bilayers. The following considerations offer a general guide to appreciate effects of solute on bilayers.

Limited Solubility of Nonpolar Solutes in Bilayers

In Chapter 5 we noted that incorporation of several lipids like cholesterol into bilayers of phospholipids occurs up to 50 mole%. Several hydrophobic solutes appear to have very limited solubility in bilayers. For example, only about 2.8 mole% cholesterol oleate and trioleoyl glycerol are incorporated into vesicles of egg phosphatidylcholine, even when premixed during the dispersal of lipids by sonication (Hamilton et al., 1983). Such observations suggest that the structure of a solute and the relative position of functional groups (Jain and Wu, 1977) have pronounced effects on the ability of a solute to perturb the organization of a bilayer. Certain solutes could preferentially interact with lipid components and form stoichiometric complexes (Boggs et al., 1986). Similar factors are apparently responsible for bilayer-to-hexgonal phase change induced by a variety of solutes like dolichols, peptides, and proteins (Killian and De Kruijff, 1986).

Partial Molecular Surface Area and Volumes

Limiting area of a solute incorporated into monolayers of phospholipids contains information not only about the molecular area but also about specific interactions. From the molecular area measurements it is possible to speculate on the orientation of the solute in the monolayer, and also obtain information about a conformational change in the components of the bilayer.

Similarly, in phase diagrams specific interactions are indicated by departure from additivity.

Effects on the Organization of a Bilayer

Incorporation of a solute in a bilayer perturbs lipid–lipid and lipid–water interactions. This brings about many secondary manifestations that are summarized in Table 6-2. One of the most dramatic effects of incorporation of solutes in lipid dispersions is manifested on the motional and the order parameters (Lee, 1977; Goldstein, 1984). While these observations suggest that a change in the organization of the bilayer has been induced by a solute, the nature of the resulting organization of the bilayer depends upon additional factors. The nature and extent of perturbations induced by solutes depends upon the structure and concentration (mole fraction) of the solute, structure of the lipid, temperature, and other environmental variables. The significance of the solute-induced perturbation of bilayer organization may be appreciated by implicit assumptions such as a solute locally stabilizes a conformational state of phospholipid molecules which ultimately stabilize defects in the bilayer organization. Except under extreme conditions, these conformational states of phospholipid molecules may be effectively analogous to those that characterize the various polymorphs. Thus, the global bilayer organization with topological continuity is retained, while local organization may be appreciably disordered, and the domains of disorder may not show in the lamellar periodicity (Jain and White, 1977; Jain, 1983). It is probably also unreasonable to expect that separate bulk phases formed at higher mole ratios of solutes would also be formed at lower mole fractions. This is because bilayers have a finite capacity to accommodate local perturbations or conformational disorder.

Effect of Solutes on Gel–Fluid Transition

Solutes in a bilayer isothermally perturb the local organization without disrupting the topography. Most solutes incorporated in bilayers alter the thermotropic transition characteristics. Almost no information is available on the effect of solutes on subtransition. This is probably because the subgel (L_c) state can not be induced in the presence of impurities, or that the solutes are squeezed out during the induction of the subgel state. In general, the effect of additives is much more pronounced on the pretransition than it is on the main transition. Since pretransition is small and exhibits considerable hysteresis, such effects have been studied only in a few cases.

Effects of a large variety of solutes on the main transition have been demonstrated by almost all conceivable techniques. In general, most nonpolar or amphipathic solutes lower T_m, decrease enthalpy, and broaden the temperature range over which the transition occurs. In some cases, the transition temperature and enthalpy increase. Even in such simple cases, the physical basis of the observed effects is not known. A detailed theory explaining all aspects of gel–fluid transition in bilayers is not available yet. Some extrapolations have been obtained from the use of the phase rules developed for transitions from bulk solid to bulk liquid. Thus, temperature-composition phase diagrams for mixed lipid systems (Lee, 1977; Ethier et al., 1983; van Dael and Cehterickx, 1984; Jain et al., 1985a) are used to obtain information about the presence of the various phases, however no mechanistic insights have come. Overextension of these rules to a protein–bilayer system can also be misleading. For example, the effect of additives on the freezing point of bulk liquids is usually interpreted in terms of Rault's law or the Clausius–Claperon equation. In its simplest form, the change in the transition temperature induced by a solute is given by:

$$\frac{\Delta H}{T} - \frac{\Delta H_m}{T_m} = R \ln \frac{X_g}{X_f}$$

which has been often approximated (but not rightfully so) as:

$$\Delta T = T - T_m = \frac{RT^2{}_m}{\Delta H_m}(X_g - X_f)$$

where T_m and T are the transition temperature in the absence and in the presence of the solute, ΔH and ΔH_m are the enthalpy of transition in the presence and in the absence of a solute, R is the gas constant, and X_g and X_f are the mole fractions of the solute in the gel and the fluid bilayers, respectively. This relationship has been used to calculate the concentration of a solute in the bilayer from the measured T and T_m.

While relationships given above appear to be partially valid for small solutes like ethanol, such interpretations are, however, dubious because ethanol induces transition to interdigitated bilayers. These relationships do not account satisfactorily for the effect of solutes like alkyl esters of tryptophan or peptides (Jain et al., 1985a) or carbocyanine dyes (Ethier et al., 1983). Inapplicability of such relationships is not very surprising. In their derivation it is implicit that the solute does not interact specifically with the components of the bilayer or with other solute molecules, that the solute does not perturb the organization of the bilayer, and that the enthalpy of transition is not appreciably altered in the presence of the solute. Since solutes also modulate the pretransition behavior and this transition disappears at low mole fractions of the solute, it is not certain whether it should be considered as a part of gel-to-fluid phase transition. In fact the enthalpy term is expected to dominate partition behavior since T is expressed on an absolute temperature scale. Therefore, interpretation of the phase-transition data of bilayers modified with solutes in terms of these equations to obtain relative affinities of a solute for the gel and the fluid phases is not valid.

Solute-Induced Formation of Interdigitated Bilayers

Several solutes at somewhat higher concentrations (typically above 10 mole% in the bilayer) induce formation of bilayers in which the acyl chains from the apposing monolayers are at least partially interdigitated. These solutes are amphipathic, therefore they are expected to be localized near the head-group region. This would induce energetically unfavorable voids between the acyl chains, and such voids can be eliminated by interdigitation. Typically, the thickness of the bilayer is reduced by about 20 Å. This corresponds to a difference of about 15 methylene residues, implying that the chains from apposing monolayers penetrate up to the seventh methylene residues.

Solute-Induced Bilayer-to-Hexagonal II Phase Transition

Several solutes induce formation of hexagonal II phase. Such changes are strongly dependent upon the nature of the lipid and the solute. For example dolichols are apparently not incorporated into PC bilayers whereas at 1–2% concentration in bilayers of dioleoyl PC they induce formation of H_{II} phase (van Duijn et al., 1986). This is probably because long dolichol chains can be accommodated only in between the rods of H_{II} phase. Since H_{II} phase is usually formed by lipids with a smaller head group compared to the cross section of the acyl chain (wedge-shaped molecules), such transitions are isothermally induced when the effective surface charge is reduced by ions. Similarly, hydrophobic solutes that increase the disorder in the acyl chains can also induce H_{II} phase presumably because the wedge shape of the lipids is enhanced.

Solubilization of Phospholipids by Solutes

Solubilization of bilayers by large amphipathic molecules like detergents is a well-known case of solute-induced bilayer-to-micelle transition. Typically, increasing concentration of detergents causes expansion of bilayers that leads to a change in the shape and size of vesicles, to leakage, then to the formation of nonbilayer structures, and ultimately to solubilization whereby disc-shaped particles and micelles are formed. In general these effects are observed when the detergent-to-phospholipid mole ratio is more than 0.5 and the absolute concentration of the detergent exceeds its critical micelle concentration. Formation of anesthetic-phospholipid micelles leading to solubilization of vesicles has also been reported (Frezzatti et al., 1986).

Chaotropic agents are known to solubilize vesicles (Oku and MacDonald, 1983a,b), and to change their electrophoretic mobility (Tatulian, 1983) and aggregation behavior (Pigor and Lawaczeck, 1983). In a few cases it has been noted that at high concentration of small solutes the bilayer is disrupted and vesicles of different size are formed. As discussed in Chapter 4, formation and stability of vesicles depends upon the chain composition and the surface charge. Indeed, by incorporation of charged amphiphiles or by incorporation of single-chain amphiphiles the vesicle size changes significantly (Hauser, 1985). These effects would have significant effects on the processes that depend upon the surface area, membrane composition, and shape of the target vesicle.

BIOMEMBRANE AS A TARGET FOR DRUGS

A wide variety of metabolites and drugs find their way into the lipid bilayer of biomembranes. This happens because all solutes in an organism come in contact with a variety of membranes during their absorption, transport, metabolism, and clearance. Most of these solutes do not have any major effect on the organization of biomembranes because their equilibrium concentration in the membrane is generally insignificant. However, the local and transient concentration of many of these solutes can be occasionally high enough that many of the effects induced by nonspecific amphipathic solutes may be significant. Such effects include a change in the permeability, osmotic swelling and lysis, fusion, transbilayer movement, shape change, vesiculation and spike formation, and modulation of protein-mediated catalytic and transport functions. A molecule as simple as ethanol, for example, can produce effects ranging from intoxication to alcoholism. Other chemotherapeutic agents that perturb functions of bilayers include anesthetics, steroids, and anticancer agents (Goldstein, 1984; Tritton and Hickman, 1985). In many cases the nonspecific effects of solutes have little primary pharmacological and physiological significance, however, it is likely that the nonspecific effects of solutes on bilayer functions are probably significant in eliciting many of the side effects exhibited by drugs.

Anesthetics

An ideal general anesthetic should cause loss of sensation, amnesia, and muscle relaxation without any adverse effect on cardiac functions. The locus of action of general anesthetics is believed to be brain cortex, although brain stem has also been implicated. Ultimately, anesthetics reduce or abolish the activity of a protein or proteins required for transmission of impulse in nerve fibers (Chapter 12), however, the primary locus of action of anesthetics remains to be elaborated. The phenomenological characteristics of the anesthetic action in an organism include: the effects of different anesthetics are additive; their effectiveness is correlated with lipid solubility as well as polarity; the anesthetic effect is reversed by high pressure; and lower concentration of anesthetics are required for victims of hypothermia (cold).

It is generally believed that the anesthetic activity of lipid-soluble solutes depends on perturbation of some yet unknown feature of bilayer organization (Janoff and Miller, 1982; Chin and Goldstein, 1985) or by a direct interaction with a protein in the nerve membrane (Richards, 1978; see also Chapter 12). Anesthetics in general lack the structural specificity that would be expected of drugs that bind with stereochemical selectivity to specific receptor sites. The functional specificity of anesthetics is also limited: Different anesthetics also exhibit other effects, and even the nature of anesthesia could be qualitatively different. Anesthetics are not chemically altered during the course of their action. Therefore by implication the primary target of action of anesthetics, narcotics, and alcohols is believed to be the bilayer of biomembranes. Moreover, the lipid phase has also been found to be a locus of adaptive changes associated with the tolerance and physical dependence that occurs with the chronic use of ethanol.

One of the most notable features of anesthetic–bilayer interactions is the observation that a small mole fraction of the anesthetic in the bilayer gives rise to a large change in some yet unknown property of the bilayer in membranes of nerve cells and synaptic junctions. Coupling between the interaction of an anesthetic and a change in the function of a specific membrane protein could be via a change in ion permeability (Bangham et al., 1980; Vassort et al., 1986), a perturbation of the dielectric constant of the bilayer (Hayden et al., 1981), an isothermal shift in the gel–fluid transition (Hill, 1974; Jain et al., 1975), a solute-induced phase separation (Jain et al., 1975; Trudell, 1977), or a modification of the annular boundary lipid (Lee, 1976). According to these hypotheses, perturbations in a property of the bilayer ultimately influence one or more protein functions. How this is achieved is still a matter of conjecture. Some of the key observations related to the phenomenon of anesthesia are summarized below.

In their structural complexity the general anesthetics range from xenon, cyclopropane, halogenated hydrocarbons (chloroform, halothane), diethyl ether, alkanols, to steroids like alphaxalone. The anesthetic potency of most general anesthetics is correlated with their lipid solubility (Fig. 6-3), however not all lipid-soluble solutes are anesthetics. Incorporation of lipid-soluble solutes changes the phase properties of bilayers. Such perturbations accompany an increase in volume, therefore be reversed by increasing pressure. Thermodynamic characterization of the anesthetic binding site suggests that anesthetic molecules partition in a relatively hydrophilic and organized (like gel bilayer) environment (Katz and Simon, 1977). While theories of anesthesia based on phase transition can account for pressure reversal of anesthesia, several obvious predicitons are not readily accommodated. For example, a decrease in T_m induced by solutes at their anesthetic concentration is about 0.5°C. It would imply that heating by 0.5°C should cause anesthesia and cooling should reverse it. This extrapolation is probably not valid because anesthetic-induced perturbations are isothermal and most probably local. Also the anesthetic and thermally induced phase transitions appear to differ in some other respects (Koehler et al., 1980).

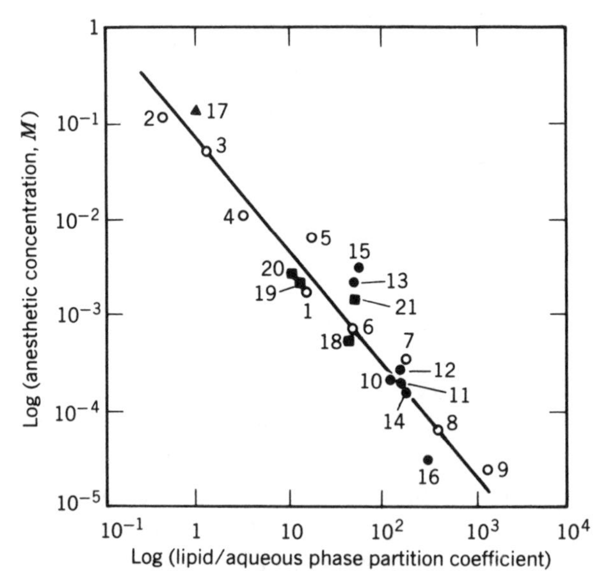

Fig. 6-3. The correlation between anesthetic potency and their partition in phosphatidylcholine bilayers. Open circles represent tadpoles, squares are newts, and triangles are frogs. Anesthetics are: 1, benzyl alcohol; 2, ethanol; 3, propanol; 4, butanol; 5, pentanol; 6, hexanol; 7, heptanol; 8, octanol; 9, nonanol; 10, halothane; 11, methoxyfluorane; 12, isoflurane; 13, fluoroxene; 14, pentobarbital; 15, thiopental; 17, acetone; 18, cyclopropane; 19, xenon; 20, carbon tetrafluoride; 21, sulfur hexafluoride. (From Janoff and Miller, 1982.)

There are several other interesting aspects of anesthesia that deserve a closer scrutiny than is given by the various theories. Concentration dependence of the effect of many anesthetics is typically biphasic. At low concentrations several anesthetics exhibit stimulatory effects. For example low levels of ethanol or diethyl ether cause stimulation and excitability (Laycock, 1953). Also there are many lipid-soluble compounds that are not potent anesthetics, and in many cases anesthetic potency does not correlate with lipid solubility (e.g., hexanol is more potent than hexane). Similarly, steroidal anesthetics exhibit considerable stereospecificity. Such aspects of anesthetic potency do not appear to correlate well with the ability of anesthetics to cause membrane expansion, to disorder the lipid bilayer and change the transition temperature, or to induce a phase separation. One of the more likely possibilities, which has not received any consideration, is that anesthetics dissipate lateral phase separation.

7 Lipid–Protein Interactions in Membranes

They sought it with thimbles,
They sought it with care;
They pursued it with forks and hope.

LEWIS CARROLL
The Hunting of the Snark

Proteins constitute up to 75% by weight of isolated cell membranes (Table 2-4). The ratio of proteins to lipid increases in the order (for rat tissues): myelin (0.23), platelets (0.7), erythrocytes (1.1), liver cells (1.4), nuclear envelope (1.6), sarcoplasmic reticulum (2), and inner mitochondrial membrane (3.2). This is approximately the order in which the level of specific biochemical functions associated with these membranes also increases. Direct evidence for the association of proteins with a bilayer matrix in biomembranes comes from a variety of observations summarized below:

1. In freeze-fracture electron micrographs the number and relative distribution of particles is related to their protein content, and they are influenced by treatment with proteases.
2. As shown by histochemical, biochemical, and analytical methods several enzymic activities are associated with specific membrane fractions, and these activities are modified by lipases as well as proteases and protein reagents.
3. Several membrane functions are genetically regulated and specific genes responsible for these functions have been isolated and cloned.
4. The number and relative proportion of protein species present in biomembranes appear to vary with its metabolic state. For example, more than 40 different proteins in inner mitochondrial membrane constitute over 70% of its mass. On the other hand only two proteins in

retinal rod discs constitute 85% of the total proteins and about 30% of the total weight of the membrane.

5. As the proportion of proteins in membranes increases the maximum membrane surface area (calculated from limiting area of extracted lipids in monolayers) occupied by lipids decreases. This is in the order: myelin (100%), endoplasmic reticulum (83%), sarcoplasmic reticulum (80%), outer membrane of mitochondria (72%), erythrocyte (67%), and inner membrane of mitochondria (40%). For the ratio of lipid/protein in these membranes see Table 2-4.
6. With photoactivable affinity labels many membrane lipids are also labeled along with the proteins.

Such observations attest to a functional significance of membrane proteins. Lipids also assist proteins and participate in many of these functions, because delipidation or disruption of bilayer organization drastically alters functions of many membrane proteins. In this chapter we elaborate aspects of lipid–protein interactions that relate to functions of membrane proteins. Compared with the behavior of water-soluble proteins, there are several unusual features of membrane proteins: A significant or substantial part of membrane proteins is in a nonaqueous environment; interaction of most proteins with lipids is noncovalent (however see below), and the stability of proteins in a bilayer is influenced by the nature and organization of lipids. Thus, the presence of proteins in bilayers of biomembranes impose several structural and organizational constraints, which require unique methodologies for their investigations. In their scope these methods range from analytical, genetic, and immunological to electrophysiological and reconstitution of semisynthetic membrane-enclosed functional structures. Some of these are discussed below and others are discussed further in subsequent chapters.

SOLUBILIZATION OF MEMBRANE PROTEINS

The *grind-and-find* techniques of classical biochemistry had very little success in the investigation of membrane proteins, because most of these methods for characterization of macromolecules rely on the physicochemical properties of molecularly dispersed soluble preparations. Disruption of tissues, cells, and organelles leaves the membranes in the pellet. Before *solubilization* of membrane proteins, it is often desirable to separate the membrane from different organelles, which is usually done by density gradient centrifugation (Chapter 2). Treatment of isolated membranes with salts and chaotropic agents (urea, guanidine, iodide, iodosalicylate) disrupts ionic interactions, and only weakly associated proteins are solubilized (Fig. 7-1). In some cases it is possible directly and selectively to transfer one or more membrane proteins to liposomes without major disruption of the membrane (Bouma et al., 1977). In most cases membrane must be disrupted to dissociate their proteins. Sonication disrupts membranes, and the fragments reseal into vesicles because the *frayed* edges cannot remain exposed to the aqueous phase. Solubilization of membrane lipids by organic solvents (chloroform + methanol 2 : 1; butanol, pentanol, 2-chloroethanol) generally causes irreversible denaturation of proteins, and this method has been successful only for isolation of very hydrophobic lipoproteins and lipids (Tenenbaum and Folch-pi, 1963; Martinek et al., 1986). However, such preparations are useful for amino acid analysis and may be for primary sequence studies.

Use of detergents to disrupt hydrophobic interactions between lipids and proteins requires careful evaluation of several factors (Helenius and Simons, 1975; Tanford and Reynolds, 1976; Lichtenberg et al., 1983). Two competing forces are at work: First, partitioning of detergents into bilayers where detergent replaces phospholipids in the vicinity of the protein and provides a stable matrix. Second, solubilization of lipids and proteins by detergents.

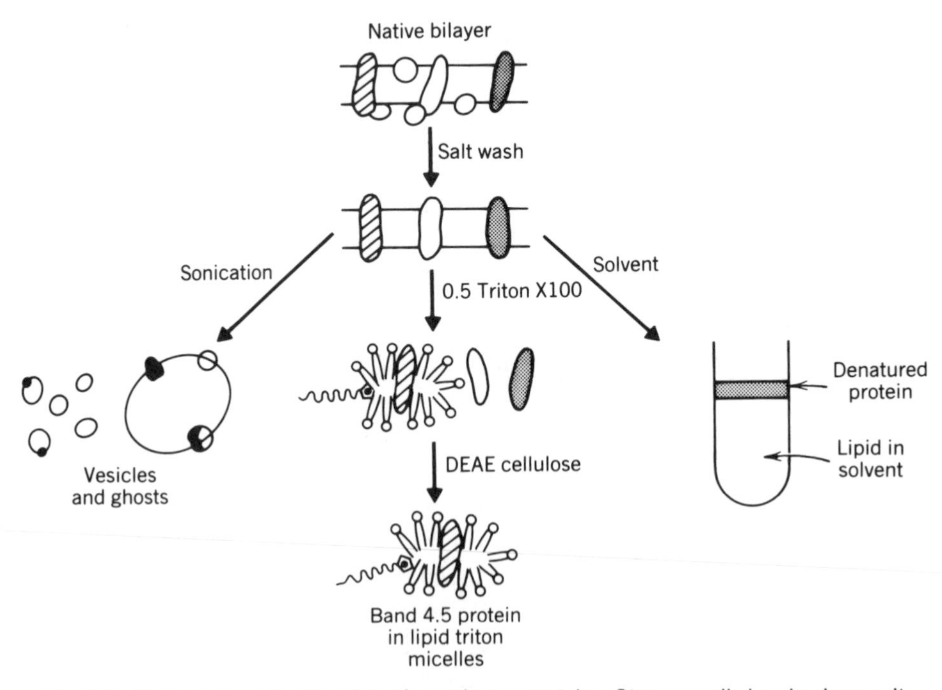

Fig. 7-1. Methods for solubilization of membrane proteins. Steps usually involved are salt wash (which removes electrostatically bound proteins) followed by sonication or detergent treatment, or solubilization in an organic solvent.

The organization of bilayer is disrupted when a critical mole fraction of detergent is incorporated.

Successful solubilization of membrane proteins by detergents is still a matter of trial and error (Womack et al., 1983). Criteria for successful solubilization include not only retention of the catalytic or binding function, but also adequate recovery, no interference with the assay, stabilization of the protein in detergent, ability to remove detergent, and reconstitution of the activity. During solubilization, detergent molecules replace native phospholipid molecules around the protein. Ionic detergents like sodium dodecyl sulfate are not very useful for solubilization of membrane proteins because their interaction is predominantly ionic. The mole ratio of SDS to proteins remains remarkably constant in mixed micelles such that these dispersions can be used for determination of molecular weight on the basis of their mobility in an electric field (electrophoresis). For such reasons it appears that strongly ionic detergents denature proteins whereas the zwitterionic or nonionic detergents do not. Detergents like cholate appear to stabilize disks of the bilayer containing proteins. When the detergent concentration is above its critical micelle concentration (cmc) for the mixed micelles, proteins and lipids disperse in micelles. If cmc is low, the bilayer is disrupted at low detergent concentration, the mixed micelles have a substantial proportion of native lipid, and removal of the detergent by dialysis or gel filtration becomes difficult. Detergents of high cmc are preferred because essentially all the lipid around the protein can be replaced by detergent molecules without formation of micelles of detergent alone.

Characteristics of detergents useful for membrane study are summarized in Table 3-5. In choosing a detergent for a specific purpose it is necessary to keep in mind that the detergent has to be removed ultimately. A complete removal of detergents like Triton X-100 by dialysis can take weeks. Removal of the detergent by gel filtration or equilibrium with Bio-Beads appears to be a preferred method. Transmembrane proteins from membranes rich in cholesterol and phospholipids can be fractionated by a temperature-induced bulk-phase separation. For example, dispersions of Triton X-114 are homogeneous at 0°C but separate into the aqueous and the detergent phase at 20°C. Hydrophilic proteins partition into the aqueous phase whereas hydrophobic transmembrane proteins are found in the detergent phase (Pryde, 1986).

Proteins of Erythrocyte Membrane

Membranes of erythrocytes (Fig. 4-9) have been examined extensively for several functional protein components. The protein profile of erythrocytes is shown in Fig. 7-2. The network of erythrocyte proteins is made up of spec-

Properties of Main Proteins Present in Human Erythrocyte Membrane

Protein	Band	MW × 10^3	Carbohydrate Content (%)	% of Total Protein	In Membrane	No. of Peptides per Ghost × 10^3
Spectrin	1	200–250	n.d.	15–25	Inner	216
Myosin-like peptides	2	235	—	12–15		235
Anion exchanger	3	90–95	7–9	25	Spans	940
Glucose transporter	4	72–78	7	4–8	Inner	180–240
Actin	5	43	n.d.	4.5	Inner	359
Glyceraldehyde-3-P dehydrogenase	6	36	n.d.	5	Inner	540
PAS-1 (Glycophorin A) + PAS-2 (MN protein)	7	>29	55	6	Spans	500

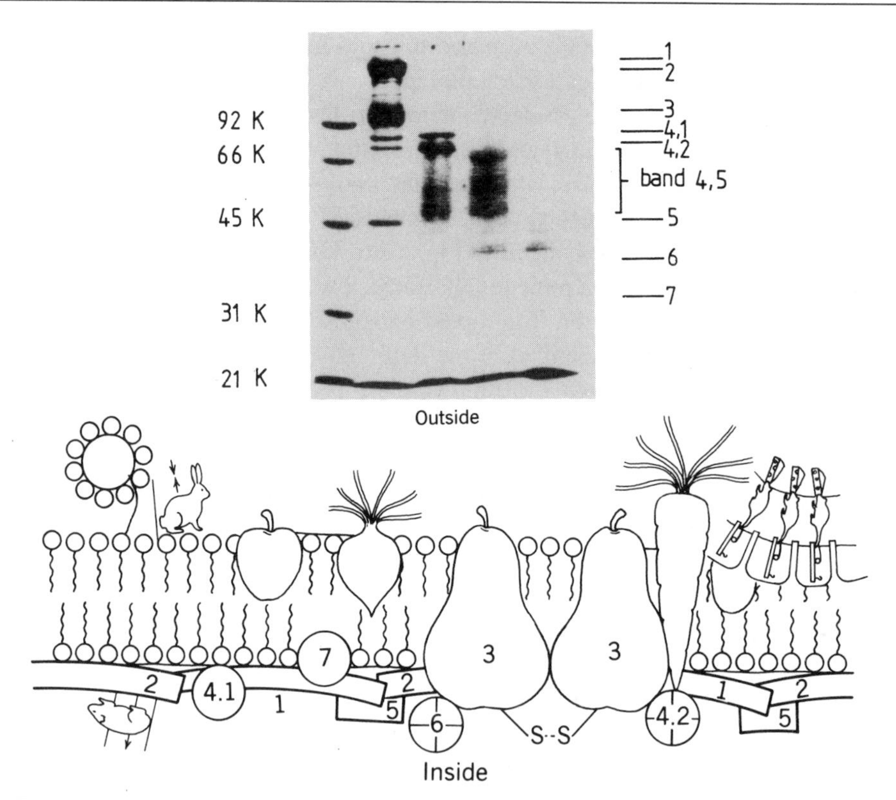

Fig. 7-2. (Top) Properties of the main proteins of erythrocyte membrane, (middle) their gel electrophoresis pattern (from left, molecular weight markers, intact red cell membrane, salt-washed membranes, solubilized in Triton X-100, cholate-solubilized membranes), and (bottom) the distribution of these and other components in the membrane. (The middle panel is from Carruthers, 1985.)

trin (bands 1 and 2), ankyrin (band 2.1), actin (band 5), and bands 4.1a and 4.1b. These proteins are readily solubilized by treatment with salt, but the skeleton remains intact during extraction of erythrocytes with Triton X-100. Band 3 and glycophorin are not solubilized by salt, but they are readily solubilized by detergents.

CHARACTERIZATION OF ISOLATED MEMBRANE PROTEINS

As shown in Fig. 7-2 simple membranes, like that of erythrocytes, contain several proteins. Careful analysis suggests that there may be well over 50 different proteins in erythrocyte membranes, however less than 10 can be normally visualized by one-dimensional gel electrophoresis. Membrane proteins solubilized in detergents can be sometimes isolated by salt precipitation and gel filtration in detergent-containing buffers. Isolation of specific protein components is generally achieved by affinity chromatography, immunoprecipitation, and fractionation after specific labeling. Use of monoclonal antibodies offers considerable promise for purification of membrane proteins (Venter et al., 1985). With appropriate corrections, analytical ultracentrifugation and gel filtration have been used to determine molecular weights of solubilized membrane proteins (Reynolds, 1985). On the other hand the radiation target inactivation method has been quite successful in determining the size of functional proteins in intact membranes. Only recently it has been possible to demonstrate that crystals of monodispersely solubilized membrane proteins can be grown in detergents like C_8E_5 (Zulauf, 1985). The three-dimensional crystals are suitable for X-ray analysis. Crystals of membrane proteins such as bacteriorhodopsin, cytochrome oxidase, conjugin, porin, and the photosynthetic reaction center have been obtained.

Determination of primary sequence of membrane proteins is somewhat tricky. Proteolysis of detergent-solubilized proteins is difficult and therefore other methods have been developed for fragmenting membrane proteins. Chemical fragmentation of proteins by cyanogen bromide dissolved in formic acid + methanol (9 : 1) has been used successfully in many situations. Detergents also interfere with the labeling procedure and hydrophobic peptides can abruptly precipitate during chemical treatments. One possible resolution of this problem is to use peptides immobilized on solid support whereby the bound peptide remains *molecularly* dispersed.

Cloning of genes of membrane proteins has become a method of choice because sequencing procedure is fast, convenient, and the nature of the sequence does not influence the physical properties of the gene. The major success stories include sequences of the acetylcholine receptor and the sodium channel. It is also possible to express these gene products in membrane of unrelated hosts such as oocytes, and thus study their functions in naturally reconstituted but an unrelated environment.

Reconstitution

Incorporation of proteins into bilayers is necessary for isolation, identification, elucidation of the mechanism of their action and relationships, and elaboration of the role of membrane components in cellular functions (Miller, 1986; Etemadi, 1985; Jain and Zakim, 1987). Questions pertaining to secretion of proteins and interaction of signal sequence are also related. In principle, reconstitution is energetically favorable because hydrophobic effect promotes spontaneous incorporation of proteins into the bilayer. However, a successful reconstitution protocol is still an art, because several competing processes can lead to formation of unwanted end products, such as an irreversibly denatured protein. Ideal reconstitution should give functionally active protein incorporated asymmetrically into a topologically continuous bilayer or vesicles, but it should not contain any residual impurities like detergents.

The exact protocol for reconstitution obviously depends upon the nature and the state of dispersion of the protein (Fig. 7-3). As discussed in the preceding chapter, predominantly hydrophilic solutes that remain in monomeric form in the aqueous phase in the absence of detergents can be spontaneously inserted into preformed bilayers or by adding them to the aqueous phase during the dispersal of lipids. In many cases, such spontaneous insertion is not observed unless specific lipid molecules are present as trace impurities. Apparently, these additives serve as specific ligands for anchoring the protein, whereas in other cases they provide defect sites for incorporation of the protein in the hydrophobic region or for promoting fusion of a bilayer with protein sheets (Jain and Zakim, 1987). In most cases spontaneous insertion of proteins is unidirectional or asymmetric, as would be expected for amphipathic molecules which cannot undergo transbilayer movement.

In principle, the protocols for the incorporation of larger hydrophobic proteins into bilayers are also the same, however the exact protocol must be developed to accommodate their dispersal characteristics. Large hydrophobic proteins can be kept molecularly dispersed only in the presence of detergents, otherwise they aggregate into large sheet-like aggregates (in favorable cases) or they denature irreversibly and form a precipitate. The simplest protocols for reconstitution consist of mixing lipids and proteins in a detergent solution and then removing excess detergent on sucrose-density gradi-

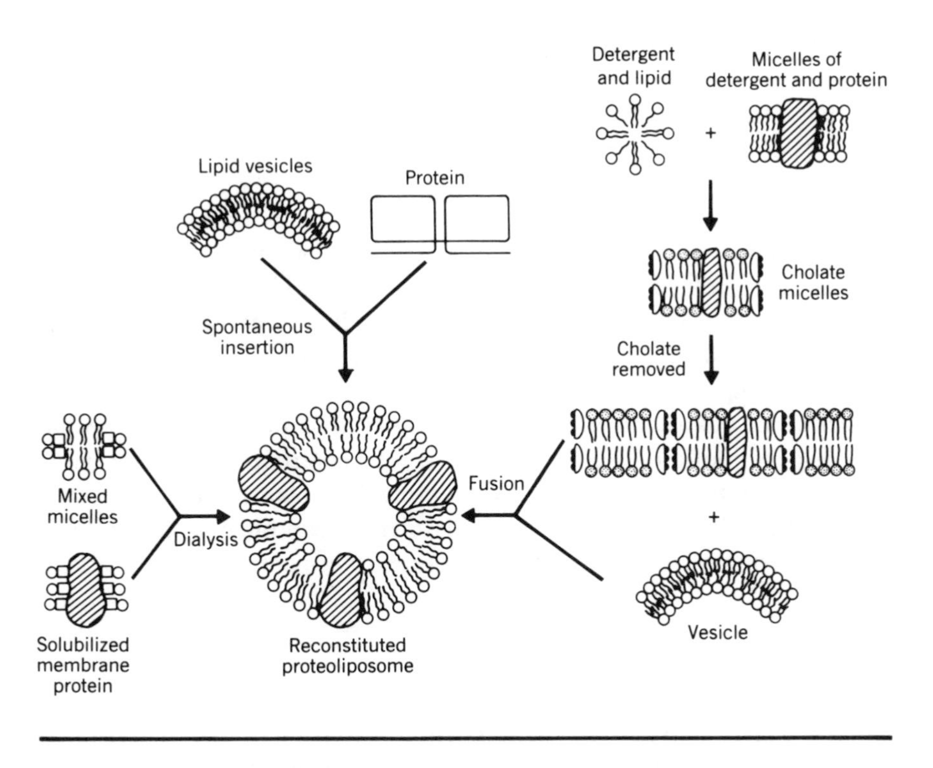

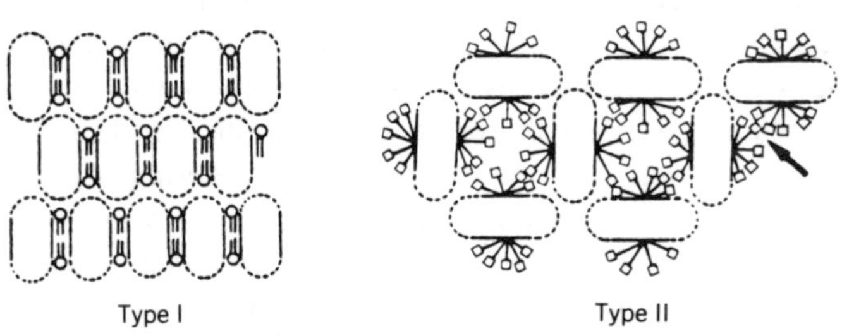

Fig. 7-3. (Top) Three methods for reconstitution. From left: Lipid and proteins solubilized in detergents are mixed and then the detergent is removed by dialysis, dilution, or gel filtration. Delipidated proteins can sometimes be spontaneously incorporated into a preformed bilayer. Partially delipidated proteins with a small amount of detergents sometimes form discs that can fuse with preformed vesicles. Other methods for reconstitution (not shown) involve exchange of lipids. (Bottom) Two types of crystals of membrane proteins are shown: Stacks of two-dimensionally crystalline membranes ordered in the third dimension; or a membrane protein crystallized with detergents bound to its hydrophobic protein surface.

ent, by salt precipitation, by dialysis, or by gel filtration during which vesicles of varying size and composition are formed (Eytan, 1982; Semenza et al., 1984; Levitzki, 1986). Such vesicles retain a small amount of detergent, they are often leaky, and generally most of the protein is oriented symmetrically. Under appropriate conditions phospholipid exchange proteins or enzymes of phospholipid metabolism can be used to alter the lipid composition in membranes. Recently, it has been possible to incorporate essentially detergent-free proteins into preformed vesicles of well-defined composition and morphology (see Jain and Zakim, 1987). In this procedure, dispersed aggregates of detergent-free sheets of a protein are mixed with vesicles containing impurities. Fusion of the membrane aggregates with the bilayer of vesicles leads to asymmetric incorporation of the protein. The composition and dispersity of the resulting vesicles can be controlled satisfactorily.

Successful reconstitution of several proteins in planar bilayer lipid membrane (BLM) has been achieved by dissolving the protein in the lipid solution, by fusing vesicles with BLM, or by apposing membrane fragments with BLM.

Conformation of Membrane Proteins

Since a bilayer provides an environment of varying polarity, the regions of proteins that interact with lipids would seek complementary polarity; hydrophobic regions would interact with acyl chains, and the polar regions with polar groups or the aqueous phase. In transbilayer proteins the segments and domains spanning the bilayer would be about 30 Å in length. Thus, the tertiary conformation of membrane proteins is determined by the unique constraints of its environment. Proteins derive their polarity from amino acid side-chains. The free energies for transfer of amino acid side-chains from water to bilayer are not known. However, such values are expected to be approximately related to the free energies for transfer from aqueous to organic bulk phase, the $\delta\Delta F$ values summarized in Table 7-1 or equivalent values are often used. Obviously this is only an approximation that has been quite useful.

In general, polypeptides that contain $<30\%$ hydrophobic amino acids exist in aqueous solutions as monomers. A higher proportion of hydrophobic amino acids is found in multi-subunit proteins. This is because only a certain fraction of the amino acid residues can be accommodated in the interior of a folded protein. For hydrophobic interactions among the subunits only patches of apolar residues are necessary. On a very gross level one might expect that membrane proteins have a higher proportion of apolar amino acids. This appears to be the case for the average of several membrane proteins compared with the average of several water-soluble globular pro-

TABLE 7-1 Transfer Free Energies (kcals/mole) for Amino Acid Side-Chains in α-Helical Polypeptides[a]

Amino Acid	(Abbr.)	Hydrophobic	Hydrophilic	Water-Oil
Phe	F	−3.7		−3.7
Met	M	−3.4		−3.4
Ile	I	−3.1		−3.1
Leu	L	−2.8		−2.8
Val	V	−2.6		−2.6
Cys	C	−2.0		−2.0
Trp	W	−4.9	3.0	−1.9
Ala	A	−1.6		−1.6
Thr	T	−2.2	1.0	−1.2
Gly	G	−1.0		−1.0
Ser	S	−1.6	1.0	−0.6
Pro	P	−1.8	2.0	0.2
Tyr	Y	−3.7	4.0	0.7
His	H	−3.0	6.0	3.0
Gln	Z	−2.9	7.0	4.1
Asn	B	−2.2	7.0	4.8
Glu	E	−2.6	10.8	8.2
Lys	K	−3.7	12.5	8.8
Asp	D	−2.1	11.3	9.2
Arg	R	−4.4	16.7	12.3

[a] Values are given for the hydrophobic and hydrophilic components of the transfer of amino acid side-chains from water to a nonaqueous environment of dielectric 2. The hydrophobic term is based on a treatment of the surface area of the groups involved. The hydrophilic term principally involves polar contributions arising from hydrogen bonding interaction. Also included in the hydrophilic term is the energy required to convert the charged side-chains to neutral species at pH 7. (From Engelman et al., 1986.)

teins (Capaldi and Vanderkooi, 1972). However, such comparisons are not meaningful because there are significant exceptions to it; the proportion of amino acids in contact with acyl chains can be relatively small, the globular proteins have a hydrophobic interior, and such information does not necessarily lead to any significant insights. It is probably pertinent to note that membrane proteins have lower disulfide content, and polar groups in many of the membrane proteins are derivatized (amidated carboxyl groups). Some membrane proteins have covalently attached fatty acids (see below), which could serve a range of functions.

The functions of membrane proteins depend not only on their structure and conformation but also on their gross orientation in a topologically intact environment. Gross organizational constraints operating on membrane proteins are similar to those operating on lipids, that is a restricted hydrophobic region of a membrane protein is incorporated into the bilayer, while the more hydrophilic sequences extend into the aqueous environment. Several predictive methods for elaborating the structural features of proteins have been suggested. These include smoothed hydrophobicity values (Hopp and Wood, 1981), probable α-helix (Fig. 7-4) and β-sheet conformations (Chou and Fasman, 1978), membrane helix potential (Rao and Argos, 1985), hydrophobicity plots (Kyte and Doolittle, 1982), and peptide turn regions (Paul and Rosenbusch, 1985). The free energies for transfer of amino acid side chains are summarized in Table 7-1. This scale for hydrophobicity of amino acids provides a reasonable prediction of the folding pattern for transmembrane α-helices (Engelman et al., 1986).

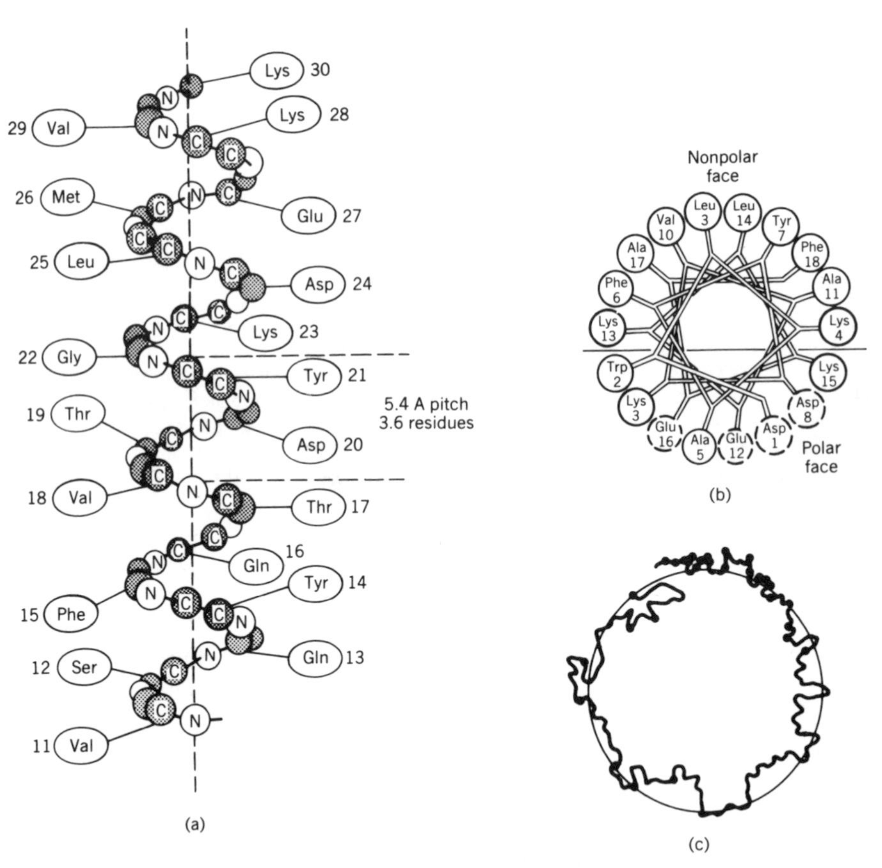

Fig. 7-4. Helical arrangement (A) of a peptide segment whose one face is hydrophobic. The axial projection (B) of this amphiphilic helix will have a polar and a nonpolar face. This arrangement is well suited for apolipoproteins which surround the edges of a bilayer disc (C) to form a stable lipoprotein particle. Similar amphiphilic surfaces can be formed with β-sheet. Such surfaces can interact with the bilayer interface, and the resoluting microinterface could be desolvated without significant penetration of the peptide into the bilayer. If the helical segment is hydrophobic on all sides as with a stretch of 20 amino acids in glycophorin (D), this segment is inserted through the bilayer as a transmembrane helix. See also Fig. 7-5 and 7-6.

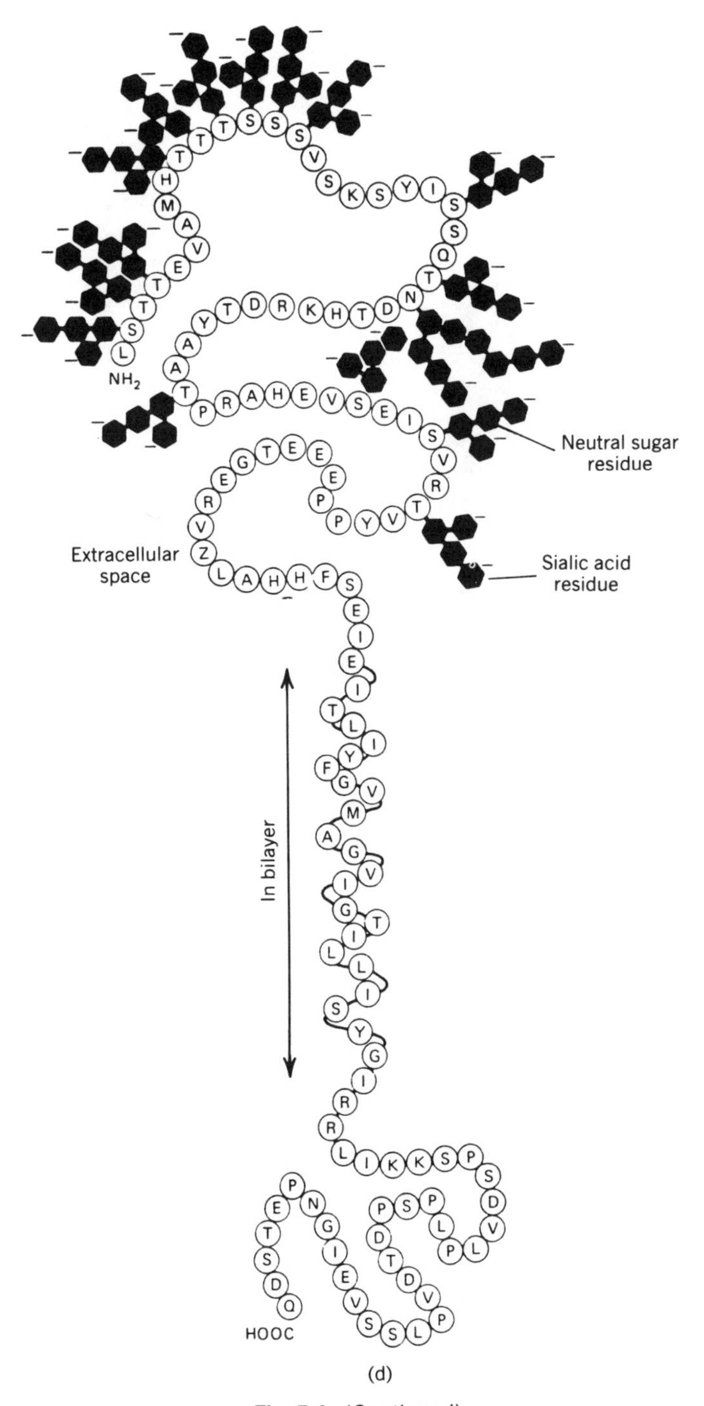

Fig. 7-4. *(Continued)*

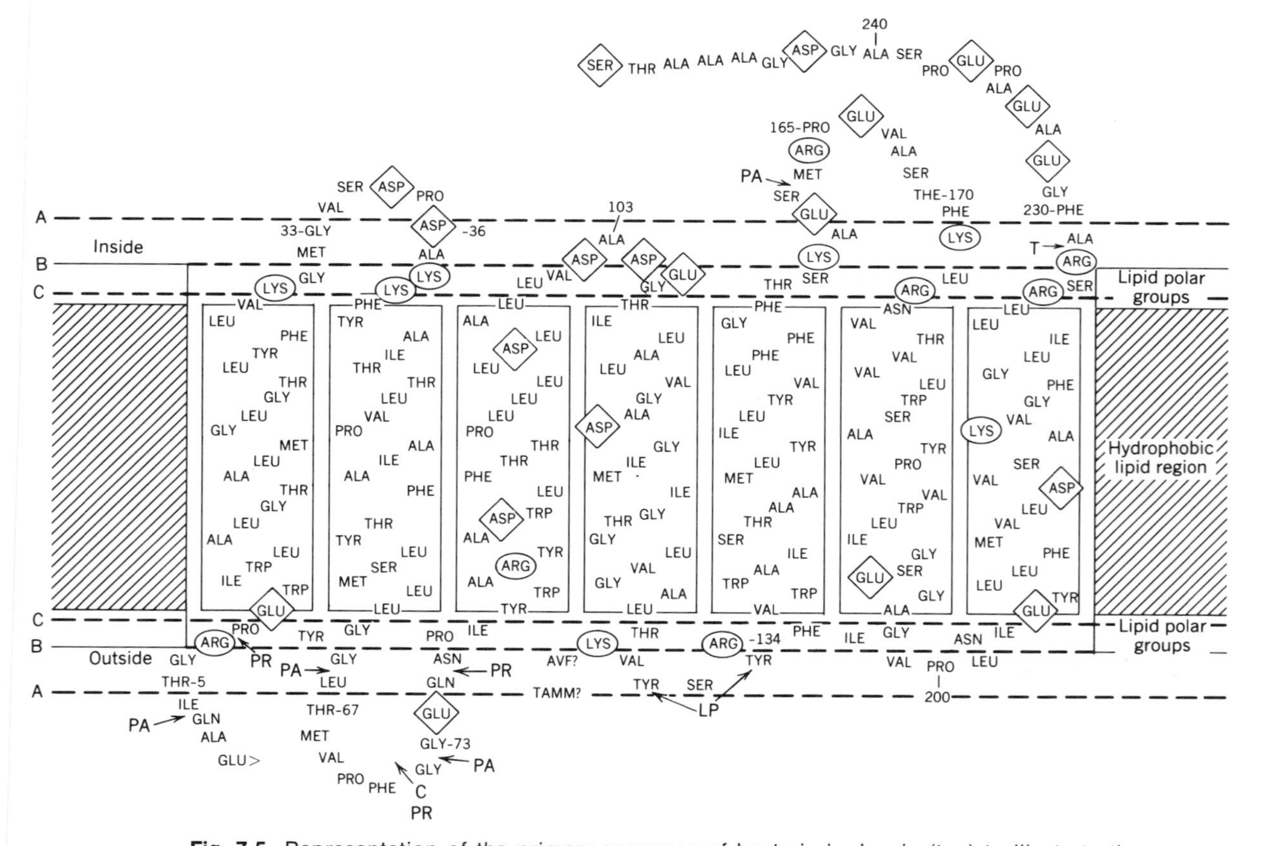

Fig. 7-5. Representation of the primary sequence of bacteriorhodopsin (top) to illustrate the seven helices and the extramembrane segments. The hydropathy plots (facing page) show seven peaks corresponding to the seven helices. (From Engelman et al., 1986.)

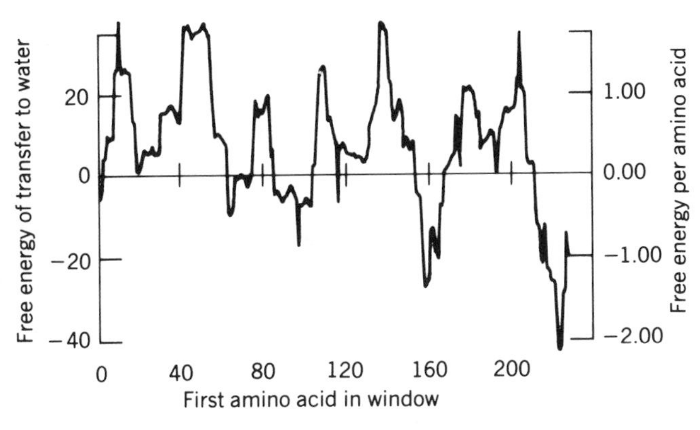

Fig. 7-5. *(Continued)*

The folding pattern of a peptide chain through a bilayer determines not only the gross conformation but also sidedness and accessibility from the aqueous compartments. In the aqueous phase peptide chains fold in a variety of secondary structures such as helices, sheets, bends, and turns (Capaldi et al., 1983). In an apolar environment the exposed groups should be nonpolar, and such an arrangement is best accommodated in a helical conformation. However, depending upon the primary sequence, β-sheet structure (for channels formed by aggregates) or amphipathic helix (for the formation of particles like serum lipoproteins) may be preferentially formed. The polypeptide backbone is relatively polar, but its effect is essentially nullified by folding such that the amide groups are hydrogen bonded (about −2.1 kcal/mole) and shielded (about −2.1 kcal/mole) from the aqueous phase. In a helical conformation the side-chains are exposed to the outer surface, and thus the helix content of membrane proteins is found to be reasonably high.

Distribution of the polar groups on the surface of a helix can be diagnostic for its environment. If the polar groups appear only on one face of the helix, it would imply that only this face is in contact with the nonpolar environment. If the helical segment spans across the bilayer, not only a majority of the side-chains will be nonpolar, but also the length of the helix would be about 30 Å, that is, about 19 residues. The hydropathy plots based on this concept have proven to be diagnostic in predicting the topography of protein segments (Kyte and Doolittle, 1982; Rao and Argos, 1985). In its simplest form hydrophobic helices anchor many digestive enzymes to membranes of gastrointestinal mucosa (Semenza, 1986). Probably a hydrophobic helix formed by the signal peptide could also promote insertion of the rest of the protein though the bilayer (Briggs and Gierasch, 1986). With the help of the so-called hydrophobicity plot shown in Fig. 7-5, only with the primary sequence data, it is possible to identify the segments that probably span the bilayer. A moving average of the hydrophobicity of 19 amino acid segments of a protein is plotted. The most hydrophobic segments that could span the bilayer show a minimum, that is, most negative-free energies for transfer into apolar environment. Although, the free energy values may depart considerably from the absolute values, the relative contribution gives a meaningful estimate. Also the residue at the peak is the one that will be in the middle of the helical segment. If more than 19 residues form the hydrophobic segment, it probably means that the helix is tilted; however, such cases are rare. Another feature that is often noted is the presence of several polar residues at the end of the helical segment, which could anchor the helix in the bilayer. Other patterns of folding like β-sheet could also give rise to hydrophobic surfaces on proteins. Such amphipathic surfaces could electrostatically interact with the bilayer interface and then promote desolvation of the protein-bilayer microinterface. The general significance of such hyrophobic interactions involving little or no penetration of the protein into the bilayer has not been widely appreciated yet.

As summarized in Fig. 7-6 the proportion of the helical segments lodged in the bilayer varies considerably in membrane proteins. Operationally, proteins can be divided into three categories: adsorbed, anchored, and trans-

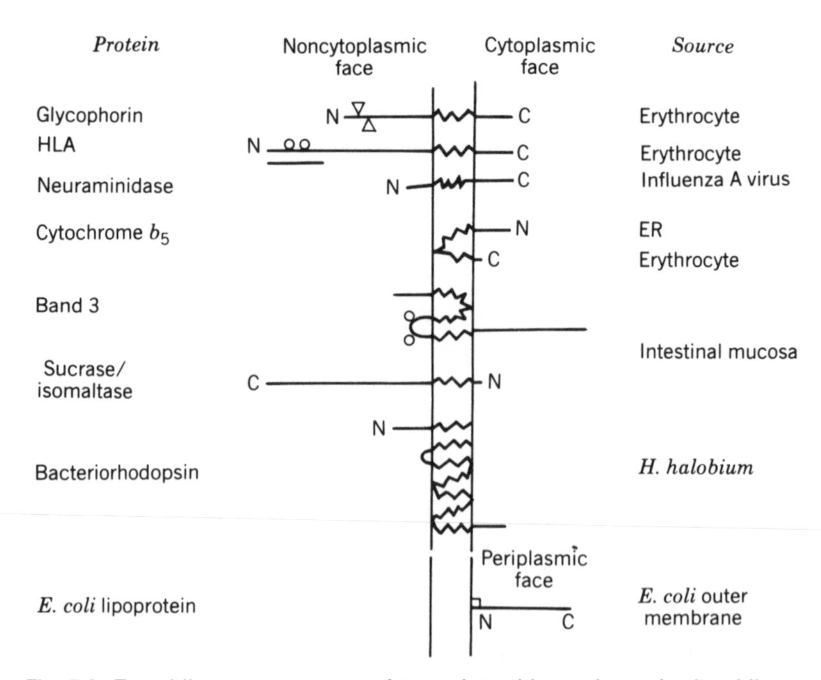

Fig. 7-6. Transbilayer arrangement of several peptides and proteins in a bilayer.

membrane (Jain and Zakim, 1987). Electrostatically adsorbed proteins can be washed easily with salts. Some proteins have a covalently linked acyl chain (see below) which may anchor the protein to the bilayer. Some proteins (like phospholipase A2 and lectins) may be anchored by specifically binding to the head group of a phospholipid or by desolvating the microinterface. In many proteins like glycophorin or M13 coat protein only one segment of about 20 amino acids spans the membrane, and a large part of the protein remains on one surface. On the other hand, the proteins like bacteriorhodopsin have seven transbilayer helical segments, and thus about 80% of the amino acid residues are in the bilayer. The helical arrangement of the peptide chain in bacteriorhodopsin (Fig. 7-7) could be confirmed by X-ray diffraction (Henderson and Unwin, 1975). The peptide regions traversing a bilayer are apparently not maintained in active configuration by folding of the remainder of the peptide sequence. In fact, large fragments of membrane proteins can be removed by proteolysis, leaving just the hydrophobic sequences, which retain functional integrity. Additional factors responsible for folding of the peptide chain in conformations other than transmembrane α-helix are not understood yet.

Topography, Sidedness, and Orientation

Functional integrity of membrane proteins is due to interaction of specific protein segments with the bilayer. Thus, it is not the high content of hydro-

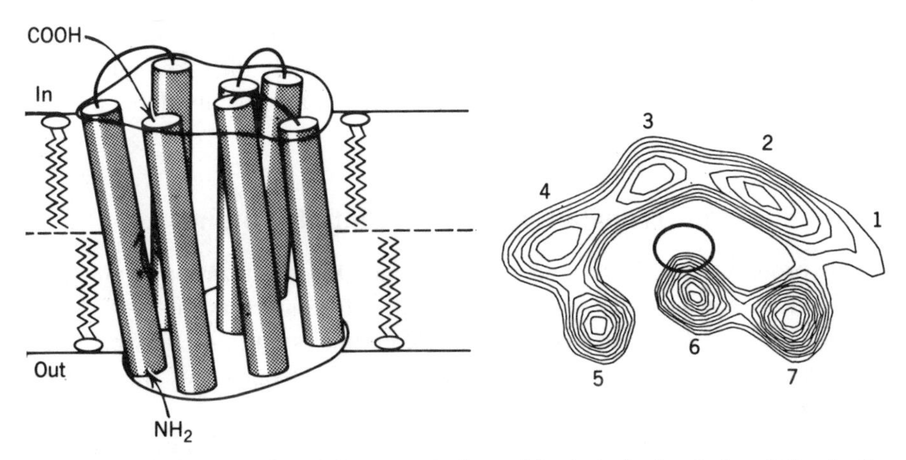

Fig. 7-7. (Left) Topography of the seven helices of bacteriorhodopsin in relation to the lipid bilayer. (Right) The position of retinal residue and the topography of the seven helices obtained from neutron diffraction. (From W. Stoeckenius, *Trends Biochem. Sci.* 483–486, 1985.)

phobic amino acids, but the disposition of such residues that determines the integration of a peptide sequence in the membrane. One of the consequences of the folding of a membrane protein would be its asymmetric orientation. The functional asymmetry of a biomembrane is absolutely critical for the survival of the cell. This is due to the asymmetric orientation of membrane proteins which are incorporated as such during their biosynthesis. Similarly, in many reconstitution procedures the proteins are inserted asymmetrically. During disruption of cells the orientation of the membrane in the vesicles can be right-side-out or inside-out, depending upon the conditions, and some scrambling is not uncommon.

Asymmetric orientation of membrane proteins can be demonstrated by a range of techniques including freeze-fracture electron microscopy, covalent labeling, treatment with proteases, and labeling with specific inhibitors or ligands. Crossed immunoelectrophoresis (Owen and Kaback, 1978) appears to be a method of general applicability for determining distribution of proteins on a membrane and for establishing topography of the vesicle preparations. From the antibody serum against whole membrane, the antibodies against specific membrane antigens are removed by progressive adsorption. The relative efficiency of adsorption reflects the relative degree at which the antigen is expressed in the membrane preparation; the antibodies for the antigens on the inner surface are adsorbed much less effectively.

Some generalizations are beginning to emerge from the study of the asymmetry of membrane proteins. The biosynthetically established absolute asymmetry is maintained through the life of the protein. Although, proteins exhibit varying degrees of rotational and translation movement, transbilayer movement is nonexistent. Most of the protein in biomembranes is generally associated with the cytoplasmic surface, and generally the carboxylic end is also on the cytoplasmic side (Fig. 7-6). In membrane glycoproteins the carbohydrates are on the external surface.

Rotational and Translational Movement of Proteins

Many proteins in a bilayer appear to modulate the organization of phospholipids. This can be seen even in the motional parameters of C—D bonds (Meier et al., 1986). There is considerable confusion in this field which may be resolved in the near future with a general acceptance of the recently developed protocols.

Proteins in membrane can rotate and diffuse laterally (McCloskey and Poo, 1984). The measured rotational times are in the range of 0.1–100 msec, and the lateral diffusion coefficients are in the 10^{-9} to 10^{-13} cm^2/sec. Only a small part of this difference is due to a difference in the size of these proteins. According to the arguments presented in Chapter 4, the lateral diffusion coefficient shows a weak dependence on the radius of the diffusing particle. The experimental data appear to be more consistent with a relationship to the square root of the molecular weight. Lower values of the lateral diffusion coefficient suggest that such slow-moving components may be anchored to the cytoplasm. Indeed, not all proteins in a membrane have the same mobility. Also not all atoms or domains of the same protein exhibit the same motional properties. Freeze-fracture electron micrographs exhibit nonrandom distribution of particles; patching and capping of membrane-localized antigens can be demonstrated under a variety of functionally meaningful conditions. For example, rotational diffusion of rhodopsin in rod outer segment membrane occurs once every 0.02 msec, whereas bacteriorhodopsin rotates once every 20 msec in purple bacteria. Even in the same membrane the lateral diffusion coefficient of different proteins is found to be considerably different (Weiss et al., 1986).

Many examples of nonrandom distribution of membrane proteins over the cell surface have been reported (Almers and Stirling, 1984). Such constraints to lateral diffusion apparently arise from the following factors:

1. The intracellular network of filaments and tubules (cytoskeleton) appears to dictate the external architecture of the cell. Anchoring of the membrane protein to microfilaments on the cytoplasmic surface (a cytochalasin-sensitive process) has been implicated for the generation of microdomains in cell membranes (Geiger, 1983). Many of these processes are not energy requiring and are not inhibited by colchicine and cytochalasin. A cytoplasmic network of spectrin and actin is attached to membrane via anchyrin, which also binds to the anion channel in erythrocyte membrane (Heast, 1982). Indeed the lateral diffusion coefficient for band 3 increases from 10^{-11} to 10^{-8} cm^2/sec when spectrin is removed by salt.
2. Aggregation and crystal-like arrangement of proteins in the membrane has been observed for bacteriorhodopsin, gap junctions, synapses, and other regions of morphological specialization. Clustering of acetylcholine receptors on muscle cells is apparently influenced by molecules of the extracellular basal lamina. In reconstituted membranes the lateral diffusion coefficient decreases significantly as the lipid/protein ratio decreases.
3. In reconstituted systems proteins have been found to aggregate or distribute nonrandomly. Such effects depend upon the phase properties of the lipid. Nonrandom distribution of receptors for endocytosis and irregularities in the surface of cells have been observed in the absence of any constraint for lateral diffusion. This is directed by the membrane flow.

4. Specific interactions between different proteins (as in the electron transport chain in mitochondria), and mechanical factors that modulate radius of curvature and therefore the packing arrangement (these situations are encountered in folds, microvilli, secretory vesicles) could also promote localization and clustering of membrane proteins.
5. In some cases the plasma membrane appears to be divided into large domains containing many different proteins, and some of these proteins cannot diffuse across the domain boundaries. The fact that morphologically distinct domains can be distinguished on a large variety of cells (sperm, rod cells, epithelial cells) suggest that lateral diffusion barriers exist in several membranes.

Nonrandom distribution of proteins in cell membranes is indicated by a variety of observations. Functional significance of this remains to be established.

Proteins with Covalently Attached Lipids

Membrane proteins from a variety of sources (animals, bacteria, viruses) contain covalently attached fatty acids (Marinetti and Cattieu, 1982; Magee and Schlesinger, 1982; Towler and Glaser, 1986) or phospholipids (Low and Kinkade, 1985; Ferguson et al., 1985) including phosphatidyl inositol (Low and Kinkade, 1985; Cross, 1987). Proteins that are known to be acylated include virus glycoproteins and hemagglutinins, myelin proteolipoproteins, transferrin receptors, and nonglycosylated xanthin oxidase. Most of these are localized on the cytoplasmic surface (Wilcox and Olson, 1987). Several acylated proteins (e.g., myelin lipoproteins) apparently contain substantial lipophilic component so that they are soluble in bulk organic solvents. Properties of several synthetic acylated proteins anchored to vesicles have also been reported (Babbitt and Huang, 1985).

Covalent attachment of lipids is a post-translational event that is apparently designed to provide a lipophilic anchor for orientation of a protein in relation to its environment. At least three different modes of attachment have been found. In the envelope of *Escherichia coli* fatty acids are attached to amino groups (Hentke and Braun, 1973). In the peptidoglycan matrix of gram-negative bacteria, diacylglycerol component is attached through thioether linkage to the side chain of the amino-terminal cysteine of a small peptide component. In addition, a fatty acid is amide-linked to the α-amino group, so the amino terminus is akin to a triglyceride. The same kind of structure has been found in the amino-terminal region of penicillinases in gram-positive bacteria. In *Trypanosoma brucei* dimyristoylglycerol phosphate is attached to a sugar moiety linked to a protein. In most cases two acyl chains are esterified to hydroxyl groups on the peptide chain. Susceptibility of these bonds to methanolic potassium hydroxide and to hydroxylamine appears to be rather sensitive to the site of acylation and denaturation of the protein.

Studies with viruses and red blood cells show that many of the acylated proteins are associated with their membranes. All membrane proteins are not acylated, which suggests that acylation serves specialized functions. The frequency and specificity of acylation of membrane proteins in eukaryotic cells (Olson et al., 1985; Magee and Courtneidge, 1985; Towler and Glaser, 1986) suggests that a substantial number of membrane proteins are acylated, however, incorporation of palmitate versus myristate in these proteins appears to be very specific. Although the functional significance of covalently attached fatty acids is not clear, several interesting proposals may be considered. These range from a role as hydrophobic anchors, as a fusogen for uptake and secretion, and as a modulator of protein conformation that brings about a proper alignment of the protein in the bilayer for catalytic, transport and recognition functions.

ROLE OF LIPIDS

Lipids form a two-dimensional matrix in or on which proteins are present in biomembranes. On a morphological level this matrix serves as an asymmetric barrier; it provides a lattice for vectorial organization of functionally related and unrelated molecules, it provides an amphipathic environment, and it provides a two-dimensional surface for diffusion of substrates, products, and ligands. On a molecular level lipids modulate the structure and functions of proteins, and the opposite is probably also true. The role of lipids in the function of membrane proteins is probably most dramatically demonstrated by the fact that most of the membrane functions are lost when lipids are removed or degraded by appropriate phospholipases. In fact, meaningful progress in the study of membrane proteins has been possible only after methods to incorporate isolated proteins into bilayers of well-defined composition have been worked out.

Requirement for Specific Lipids

Reconstitution of membrane proteins into bilayers of well-defined composition has permitted examination of the specificity of lipid for functional reconstitution. It may be emphasized that in reconstituted systems, it is not always possible to assure that all the lipid has been exchanged, or that all the detergent has been removed, or the reconstituted vesicles are monodisperse. A near-absolute requirement for phospholipids of specific head group has been shown only in two cases: β-hydroxybutyrate dehydrogenase (Fleisher et al., 1983) and pancreatic signal peptidase (Jackson and White, 1981). Duramycin, a peptide antibiotic, also apparently interacts only with phosphatidylethanolamine (PE) in membranes (Navarro et al., 1985). The idea of solvation of proteins by phospholipids also implies specific interaction between lipids and proteins (Devaux and Seigneuret, 1985). In these and other cases the implied specificity could be due to a requirement for charge or organizational features often necessary for reconstitution or for optimal functioning of the reconstituted protein. Indeed, lipid molecules in the environment near the interface with a protein molecule could experience unique organizational and conformational constraints.

The phospholipid molecules in bilayers could provide a specific microenvironment to proteins. Relevant evidence comes from two types of studies: boundary lipids and Arrhenius plots. From the activation of delipidated proteins, it is possible to calculate the number of phospholipid molecules required to activate or functionally reconstitute a protein molecule. For example, about 70 phospholipid molecules are required for activation of Na,K-ATPase, 35 for Ca-ATPase, 50–60 for cytochrome oxidase, 11 for myelin proteolipid, and 8 for mellitin. These numbers apparently correspond to the square root of molecular weight, i.e., they could form a monomolecular ring around the periphery of the protein embedded in the bilayer. In many cases it appears that the protein–protein contact reduces the catalytic activity.

Evidence for immobilization of a fraction of lipid in bilayers for proteins comes from Raman, fluorescence, and electron spin resonance (ESR) studies (Devaux and Seigneuret, 1985), whereas by deuterium NMR no such immobilization could be shown. Fourier transform infrared spectroscopy stearoly-oleoyl PC (SOPC) + DPPC vesicles containing Ca-ATPase suggests that SOPC is disordered preferentially by the protein (Jaworsky and Mendelsohn, 1985). It appears that this discrepancy is due to the residence time of lipid molecules in the boundary of the protein. The residence time for boundary lipid is thought to be about 1 μsec, which is considerably faster than the turnover time for most membrane enzymes, which is typically <1000/sec. The correlation time for chain motion (about 20 nsec) is also not drastically (less than three-fold) altered in the presence of proteins. Recent studies also suggest that the relative specificity for the various lipids for a protein in bilayers is less than 10-fold. By assuming a cylindrical shape, and the radius proportional to the square root of the molecular weight of the protein, it can be computed that often the total boundary lipid can be accommodated as a single layer around the protein (Marsh and Seddon, 1982).

Selective microenvironment of proteins in biomembranes is also indicated by breaks or discontinuities in Arrhenius plots; even in the same membrane different functions exhibit breaks at different temperatures, or no breaks at all (Esfahani et al., 1977; Eze and McElhaney, 1987). On the other hand soluble enzymes are known to show breaks in Arrhenius plots. Generally, the activation energy is higher at temperatures below the breaks, which is consistent with the suggestion that a less disordered lipid environment hinders the motional freedom of the bound protein. Also in fatty acid auxotrophs of *E. coli*, the breaks or discontinuities are observed at temperatures corresponding to the transition temperature of the lipid bilayer. Interpretation of these observations is tenuous because the origin of discontinuities in the Arrhenius plots could be from a variety of factors, ranging from a change in the conformation or the state of aggregation of the protein, a change in the affinity or the turnover of the protein, or a change in the rate-limiting step, or a phase change or phase separation, or a change in the solvent structure or the p*K*a of a critical ionizable group. Even a possibility of experimental artifacts is implied by the fact that rather sharp breaks (over 1–3°C) are observed in the kinetic data, whereas the corresponding sharp cooperative thermotropic transitions have not been detected in biomembranes.

Effects of Lipids on Catalytic Properties of Proteins

Enzymes localized in a membrane display all the modes of regulation described for soluble enzymes: induction, allosteric modulation by ligands (substrates, inhibitors, activators), and post-translational modification. While some of these modes are further modified by solubility and accessibility of the substrate and products as regulated by the membrane, the amphiphilic environment of the bilayer offers additional possibilities: anisotropy of organization; regulation of the conformational change, aggregation, and disaggregation; and vectorial coupling with ligand binding, chemical reaction, and electrical field. Thus, regulation of membrane proteins can occur by changes in composition, organization, conformation, and the catalytic step. Such changes can be induced by external perturbations under certain pathological conditions, and probably by genetic regulation.

The precise nature of lipid–protein interaction is not known in any system. Phospholipids do not appear to act as cofactors. Similarly, activation of an enzyme by a structurally specific head group of a lipid has been shown only for β-hydroxybutyrate dehydrogenase (see below). In most other cases where such claims have been made, they appear to be due to a requirement for a specific phase property of the bilayer required for incorporation or reconstitution. Some of the functional properties illustrating the molecular basis of the role of lipids are given below, and others will be discussed in subsequent chapters.

Cytochrome b_5: In mammalian endoplasmic reticulum it is functionally associated with cytochrome b_5 reductase (MW 700 kD), which catalyzes metabolism of endogenous and xenobiotic compounds (Wislocki et al., 1980). On the cytoplasmic side the reductase catalyzes the transfer of electrons from NADH to the covalently attached heme in cytochrome b_5, and phospholipids are required for optimal activity. When isolated by detergent solubilization it contains a hydrophobic segment of 44 residues, and the rest of the molecule is polar (Spatz and Strittmatter, 1971). By trypsin treatment the hydrophobic segment can be removed and the water-soluble fragment is functionally active, but it cannot be incorporated into bilayer. The role of phospholipids is to promote coupling of cytochrome b_5 with the reductase, although an active complex can be produced at higher concentrations of the components in the absence of phospholipids (Muller et al., 1984). The phase state of bilayer has little or no effect on the activation energy or the catalytic parameters.

This is an example of stalked membrane protein. Many digestive proteins (e.g., sucrase-isomaltase, aminopeptidase) in the intestinal brush border also appear to be anchored through a segment located in the amino-terminal region (Semenza, 1986).

Glycophorin is a glycoprotein from erythrocyte membranes (Fig. 7-4d) that serves as the MN blood group determinant and as the receptor for phytohemagglutinin and influenza virus. The carboxyl end, which is localized on the cytoplasmic side, contains a stretch of six charged amino acids followed by 23 hydrophobic amino acids (Segrest et al., 1973). This hydrophobic segment could span the bilayer to anchor the carbohydrate-containing amino-terminal region on the erythrocyte surface. Other properties of this protein are discussed in the next chapter.

Phospholipase A2: Phospholipase A_2 (PLA) is a small protein (MW 14 kD) from venoms and pancreas, and is probably also present in all cells. PLA from pig pancreas has been characterized extensively and its crystallographic structure is known at 1.7 Å resolution. The following discussion relates largely to the pancreatic enzyme. This esterase is specific for the *sn*-2-acyl group in glycero-3-phospholipids. The catalytic action of PLA from a variety of sources is considerably more pronounced on lipid–water interface compared with that on a monomeric substrate (Verheij et al., 1981). Catalysis in the interface according to the scheme in Fig. 7-8 is observed only under certain conditions; for example, micelles with specific detergents in certain

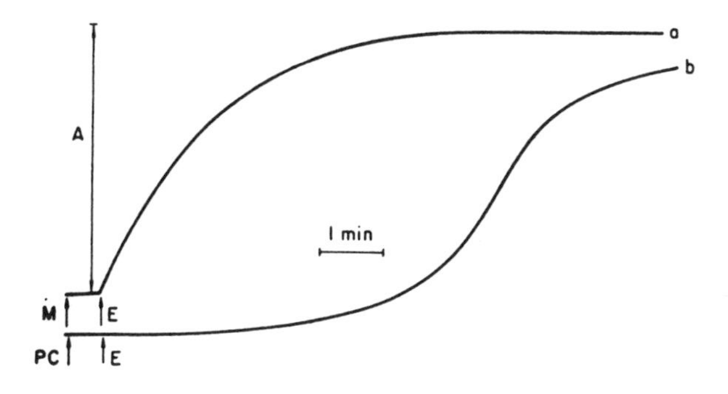

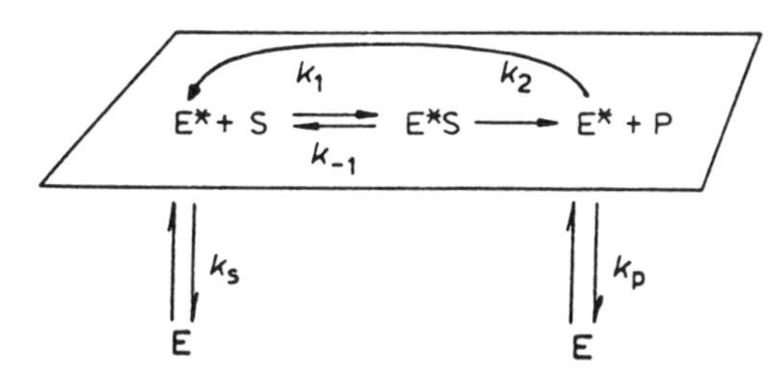

Fig. 7-8. (Top) Reaction progress curves for the hydrolysis of (a) dimyristoylphosphatidylmethanol and (b) dimyristoylphosphatidycholine vesicles by pig pancreatic phospholipase A2. (Bottom) A Kinetic scheme for interfacial catalysis by phosphlipase A_2. The enzyme in the aqueous phase binds to the interface, where it undergoes the catalytic cycle. If the enzyme desorbs (the intervesicle hopping mode of hydrolysis) occasionally from the interface the reaction progress curve of type b is obtained, where the latency is due to formation of a critical mole fraction of the product. If the bound enzyme does not desorb (as in the scooting mode of hydrolysis) then the first-order curve of type a is obtained. (From Jain et al., 1986a.)

proportions, monolayers at certain surface pressure, and bilayers only in the presence of certain additives. Such factors appear to modify the binding of the enzyme to the interface (E to E* step). The catalytic turnover by E* (monomeric PLA in the interface) is postulated to occur by the reaction scheme shown in the box. Thus, the average residence time of the enzyme in the interface determines its catalytic efficiency. Indeed, direct binding of PLA to the substrate interface corroborates such conclusions (Jain et al., 1986a–d). For example, as shown in Fig. 7-8, the reaction progress curve for the action of pig PLA on vesicles of anionic phospholipids is first order, as would be expected if all the enzyme is in the interface. In the presence of excess vesicles, one enzyme molecule per vesicle hydrolyzes all the substrate molecules available in the outer monolayer of the vesicle, and the bound enzyme does not exchange with excess substrate vesicles. Vesicles are not disrupted nor their contents released when their outer monolayer is completely hydrolyzed or when excess enzyme binds to vesicles.

On the other hand as shown in Fig. 7-8, the reaction progress curve for the hydrolysis of phosphatidylcholine (PC) exhibits a long latency phase followed by a steady-state phase. As confirmed by direct binding studies PLA does not bind to PC vesicles unless a critical mole fraction of negative charge is introduced by the products of hydrolysis, lysoPC, and fatty acid. Such observations suggest that the binding of PLA to the substrate interface occurs predominantly via an anion-binding site. However, additional factors related to the phase properties are also at work. Binding and catalytic action of PLA on anionic vesicles occurs at all temperatures at, below, and above the phase transition. In PC vesicles the effect of the low mole fractions of the products is optimal near the phase-transition temperature. The interpretations of this behavior is that in PC vesicles, containing the products of hydrolysis at low mole fractions, the anionic additives phase-separate to create a critical anionic charge density that promotes binding and increases the residence time of the enzyme in the interface. Such phase separation is apparently dissipated by additives like alkanols and general and local anesthetics which prevent binding of PLA also inhibit the rate of hydrolysis (Jain and Jahagindar, 1985). Anions in the aqueous phase also reverse binding of the enzyme. The interfacial catalysis by PLA therefore occurs without significant penetration of PLA in the bilayer; at the same time binding of the enzyme to the interface is very strong such that the bound enzyme remains in the interface for several thousand catalytic turnover cycles, and the estimated K_d is less than 10^{-13} M.

Binding of phospholipase A2 to the bilayer is accompanied by desolvation of the lipid–bilayer microinterface. Since this is also a very slow step (about 0.2/sec), the catalysis in the hopping mode, where the enzyme adsorbs and desorbs during each catalytic turnover cycle, is very slow, as is the case with prophospholipase, which does not desolvate the microinterface. On the other hand, the enzyme bound with desolvation of the microinterface remains in the interface for several thousand catalytic turnover cycles. Thus, interfacial activation of the enzyme is due to the factors that increase the residence time of the enzyme in the interface.

β-Hydroxybutyrate Dehydrogenase: This lipid-activated enzyme (Fleischer et al., 1983) is associated with the inner membrane of mammalian mitochondria, where it is accessible only from the matrix. During fatty acid metabolism in liver hydroxybutyrate dehydrogenase (BDH) converts acetoacetate to D-β-hydroxybutyrate, and in peripheral tissues it catalyzes the reverse reaction. It can be solubilized by phospholipase A2 as well as by cholate and salt treatment. The soluble enzyme is catalytically inactive, and the activity in both directions can be restored completely by PC but not with any other class of lipid. Lysophosphatidylcholine and its analogs, as well as short-chain phospholipids in monomer form, activate the enzyme, and the activity is lost above cmc. A more stable lipid protein complex is formed with unsaturated phospholipids in a bilayer. Optimal activation of the enzyme requires a careful control of lipid composition; a mixture of PC + PE + an anionic lipid appears to be as good as the mixture of mitochondrial lipids (Fig. 7-9a). This requirement for mixture of lipids is probably for insertion of the protein rather than its activation which follows binding of PC. The enzyme reconstituted with synthetic lipids shows discontinuities in Arrhenius plots (Fig. 7-9b) at the same temperature as the T_m for the lipid mixture; the activation energies are about 16 and 27 kcal/mole in the liquid crystalline and the gel phases, respectively. During reconstitution the apoenzyme must be incubated with lipid vesicles, and in the reconstituted vesicles the enzyme is localized only on the side to which it is exposed during reconstitution. BDH appears to be active in tetrameric form (subunit MW 30 kD), however, PC is not required for tetramer formation. Studies with synthetic phospholipids show that a substantial modification of the head group region can be tolerated (Issacson et al., 1979): D- and L-isomers, octadecyleicosylphosphorylcholines, and phosphono analogs activate; whereas the phosphinate analog, with both substituted oxygens missing, does not activate. The enzyme acti-

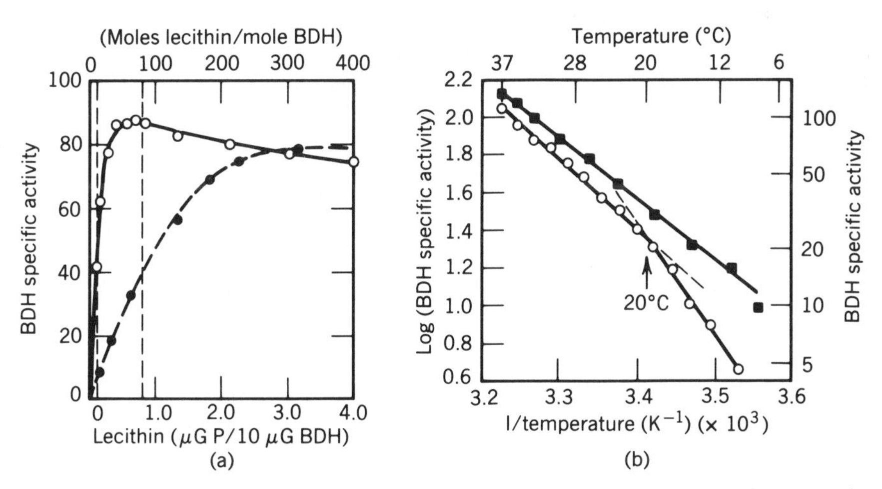

Fig. 7-9. (Left) Titration curves for activation of apo-β-hydroxybutyrate dehydrogenase by mitochondrial phospholipid (open circles) and mitochondrial PC (closed circles). The amount of PC required for half-maximal activity of BDH is 7 and 80 moles. PC per moles of BDH. (Right) Arrhenius plot of BDH activated with mitochondrial phospholipid (filled squares) or a mixture of synthetic lipids (open circles) with transition at 19°C. (From Fleisher et al., 1983.)

vated with various phospholipids has the same ordered sequential reaction mechanism in which binding of NADH is the first step. The effect of lipids is primarily on the affinity of the enzyme for NADH. Although the precise mechanism of activation of BDH by phospholipids is not known, data at hand suggests that this enzyme probably does not penetrate into the bilayer (Jain and Zakim, 1987).

In malate quinone reductase from *Mycobacterium* (Imai, 1978) and pyruvate oxidase from *E. coli* (see below) also, the role of lipids may be related to the coenzyme binding.

C_{55}-Isoprenoid Alcohol Phosphokinase: The product of ATP-dependent phosphorylation by this enzyme serves as a lipid carrier in bacterial cell wall synthesis. The kinase from *Staphylococcus aureus* has 58% hydrophobic amino acids (MW 17 kD), out of which about 30% are leucine and isoleucine. It is not water soluble and it dissolves in *n*-butanol; it can be readily renatured after solubilization by sodium dodecyl sulfate; and lipids are required for activity. Optimal reconstitution is attained with a large molar excess (about 1000:1) of lipids, and lower activity is observed both below and above this ratio (Gennis et al., 1976). The disordered phase state of lipid rather than its chemical nature appears to be of primary importance for activation. Besides fatty acids and detergents, numerous phospholipids serve as activators in the decreasing order: lyso-PE, lyso-PS > PC, PI > PG, CL > PE, PS. In PC series the activity decreases with the chain length and increases with the degree of unsaturation in the decreasing order: 8:0 > 12:0 > 18:1 > 14:0 >> 16:0 and 18:0, which are active only at temperatures near T_m. How the phase state of a lipid aggregate influences the catalytic activity remains to be established. However, the high (about 25%) helix content suggests that the arrangement of transmembrane helix could modulate the catalytic function or accessibility of lipid soluble substrate.

Bacteriorhodopsin is the major component of the purple membrane of purple bacteria, *Halobacterium halobium*. As discussed later (Chapter 13) it converts light energy into the proton gradient. About 80% of the 247 amino acids are hydrophobic, and are present in seven transbilayer α-helices (Fig. 7-5). The contiguous helical segments in the sequence are adjacent in the structure (Fig. 7-7) suggesting a simple folding pathway. Out of the 27 charged residues 9 are buried in the interior of the membrane and are probably ion-paired, and probably line the proton channel (Engelman et al., 1986). The coupling between the function of the proton channel and the photon-induced isomerization of retinal is not established. The role of lipid is to provide an impermeable barrier to retain the proton gradient, and probably to stabilize the arrangement of the helix.

Pyruvate Oxidase is a water-soluble enzyme that can be removed from *E. coli* by sonication. It has four identical subunits (MW 270 kD) and shows no unusual preponderance of hydrodrophobic amino acids (Cunningham and Hager, 1971; Blake et al., 1978). Binding of lipids or binding to membranes is not required for activity, however *V*m increases 10- to 50-fold in the presence of detergents and lipids; phospholipids, monoolein, fatty acids, and detergents are equally effective. Trypsin treatment activates the enzyme and the modified enzyme is not activated by lipid. Lipid activation requires several minutes, and optimal activation is achieved in the presence of substrate, cofactors, and excess lipids. The effect of lipids is both on *K*m and *V*m, however, the only clear difference is seen in Hill plots. Thus, lipids probably act as allosteric effectors for pyruvate oxidase, whose physiological function is to transfer reducing equivalents to the electron transport chain located in the membrane.

UDP-Glucuronosyltranferase: In liver endoplasmic reticulum it (MW 217 kD, four subunits, three of which are nonidentical) is involved in metabolism of lipid-soluble endogenous and xenobiotic substances. They are activated by perturbation of the membrane organization by sonication, detergent, or phospholipase A2 treatment. Besides solubilizing the substrate, phospholipids are required for catalytic activity (Hochman et al., 1981). The purified delipidated enzyme can be activated up to 700-fold at *V*m by lysophosphatidylcholines in the order $14:0 < 16:0 < 18:0 << 18:1$. The effect of lipid bilayers on the kinetic parameters is rather complex (Hochman and Zakim, 1984). Phospholipids stimulate the rate of glucuronidation catalyzed by the GT_2 form of UDP-glucuronyltransferase by modulating the binding of the substrate.

The behavior of the enzyme during reconstitution to vesicles of diacylphospholipids is also curious (Scotto and Zakim, 1985; Jain and Zakim, 1987). Delipidated enzyme added to vesicles is activated, but the binding is reversed during density gradient centrifugation. However, in the presence of a small amount of impurities (cholate, fatty acid, cholesterol) a stable complex is formed. Besides their effect on *V*m, lipids in the gel phase change the affinity for UDPGA for the substrate; the double-reciprocal plot becomes nonlinear in the presence of lipid. Bilayer in the gel phase also mediates allosteric activation by UDP-*N*-acetylglucosamine. In the liquid crystalline phase, UDPGT exhibits linear double-reciprocal plots, and is not activated by UDP-*N*-acetylglucosamine. These effects are similar to those observed in intact microsomes in the presence of detergent. Such observations suggest that the catalytic and regulatory parameters of UDP-GT are modulated by the phase properties of the lipid.

8 Glycoproteins and Glycolipids

Everything is incredible,
if you can skin off the crust
of obviousness our habit put on it

ALDOUS HUXLEY
Point Counterpoint

The plasma membrane plays a key role in all expressions of the social behavior of a cell. Obviously, structures on the outer surface of a cell must be important in such processes that involve *recognition* at the surface, *transduction* across the barrier, and *translation* in the cytoplasm of the external signals and stimuli. The outer surface of cell membranes is covered with a substantial amount of oligosaccharides that are covalently linked to lipids and proteins. Modulation of the surface charge, adhesion of surfaces, and recognition of foreign molecules are among the major functions of these glycoconjugates.

Typically, carbohydrates make up 0–10% of the membrane by weight, but they are always associated with lipids and proteins. Certain membrane components may contain as much as 70% carbohydrate by weight. Oligosaccharide residues are found associated with the outer surface of membranes of all cells. They are often distinguishable by electron microscopy as a feltwork of fine filaments or *fuzz* that follows the extracellular contours of the plasma membrane. Many functions of cells require intercellular communication, which involves binding or recognition of these surface carbohydrates. Thus, glycoconjugates stand out not only as antigens but also as receptors for hormones, transmitters, and toxins. Their roles in cell–cell recognition and transduction of cell-surface events into cytoplasmic response and protection of glycoprotein conjugates from proteolysis are probably of general biological significance.

Many types of mammalian cells also contain an extraneous surface coat (Fig. 2-1) consisting of mucopolysaccharides. These coats, sometimes referred to as the glycocalyx, function as cell walls or pellicles which undoubtedly play important roles in the normal physiology of cells, although they offer little resistance to the transport of small molecules. It provides stability against turger pressure; denuded cells lyse readily. There is no distinct boundary between plasma membrane and these extramembranous structures. Glycocalyx is often destroyed during conventional membrane isolation procedures, and in many cases the cell wall can be removed by hydrolytic enzymes without altering the viability of the cell and integrity of most membrane functions.

IDENTIFICATION OF GLYCOCONJUGATES ON CELL SURFACE

The morphological and histochemical evidence for distribution of carbohydrates on the external surface of plasma membranes has been shown by dye stains like acid fuchsin on periodate-treated cells, or by the electron-dense reagents like cationized ferritin or colloidal hydroxides of iron, thorium, and lanthanum. Ruthenium red is also specific for carbohydrates and can be made electron dense with osmium tetroxide treatment. These stains do not distinguish between the surface coat of mucopolysaccharides and the glycoconjugates associated with the plasma membrane. Treatment of cells with glycosidases or with mucolytic enzymes like hyaluronidase abolish many of the staining reactions. Similarly, anionic groups on the sialic acids, uronic acids, sulfatides, and mucopolysaccharides can be identified with histochemical techniques coupled with digestion with specific glycosidases and by electrophoretic methods (Hughes, 1976). Radiolabeled or fluorescent or ferritin-tagged lectins and antibodies are somewhat more specific for identifying membrane glycoconjugates. Although these agents may perturb the normal organization of membrane components, they are useful in determining the distribution of specific carbohydrates on the surface of intact and viable cells.

The staining methods have been useful in providing information about the role of surface carbohydrates in cells. Membrane carbohydrates are found on the outer surface of a large variety of cells and between cells at junction complexes. The density of glycoconjugates on cell surface is often high. For example, on the erythrocyte surface the average distance between polysaccharide molecules is about 5 nm, and on the average these chains reach out about 10 nm from the bilayer surface. It appears that glycoproteins up to 100 nm long are present on the surface of many types of cells. In some cases changes in the size and shape of carbohydrates, or the relative distribution of glycoconjugates in response to metabolic or physiological change, or internalization of membrane receptors have been noted. Such observations suggest an active turnover of glycoconjugates on the cell surface. However, detailed molecular resolution or quantitative estimation of the carbohydrate content is not possible from the staining techniques.

Isolation and Characterization

Membrane carbohydrates occur as glycoconjugates, therefore, methods for their isolation are the same as those for the corresponding lipids and proteins but with some important differences. The glycolipids are more polar than phospholipids, therefore extraction and purification procedures for glycolipids require use of more polar solvents. Properties of glycoproteins are not altered as much because of conjugation. However glycoproteins do not bind as much anionic detergent, and therefore their electrophoretic mobility in sodium dodecylsulfate (SDS) is lower, thus giving a higher estimate of their molecular weight. Similarly, glycoproteins sediment slower on a sucrose gradient containing anionic detergent. Specific procedures for purification of glycolipids and glycoproteins utilize carbohydrate binding specificity of lectins, such as affinity chromatography or immunoprecipitation of the lectin complexed with the carbohydrate.

Isolated carbohydrates are readily quantitated by colorimetric assays. Neutral sugars (total hexose) react with anthrone to form a chromophore that absorbs at 620 nm. Bound sugars like galactose can be cleaved by enzymes like galactose oxidase, and the released product reacts with *o*-toluidine and peroxidase to yield a product that absorbs at 425 nm (Avigad et al., 1962). Identification of individual sugars is done by paper or by thin-layer chromatography coupled with group-specific color reagents. An automatic sugar analyzer employing an orcinol reagent in conjunction with a high-pressure liquid chromatograph has been described (Marsh et al., 1977).

Structure of Oligosaccharides from Membranes

Membrane oligosaccharides are derived from only nine different pentoses and hexoses, as shown in Fig. 8-1. Six of these are simple sugars: galactose, glucose, mannose, fucose, arabinose, and xylose. The other three, glucosamine, galactosamine, and *N*-acetylneuraminic acid (sialic acid), are complex derivatives of the corresponding hexoses. For example, the term sialic acid denotes one of the 20 naturally occurring derivatives of neuraminic acid. The amino group of neuraminic acid is derivatized with acetyl or glyco-

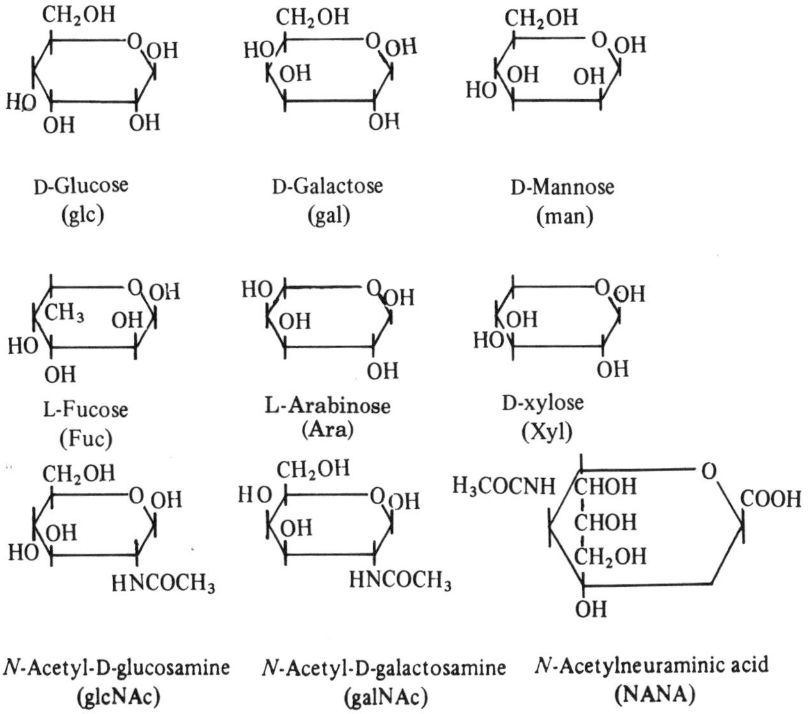

D-Glucose (glc) D-Galactose (gal) D-Mannose (man)

L-Fucose (Fuc) L-Arabinose (Ara) D-xylose (Xyl)

N-Acetyl-D-glucosamine (glcNAc) *N*-Acetyl-D-galactosamine (galNAc) *N*-Acetylneuraminic acid (NANA)

Fig. 8-1. Structures of the nine different carbohydrate residues that are found in oligosaccharide chains of glycoproteins and glycolipids.

syl residues, and the hydroxyl groups may be methylated or esterified with acetyl, lactyl, sulfate, or phosphate groups. In glycoconjugates, sialic acids are often linked to galactose or *N*-acetylgalactosamine or sialic acid (Schauer, 1982). The sugar residues are generally present as oligosaccharides in glycolipids and glycoproteins on the outer surface of a cell. Thus, the outer surface of an erythrocyte contains about 20 million sialic acid (pK_a 2) residues on terminal positions. Many similarities are found in the structure of the sugar chains conjugated to lipids or to proteins, especially in the terminal moieties of the more extended structures (Rauvala and Finne, 1979). This suggests that the role of surface carbohydrates is distinct from the specific functions of individual membranes components to which they are conjugated.

Glycolipids

There are two major types of glycolipids (Wiegandt, 1985): Glycogycerolipids (Fig. 2-6) and glycosphingolipids (Fig. 2-8, 8-3) can constitute up to 50% of the membrane lipids. Other minor classes of glycolipids (Fig. 8-2) also

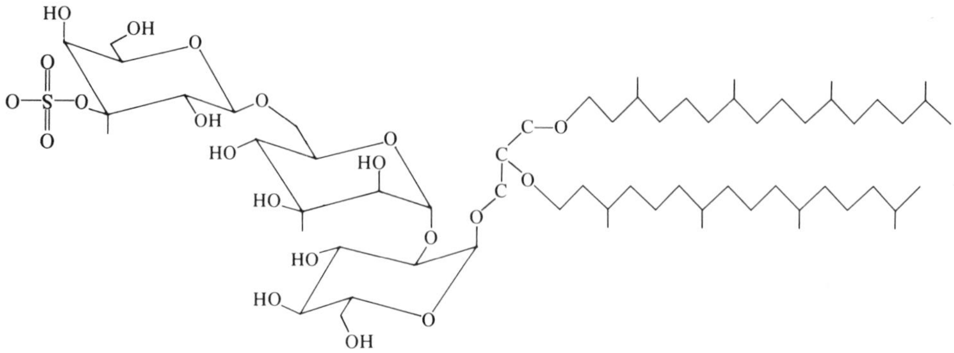

Sulfoglycolipid
(*Halobacteria*)

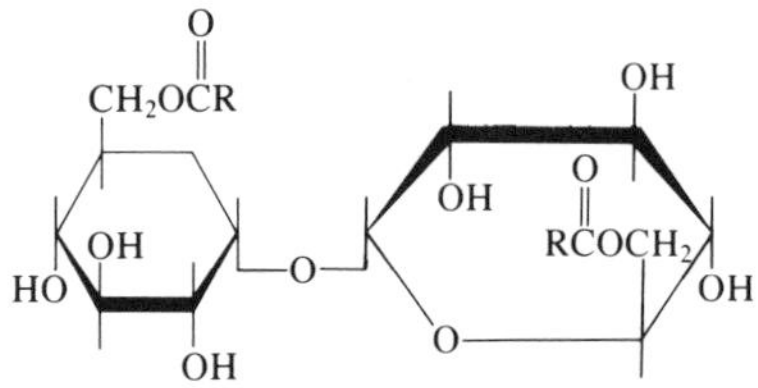

Cord factor
(*Corynebacterium*)

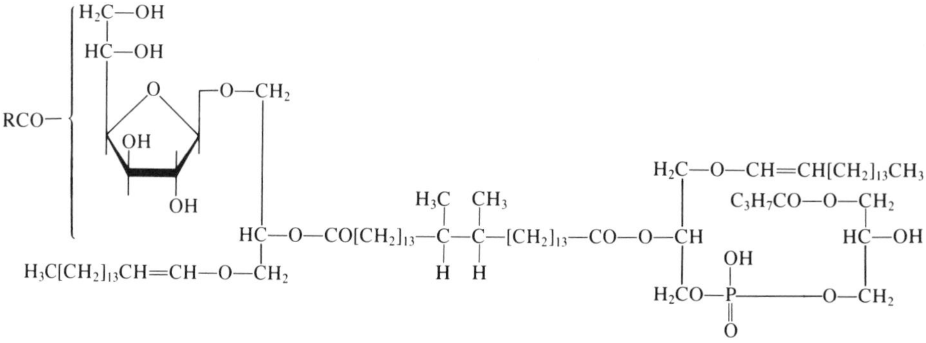

Diabolic acid
(*Butyrivibrio S2*)

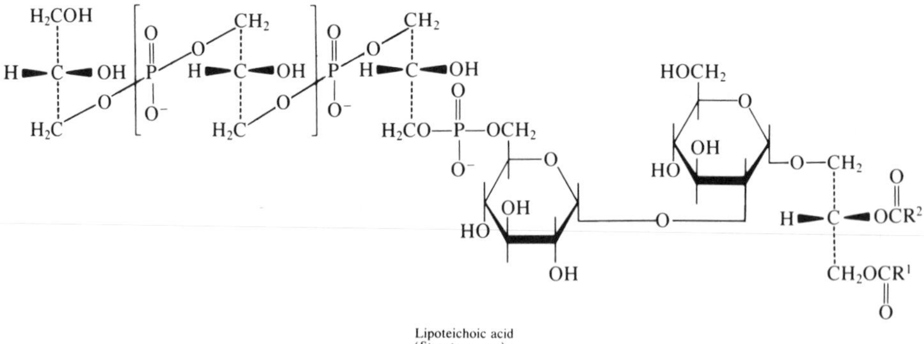

Lipoteichoic acid
(*Streptococcus*)

Fig. 8-2. Structures of some unusual glycolipids that are mostly found in bacterial membranes.

L-4-AraN $\overset{1}{—}$ (P) — D-GlcN$_p$ $\xrightarrow{\beta 1,6}$ D-GlcN $\overset{1}{—}$ (P) $\overset{1}{—}$ D-GlcN

Lipid A
(*Chromobacterium*)

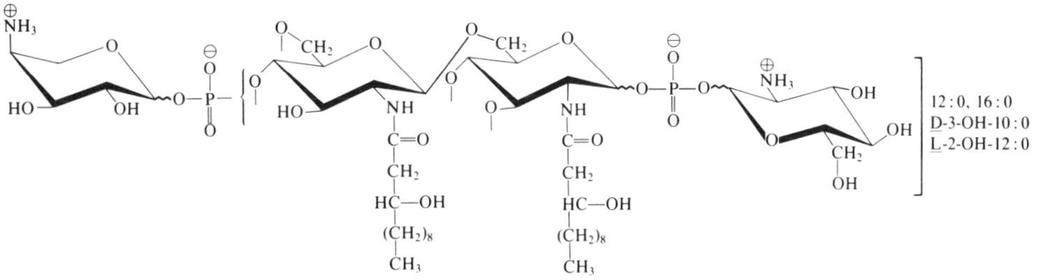

Fig. 8-2. (*Continued*)

include polyprenol phosphate glycoside, steryl glycoside, lipopolysaccharides of gram-negative bacteria, wax esters of mycobacteria, acylated carbohydrates, and several derivatives of erythritol, ethylene glycol, or glyceric acid. Many of the minor glycolipids are probably not the normal membrane components, but serve metabolic, regulatory, and structural functions, and appear only transiently in certain membranes.

Glycolipids are often classified according to their carbohydrate structure. Glycoglycerolipids (Fig. 2-6) include several classes of glycoconjugates that occur largely in bacterial and plant membranes. The sugar residue generally contains sialic acid, uronic acid, sulfate, phosphate, phosphono, or carboxyl groups. The hydrophobic moieties are generally 1,2-diacylglycerol or alkylacylglycerol residues.

Glycosphingolipids (Fig. 2-8, 8-3) in which the hydrophobic moiety is ceramide, are mostly found in animal plasma membranes and they are abundant in nervous tissues. The lipophilic portion of a fatty acid is attached by an amide bond to a sphingosine base. Ceramide shows microheterogeneity, and there is considerable variability in the composition of the three structural moieties (polar head group, backbone, chains) in glycolipids (Fig. 8-3). For example, sphingosine is found in glycosphingolipids from animal membranes. Dihydrosphingosine (sphinganine) is found in several eukaryotes. Phytosphingosine (4-D-hydroxysphinganine) is the principal base in sphingolipids derived from plants and fungi. The acyl chains (Fig. 2-4a) generally have more than 20 C-atoms and few double bonds. Hydroxy acids are also found in sphingolipids from many sources.

The carbohydrate moiety in glycosphingolipids is attached at the primary hydroxyl group of ceramide. Glycosphingolipids containing *N*-acetylneuraminic acids as the last residue are called gangliosides. Several different

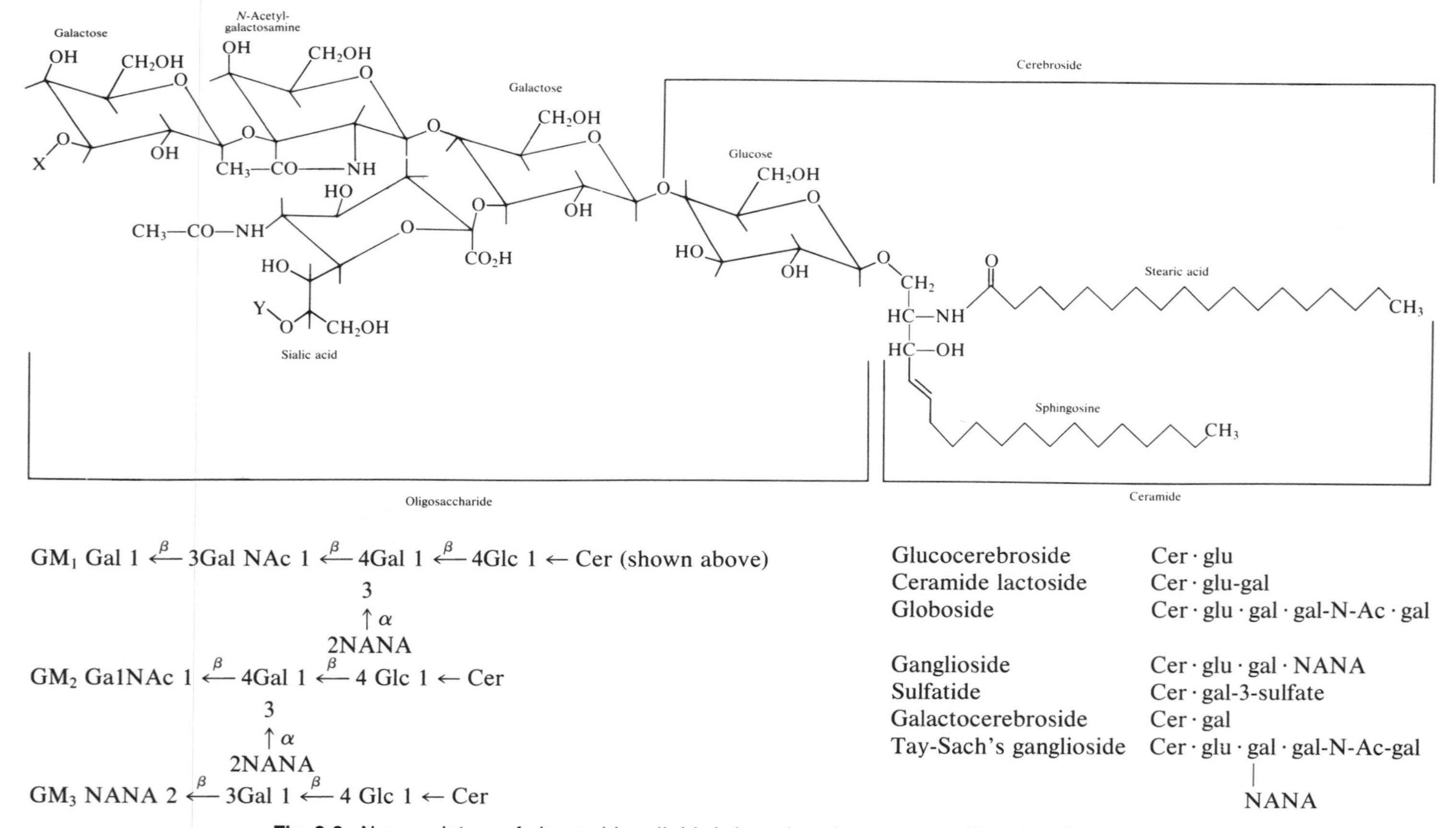

Fig. 8-3. Nomenclature of glycosphingolipids is based on the sugar substituent on the ceramide (see also Fig. 2-8).

types of oligosaccharide moieties with one to over 50 sugar residues have been found. It may be noted that in all such cases the reducing end of the terminal sugars is not free but forms a glycosidic bond with the next residue. Some microheterogeneity in the structure and function of glycosphingolipids arises because the oligosaccharides may be present in different stages of glycosylation, and also because different acyl chains may be present in the hydrophobic portion.

Biosynthetic incorporation of sugar residues in glycolipids and glycoproteins is mediated by glycosyltransferases to ceramide from UDP-sugars (Gal, Glu, *N*-acetylgalactosamine, *N*-acetylglucosamine) or CMP-sugars (sialic acid, L-fucose). The glycosyltranserases probably exist in Golgi membrane. These are generally membrane-bond enzymes that seem to require lipids for activity. Their specific activity toward glycolipid substrates is considerably lower than it is for the glycoprotein substrate. It is, however, not clear that all glycosyltransferases have distinct specificity for lipid or protein substrates. The turnover of glycolipids is not controlled only by induction of hydrolases and transferases; activator and transfer proteins are also known to play crucial roles. The glycolipid-degrading enzymes include not only lipolytic esterases and amidases (Fig. 2-10) but also sialidases and glycosidases. The genetic deficiency of glycolipid-degrading enzymes or their activators leads to accumulation of glycolipids and is responsible for a variety of hereditary diseases or gangliosidoses summarized in Table 2-6. Gangliosides are prominent components of nervous tissues, however their accumulation can be fatal. Developmental increase in the quantity of gangliosides has been noted during periods of rapid growth of dendrites, axons, and neuronal interconnections. Thus, disorders of ganglioside metabolism are associated with brain disorders. Changes in ganglioside levels have also been noted during growth, transformation, drug metabolism, and uptake of exogeneous gangliosides.

GLYCOLIPIDS IN AQUEOUS DISPERSIONS

The properties of glycolipids in aqueous solution attest to their amphipathic character. They are readily hydrated and dispersed in excess water. The lateral packing characteristics of sphingomyelin (SM) are similar to those for glycerophospholipids, however, subtle differences have been noted. For example, T_m for gel–fluid transition of SM dispersions is higher, presumably due to intermolecular H-bonding. In the gel state of SM dispersions, the acyl chains are partially interdigitated (Levin et al., 1985). Since this behavior is expected from the naturally occurring SM, it would imply that a local order is created by interdigitated chains. However, such intermolecular interactions do not appear to interfere with the interaction of cholesterol codispersed with SM (Estep et al., 1979).

Glycosphingolipids disperse in water to form bilayers, which exhibit a thermotropic chain-melting transition (Thompson and Tillack, 1985; Maggio et al., 1985). The transition temperatures are about 20–40°C higher than those for the corresponding phospholipids of the comparable chain length. The enthalpy and T_m decrease with the increasing complexity of the oligosaccharide chain or the presence of anionic sugar residues. A direct correlation is found between the area of cross-section of glycosphingolipids and the enthalpy or T_m of their aqueous dispersions. This indicates that the intermolecular spacing is an important determinant in establishing the thermotropic properties.

Because of their relatively large polar head group, most gangliosides form nonbilayer structures like micelles when dispersed alone in aqueous solution, and critical micelle concentration (cmc) is estimated to be less than 1 n*M* for most naturally occurring species. In dilute solutions, gangliosides form oblate ellipsoid micelles of MW 200 kD to 500 kD depending on the ratio of C_{20} and C_{18} sphingosine and the nature of the sialo-oligosaccharide. The tendency to form micelles of smaller aggregation number increases with increasing number of sialic acid residues. Also the increase in chain length and unsaturation appears to increase the size of the micelle. Such observations suggest that the micelle size and their dispersity depends upon the concentration as well as the structure of gangliosides (Corti et al., 1980). At low water content, some gangliosides form a hexagonal phase in which about 20% (w/w) of the water is incorporated between the cylinders. The motional freedom of GM1 and GD1 changes significantly in the 10–60°C range (Hinz et al., 1981), even though there is no evidence of a phase transition. The rate of transfer of ganglioside monomers from their micelles to vesicles of phospholipids is relatively slow, with a half-time of more than 30 min at room temperature, and it is much more rapid above 45°C (Felgner et al., 1983). It implies that the lifetime of ganglioside micelles at room temperature is considerably longer than that of micelles of single-chain amphiphiles. It is quite likely that the aqueous dispersions of gangliosides may be in a lyotropic equilibrium of several aggregates.

Gangliosides transferred from their aqueous micellar dispersions are exclusively incorporated in the outer monolayer of phospholipid vesicles, and the rate of transbilayer movement is very slow. When codispersed with phospholipids, glycolipids readily form bilayers. The thermotropic transition data suggest that glucosylceramide does not mix ideally with dipalmitoyl phosphatidylcholine (DPPC) in bilayers. Similarly, gangliosides are not uniformly distributed in the bilayers of binary codispersions as reflected in their binding characteristics to lectins, permeability properties, and fusion properties. Acidic groups on oligosaccharides of gangliosides also bind divalent ions and thus influence the properties of bilayers in which they are present. Due to their molecular and organizational properties, gangliosides in bilayers occupy a larger surface area. This would disorder the lipophilic region and lead to lateral-phase separation, which could be lytic and promote fusion.

An appreciation of the biological role of glycolipids and glycoproteins has evolved from the study of their immunological properties, their ability to bind lectins, and from the use of inhibitors and activators of their metabolism (Brady, 1986). The functional characteristics of glycolipids arise from their carbohydrate moieties, although the hydrophobic portion may enhance the biophysical characteristics of the carbohydrate residues by a conformational change, by the orientational effects, or by altering their metabolism. One of the major biological functions of glycosphingolipids is that they are antigenic, as for blood groups, serum, and milk. Antibodies to gangliosides have therefore been used as markers for the specific immunological identification of cell types of the immune and nervous system. Monoclonal antibodies of glycosphingolipids have been detected as prominent cell-surface antigens in certain tumor tissues such as colon carcinoma and human melanoma. The antigenic determinants reside in the nonreducing terminus, thus antigenic polymorphism can result from variation in the structure of the terminal residues. The antigenic carbohydrates of lipids are also sometimes found in glycoproteins. Thus, antigenic expression is controlled by glycosyltransferases which insert or remove specific sugar residues. The antigenicity of a carbohydrate residue is also determined by the environment. Thus, compared with glycolipids, the glycoproteins with the same antigen bind better to the antibodies. However, glycolipids dispersed with inert lipids like PC and cholesterol also exhibit apparently higher antigenicity and affinity for the antibodies (Yogeeswaran, 1980).

Some of the glycolipids are known to serve as receptors for toxins and viruses, while others appear to play a role in transformation. The glycolipids are not major structural components, and thus on a mole basis these complex lipids are minor components of plasma membranes. Antisera to gangliosides exhibit a number of biological actions: They stimulate mitosis in rat thymocytes, decrease cell growth in fibroblasts, and inhibit the expression of transformed phenotypes in virus-transformed cells. Levels of glycosphingolipids in membranes appear to change in association with cellular differentiation and oncogenic transformation. Most of the glycosphingolipids are immunologically active both in heptenic reactivity and in antibody-producing potency.

GLYCOPROTEINS IN MEMBRANES

Affinity chromatography on lectin columns has been used for specific separation of glycoproteins from solubilized membranes. Glycolipids apparently are not retained because their affinity for detergent micelles is much higher. Bound glycoproteins are usually eluted with appropriate sugars. The carbohydrate residues on glycoproteins are complex and the sugar specificity of lectins is broad, often many glycoproteins are therefore retained on affinity columns of most lectins.

Glycoproteins exhibit anomalous mobility on polyacrylamide gels, depending upon the gel concentration and the degree of cross-linking. As the percent concentration of polyacrylamide increases, the apparent molecular weight of a glycoprotein decreases. For example, the apparent molecular weight of glycoprotein is found to be 92 kD in 5% gel and only 55 kD in a 12.5% gel (Segrest et al., 1971). Usually, the molecular weight of a glycoprotein can be obtained reliably with gels at high concentration of acrylamide and a high degree of cross-linking (Glossman and Neville, 1971).

Histochemical techniques suggest that oligosaccharide residues of membranes are localized on the outer surface. Carbohydrate residues in membrane glycoproteins are readily accessible to reagents (such as enzymes, lectins, antibodies, toxins, polar reagents) from the aqueous phase, and these residues, along with a peptide fragment, are readily removed by proteases. Such observations show that the carbohydrate moieties do not play an important role in integrating proteins into the hydrophobic region of membranes.

Several reagents have been useful for the study of carbohydrates at the cell surface. Neuraminidase selectively removes sialic acid residues from the cell surface. Galactose oxidase can be used for selective labeling of galactose residues. Similarly, periodate oxidation of vicinal dihydroxyl groups is useful for labeling of the sugar residues (Liao et al., 1973) because the resulting carbonyl function can be reduced by tritiated borohydride. Modification of proteins (as by iodination of tyrosine, protein kinase-mediated phosphorylation of serine and threonine, and reactions of amino groups with a variety of polar reagents) has been used for labeling of glycoproteins. One of the major problems encountered with the use of enzyme reagents is due to *crypticity* of the target residue; i.e., often such groups are not accessible in intact cells, and they become accessible when membranes are perturbed by lysis, or by treatment with trypsin and neuraminidase. Restricted accessibility of oligosaccharides is more common with glycolipids than with glycoproteins. The origin of crypticity is not known, however, lateral segregation of ligands in the plane of the membrane is one of the more likely explanations.

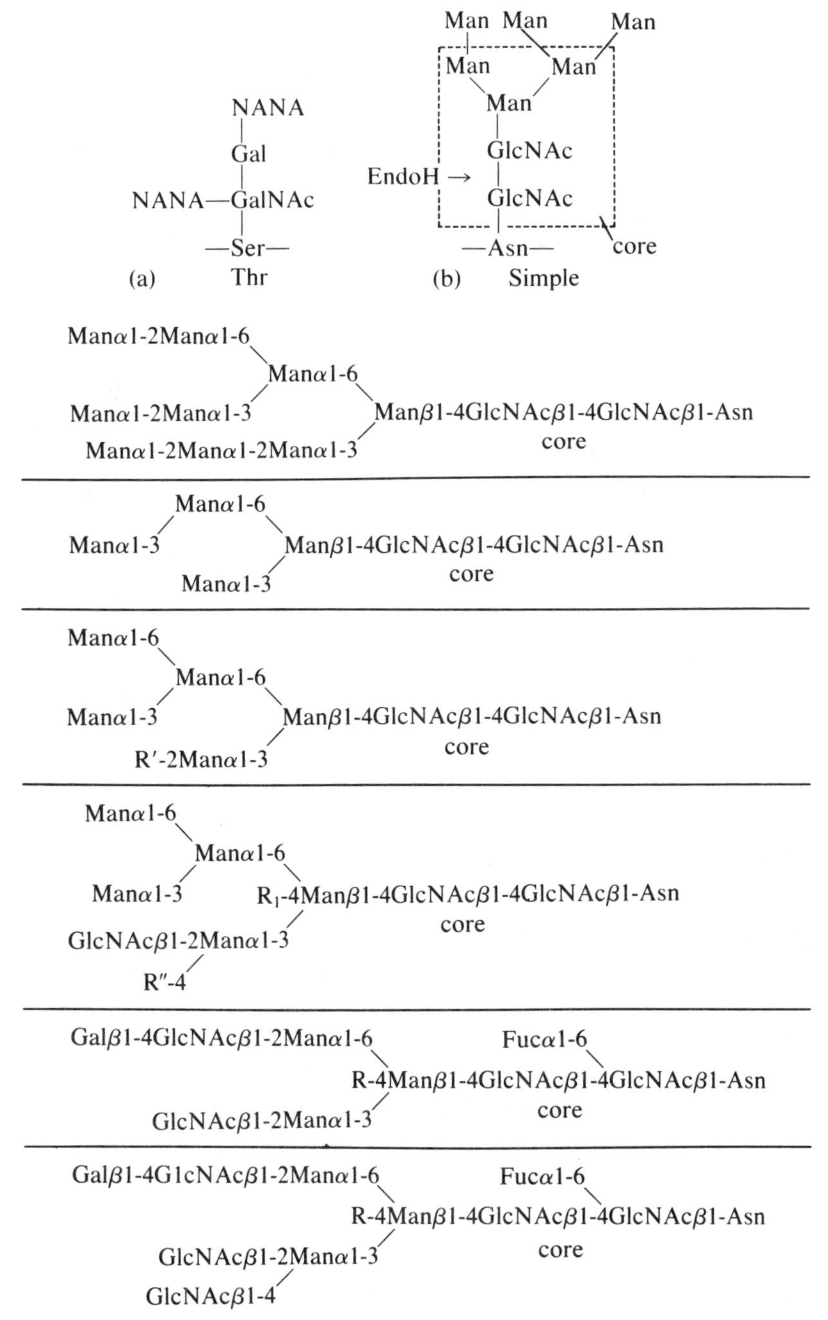

Fig. 8-4. Carbohydrate chains attached to proteins are either *O*-linked or *N*-linked. These chains are generally rich in mannose and appear to have a common core structure. The site of cleavage by endoglycosidase H is also shown.

Conjugation of oligosaccharides with proteins occurs by *O*-glycosylation to serine, threonine, hydroxylysine, and hydroxyproline, or by *N*-glycosylation to asparagine (Fig. 8-4). Detailed structures of six major classes of carbohydrate chains that are *N*-linked to proteins of cell membranes and serum are also shown in this figure (Brisson and Carver, 1983). These carbohydrates have a common core and well-defined three-dimensional structures with relatively few degrees of freedom. The flexibility of the man-α1-6-man-β1-arm is also restricted by interactions with other parts of the molecule. However, it appears unlikely that the oligosaccharide chain could alter the gross conformation of a glycoprotein.

A single protein may contain oligosaccharides linked by both *O*- and *N*-glycosidic bonds, and more than one type of linked structure at the same site on a protein. This microheterogeneity arises largely from post-translational modifications by glycosyltransferases. The final structure of oligosaccharides presumably depends on the specificity of the enzyme and the conformation of receptor proteins. The *O*-glycosyltransferases are membrane-localized in Golgi facing the lumen, and they utilize sugar nucleotide substrates. Biosynthesis of *N*-glycoconjugates occurs via phosphorylated polyisoprenoid intermediates whose synthesis involves both cytoplasmic and membrane-associated (endoplasmic reticular) enzymes. The preassembled oligosaccharide unit is transferred to asparagine while the peptide chain is still being translated. Further modification of the oligosaccharide occurs in the Golgi (Krag, 1985).

FUNCTIONS OF GLYCOCONJUGATES

A large variety of glycoconjugates are present in cells and not all of these are on the outer surface of the plasma membrane. The genesis of organelles also gives rise to certain carbohydrate-containing components. For example, the membrane of lysosomes is derived by pinching-off membranes of Golgi or smooth endoplasmic reticulum. Thus many of the lysosomal acid hydrolases are glycoproteins, e.g. arylsulfatase A and B, acid phosphatase, β-galactosidase, β-glucuronidase, and β-*N*-acetylhexosaminidase. The calcium-binding glycoprotein of mitochondria also raises some interesting questions about its origin.

Glycoproteins and Glycolipids as Antigens

Numerous biological functions of glycolipids are intertwined at several levels of organization and metabolism. Their ability to bind cations and proteins, asymmetric distribution, and segregation add another dimension of complexity to their putative functions. Presence of substances with blood group activity has been shown on the surface of almost all types of cells. The oligosaccharide determinants for many antigens have been identified. The A, B, and H determinants are located on glycolipids, whereas the MN antigen is on a glycoprotein. Histocompatibility antigens, which establish the acceptance or rejection of transplanted tissues (HL-A antigen), have also been identified as glycoproteins.

Glycolipids, like gangliosides, have also been implicated as receptors for a variety of viruses, interferons, growth factors, hormones, toxins, and bacteriophages. Although the mechanism of receptor–effector coupling is not known in any of these situations, it appears that their physiological effects are most probably due to a change in the organization and distribution of the receptors and other membrane components including the lipids (Crook et al., 1986).

Glycophorin A

This MN antigen is a major component of erythrocyte membrane (Segrest et al., 1971, 1973) with no known function. There are one million copies of it per erythrocyte. It consists of a single polypeptide chain of MW 31 kD and about 22 kD oligosaccharides. It is present as a dimer in the membrane. Residues 12 through 34 are nonpolar and appear to form a transmembrane helix which is braced by polar groups like Glu-9 and Glu-11, and Arg-35, Arg-36, Lys-39, and Lys-40 (Fig. 7-4d). The five amino-terminal residues determine the MN-blood group antigenicity. The amino terminus, which is exposed to the external surface, contains 15 *O*-glycosidically-linked serine and threonine oligosaccharides with sialic acid, *N*-acetylgalactosamine, and smaller amounts of *N*-acetylglucosamine, galactose, and mannose as tri- and tetrasaccharide units (Adamany et al., 1983).

Inhibitors of Glycosylation

The role of cell-surface glycoconjugates is implicated by the effect of inhibitors of their biosynthesis. Bacitracin chelates to phosphate via zinc, and thus ultimately causes a shortage of monophosphopolyprenol required for biosynthesis of peptidoglycans, sterols, and ubiquinones. Tunicamycin, a metabolite from *Micrococcus lyso superificus*, inhibits peptidoglycan biosynthesis, *N*-glycosylation of proteins, and cross-linking of peptidoglycans to teichoic acids and teichuronic acids. All these pathways inhibited by tunicamycin have a common step: translocation of *N*-acetylhexosamine-1-phosphate from UDP-sugar to monophosphopolyprenol. As an inhibitor, tunicamycin acts as a structural analog of the substrate as well as the hydrophobic product by binding irreversibly to the translocase. Inhibitory effect of tunicamycin has been observed in a variety of organisms, and the range of processes affected (development, histogenesis, metastasis, and function of certain receptors) attests to the diversity and significance of glycoconjugates (Olden et al., 1982).

Antibodies against Liposomal Antigens

Among the potential antigens as head-group substituents of membrane lipids, naturally occurring antibodies are found only against gangliosides and to a lesser extent against cardolipin and sphingomyelin. Several lipid species like Frossman antigen, lipid A, and lipopolysaccharide from *Salmonella* (Fig. 8-2) exhibit strong antigenicity. Similarly, proteins incorporated into the bilayer (such as glycophorin, glucose carrier, myelin basic proteins, viral antigens, histocompatibility antigens) act as antigens and bind to their antibodies. Nearly all phospholipids form anti-liposome antibodies (Alving, 1984). Many of such antibodies do not appear to have any ill effect on rabbits or humans; however, interaction of such antibodies with antigens disrupts cells and liposomes, and releases trapped solutes. Therefore, this reaction could limit utility of liposomes as drug carriers. Complement activation also involves binding of an antibody, which ultimately leads to binding of nine complement components to cell membranes, and the damage is due to increased permeability and lysis resulting from the formation of channels.

The antibodies tagged with a fluorescent label or radiolabel have been used as membrane markers for purification and assay of membrane components, for immune damage, and for monitoring mobility of antigens on cell surfaces. The dynamics of antibody interaction with surface antigens is complex, as it depends upon the size, type, density, and lateral mobility of antigens in the bilayer.

Cell Adhesion

Adhesion of cells to each other and to the extracellular matrix provides the basis for morphogenesis and development, yet its mechanism remains elusive. Organ-specific cell adhesion molecules, which are transmembrane proteins, have been shown to express during development (Edelman, 1984).

LECTINS

Specific carbohydrate-binding properties of lectins and other proteins that bind to carbohydrate residues (like glycosidases, glycosyltransferases, fibronectin, cell adhesion molecules) have been used not only as histochemical markers but also for a variety of studies, including agglutination of cells, labeling of surface carbohydrates, and purification of glycoproteins on the affinity columns of lectins. Their ability to interact with surface carbohydrates also initiates secondary processes like cell division, agglutination, fusion, patching, and capping arising from redistribution of receptors. Many lectins are cytotoxic because they interfere with metabolic activity (Hughes, 1976; Brown and Hunt, 1978). It should be noted that the responses of lectins are not necessarily related. For example, there is no correlation between mitogenesis induced by concanavalin and its ability to agglutinate transformed cells.

Lectins are often present in plant seeds, however, lectin-like molecules with avidity for sugar residues have been found in other organisms, including mammals. The properties and sugar specificity of some lectins are summarized in Table 8-1. Lectins are usually oligomeric, and the binding site on each subunit usually binds a mono- or disaccharide residue. In an oligosaccharide chain these are usually the last residues; however, penultimate residues also contribute to the binding specificity.

Several proteins of cell surface also bind or react with carbohydrate residues (Neufeld and Ashwell, 1980). Their potential roles include binding of circulating glycoproteins for endocytosis and clearance, specific cell adhesion, and covalent modification of carbohydrates. Adhesion of cells is an energy-requiring process that is inhibited by drugs and metabolic inhibitors (azide, diamide, cytochalasin, and lower temperature).

Interaction of Lectins with Receptors in Bilayers

Lectins can bind to sugar residues in glycoproteins and glycolipids. Since each lectin molecule has two or more binding sites, lectins cross-link receptors on adjacent cells, and thus lead to aggregation or agglutination of the receptor-containing particles. Lectin-mediated agglutination depends upon the receptor density on the membrane. Other factors, including membrane fluidity, do not appear to influence agglutinability. However, there are other factors that also regulate agglutination. Key observations are: (i) Agglutination is profoundly affected by temperature and metabolic inhibitors. (ii) With certain lectins, transformed cells agglutinate more readily than the normal

TABLE 8-1 Characteristics of Some Lectins

Lectin	Common Name	Approx. MW	Subunits	Blood Group Specificity	Simple Inhibitory Sugars (D-configuration)
Agaricus bisporus	Mushroom	58000	?	—	?
Bandeiraea simplicifolia	—	114000	4	B	α-Gal
Concanavalia einsformis	Jack bean, con A	50000	2	—	α-Man, Glc
Lens culinaris	Lentil, LCA	42000, 69000	2	—	α-Man
Limulus polyphemus	Horseshoe crab lectin	400000	18	—	Sialic acid
Lotus tetragonolobus	Lotus A, B, C	120000, 58000, 120000	4,2,4	H	Fuc
Phaseolus lunatus	Lima bean lectin	269,000, 138000	8,4	A	GlcNAc, Man
Phaseolus vulgaris	Red kidney bean I	138000	4?	—	GalNAc
	II	98000–138000	4?	—	
Phytolacca americana	Pokeweed mitogen	32000	—	—	?
Ricinus communis	Ricin I	120000	4	—	Gal
	II	60000	2		Gal
Robinia pseudoacacia	Robin	100000	—	—	Man, GlcNAc
Solanum tuberosum	Potato lectin	92000	2	—	$(GlcNAc)_2$
Triticum vulgaris	Wheat germ, WGA	23000	2	—	$(GlcNAc)_2$
Ulex europaeus	UEA I	170000	—	H	Fuc
	II	?	—	H	$(GlcNAc)_2$
Vicia graminea	—	—	—	N	GalNAc

cells, although both types of cells can be agglutinated at higher lectin concentration. On the other hand, normal cells agglutinate more readily with other lectins. (iii) Even closely related lectins with the same carbohydrate specificity often differ both qualitatively and quantitatively in their response. (iv) Agglutination may take place selectively among several cell types that bind almost identical numbers of lectin molecules. (v) The sequence of lectin-induced events involving binding and redistribution of surface receptors is remarkably similar to those observed for the binding of certain hormones, toxins, and antibodies (cf. Chapter 15).

These observations suggest that not only the affinity, density, and selectivity of lectin receptors, but also some property of the matrix in which the receptors are localized determine the agglutination behavior. Indeed, such a property of the membrane matrix may also be responsible for internalization of occupied receptors mediated by membrane flow, microvilli formation, and rearrangement of cytoskeletal elements.

Experiments with reconstituted systems have yielded significant insights into possible modes of agglutination. Glycolipids and glycoproteins incorporated in liposomes retain their ability to bind lectins (Grant and Peters, 1984). Thus, influenza virus, antisera to MN blood group determinant, wheat germ agglutinin, and phytohemagglutinin agglutinate vesicles containing glycophorin. While agglutination does not appear to depend significantly upon the lipid composition or size of liposomes, it can be inhibited by hapten inhibitors of lectins. There appears to be some interaction between glycoproteins present in the same membrane. Band 3 of the erythrocyte membrane binds specifically to concanavalin A (Con A). Vesicles containing band 3 and glycophorin agglutinate considerably less well with MN antisera, while antibody binding and Con A-mediated agglutination remain normal.

Vesicles containing glycolipids also agglutinate. For example agglutination of GM1-containing vesicles by RCA1 (*Ricinus communis* agglutinin-1) is quite sensitive to the ganglioside concentration in the range 2–10% (Surolia et al., 1975). Agglutinability of these vesicles also depends upon the lipid composition. Similar results have been reported for vesicles containing galactosylceramide or lactosylceramide, and synthetic lipids containing specific carbohydrate groups. In all such cases the specificity of the lectin for the sugar remains the same. One of the interesting features of the studies with glycolipid-containing vesicles is that a certain minimum density of receptors is required for supporting agglutination (Grant and Peters, 1984; Rando and Bangerter, 1979). Typically, the glycolipid concentration must exceed about 4% (Fig. 8-5). Apparently, this effect is related to the statistics of agglutination, where the concentration of receptors per aggregate obeys a higher order (power) relationship (Von Schulthess et al., 1976). The extent of agglutination (S_m) of m vesicles, each containing n receptors, is given by:

$$[S_m] = K_1, K_2 \cdots K_{m\ 1}\, n_b(n - n_b)^{m\ 1}(S_1)^m$$

where K_i is the equilibrium constant for the binding of the $(i + 1)$th vesicle, and n_b is the number of bound receptors per vesicle. This equation predicts that the agglutinability of vesicles containing receptors varies as the surface concentration of receptors is raised to a power equal to the number of vesicles in the aggregate.

Binding of lectins to monomeric sugars in the aqueous phase is weak (K_d are typically >100 mM). This reaction appears to be a first-order process whose activation energy is dominated by the enthalpy term (Khan et al., 1986). K_d for lectins bound to receptors in vesicles is often in the submicro-

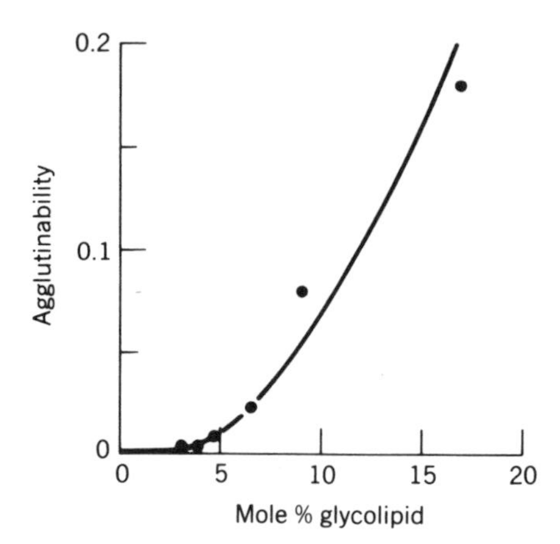

Fig. 8-5. The threshold effect that is observed during agglutination of vesicles and cells treated with lectin. The vesicles contained lactosyl ceramide, and agglutination was induced by RCA_1. (From Grant and Peters, 1984.)

molar range, and it changes with the nature of lipid (Grant and Peters, 1984). In fact, binding of lectins to glycoconjugates in vesicles is often strong enough that the bound lectin is not removed by washing. In contrast to the agglutination behavior, the extent of binding of RCA1 to GM1 in egg PC vesicles increases linearly with the receptor density, whereas the affinity for binding does not change with the receptor density (Surolia et al., 1975). The number of accessible binding sites is very sensitive to the position of sugar from the interface and the composition of vesicles (Surolia and Bhachawat, 1978). These experiments suggest that the influence of the bilayer matrix on lectin binding drops off very sharply as the length of the oligosaccharide head group increases, and the binding affinity is fairly insensitive to host matrix. Similar behavior has been observed not only for other lectins but also for binding of peptide hormones and antibodies (Grant and Peters, 1984). Other interesting features of lectin binding to receptors in vesicles include: the binding curves are sigmoidal; the time course for binding to bilayer is slow, with a half-time of several minutes compared with about a second for soluble sugars. Such observations suggest that binding of lectins to their receptors localized in the bilayer is a complex process probably involving a lateral rearrangement and clustering of the complex (Peters and Grant, 1984).

9 Ionophores: Models for Channels and Carriers

"Undoubtedly," said Wimsey, "but if you think that this identification is going to make life one grand, sweet song for you, you are mistaken . . . . Since we have devoted a great deal of time and thought to the case on the assumption that it was murder, it's a convenience to know that the assumption is correct."

DOROTHY L. SAYER
Have His Carcase

The permeability coefficients for the transfer of anions (chloride) and cations (Na, K) across bilayers are less than 10^{-11} cm/sec. This is largely due to a very low solubility of ions in the bilayer. The selective and enhanced transport of polar and ionic solutes across biomembranes therefore requires modification of the bilayer. Proteins facilitate transmembrane transport of polar solutes by one of the following mechanisms.

1. Destabilization or a phase change in the bilayer is facilitated by a variety of solutes, including proteins (cf. Chapter 6), such that the permeability barrier is transiently or permanently breached (pores or cracks). such an anomalous increase in the permeability of solutes is observed at the phase-transition temperature or in the presence of lipid-soluble solutes including surface-active proteins like mellitin and glycophorin. We have not considered this possibility further.
2. Translocation by a carrier involves a stoichiometric binding of the solute followed by translocation without any direct contact between the transmembrane aqueous compartments. Transmembrane proteins that expose the solute binding site alternately to the two aqueous compartments (conformational change) also satisfy this condition.
3. In channels proteins provide polar pathways that connect the two aqueous compartments. The polar pathway provided by the channel

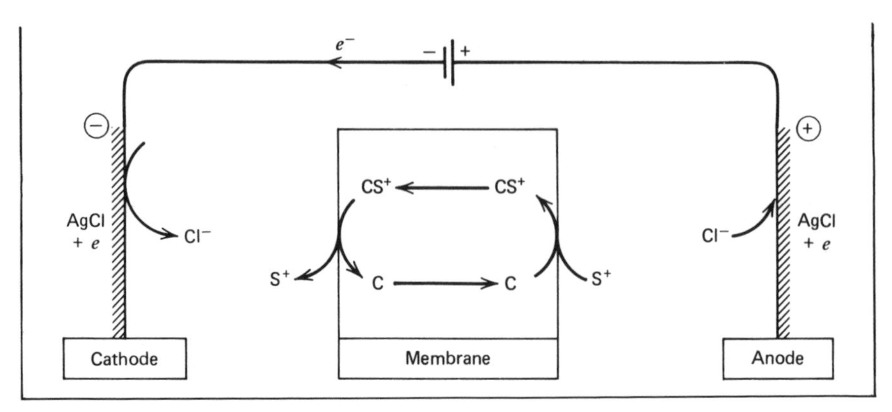

Fig. 9-1. The pathway for the flow of ions and electrons when a current pulse is applied across a membrane containing a cation-carrier that forms a charged complex in the bilayer. The reversible electrodes facilitate charge equilibrium between the chloride ions in solution and the electrons in the metallic conductors. Such reactions at the electrode interfaces, membrane interfaces, and in the battery compensate each other and maintain electroneutrality in the two aqueous compartments.

may or may not be filled with water, it may not be contiguous, but it should be accessible simultaneously from both aqueous compartments. Also a channel may always remain open, it may fluctuate between several conducting and nonconducting states, or it may be gated by membrane potential or by binding of regulators like hormones and neurotransmitters.

Our current understanding of facilitated ion transport in biomembranes is based largely on the study of ionophores. The term "ionophores" generally refers to small naturally occurring or synthetic peptides that facilitate transport via discrete sites by a carrier or channel mechanism. A conceptual framework for ion binding and selectivity, kinetics of translocation, single-channel conductance, and gated channels has developed from the study of ionophores. The electrical analogs of the molecular processes involved in transport of ions (conductance) across membranes are shown in Fig. 9-1. In such a scheme the current flowing through the membrane depends upon the driving force (electrochemical potential of the *battery*, which is the sum of the applied potential and the transmembrane gradient), and the selectivity of the barrier, which is related to ion binding, turnover, and conductance. The conductance pathways may be electrically or chemically gated as if a switch were operating in the circuit.

CARRIER-MEDIATED TRANSPORT

A carrier mechanism for translocation of solutes invokes a site that stoichiometrically binds the permeant, and the free as well as the occupied site appears alternately at the two membrane interfaces by diffusion, or rotation, or conformational change. Thus, the two aqueous compartments remain osmotically separated during the translocation cycle. A schematic representation of the minimum number of steps required for transport of solute (S) by a carrier (C) is given by:

$$\begin{array}{ccccc} & SC & \rightleftarrows & SC & \\ S \nearrow & \uparrow\downarrow & & \uparrow\downarrow & \searrow S \\ & C & \rightleftarrows & C & \end{array}$$

This minimal kinetic scheme (also shown in Fig. 9-1) is valid for transport of both ionic and neutral solutes. Some of the consequences of ion transport implicit in this scheme are examined in this chapter.

Ion Binding and Selectivity

Selectivity of carrier-mediated transport arises from the equilibrium binding specificity for the reaction:

$$S + C \rightleftarrows SC$$

which can occur in the aqueous phase or in the interface. According to this reaction the carrier or ionophore can be viewed as a complexing agent, hence the term *complexon* has been suggested. The free energy for binding of an ion to a site arises largely from electrostatic interactions. The electrostatic free energy (in kcal/mole) for binding of a cation of radius r to the ligand of radius r_L is given by the Born (1920) equation as $-322Nq/(r_L + r)$, where N is the coordination number of the ligand and q is the electronic charge. In this derivation it is assumed that the interaction occurs between two nonpolarizable monopolar point charges, each at the center of an incompressible sphere. For nonpolarizable point sites $N = 1$ and q is an integer number ($=1$ for monovalent cations). For dipolar sites the electrostatic energy term is more complex, and it is often approximated. The relationship given above is also applicable if the dipole separations are large compared to r_L.

The overall process for transfer of an ion from the aqueous phase to the carrier can be considered to occur in two steps: Ions are desolvated, and then the naked ion coordinates with ligands in the carrier. If the equilibrium constant for binding of an ion to the carrier is very large, the problem of ion selectivity reduces to a problem of assessing competition for solvation and binding to ligands. In this approach the terms arising from the change in entropy and strain energy differences due to a conformational change in the carrier are assumed to be the same for the two ions that compete for the binding site. Thus, for two ions of radii r_1 and r_2 competing for the ligand of radius r_L, the overall difference in the free energy is given in terms of ion–water and ion–site interactions (Ling, 1962; Eisenman 1962; Wright and Diamond, 1977; Eisenman and Horn, 1983):

$$\delta\Delta F_{21} = \left(\frac{-322}{r_A + r_2} - \frac{-322}{r_A + r_1}\right) qN - (\Delta F_2 - \Delta F_1)$$

The terms ΔF_1 and ΔF_2 are the free energies of hydration of ions 1 and 2. The equilibrium constant K for the competing binding reaction, which will give the ratio $K = (CS1)/(CS2)$, is given by:

$$\delta\Delta F_{21} = -RT \ln K = 1.36 \log K \qquad \text{(at 25°C)}$$

These expressions have been solved completely to obtain selectivity sequences for a variety of ions (Fig. 9-2) because the sign and magnitude of the $\delta\Delta F_{21}$ term would change with the radius of the ligand, r_L. Thus, depending upon the value of r_L, the binding preference of a ligand to the ions can change. Consider two extreme cases for alkali metal ions:

1. When r_L is large (i.e., low field strength), the electrostatic free-energy term for the ligand–ion interaction would be small. The $\delta\Delta F_{21}$ term would be dominated by the difference in the standard free energies for hydration of the cation which are: H, −192.7; Li, −54.3; Na, −30.4; K, −12.7; Rb, −7.7; Cs, 0.0 kcal/mole. For ligands of large r_L (low-field strength), the selectivity sequence therefore would be:

   $$\text{Cs} > \text{Rb} > \text{K} > \text{Na} > \text{Li}$$

   In this case the smallest cation that has highest affinity for water will have the most unfavorable $\delta\Delta F_{12}$ value, so that the relative affinity will decrease with decreasing ionic radius.

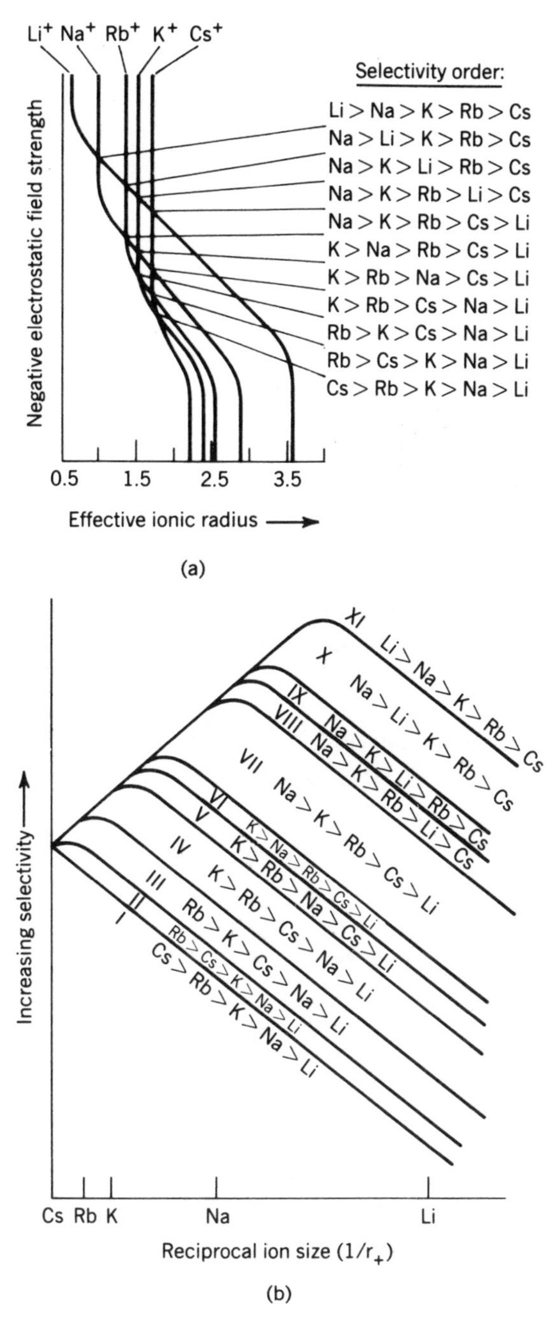

Fig. 9-2. (a) Selectivity isotherms for alkali metal cations are generated from the consideration of electrostatic field strength of the ligands and ionic radii. (b) The topology of the selectivity isotherms is more obvious in the plots of this type. (From Eisenman and Horn, 1984.)

2. When r_L is small (ligand of a high-field strength), the $\delta\Delta F_{12}$ is controlled by the electrostatic free-energy term. The smallest cation will have its center of charge nearest to the ligand and will therefore experience the largest attractive force, which for a strong site outweighs the fact that it also has the highest free energy of hydration. Thus, for a ligand with small r_L (high-field strength), the selectivity sequence would be:

$$\text{Li} > \text{Na} > \text{K} > \text{Rb} > \text{Cs}$$

that is, the affinity will decrease with increasing radius of the ion.

As r_L is varied continuously from a large to a small value, selectivity proceeds through the transition sequences shown in Fig. 9-2 for monovalent alkali metal cations, as well as for other classes of ions. It may be noted that only a few selectivity sequences are predicted: 11 out of a possible 120 permutation (= 5!) sequences for the five monovalent cations (see Table 9-1 for other ions). Each sequence corresponds to the situation where a selectivity optimum occurs as a monotonic function of the ion (solute) size (Fig.

TABLE 9-1 Equilibrium Selectivity Sequences

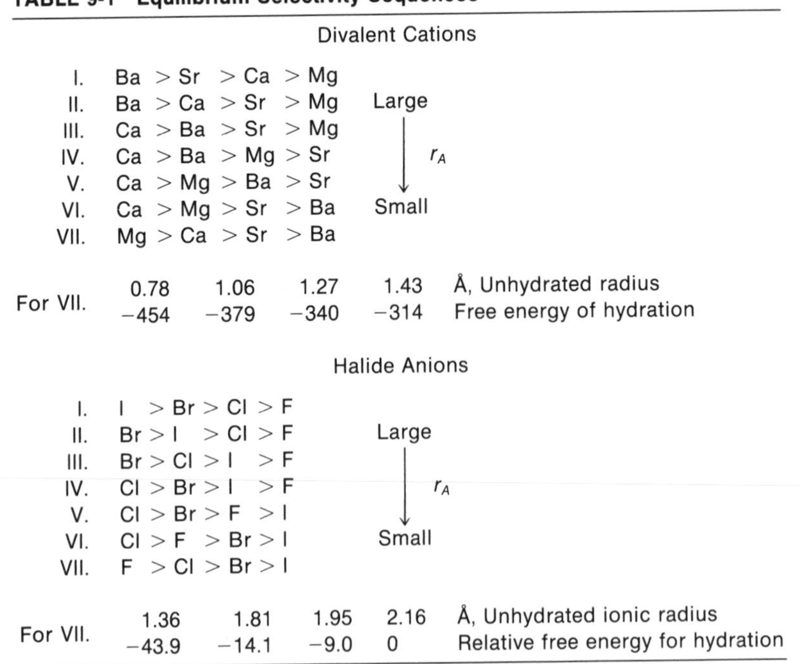

Divalent Cations

	Sequence	r_A
I.	Ba > Sr > Ca > Mg	
II.	Ba > Ca > Sr > Mg	Large
III.	Ca > Ba > Sr > Mg	↓
IV.	Ca > Ba > Mg > Sr	↓
V.	Ca > Mg > Ba > Sr	↓
VI.	Ca > Mg > Sr > Ba	Small
VII.	Mg > Ca > Sr > Ba	

For VII.	Mg	Ca	Sr	Ba	
	0.78	1.06	1.27	1.43	Å, Unhydrated radius
	−454	−379	−340	−314	Free energy of hydration

Halide Anions

	Sequence	r_A
I.	I > Br > Cl > F	
II.	Br > I > Cl > F	Large
III.	Br > Cl > I > F	↓
IV.	Cl > Br > I > F	↓
V.	Cl > Br > F > I	↓
VI.	Cl > F > Br > I	Small
VII.	F > Cl > Br > I	

For VII.	F	Cl	Br	I	
	1.36	1.81	1.95	2.16	Å, Unhydrated ionic radius
	−43.9	−14.1	−9.0	0	Relative free energy for hydration

9-2). Note that in these sequences each ion, in turn, is favored, and with that an inversion of a particular selectivity sequence occurs. The number of water molecules present in the vicinity of the interacting cations and the anionic site influences the magnitude, but not the sequence, of specificity. Experimentally, only the predicted and very few other selectivity isotherms have been observed for both biological and synthetic ligands (Wright and Diamond, 1977). The ubiquitousness of selectivity sequences arises from the fact that the independent variable is r_L, whereas the variable in the electrostatic term is $r_L + r$. If $r_L > r_1$ or r_2. Any multiplication operation on the hydration energies or the electrostatic energies can always be approximately compensated for by a change in r_L. The selectivity of protons relative to alkali metal cations is also expected to depend upon the value of r_L. Binding of protons will be preferred markedly at small r_L (high field strengths), while large alkali metal cations are preferred at large r_L (low field strength).

It must be emphasized here that although these calculations are based on the variation of the field strength of a ligand by varying r_L, similar changes in the field strength can be achieved by inductive effects from the microenvironment of the ligand. A permanent or induced dipole would also give rise to field strengths just the same as ionic groups do. Therefore, these equilibrium selectivity isotherms are expected under most situations, whether the ion is bound to the ligand by a point charge or coordinated by several monopoles of dipoles (Eisenman et al., 1973). The spatial distribution of individual anionic groups also influences selectivity through the overlap of the electrostatic forces. An increasing overlap between sites leads to an increase in the effective field strength without changing the overall selectivity sequence of ions having the same charge. The effect of charges, asymmetry, and other kinetic factors that influence selectivity have been reviewed (Eisenman and Horn, 1983).

ION-PERMEATION AND SELECTIVITY IN CHANNELS

Passage of ions across channels lined with ligands for the whole length or for a small distance (which acts as a *selectivity filter*) has been modeled with two main assumptions: (i) Free diffusion through large-pored channels, where ions move continuously and independently through a relatively structureless medium. In this treatment the pathways for ionic movement are symbolized by fixed conductors, and the thermal driving force on ions is expressed as gradients of electrochemical potential in series with the conductor. For a thin homogeneous membrane in which the electric field would be constant from one interface to the other, reversal potential or the zero-current potential is given by Goldman–Hodgkin–Katz equation:

$$V_m = \frac{RT}{nF} \ln \frac{\sum_{i=1}^{n} P_i a_i'}{\sum_{i=1}^{n} P_i a_i''} = \frac{0.059}{n} \log \frac{a_i'}{a_i''}$$

in which P is the voltage- and concentration-independent permeability for the ion. (ii) Rate theory: According to this treatment ions hop stepwise across a series of discrete energy barriers. The fundamental concept is that an ion from the energy well enters into an activated or transition state of high potential energy from which it can jump across or fall back as shown in Fig. 9-3. The rate constant, k, for the reaction is exponentially related to the

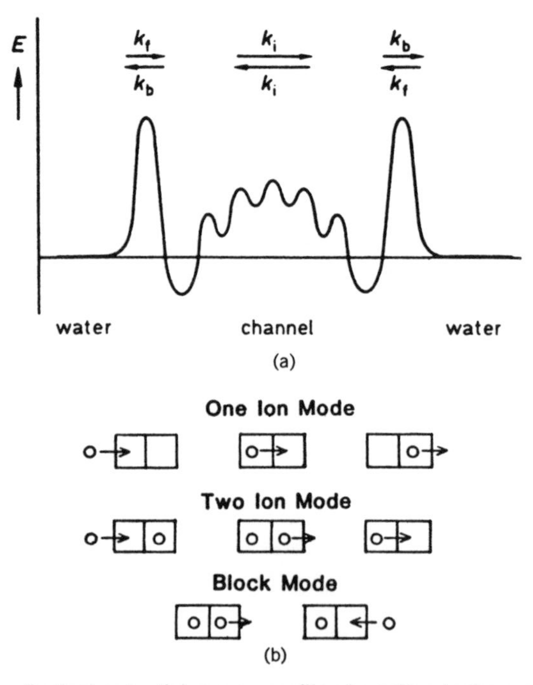

Fig. 9-3. (a) A hypothetical potential-energy profile of a cation in the gramicidin channel. Such a profile results from a superposition of a broad dielectric energy barrier and local potential variations caused by interactions with the ligands in the channel. The energy barriers at the opening imply energetically unfavorable transition states between the completely hydrated and the ligand-associated states of the ion. k_i is the rate constant for the ion jump across the central barrier; k_f and k_b are the on and off rate constants for ion binding when the second ion binding site is not occupied. (From Lauger, 1985.) (b) A schematic representation of the two-site model for a channel when only one ion moves across (top), when two ions move in the same direction (middle), when two ions move in the opposite direction (bottom). (Courtesy of S. Hladky.)

height of the barrier (the activation free energy) as:

$$k \propto \exp \frac{-(Ea)}{RT}$$

The rate theory has been useful for describing mechanisms of chemical and enzymatic reactions, and some of the salient features are now being adopted to describe kinetics of ion permeation across channels (Levitt, 1986). The passage of ions across a channel may occur through one barrier and two wells, or through two barriers and three wells, and so on corresponding to a reaction sequence $C + S^0$ to CS^1 to CS^2. . . . CS^n to $C + S^i$, where a solute is translocated from outside (S^0) to inside (S^i) through several intermediate states (S^1 through S^n). According to the rate theory each well represents a binding site, and each barrier is characterized by an activation energy term that provides an exponential term that may or may not be independent of the electric field. Obviously, more barriers provide more exponential expressions to simulate data, and the ability to resolve such rate constants depends upon the level of noise in the data. Although, the kinetic proof for a model is always circumstantial, such treatments can be useful for quantifying data, for developing models and eliminating alternatives, and of course for polarizing dialectics. The rate theory also has some serious limitations: It is valid only for barrier heights much greater than about 5 kT, and the model becomes analytically intractable for any but the simplest barrier structures. Other suggestions for modeling a channel have also been made (Cooper et al., 1985).

CARRIERS FOR CATIONS

Many organisms, mainly spore-forming bacteria, produce ionophores that increase ion permeability in membranes of target organisms. Ion binding and translocation characteristics of a variety of ionophores have been studied extensively (Hladky, 1979; Hladky and Haydon, 1984; Lauger et al., 1981; Easwaran, 1985). To function as a carrier an ionophore must undergo the reaction cycle (Fig. 9-1) which includes: bind ions to form CS complex, translocate CS complex across the bilayer, and couple the flux of ions with their electrochemical gradient.

Binding of ions to ionophores like valinomycin can be demonstrated by a variety of techniques. The structure and conformation of free and complexed valinomycin $(\text{L-Val-D-Hylval-D-Val-L-Lac})_3$ has been established in solution and X-ray diffraction analysis. In all these studies the selectivity ratio for K/Na is very high, however the exact value depends upon the medium. Valinomycin adopts different conformations depending on the polarity of the solvent and the nature of the cation. The conformation of free- and K-valinomycin (1 : 1 complex) is the same in nonpolar solvents. As shown in Fig. 9-4, the core of the bracelet structure contains a K ion coordinated with six ester carbonyl groups. The backbone of the molecule is folded into six loops which are held together by six hydrogen bonds between NH and C=O groups. Binding of the K ion is through dipoles, and the complex has a net positive charge because valinomycin is neutral. In crystals, this charge is compensated by anions, which are also present in the crystal lattice; the conformation of the complex does not depend upon the nature of the anion.

Similar structure is apparently assumed by a variety of other ionophores (Table 9-2 and 9-3; structures given in Fig. 9-5). These ionophores have a backbone in which oxygen atoms of functional groups (like hydroxyl, ester, ether, amide, carbonyl) are arranged strategically such that they form a polar cavity in the middle of the bracelet; a cation of appropriate size can fit snugly

TABLE 9-2 Properties of Some Carrier-Type Ionophores[a]

Ionophore	MW	pK_a	Selectivity Sequence	K/Na	Stoichiometry	Remarks
Valinomycin	1110		III, M	17,000	CS^+	CS > Ag > Tl > NH_4 > Na
Enniatin A	681		V, M	3.2	CS^+	Also forms C_2S^+ and $C_3S_2^+$ complexes
Enniatin B	639		III, M	2.8	CS^+	III for D
Enniatin C	681		III, M	2.2	CS^+	
Beauvericin	781		I, M	10	CS^+	
Nonactin	736		IV, M	16	CS^+	
Monactin	750		IV, M	18	CS^+	
Dinactin	764		IV, M	5.7	CS^+	
Trinactin	778		IV, M	1.3	CS^+	
Tetranactin	792		IV, M			
Dicyclohexyl-18-crown-6	373		IV, M	85	CS^+	
Monensin	670	7.95	VIII, M	0.1	C^-S^+	Also forms C_2S^+ and $C_3S_2^+$ complexes
Nigericin	724	8.45	V, M	45	C^-S^+	N
Grisoryxin	708					
Dianemycin	867	7.85	VIII, M	0.5	C^-S^+	N
X-206	774	8.30	V, M	20	C^+S^+	N
X-537 A (lasalocid)	590	5.80	I, D	3	C^-S^{2+}	N, I for M^+, also carries dopamine
A-23187 (calcimycin)	523		VI, D		C_2S^{2+}	N, XI for monovalent cations

[a] C, Carrier; M, Monovalent cations; D, divalent cations; N, electrically neutral ion carriers that mediate cation–proton exchange; S, solute or permeant.

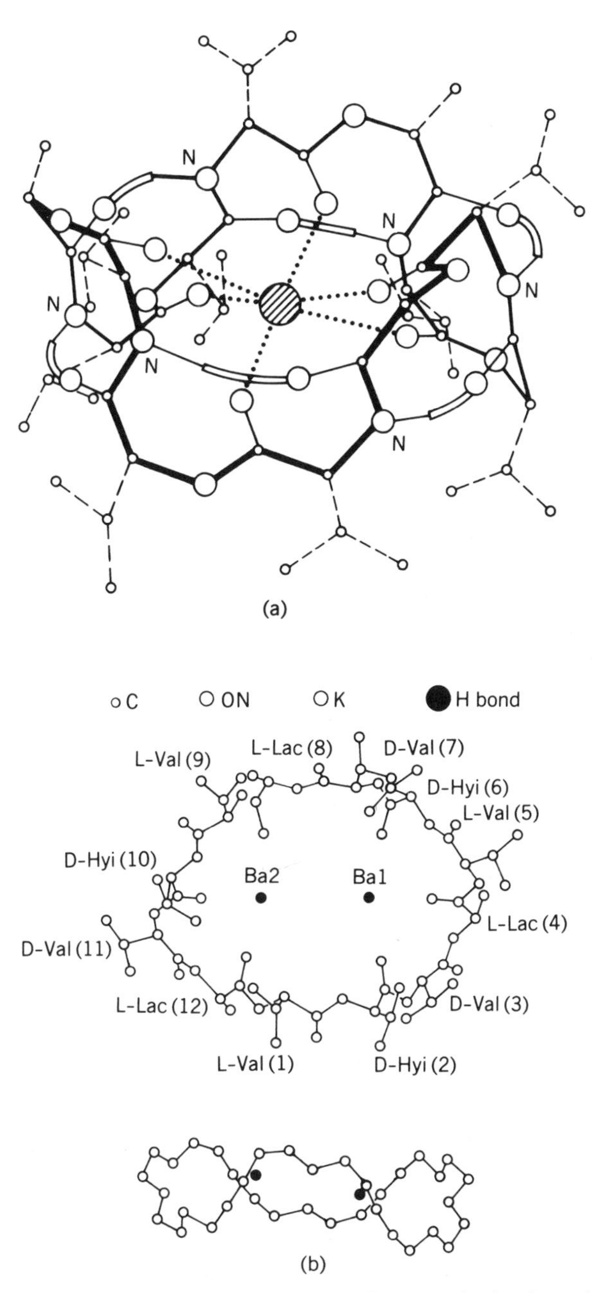

Fig. 9-4. (a) Structure of valinomycin · K^+ complex. The complex is a bracelet 4 Å high and 8 Å in diameter. The backbone loops up and down the plane of the bracelet, stabilized by intramolecular hydrogen bonding, so that the three ester carbonyls focus centrally on the top and three others centrally on the bottom to define a sphere of radius 1.4 Å. This can accommodate a K ion (1.33 Å) or Rb ion (1.45 Å) or Cs ion (1.69 Å) with a little stretching. The backbone is too rigid to permit a snug fit of small ions like Na (0.95 Å) or Li (0.60 Å). (From Shemyakin et al., 1969.) (b) Top and side views of valinomycin complexed with barium. (From Easwaran, 1987.)

TABLE 9-3 Single-Crystal X-Ray Structure Data for Some Ion–Macrocycle Complexes

Complex (Reference)	Backbone	Orientation of Polar Groups	Shape (diameter)	No. and Arrangement of Oxygens	Range of M-0 Distance
Enniaton-KI	Ring has 18 atoms made up of both peptide and ester links	Both peptide and ester groups are in trans form and in plane	Disc shaped; anions do not occupy definite position (15 Å × 6.5 Å)	Octahedral coordination by six oxygens arranged 1.5 Å above and below the mean molecular plane	2.6–2.8
Nonactin-KCNS	Ring has 32 atoms made up of ester and ether oxygen	All the oxygen functions are centrally directed	Approximately spherical; formed by foldings like seam of tennis ball; no fixed position for anion (d = 12 Å)	Eight oxygens (four from carbonyl and four from THF) in approximately cubic coordination	2.73–2.88
Valinomycin-$KAuCl_4$	Ring has 36 atoms made up of peptide and ester linkages	Ring is internally hydrogen bonded; the nitrogen of each of the six valine residues linked to the ester carbonyl oxygen three residues removed (O · · · H 2.8 to 3.0 Å)	Approximately spherical; anion occupies definite position in the complex (15 Å × 12 Å)	Six carbonyl of valine valine residues in octahedral coordination	2.7–2.8
Nigericin-Ag[a]	A linear chain which can cyclize by a pair of H bonds to form a 17-membered ring	Three THF rings[b] are approximately planar and their oxygens are centrally directed	Disc shaped (irregular) (9–15 Å)	Five oxygens coordinate; carboxyl group participates	
Monensin-Ag	A linear chain which can cyclize by a pair of H bonds to form 17-membered ring; H bond distances 2.5–2.9 Å	Three THF rings are approximately planar and their oxygens are centrally directed	Disc shaped (irregular) (9–15 Å)	Six oxygens coordinate; carbonyl group does not participate	2.4–2.6
$[X\text{-}537A]_2\text{-}Ba \cdot H_2O$	A linear chain	Centrally directed	Dimer (disc)	Nine oxygens	2.8–3.0
$X\text{-}206 \cdot Ag^+$	A linear chain	Centrally directed	Irregular, 11–15 Å		

[a] Isomorphous with corresponding Na complex.
[b] THF, Tetrahydrofuran.

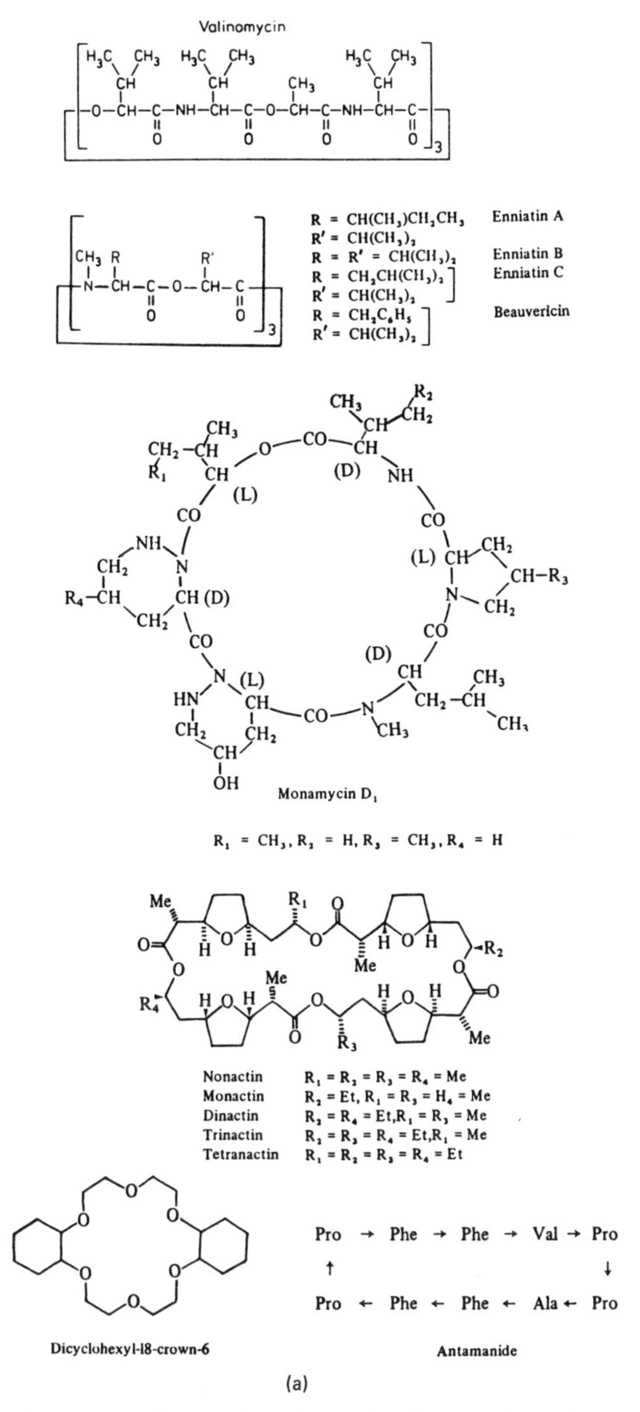

(a)

Fig. 9-5. Structures of carrier-type ionophores that have cyclic and open structures. Names of individual ionophores are given under each structure.

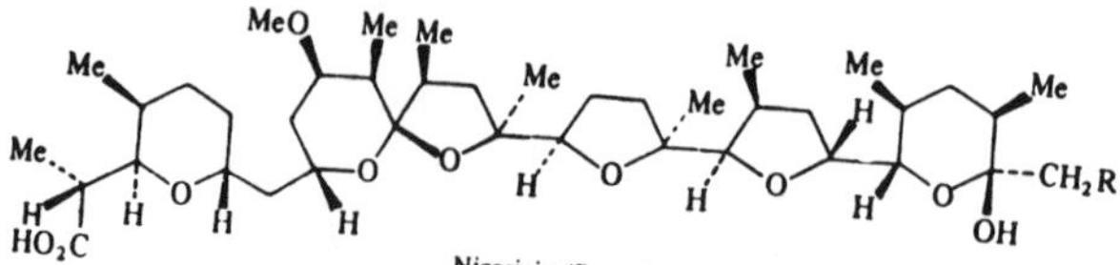

Nigericin (R = OH); Grisorixim (R = H)

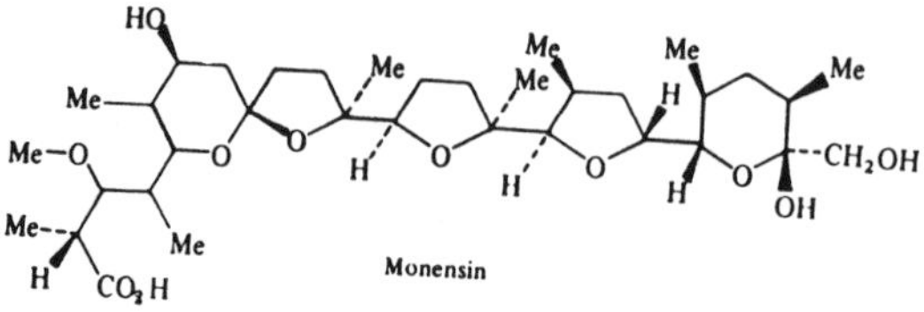

Monensin

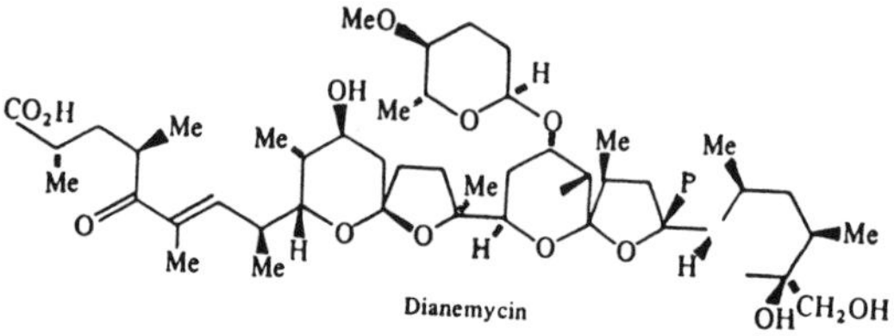

Dianemycin

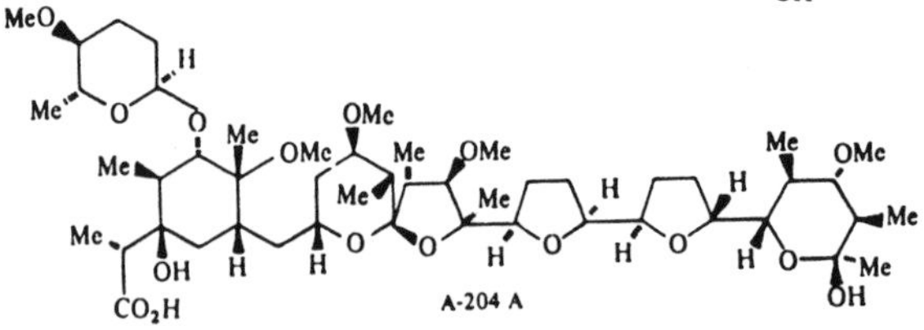

A-204 A

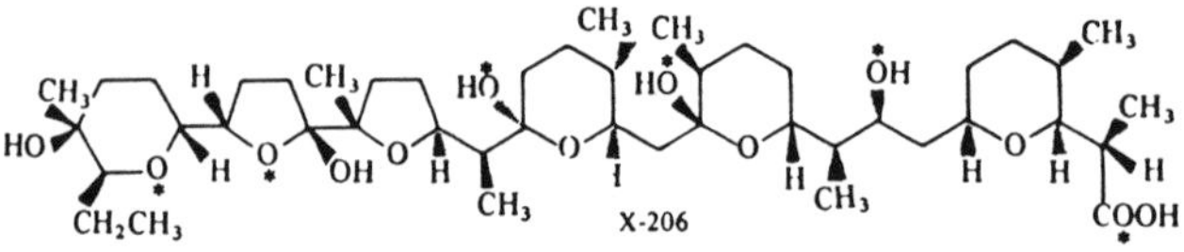

X-206

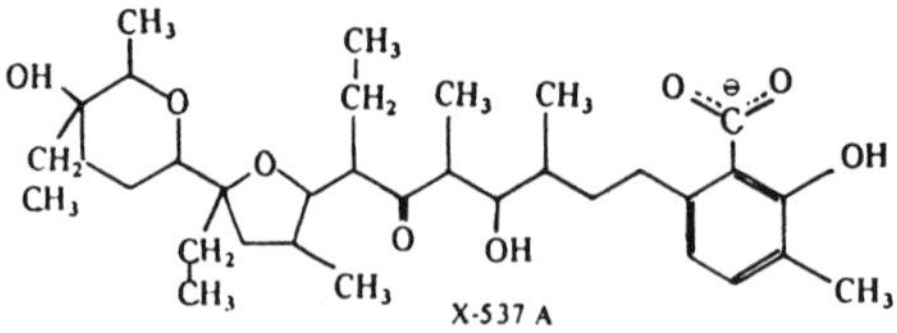

X-537 A

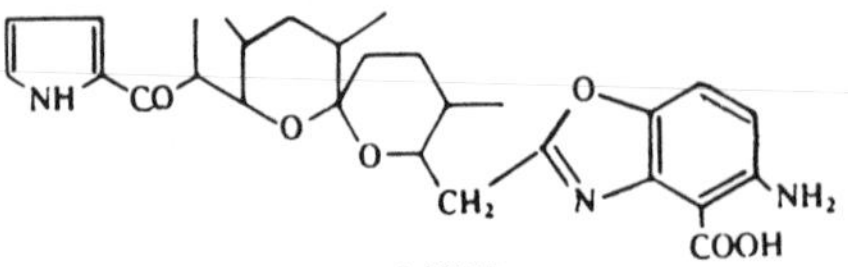

A 23187 (Calcimycin)

(b)

Fig. 9-5. (*Continued*)

in this cavity. The polar groups orient toward the center of the complex, while the exterior contains nonpolar groups. The lipid solubility of the complex is partially due to effective shielding of the polar interior and also due to compatibility of the hydrophobic exterior with solvents of low dielectric constant.

The conformation of valinomycin and its ability to bind ions is intricately related (Easwaran, 1985, 1986). Conformation of valinomycin depends both on the nature of the solvent and to a lesser extent on the nature of the cation. In nonpolar solvents valinomycin exhibits a closed conformation similar to that of the K complex, whereas in polar solvents an open conformation with a smaller number of hydrogen bonds predominates (Fig. 9-4). Transition between these conformations takes place on the time scale of 10–100 nsec. The group IA alkali metal cations form a 1 : 1 complex with valinomycin. In the Li and Na complexes the cation is located on the periphery of the molecules held by the D-Val ester carbonyls. In complexes with K, Rb, and Cs the bracelet conformation predominates with the ion located in the middle. This bracelet conformation is suitable for translocation across the bilayer, however valinomycin also forms several complexes that cannot be translocated across the bilayer. The 1 : 1 and 1 : 2 (valinomycin : cation) complexes with divalent cations are also formed. Whereas Ba and K ions have the same diameters, their complexes with valinomycin are very different (Fig. 9-4). The conformation of valinomycin · Ba_2 complex consists of an extended depsipeptide chain in the form of an ellipse in which two cations are localized at the two foci. In this relatively flat open conformation each barium ion is liganded to three consecutive amide carbonyls, and the two Ba ions are separated by 4.57 Å. The remaining coordination is provided by the solvent molecules. Valinomycin forms a peptide/cation/peptide type of sandwich complex with Ca and Mg ions. In these complexes cations are liganded to the three ester carbonyls of the two valinomycin molecules on the D-Val-L-Lac side of the bracelet structure.

The difference in the conformation of K and Ba complexes of valinomycin is apparently due to steric hindrance of the valyl side-chains. For example, an analog of valinomycin with less bulky side-chains, cyclo(Val · Gly · Gly · $Pro)_3$, has an identical bracelet-like conformation with K as well as with Ba ions and forms a 1 : 1 complex with both of these cations.

Translocation across Bilayers

The ability of ionophores to solubilize ions in a hydrophobic medium has been demonstrated extensively. Thus, for example, ionophores like valinomycin solubilize the $KMnO_4$ anion in organic solvents where it can be used for oxidation. Similarly cations like K can be translocated between two aqueous phases separated by a layer of an immiscible organic solvent in the presence of large polarizable anions like permanganate, picrate, and thiocyanate. The bilayer water partition coefficient of valinomycin is about 50,000; however translocation of ions across closed structures like vesicles is possible only when the counterions can leak through or translocate with the complex, or if protons are allowed to exchange in the opposite direction, as in the presence of proton carriers or uncouplers. With ionophores containing a protonated group, like monensin, an exchange of protons with cations is observed in the absence of an applied electric field.

Probably the most elegant demonstration of the ability of ionophores to mediate translocation of ions across bilayers has been obtained from the study of the electrical properties of bilayer lipid membrane (BLM) (Ciani et al., 1969; Eisenman et al., 1969; Szabo et al., 1969; Hladky, 1979; Lauger et al., 1981). As shown in Fig. 9-1, by placing appropriate electrodes in the aqueous compartments across the BLM, appropriate electrical properties can be measured readily (Fig. 9-6). The electrochemical driving force for the flow of ions across the BLM is the sum of the concentration gradient of the permeant ion and the applied electric field.

The flow of ions driven by the concentration gradient of a permeable ion is counteracted by the Nernst equilibrium *diffusion potential*:

$$V_m = \frac{RT}{nF} \ln \frac{\sum_{i=1}^{n} P_i a_i'}{\sum_{i=1}^{n} P_i a_i''} = \frac{0.059}{n} \log \frac{a_i'}{a_i''}$$

In this equation a_i' and a_i'' are the activities of the ith ion in the two compartments, and P_i is its permeability. When only one ionic species is permeable across the membrane, the expression for diffusion potential is simplified. For an ion of valence n and the concentration C_1 and C_2 in the two compartments (at 25°C, V_m in mV):

$$V_m = \frac{59}{n} \log \frac{C_1}{C_2}$$

Thus a 10-fold gradient of the permeable monovalent cations across a bilayer would create a potential of 59 mV when there is no flux of ions (in an open circuit where there is no flow of current). The polarity of this potential would be negative in the compartment which has a higher concentration of the permeant cation. Conceptually, the origin of diffusion potential is rather simple. It arises from the charge separation between ions induced by differ-

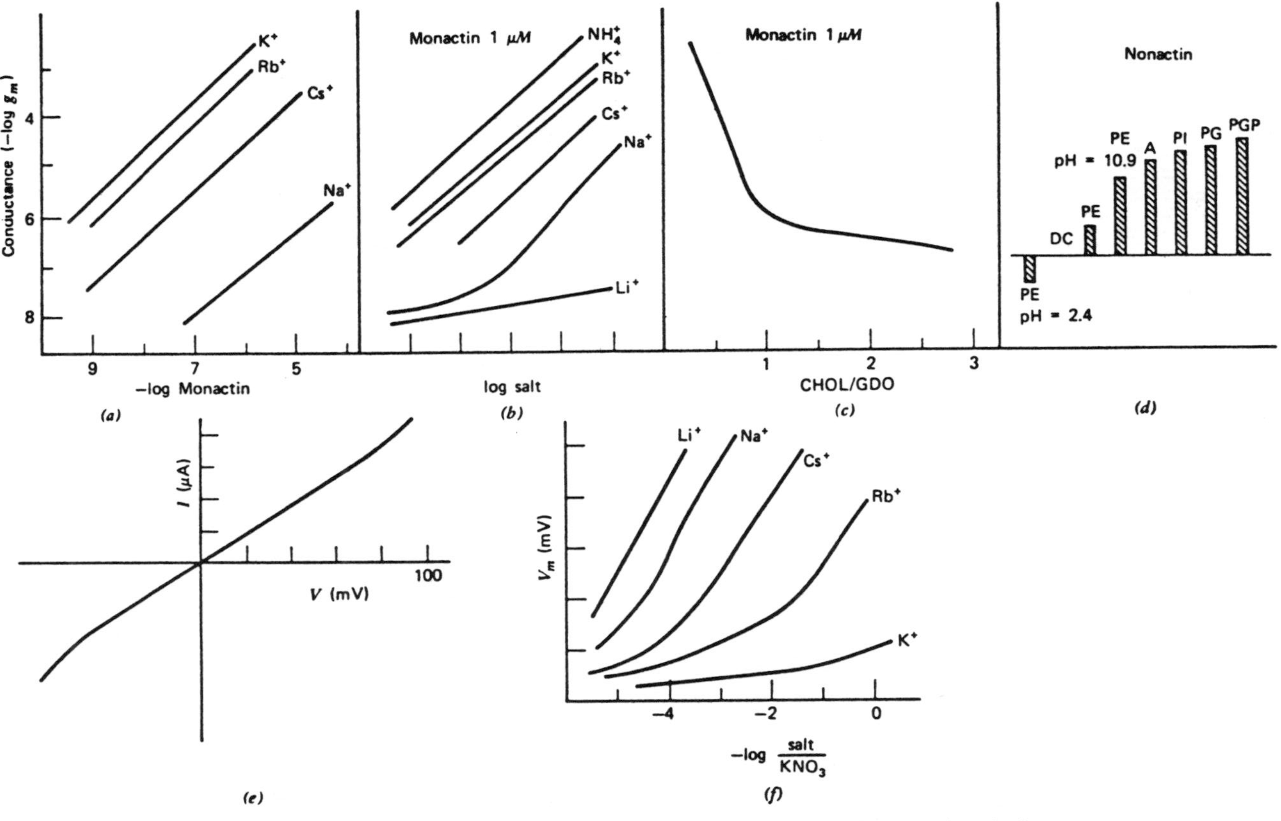

Fig. 9-6. Effect of changing the environment or the bilayer on the steady-state electrical properties of BLM. The change in the conductance is plotted against (a) the monensin concentration in the presence of several alkali metal halides, (b) salt concentration, (c) increasing cholesterol content, (d) changing phospholipid composition. *I-V* curve (e) and the diffusion potential versus cation gradient (f) also suggest that monensin translocates cations as a 1 : 1 complex, and the selectivity sequence is ammonium, K, Rb, Cs, Na, and Li.

ential permeability across the bilayer. Thus, the flow of cations toward the compartment with a lower concentration of cations would create a charge separation, until an opposite drag created by the potential compensates the flow of cations. Under these conditions an equilibrium is established and there is no net flux of ions, i.e., the current loop is not completed and only a diffusion potential develops. In the presence of ionophores like valinomycin, BLM behaves as if its permeability for K is very high compared with that for chloride. Obviously, the magnitude of V_m is considerably lower in the presence of anions that are permeable. Similarly, selectivity for translocation of two different cations can also be compared by placing these ions on the opposite sides of the bilayer. From such studies it can be shown that the K/Na selectivity for translocation by valinomycin under equilibrium conditions is >1000.

Yet another measure of ion translocation across bilayers is obtained by determining the transport of ions driven by applied electrical field when the current loop is completed (short circuit). The current flow (I) in response to an applied potential (V) is related to the proportionality constant, g, the membrane conductance:

$$I = g\ V$$

If the two media surrounding a membrane are in electrical contact (short circuit) through a pair of reversible electrodes (Fig. 9-1), the passage of cations across the membrane toward the cathode (the negatively charged electrode) would be driven by the applied transmembrane potential. Typically, the conductance of the membrane is considerably lower than the conductance of the aqueous salt solutions; the potential drop across the membrane is essentially the same as the potential applied from the electrodes. Passage of 1 ampere current is equivalent to the transfer of 1 Faraday of charges per second, i.e., equal to 1/95,500 mole of electrons per second. One mole of electrons is 6.23×10^{23} electrons. From the current–voltage characteristics (I vs. V plots), one can measure membrane conductance g as the slope whose units are S (Sieman) or mho. The reciprocal of g is often expressed as membrane resistance whose unit is ohm. Membrane *permeability* for ions can be monitored continuously with great accuracy by measuring transmembrane current in response to an applied potential. With the technology available now, currents of about 0.1 picoampere can be monitored with significant precision, which corresponds to the flow of about 10,000 ions per second.

Translocation of ions across bilayers by carriers like valinomycin can be described quantitatively under most experimental conditions by the scheme

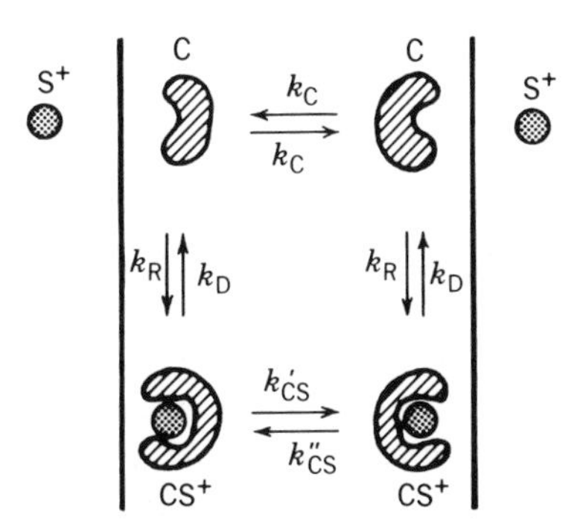

Fig. 9-7. The minimum kinetic scheme for the carrier (C)-mediated translocation of a solute (S) across a bilayer. In this model complexation of the carrier occurs in the bilayer. Models invoking complexation in the aqueous phase have been rejected because of their inconsistency with the experimental data.

shown in Fig. 9-7. As predicted by this scheme the steady-state conductance of BLM increases linearly when ionophores are added to the aqueous medium or to the lipid solution used to prepare BLM (Fig. 9-6). The increase in conductance does not depend upon the side to which the ionophore is added, apparently an equilibrium distribution of ionophore between BLM and the aqueous phase is rapidly established. For ionophores that translocate ions by a carrier mechanism, the increase in conductance shows a lower-order (generally first) dependence on the ionophore concentration, which implies that the permeating species has stoichiometry of ionophore as one. This is also consistent with the linearity (slope = 1) of log fluorescence versus log conductance plot for dansyl valinomycin. Similarly, the conductance increases linearly with the salt concentration (as predicted by the partial reaction C + S to CS), and reaches a maximum value, presumably because all the carrier in the bilayer is in the CS form. The relative conductance at equal concentrations of two different salts is a measure of relative permeability. Thus, by measuring conductance of BLM in the presence of KCl and NaCl, the K/Na selectivity for translocation by valinomycin has been found to be over 10,000. The I versus V curve for most ionophores is linear, however a departure from linearity indicates that one or more steps in the permeation process are voltage-dependent or some other type of polarization occurs.

Ion transport mediated by ionophores also depends upon the nature of phospholipids in the bilayer. These effects are of three types. The surface charge of the bilayer acts as a screen for the free and complexed ions. Thus, the factors that influence the state of ionization of the interface or of the ionophore (pH, salt concentration, ionic additives) would tend to influence membrane conductance. The partition coefficient of the ionophore and its mobility is also significantly affected by the disorder in the chain composition or the presence of cholesterol in the bilayer (cf. Chapter 5). Anomalous conductance has also been observed in the gel–fluid transition region.

The kinetic mechanism of carrier-mediated ion transport has been elaborated by relaxation methods in which the transmembrane current is monitored in response to a voltage jump, or by the charge-pulse method in which the decay of membrane potential (attained by a 1–50 nsec current pulse) is monitored (Lauger et al., 1981). The time resolution of voltage jump is about 1 μsec and it is limited by the charging time of the membrane capacitance. In the charge pulse method the relaxation of the voltage can be monitored with a resolution of about 40 nsec. The time dependence of the relaxation curves can be described by the sum of 2 or 3 exponential functions. From their amplitudes and the time constants it is possible to calculate the rate constants for the four steps in Fig. 9-7. The rate constants for the translocation of Rb by valinomycin in BLM of monoolein + *n*-decane (25°C, 1 *M* RbCl) are summarized in Table 9-3. The rate constants for loading the carrier (k_R 300,000 per mole per sec) are the same as for unloading (k_D 300,000 per sec) or for translocating (k_{CS} = 200,000 per sec) the complexed ionophore. The rate-limiting step is the transport of uncomplexed ionophore, k_C = 40,000/sec. The time required for translocation of CS is $1/k_{CS}$ = 5 μsec, compared with the time for diffusion of CS complex across 5 nm of the aqueous phase, which would be about 40 nsec. The maximum turnover rate (f) or the number of ions transported per unit time (= n/t) can also be calculated from these rate constants as:

$$f = \left[\frac{1}{k_{cs}} + \frac{1}{k_c} + \frac{2}{k_d}\right]^{-1} = n/t$$

Thus, a single valinomycin molecule is able to translocate ($f = n/t$ or the turnover number) more than 20,000 Rb ions per second, even though the binding or the stability constant for the complex in the interface k_R/k_D is only 1.5 per mole.

Although most of the experimental data are consistent with the scheme shown in Fig. 9-7, it is not clear why the association rate constant appears to depend upon the concentration of Rb ions (see Hladky, 1979, for a discussion). It may also be noted that the turnover number for most analogs of valinomycin as well as those for other ionophores are considerably smaller than those for valinomycin. It is possible that the symmetrical bracelet conformation of the K · valinomycin complex orients it such that in the bilayer the two polar ends of the bracelet open alternately at the two interfaces. Such a complex need not rotate (or do a half-somersault) in the middle of the bilayer as would be required for the translocation by less symmetrical ionophores. Indeed, the rate of turnover by valinomycin is faster by several orders of magnitude compared with that for less symmetrical ionophores.

The effect of lipid composition on the rate constants is complex (Lauger et al., 1981; Lauger, 1985). The effect of the acyl chain length on the rate constants is summarized in Table 9-4. The kinetic constant that is influenced most by the acyl chain composition is k_{CS}, which changes by a factor of 10 when the acyl chain length is increased from C_{16} to C_{22}. There is a surprisingly large effect of the acyl chain length on k_R. The origin of this effect is not known. An effect of the change in the viscosity or an equilibrium between different conformers or states of C has been implicated. The effect of increasing number of double bonds is predominantly on k_{CS}, whereas the effect of increasing the mole fraction of cholesterol is on k_R, k_{CS}, and k_S. These effects suggest that the overall effect of the change in the order in the hydrophobic region cannot be interpreted only in terms of the dielectric constant, or the microviscosity, or the thickness of the bilayer.

From the discussion in this section it is clear that the criteria used for favoring a carrier mechanism include the following: small size of the ionophore and low stoichiometry for ion binding; a lower-order (generally one) dependence of conductance change on ionophore concentration; ion selectivity that does not change significantly with the nature of the membrane; ion translocation observed in bulk organic solvents; a large difference in conductance related to the the dipolar state of the bilayer interface; the conductance and the permeability ratios are comparable and independent of salt concentration over a wide range.

TABLE 9-4 Influence of the Chain Length and the Number of Double Bonds of Lipids on the Rate Constants for the Valinomycin-Mediated Rb^+ Transport

Lipid	k_R [10^4 M^{-1} s^{-1}]	k_D [10^4 s^{-1}]	k_{CS} [10^4 s^{-1}]	k_C [10^4 s^{-1}]
Variation of the chain length				
(16:1)-MG	43	13	74	8.5
(18:1)-MG	37	24	27	3.5
(20:1)-MG	23	12	10	1.8
(22:1)-MG	24	13	7	1.1
Variation of the number of double bonds				
(20:1)-MG	23	12	10	1.8
(20:2)-MG	34	9	39	3.2
(20:3)-MG	34	5	136	9.4
(20:4)-MG	42	3	240	12

The membranes were formed from α-monoglycerides (MG) containing the following fatty acid residues: palmitoleoyl (16:1), oleoyl (18:1), 11-eicosenoyl (20:1), erucoyl (22:1), 11,14-eicosadienoyl (20:2), 11,14,17-eicosatrienoyl (20:3), and arachidonoyl (20:4). The membranes were spread from n-decane. $T = 25°C$. (From Lauger, 1985.)

CARRIERS FOR PROTONS

The magnitude of the net flux of protons across bilayers is still being debated. Reported values of permeability coefficients range from 4×10^{-9} (Gutknect and Walter, 1981; Cafiso and Hubbell, 1981) to about 10^{-4} cm/sec (Deamer and Nichols, 1985). Higher values are in part due to the presence of anionic lipids in the bilayer; however, the actual values may be near the higher end of the range mentioned above. However it may be emphasized that normally the concentration of protons is very low (10 nM at pH 8), therefore the actual charge translocation by protons would be very small.

Weak acids that dissolve in bilayers can translocate protons:

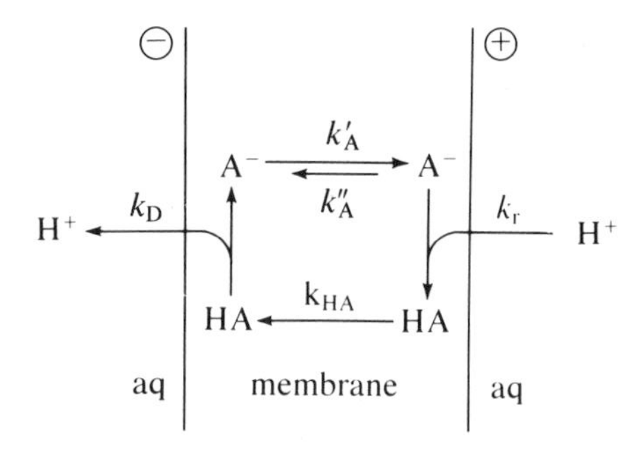

Some of the commonly used proton carriers include carbonylcyanide phenyl hydrazones and nitro-, halo-, and oxygenated phenols (Fig. 9-8). Two important factors that regulate the ability of a proton carrier are its lipid solubility and its dissociation constant K_a, and $pK_a = -\log K_a$. Optimal proton transport is achieved when pK_a is the same as the pH of the medium. This is because on either side of this value either the carrier (A^-) or the permeant (H^+) concentration is lower and therefore suboptimal.

The criteria for establishing the proton-carrying ability of a carrier are essentially the same as for ionophores. For example, in the presence of a proton carrier the equilibrium diffusion potential is generated in a pH gradient: transmembrane pH difference of one unit would give rise to 59 mV transmembrane potential, negative on the side with lower pH. The log g versus log (carrier) plots usually show a slope of 1 in most cases. The slope of 2 for bezimidazoles is ascribed to $(HA)_2$ as the charge-transporting species in BLM containing decane, and AH (slope = 1) is the proton-translocating species in BLM containing chlorodecane (Dilger and McLaughlin, 1979). These effects have been ascribed to a difference in the dielectric constant of

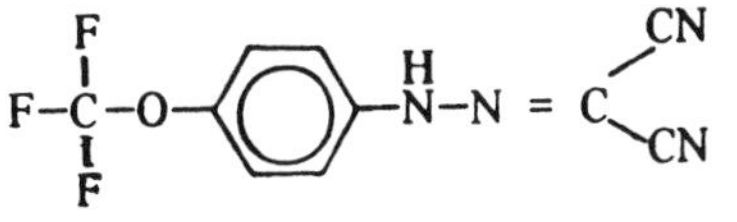

Carbonylcyanide *p*-trifluoromethoxyphenylhydrazone (FCCP)

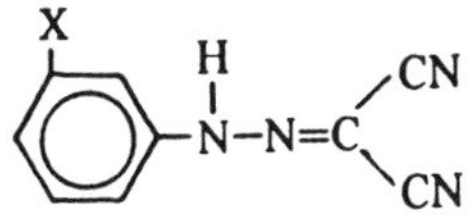

Carbonylcyanide *m*-chlorophenylhydrazone (CCCP) X = Cl
Carbonylcyanide phenylhydrazone (CCP) X = H

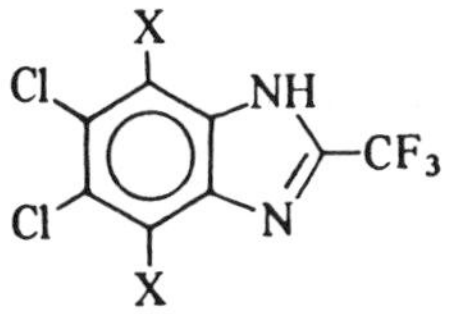

Tetrachloro-2-trifluomethyl benzimidazole
(TTFB, X = Cl)
5, 6-dichloro-2-trifluoromethyl benzimidazole
(DTFB, X = H)

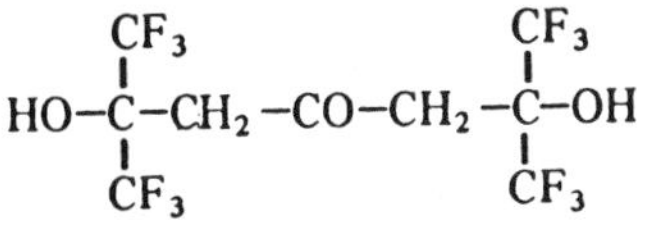

Uncoupler 1799

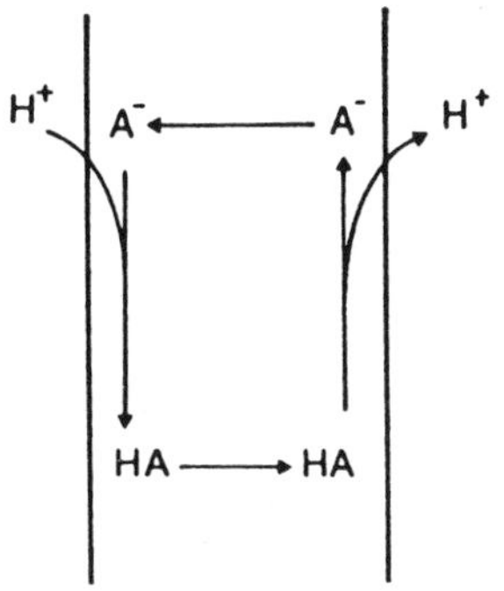

Fig. 9.8. Structures of some proton carriers (uncouplers). The pK_a for these weak acids are about 7, which makes them effective proton carriers under physiological conditions. The minimum kinetic model for translocation is also shown.

BLM. This monomer–dimer equilibrium apparently depends upon the dielectric of the bilayer. The log g versus pH curves show a maximum at the interfacial pK_a of the carrier. At any given pH, membrane conductance changes linearly with log A^-, and the conductance characteristics are not affected by other ions. The membrane current in response to an applied potential shows a time-dependent decrease, and the magnitude of the overall change decreases with increasing buffer capacity. Like ionophores that form a charged complex, the proton carriers do not transport ions across closed systems like liposomes unless the system is short-circuited by leakage or by parallel conductance pathways across the bilayer.

The disorder in the organization of acyl chains (usually considered in terms of the dielectric of BLM) has little effect on the function of proton carriers. However, the pH of the maximum conductance does shift with the nature of the polar head group. This is probably because the interfacial pK_a changes with the surface potential (Chapter 4). For example, the apparent pK_a in an anionic interface is usually greater than in a zwitterionic interface. The charged interface also influences the distribution of the conjugate base (A^-) in the bilayer. Thus, the concentration of A^- in a positively charged bilayer would be higher; however, this interface would shield protons (Hopfer et al., 1970). Relaxation experiments with CCCP in BLM of diphytanoyl PC + chlorodecane are also consistent with the conclusion that the back-diffusion of the conjugate base A^- is the rate-limiting step: $k_A = 175$ per sec; $k_{AH} = 12{,}000$ per sec (Kasianowicz et al., 1984). Similar values have been obtained for FCCP (Benz and McLaughlin, 1983). The interfacial rate constants k_D and k_R are essentially diffusion limited $>10^{10}$/mole per sec. The kinetic constants exhibit a strong dependence on the lipid composition. These results have been interpreted in terms of the dipole potential at the interface.

The protonophore function of weak acids in bilayers is in accord with their ability to uncouple oxidative phosphorylation from electron transport. As discussed in Chapter 13, the energy from the electron-transport chain is first conserved in the form of a proton gradient across the inner membrane of mitochondria, and dissipation of the proton gradient would stop oxidative phosphorylation without inhibiting the electron-transport chain.

LIPOPHILIC IONS

Several hydrophobic ions (Fig. 9-9) transport counterions across bilayers. Except for ion selectivity their behavior is essentially similar to that of uncouplers; i.e., they increase conductance, develop transmembrane diffusion potential, and dissolve counterions in nonpolar solvents. Relaxation experiments show that the current decay occurs by a single rate constant (Hladky, 1979; Lauger et al., 1981). Such experiments suggest that the exchange of ions like tetraphenylboron and dipicrylamine between membrane and bulk aqueous solution is slow in neutral bilayers. In anionic bilayers, the desorption rate constant and the translocation rate constants are about 50/sec. Adsorption of hydrophobic ions such as tetraphenylboron to BLM satu-

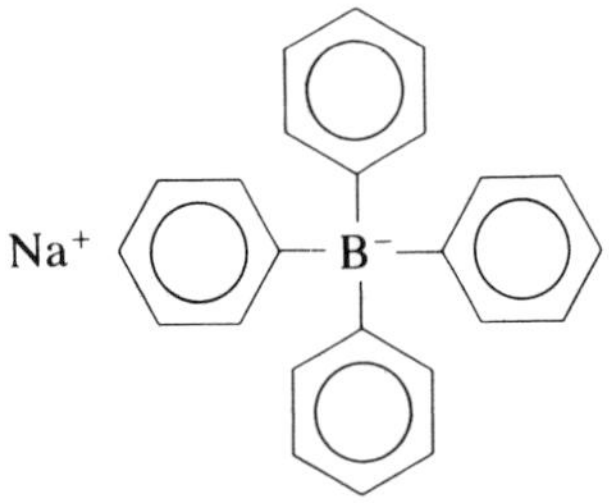

Sodium tetraphenylboron

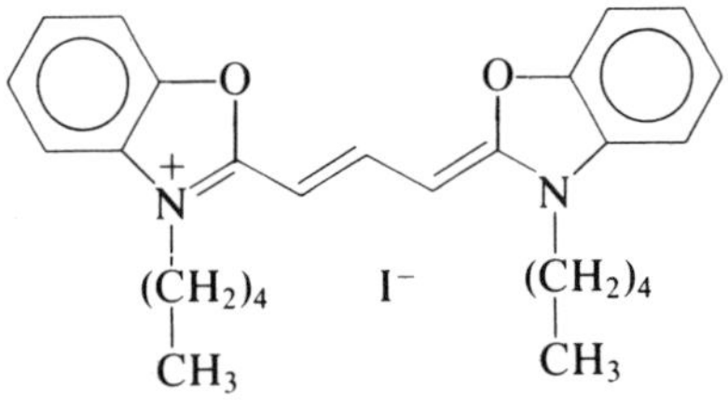

3,3′-Dipropyloxacarbocyanine iodide

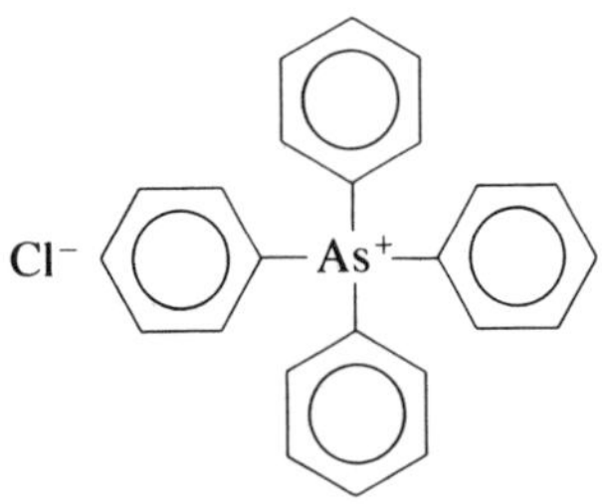

Tetraphenylarsonium chloride

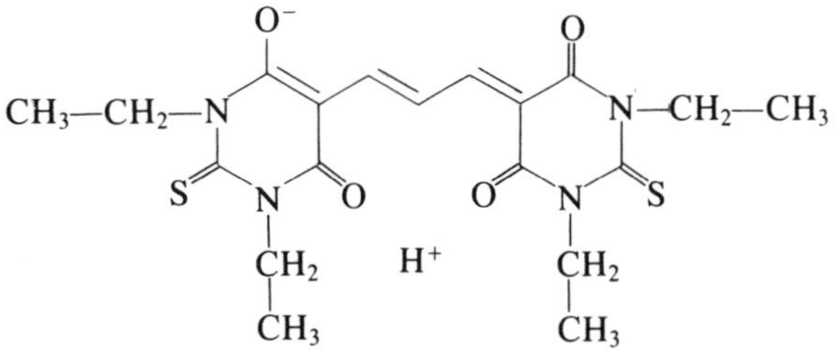

Bis(1,3-diethyl-thiobarbiturate)-trimethineoxonol

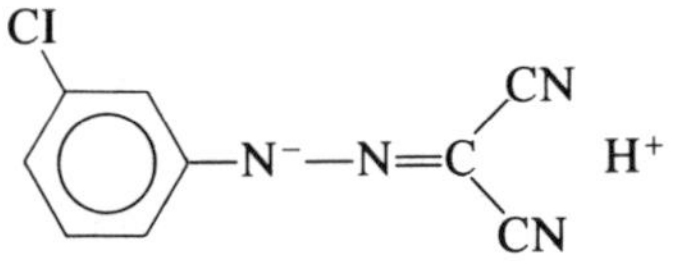

Carbonylcyanide-*m*-chlorophenylhydradrazone

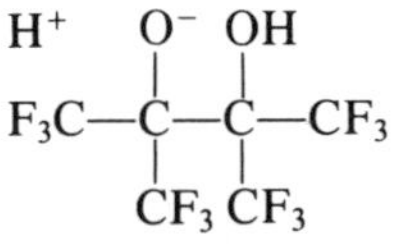

Perfluoropinacol

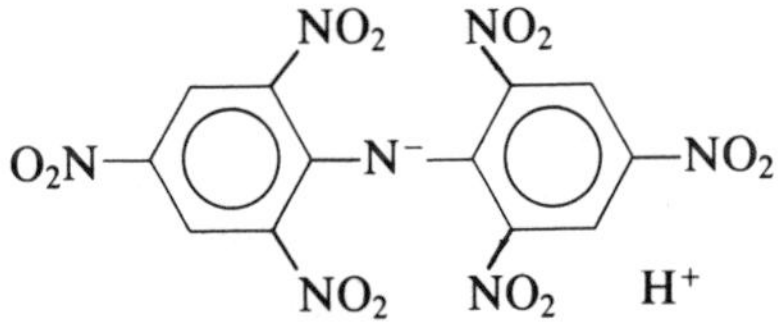

$K^{+}I^{-}_{5-9}$

Polyiodide

Dipicrylamine

Fig. 9-9. Structures of lipophilic (lipid-soluble) ions that translocate counterions across bilayer.

rates at 0.1 ion/nm^2. This is probably due to lateral electrostatic repulsion between the ions that are partially embedded in the bilayer (Tsien and Hladky, 1982).

Certain lipophilic ions have been used to monitor membrane potentials based on the following principle. Lipophilic cations can diffuse through bilayers. In the absence of lipophilic anions the cations distribute at the two interfaces (and compartments) in the ratio of surface potentials, i.e., the equilibrium potential of the ion should equal the membrane potential. At equilibrium, the electrochemical potential of the ion would be zero. The membrane potential can thus be calculated from the ratio of the concentration (C_i/C_o) of the permeant ion at equilibrium. A cationic dye like sefranin or oxonol is taken up by vesicles with inside negative potential. The change in the spectral properties of the dye is thus used to monitor membrane potentials where there is no space to insert electrodes (Bashford and Smith, 1979; however see Ritchie, 1984, and Bakker et al., 1986 for many limitations of this technique).

ION TRANSPORT THROUGH CHANNELS

Transport of solutes through channels has many of the same macroscopic characteristics that are observed with carriers. For transport of ions, both channels and carriers increase conductance, and give rise to a diffusion potential; transport is driven by an electrochemical gradient, and both exhibit an exponential relaxation when the driving force is perturbed.

Certain distinguishing features of channels may be noted. Turnover number or the number of ions passing through a channel is typically 10 to 100 million/sec compared to 10 to 50,000/sec for most carriers, and 10^{-4}/sec for a nm^2 patch of a bilayer. The activation energy for transport of ions through channels is about 3 to 7 kcals/mole, whereas values for carriers are usually between 12 to 30 kcals/mole. In the transport mediated through channels, the permeability coefficient is inversely correlated with the size of the solute rather than the lipid–water partition coefficient (for diffusion) or some unique structural feature that is necessary for binding during the transport by carriers. Thus, structurally very different solutes can pass through channels. This gives rise to certain phenomena in narrow channels. For example, at some point the pore may be so narrow that an ion and water must move in single file. Therefore, if an activity gradient of water were applied across the bilayer (for example, an osmotic gradient of impermeable solute) then the ion in the channel would be dragged against its own electrochemical gradient by the flow of water through the channel. Thus, a streaming potential develops, which is equal to the voltage that must be applied to reduce the ionic current to zero. If it is assumed that the channel consists of a water-filled region in series with a narrow single-filing region for ions, then the streaming potential (V), molality of water (M), and the molal gradients (ΔM) are related to the number of water molecules (N) constrained to move in single file with the ion as $VM/RT\Delta M$.

In very large channels the activation energy for transport is the same as the activation energy for transfer through the bulk aqueous phase. Assuming that such channels of large pore size are filled with a solution of the same specific conductivity (σ) as the bulk aqueous phase, and assuming a pore length (l), it is possible to calculate the diameter (d) of the channel from the single-channel conductance $\gamma = \sigma \pi d^2/1$. For most channels for which γ has been measured directly, a significant departure is observed from this approximation. Obviously, channels are not right circular cylinders. Unfortunately, the detailed architecture of none of the naturally occurring channels is known. However, it may be appreciated that the electrical interaction between the detailed architecture of a pore can affect the passage of ions in two ways. Long-range electrostatic forces control access of ions to the channel and modulate the gating properties. Short-range forces govern selectivity and the shape of the potential energy profile for ion permeation. For elaboration of these themes see Jordon (1984).

Probably the most distinguishing characteristic of the transmembrane channels arises from the stochastic nature of the underlying molecular process, i.e., the single-channel conductance. This is reflected at two levels. If only a few channels are present in the field of observation (as in a patch of bilayer or biomembrane), opening and closing of individual channels can be monitored (Hladky and Haydon, 1984; Sakman and Neher, 1984). From such records (Fig. 9-10), one can obtain conductance of a single open channel, the number of conducting states, sometimes also the number of the nonconducting states, average life-time of the conducting states, τ, and the voltage dependence of the population of the conducting states. When a large number of channels are present, the number of open channels varies incessantly around an average value and produces random fluctuations in membrane conductance; fluctuation analysis of this data also provides information about γ and τ (Eisenberg et al., 1984). The statistical characteristics of these conductance fluctuations reflect the dynamics of underlying molecular processes. For example, a class of channels may produce small and rapid fluctuations, whereas others may yield slow noise. Through such studies it is possible not only to obtain information complementary to that obtained by average response to a perturbation or relaxation, but in many cases through such studies it is also possible to obtain information not available otherwise. For example, different coexisting molecular mechanisms can exhibit identical average behavior, but fluctuations for each will occur with distinct characteristics. Stochastic models of channel activity have been constructed (Colquhoun and Hawkes, 1981) on the basis of certain assumptions: Channels exist in discrete "states" that have lifetimes much longer than the time needed to switch between states; the probability that a given transition will

Temperature Dependence of Gramicidin A Kinetics

	di-(18 : 1)-Lecithin/ *n*-decane		1 *M* Nacl	*V* = 135 mV		
	k_R ($cm^2 mole^{-1} s^{-1}$)	k_D (s^{-1})	$K = k_R/k_D$ ($cm^2 mole^{-1}$)	Λ (Ω^{-1})	τ (s)	$1/k_D$ (s)
at 25°C	20×10^{13}	1.6	12×10^{13}	1.2×10^{-11}	0.73	0.63
Activation energy (Kcal/mole)	20	17		23		

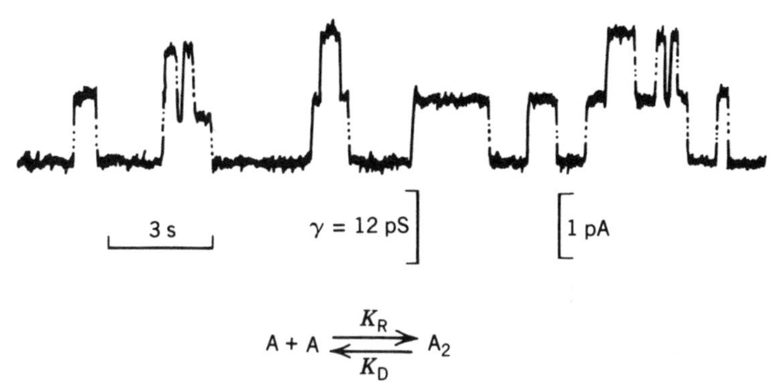

Fig. 9-10. Current steps due to fluctuation (open–close transitions) of two gramicidin A channels. Dioleoyl PC + *n*-decane bilayer in 1 *M* NaCl is polarized with 90 mV. γ is 12 pS (25°C) and it increases 1.4-fold for a temperature increase of 10°C. As discussed in the text the kinetics of gramicidin channel is best described by monomer (closed) dimer (open) equilibrium. The values of kinetic constants and the channel parameters are given in the table. Enthalpy for the association constant is 32 kcal/mole. (From Bamburg and Lauger, 1974.)

occur depends only on the channel's current state, and not on its past history; the rate of switching between two states is determined by a single rate constant, which does not change with time but may depend upon temperature and membrane potential. The transition probability (rate constant) can also depend upon the state of the channel and how long the channel remains in that state. Such a gating behavior with "memory" has been solved on the basis of a fractal description, and it has been successfully applied to model the behavior of channels (Liebowitz et al., 1987).

Channels can be blocked by protein reagents and by structurally related solutes that occlude the passage of ions. Both of these factors give rise to a relatively large decrease in permeability of each channel that is blocked. This characteristic of channels is also reflected in the single-channel and noise characteristics. Also the ability of channels to open and close can be gated or regulated by transmembrane potential, cofactors, and transmitter.

Certain anomalies in the conductance fluctuation behavior of channels has also been noted. For example, some channels do not always exhibit a stable equilibrium (''gearshift'' phenomenon) as shown for calcium-activated potassium channels (Moczydlowski and Lattore, 1983). Ca- or Na-channels exhibit ''rundown'' characteristics as the currents or activation parameters shift slowly (Fenwich et al., 1982). Similarly, the burst of the open state is seen with acetylcholine receptors. The physical basis of such phenomena is not known; however, these observations do emphasize the complexity of the underlying gating and kinetic processes.

CHANNELS OF GRAMICIDIN A

Transmembrane channels formed by dimers of gramicidin have been studied extensively, and are amongst the best-characterized ion channels. Structurally, these channels may not resemble channels in biological membrane, but they have proven to be excellent models for studying conductance, ion selectivity, water permeability, single-channel characteristics, and lipid–peptide and peptide–peptide interactions in bilayers as well as bulk media.

The biological role of gramicidin in *Bacillus brevis*, which produces it, is probably on the inhibition of RNA polymerase (catalyzes DNA transcription) during sporulation (Ivanov and Sychev, 1982). Solubility of gramicidin in water is less than 100 p*M*, however it readily dissolves in organic solvents and bilayers. At high mole ratios it appreciably perturbs the local organization of bilayers. The increase in the conductance of BLM induced by gramicidins is due to the formation of channels, and such an increase in conductance persists in frozen membranes, and the increase in conductance is not seen in thick membranes. Water and protons are also permeable through these channels. The streaming potential measurements suggest that six to nine water molecules move in single file with ions. Also the ratio of hydraulic to tracer flux is more than 5.

In the presence of very small amounts of gramicidin A, the conductance of BLM increases in discrete steps, and in the steady-state relatively significant fluctuations can be seen about the mean value (Hladky and Haydon, 1984). A typical single-channel conductance record of BLM in the presence of gramicidin is shown in Fig. 9-10. Fluctuations of this type are observed in patches of membranes in which the channels have either a low density or a low probability of opening. Discrete pulses of current originating at and returning to the zero current level are observed. The currents jump quickly to a steady level of flow, and stay at that level until they return to the zero current level. By measuring the difference in the current level between the baseline and the open channel level, one calculates the amount of current (I_s) that flows through a channel under the transmembrane potential V_m. The single-channel conductance, γ, can be calculated as the ratio I_s/V_m. Such observations suggest that these conductance fluctuations are due to opening and closing of single channels through which several million ions flow per second under the driving force of 100 mV in a 1 *M* salt solution. For example, the current above the first level is found to be an integral multiple of the basic unit, i.e., the conductance of a single open channel, γ. Histograms of single-channel conductance (Fig. 9-11a) show a range in which a particular conductance value predominates (Bradley et al., 1981). This implies that the

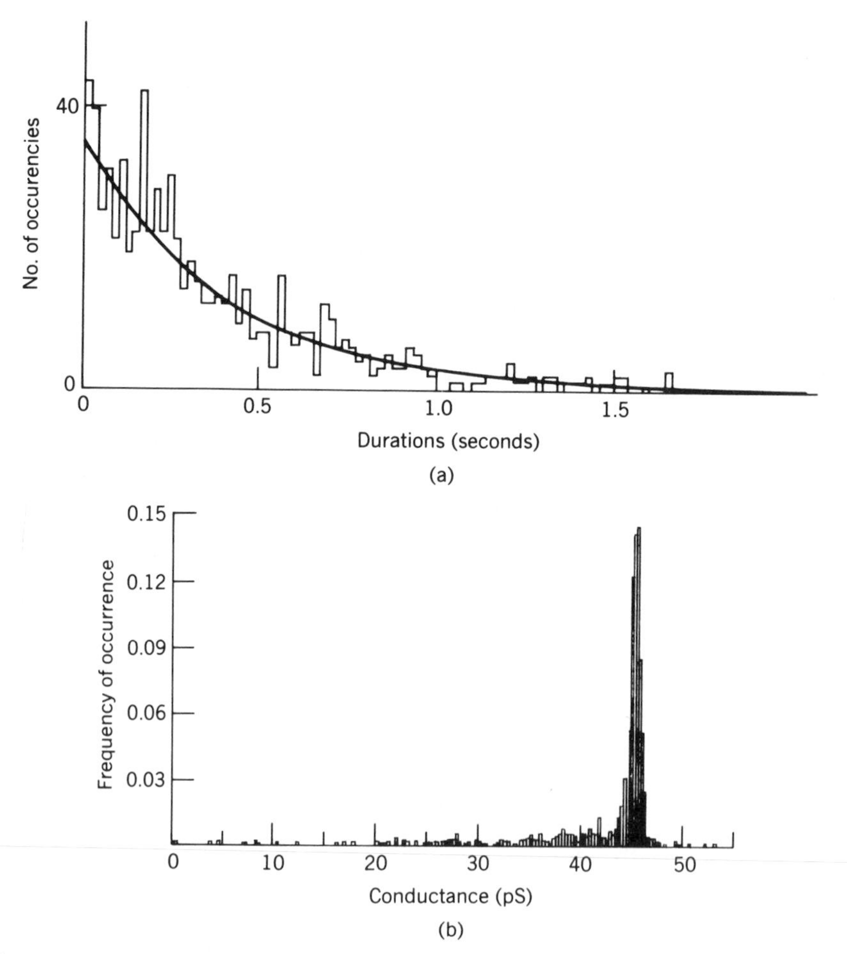

Fig. 9-11. A histogram of the frequency of occurrence of the step changes in conductance as a function of the open-time (a) and the conductance (b) of gramicidin channels. (From Prasad et al., 1982.)

channel can adopt a range of long-lived conformations. The opening of channels is due to association of two gramicidin molecules apposed in the two monolayer halves of the bilayer. As expected the lifetime of the channel is strongly dependent upon the composition and thickness of the bilayer, whereas the single-channel conductance is not.

The probability of opening and closing of channels is independent of time as well as of the number of other channels that are open or closed. Statistically, the probabilities of several channels opening simultaneously follow a Poisson distribution. The statistical distribution of the channel lifetimes, i.e., the duration they remain open, is not a single value but a range of values (Fig. 9-11a). The frequency of occurrence of channels with any given duration declines exponentially as the duration increases. This would be expected for a first-order process, such as dissociation of a dimer or a conformational change. The rate constant ($1/\tau$) for this process or the mean open time of the channel is related and obtained from such plots. Conductance of a single gramicidin channel is independent of membrane potential, i.e., the I-V curve for the open channel is linear. Similarly, the single-channel conductance does not depend upon the thickness of the bilayer, although the mean open-time for the channel decreases as the membrane becomes thicker (Table 9-5), i.e., the dimer that forms the channel dissociates less readily in thinner membranes.

Gramicidin channels are cation selective. Single-channel conductance shows a saturation effect with salt concentration (Fig. 9-12). Half of the maximum γ for different ions is obtained between 80 and 300 mM. This dependence of conductance on ion concentration suggests that only one ion is allowed in the channel at a time, and the conductance could tend to a limiting value corresponding to the 100% occupancy of the channel. At

TABLE 9-5 Single-Channel Conductance Data on Gramicidin A Derivatives

Ionophore	Lipid[a]	γ (PS)	τ (s)
Gramicidin A	DOPC	50	0.22
	GMO-HD	45	3
	Diphytanoyl P-decane	28	0.8
	PC + chol + decane	29	1.7
Malonyl bis-des-formyl-	PC + chol + decane	16	~200
O-Succinoyl des formyl-	GMA + decane	48	1.3
N-Succinoyl des formyl-	GMO + decane	19	0.22
O-pyromellityl-	GMO + hexadecane	82	2.2
N-pyromellityl-	GMO + hexadecane	No channel	—
N-acetyl-des formyl-	GMO + hexadecane	27	0.056

Adopted from Urry et al. (1985).

[a] DOPC, Dioleoyl phosphatidylcholine; GMO, glycerol monooleate; HD, hexadecane.

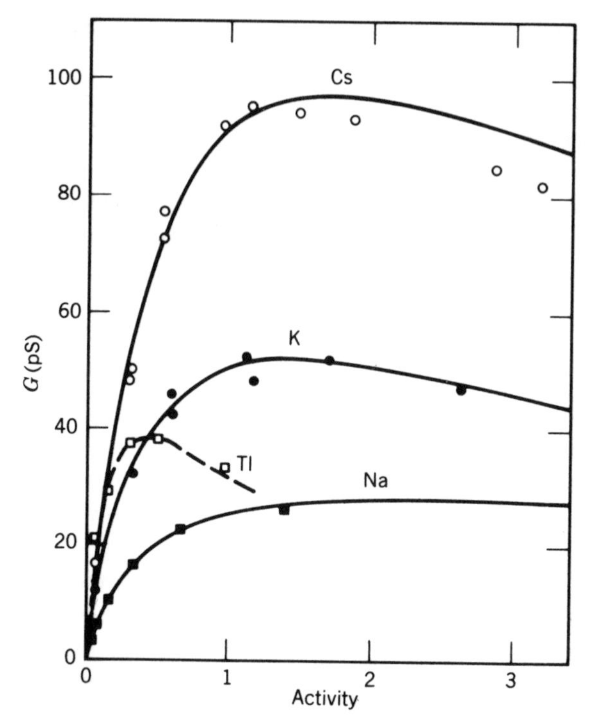

Fig. 9-12. The single-channel conductances at 50 mV versus molal activity of CsCl (from top, open circles), KCl, TlCl (open squares), and NaCl. Smooth curves are theoretical based on the model shown in Fig. 9-3. (From Hladky and Haydon, 1984.)

higher concentrations of Cs ions conductance decreases, which could be expected if a second ion could enter and block the channel (see Fig. 9-3b).

The values of τ and γ are strongly influenced by the presence of fixed charges, as well as by hydrophobic solutes like alkanols (decrease τ) and cholesterol (increase τ). In BLM of phosphatidylcholine and phosphatidic acid, the histogram for the single-channel conductance and channel lifetime for gramicidin exhibits phase separation in the presence of calcium ions (Knoll et al., 1986). At low ionic strength γ is higher in the anionic bilayer than in neutral membrane. At high ion concentration γ in different membranes approaches the same value. This is because in the electrical double layer, the surface concentration of ions reaches a limiting value even at high surface charge densities, therefore γ reaches a finite limiting value.

The statistical basis for assigning τ as the average open time of a channel is following. The mean value of the lifetime of an open channel is obtained by monitoring the fluctuations from a single channel over a long period of time. In the conductance records of type shown in Fig. 9-10, there is a statistical distribution of the channel lifetimes; i.e., the duration they remain open is not a single value but is range of values. According to the stochastic model, each time the channel opens, it is entering a distinct kinetic state. The amount of time it spends in the open state is different for each opening. The channel open time is the duration between the channel's entry into its open state, and the first instance of it leaving that state, i.e., closing. The average open time based on such a conditional probability can be obtained from a histogram of measured open times for many individual channels. As shown in Fig. 9-11a such a histogram is usually well fitted by a single exponential, and is consistent with the idea that this channel has a single open state. The time constant of this exponential is defined as mean open time of a channel. The distribution of lifetimes in the open state and τ depend upon temperature, membrane thickness, lipid composition, as well as the interfacial potential. Similar information can be obtained from the relaxation kinetics or fluctuation analysis of a large number of channels.

If a large number of channels are present in a bilayer, as shown in Fig. 9-13, significant fluctuations about the mean conductance are observed. Based on the preceding discussion, the mean conductance is related to the average number of channels that are open at any given instance. The rate constants for the formation and dissociation (or closing) of the channel can be directly obtained from the voltage-jump relaxation technique (Lauger et al., 1981). The rate of change of conductance to a change in an applied potential (relaxation), and the fluctuations about the mean value contain considerable information.

The kinetics of close to open (or monomer to dimer) equilibrium for the channel of gramicidin has been investigated by relaxation methods. The current relaxation from BLM containing gramicidin A consists of an initial current (related to the channels initially present) followed by an asymptotically increasing current corresponding to the voltage-induced shift in the equilibrium constant for channel formation. From these curves it is possible to compute not only the kinetic order of the channel formation reaction, but also the values of the association (k_R) and dissociation (k_D) rate constants. The values are in agreement with those obtained from the single-channel records. The forward reaction for the formation of the gramicidin channel is found to be second order, and the reverse reaction is first order. This is consistent with the scheme:

$$\mathrm{G} + \mathrm{G} \underset{k_D}{\overset{k_R}{\rightleftharpoons}} \mathrm{G_2}$$

The dissociation rate constant $k_D = 1/\tau$, where τ is the mean open-time of the single channel. The forward second-order rate constant k_R is several orders

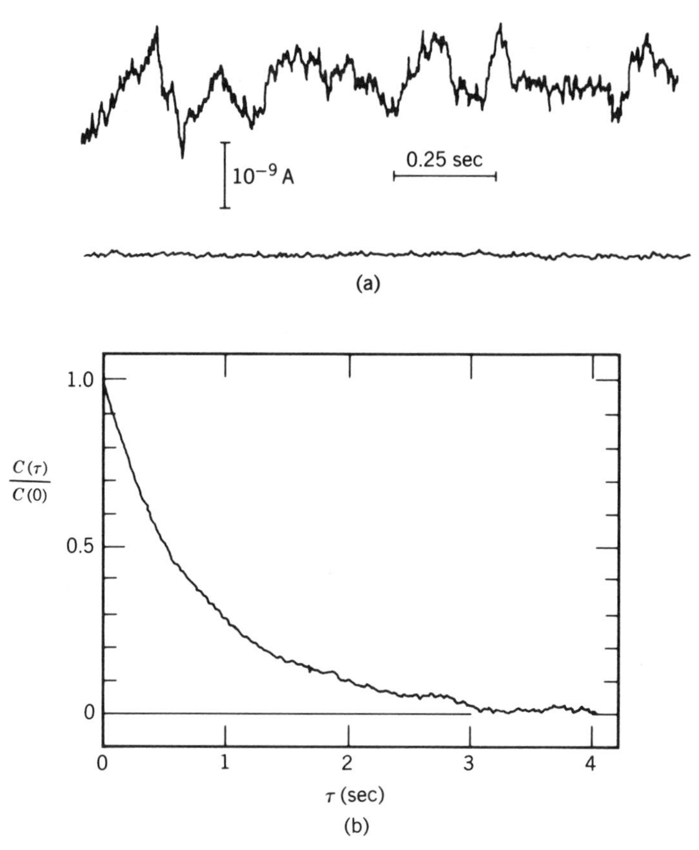

Fig. 9-13. (a) Current noise generated by gramicidin (background noise is shown by the lower trace). Such noise can be analyzed by several methods. The autocorrelation function of the current fluctuations as a function of time is shown in b. (See Lauger et al. (1981) for details.)

of magnitudes smaller than the diffusion-limiting values. This would be expected if, for example, a rearrangement of lipid around the polypeptide to make a dimer is the rate-limiting step.

At first sight the fluctuations in the steady-state conductance may appear like random noise, however with the help of autocorrelation analysis or spectral power density methods, such fluctuations can give significant information about the underlying molecular events. In a channel of this type the basic stochastic molecular process is described by the two-state equilibrium reaction:

$$\text{Close} \underset{K_D}{\overset{k_R}{\rightleftharpoons}} \text{Open}$$

The relaxation time τ^* for this process is given by:

$$\frac{1}{\tau^*} = k_d + 4\left[\frac{k_R k_d \lambda}{N\gamma}\right]^{1/2}$$

where λ is the mean steady-state conductance, N is Avogadro's number. For the analysis of conductance noise several analytical methods can be applied (Eisenberg et al., 1984), and under favorable conditions the channel parameters obtained from such analysis may be comparable to those obtained from single-channel conductance measurements. Lately, this technique is not favored because it is very difficult to rule out artifacts arising from multiple lifetimes, presence of several populations, and the effects of voltage on the gating properties of the channels.

Effect of the Structure of Gramicidin A

The effect of changing the structure of gramicidin on the single-channel parameters has been reviewed by Urry (1984). Gramicidin A is a neutral linear pentadecapeptide (Fig. 9-14) in which the amino terminus is formylated and the carboxyl terminus blocked by an ethanolamino group. Gramicidin B is Val-1-, Phe-11-, and C is Ile-1-, Tyr-11. The sequence is one of the most hydrophobic known. Similarly, alternating L- and D-amino acids permit a unique helical structure in which the polar groups line the central cavity.

The dimer that gives rise to a 28-Å-long channel of about 4 Å internal diameter is formed by the two left-handed β-helices associated head-to-head, i.e., the N-formyl ends are in the center of the bilayer (Fig. 9-14). The hole is lined preferentially with the negative ends of dipoles formed by the carbonyl oxygens of the peptide bonds. Support for this organization comes from a variety of observations (Urry, 1984; Hladky and Haydon, 1984; Lauger, 1985):

1. The steady-state conductance of bilayers shows a second-order dependence on gramicidin concentration, which suggests that two molecules of gramicidin are required for the formation of each channel. Tracer and osmotic fluxes of water display the single-file effect for a long pore, and the movements of ions and water (six to nine water molecules per ion) are coupled.
2. A symmetrical structure for the channel is implicated by symmetrical I-V curves for the most probable single-channel conductance state (Busath and Szabo, 1981).

HCO-NH-L-Val-Gly-L-Ala-D-Leu-L-Ala-D-Val-L-Val-D-Val-L-Trp-
D-Leu-L-Trp-D-Leu-L-Trp-D-Leu-L-Trp-CO-$NHCH_2CH_2OH$

(a)

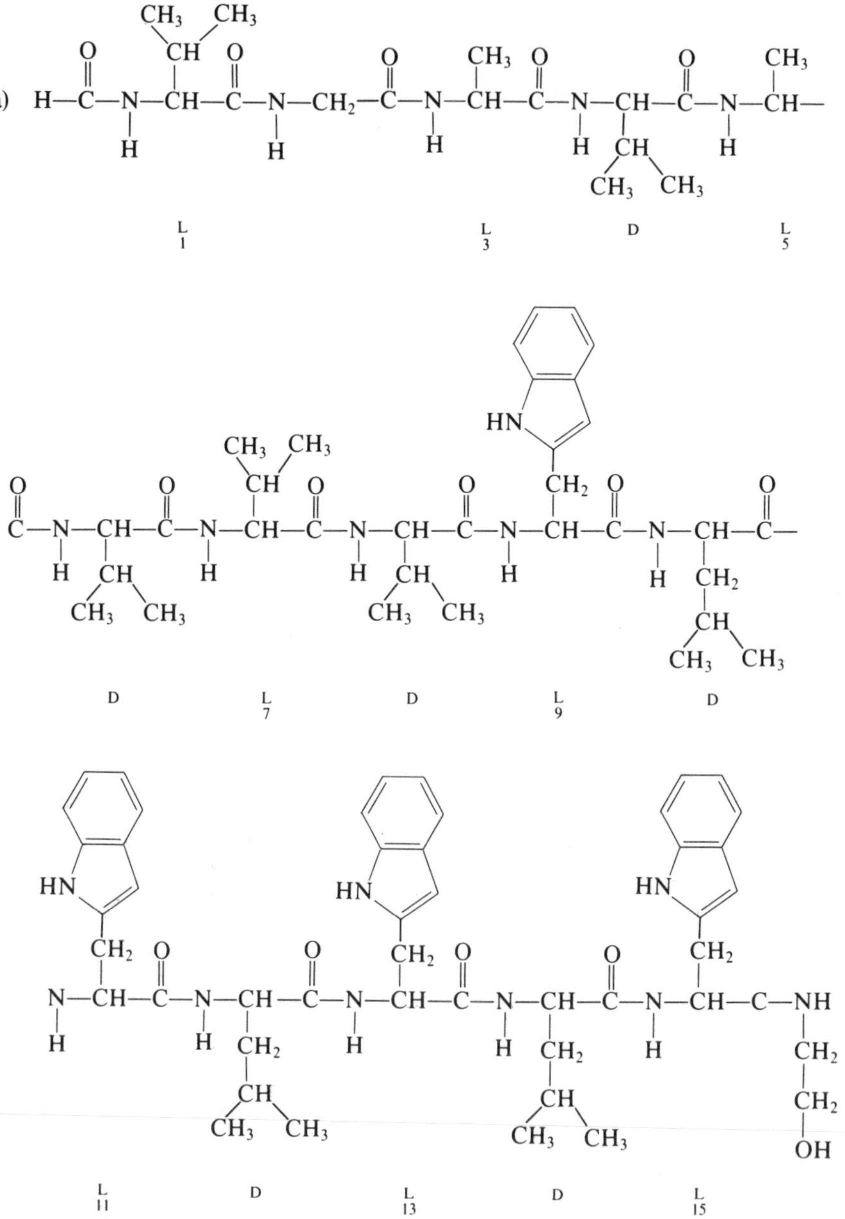

Fig. 9-14. (Top) The primary sequence of gramicidin. The channel is formed by a left-handed helical dimer viewed from the side (b) and from the top (c). (b is from Ivanov and Sychev, 1982; c from Lauger, 1985.)

(b)

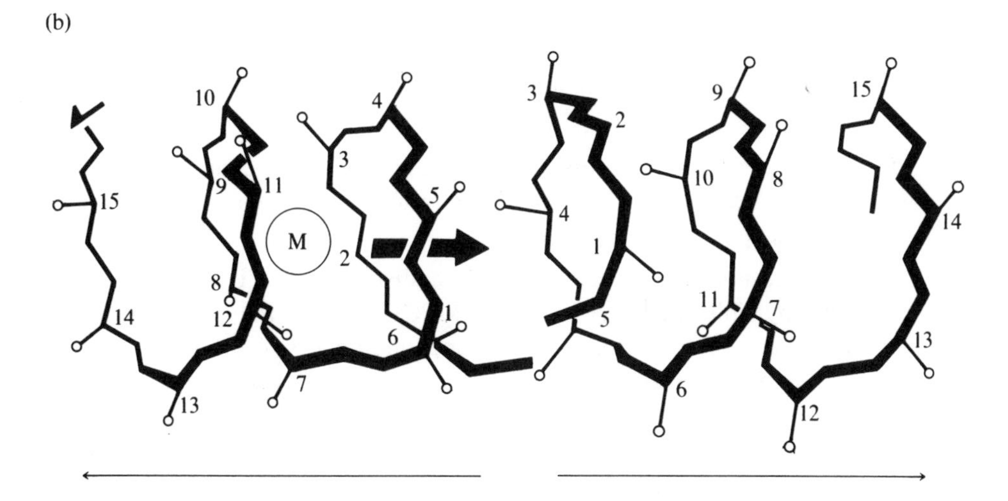

(c)

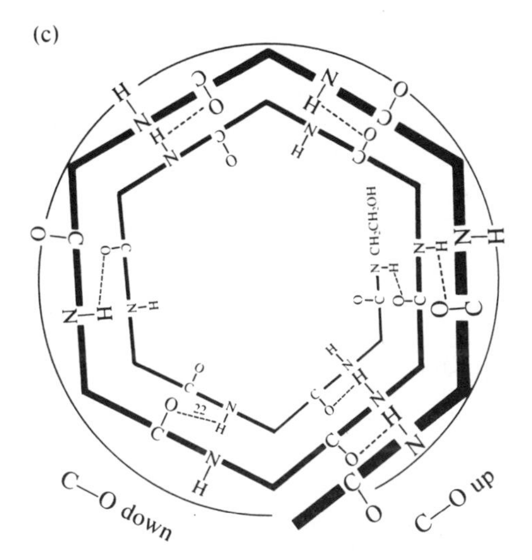

Fig. 9-14. (*Continued*)

3. The single-channel conductance properties of gramicidin analogs (Table 9-5) suggest that a change in positions 9 or 11 or on the ethanolamine end (which are at the interface) do not significantly influence τ and γ. On the other hand a change in the formyl end (located in the middle) causes a dramatic decrease in γ and τ. This is the region of the molecule that comes in contact with each other during the association in the bilayer. It is also relevant to note that the malonyl dimer of gramicidin A forms a channel with a lifetime of about 200 sec.

4. The channels from *O*-pyromellityl gramicidin A can be formed only when the ionophore is placed on both sides of the membrane. This is because the negatively charged substituent decreases the permeability of the peptide through the bilayer.
5. Hybrid channels consisting of neutral and negatively charged gramicidin have been seen when both the peptides are present.
6. Circular dichroism (CD) and nuclear magnetic resonance (NMR) studies on gramicidin in vesicles and in codispersions of lyso-PC and dimyristoyl phosphatidylcholine (DMPC) suggest that a significant change in the conformation of the peptide occurs in the presence of bound cations. Ion-induced chemical shifts show that carbonyl oxygens centered around residue 11 are involved in ion binding. Moreover, such an ion-induced chemical shift is not observed with D-Val-8 (Urry et al., 1983).

Taken together, these observations show that the gramicidin channel is formed by interaction of the amino terminii in the middle of the bilayer which involves association of a left-handed, single-stranded β-helix with 6.3 residues per turn (Fig. 9-14). This 28-Å-long structure also predicts relatively rapid on–off kinetics for the ion, and is consistent with two ion-binding sites, one close to each end of the channel. The gramicidin channel with a diameter of 4 Å is selective only for monovalent cations, and it does not permit passage of anions. The basis for this selectivity is to be found in the nature, tilt, and position of the coordinating atoms, which may only partially desolvate the ion. The NH hydrogen carries a +0.17 charge and becomes available for coordination of an anion only in an energetically unfavorable conformation. On the other hand carbonyl oxygens with a −0.4 charge are oriented such that they line the channel and thus create a negative potential for cations. The radius of the channel is optimal for hydrated Rb^+ and Cs^+ and the selectivity sequence for transport is Tl (8.2), Cs (5.8), Rb (5.5), NH_4 (4.6), K (3.9), Na (1), Li (0.33). This sequence is not identical but is of the same order as that for the mobilities of these ions in bulk aqueous medium: Li (0.77), K (1.47), Na (1), Ca (1.54), H (7), NH_4 (1.47), Tl (1.49). A channel of 4 Å diameter would not be able to discriminate against chloride, Ca, and urea. If the rate of transport were proportional to the binding of ions to sites, the transported species would be partially desolvated. Larger ions can be more easily desolvated and therefore would have higher affinity for the site. Under such conditions selectivity would be strong, and no flux of protons (smallest ion with strongest solvation) would be expected. The rate-limiting step for the transfer of ions is also weakly potential dependent. The rate of transfer is reduced by sucrose in the aqueous phase, even though sucrose cannot enter the pore. Thus the rate-limiting step occurs at or just outside the mouth of the pore (Andersen, 1983).

Divalent ions like Ba and Ca (r = 0.99 Å) are not transported across the gramicidin channel, yet they bind to the channel and interfere with the transport of monovalent cations. This is probably because the proximity of the lipid barrier during the translocation step would bring into play a z^2 term for the divalent cation in the Born solvation energy expression. Thus, the activation energy for translocation would increase from about 3 kcal/mole to about 12 kcal/mole. For monovalent cations the rate-limiting step is the transfer of ion from water to the mouth of the channel. By measuring the rate of spin lattice relaxation of monovalent cations, it has been possible to calculate several equilibrium and rate constants that are involved in ion transport (Urry, 1984). These results suggest that the height of the interfacial barrier and that of the transmembrane barrier is approximately 6–7 kcal/mole. The nature and the number of the rate-limiting barriers, i.e., the nature of the interactions they represent, is yet to be resolved satisfactorily.

Interaction of an ion with the ligands in a channel creates a series of potential energy minima and maxima along the transport path. The dielectric interaction of the ion with the lipids phase creates a broad energy maximum in the middle (Fig. 9-3a). This is superimposed over the two minima at the two interfaces corresponding to the two ion-binding sites. This two-site, three-barrier model, which is characterized by five rate constants, describes essentially all the major experimental observations. Several models invoking three-barrier and two or four ion-binding sites (Eisenman and Horn, 1983; Sandblom et al., 1983) or the "liberation mechanism" have been proposed to account for ion selectivity in the gramicidin channel. The need for such models is based on the observation that tetramethyl ammonium ions lead to a voltage-dependent block, whereas the tetraethylammonium ion can compete only for outer sites and cannot enter the channel.

Other Channel-Forming Ionophores

Several synthetic and naturally occurring molecules have been shown to form ion-conducting channels (Fig. 9-15). The polyene antibiotics are of special interest (Bolard, 1986). Their alicyclic structure consists of a polar and a nonpolar face along the length of the molecule. They form channels by aggregation of six or more monomers only in a bilayer containing sterol. Apparently such aggregation cannot occur in membranes that do not contain sterols. This also explains the growth-inhibiting and lytic effect of the polyenes only in organisms containing sterol. The effect of these antibiotics is

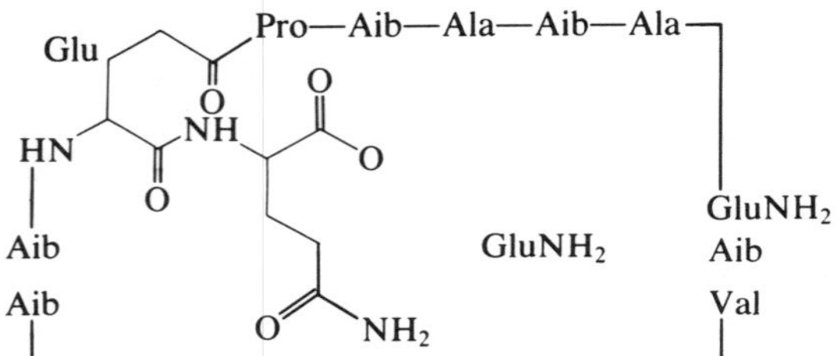

Alamethicin

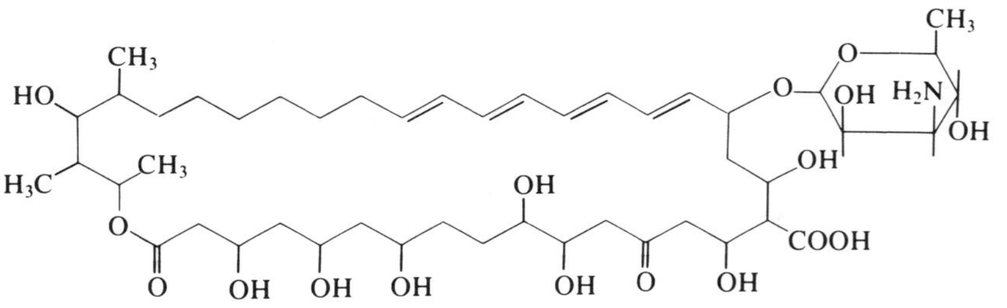

Nystatin

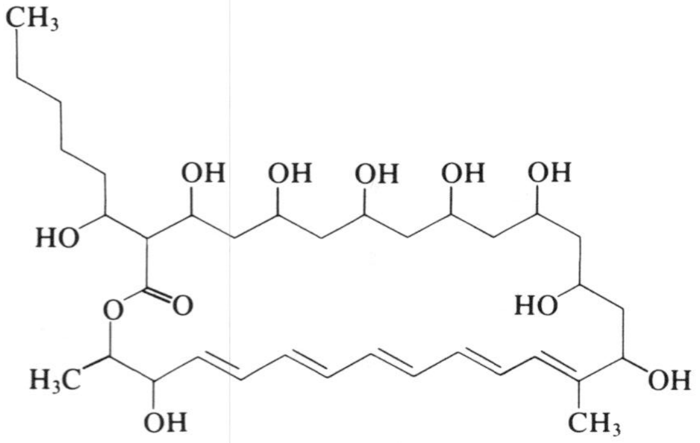

Filipin

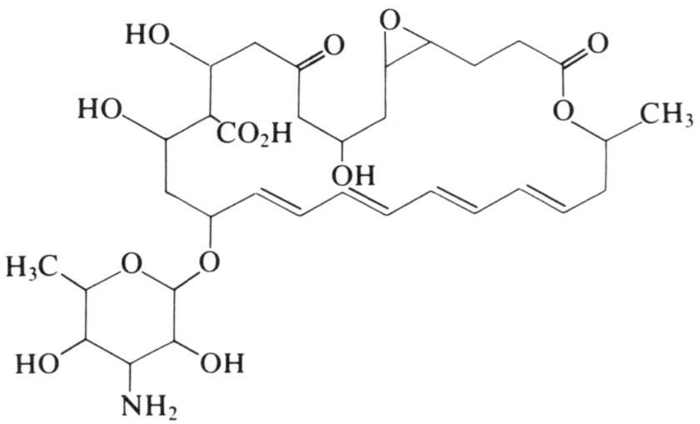

Pimaricin

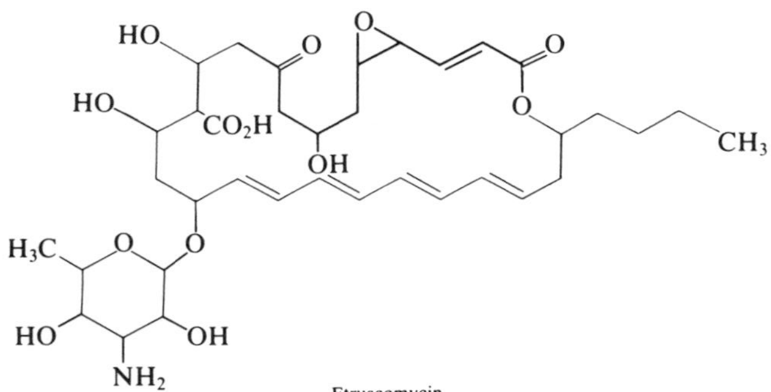

Etruscomycin

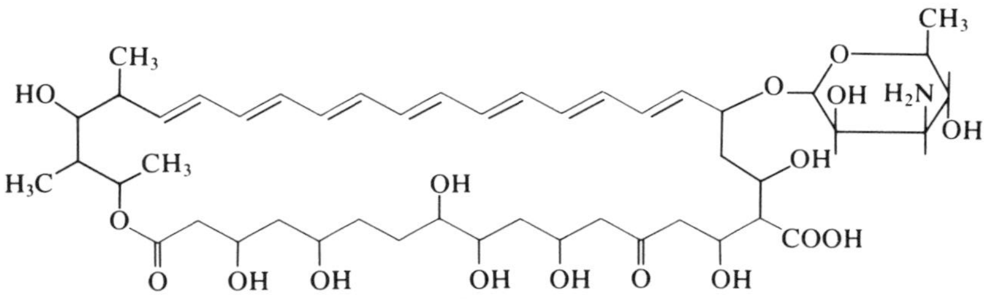

Amphotericin B

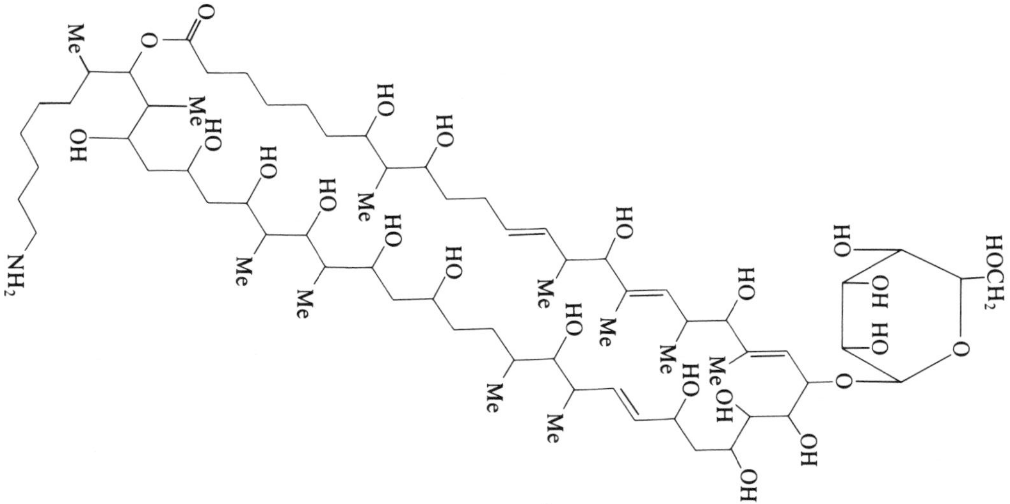

Monazomycin

Fig. 9-15. Structures of some compounds that form channels across bilayers. The polyenes like filipin form channels in sterol-containing membranes, whereas alamethicin and monazomycin form voltage-gated channels.

also blocked by excess sterol. Most polyene antibiotics appear to interact with sterols with 1 : 1 stoichiometry. The channels of polyene antibiotics like amphotericin are permeable to solutes of less than 4 Å diameter. Single-channel conductance for amphotericin channels is 0.4 pS, and for nystatin channels it is 0.07 pS (Ermishkin et al., 1977).

VOLTAGE-GATED CHANNELS

Channels formed by compounds like *alamethicin* and *monazomycin* exhibit many of the characteristics of channels like gramicidin, but they are also *voltage-gated*, i.e., the kinetics of their opening and closing depends upon the membrane potential. Several naturally occurring channels are voltage-gated (Table 9-6) and some of them, like alamethicin, exhibit several open states. The two states of a voltage-gated channel can be represented by an equilibrium:

$$\mathrm{C} \underset{\beta}{\overset{\alpha}{\rightleftharpoons}} \mathrm{O}$$

where the voltage-dependent rate constants α and β describe the molecular transitions between the two states by the Boltzman distribution function of probabilities. In voltage-gated channels the transmembrane potential determines not only the magnitude of the current passing through the channels but also the magnitudes of the rate constants and therefore the equilibrium constant for the reaction given above. For a simple two-state model the ratio of open (n_o) and closed (n_c) channels as a function of membrane potential, V_m, is described by

$$\frac{n_o}{n_c} = \exp\left[\frac{Q(V - V_o)}{kT}\right]$$

where Q is the product of the charge translocated and the fraction of the membrane thickness travelled by the charge when the channel changes from one state to the other. V_o is the potential at which the number of open channels is equal to the number of close channels. The term $Q(V - V_o)$ is the difference of potential energy of the two states. The model on which this equation is based assumes that a change in the state of a channel induced by membrane potential occurs due to a movement or displacement of charges. If a singly charged particle or dipoles are displaced, an e-fold change in conductance must be associated with a potential change of kT/e, which at room temperature is about 25 mV. If an e-fold change is observed with a shift

TABLE 9-6 Comparison of Single-Channel Conductance in Lipid Bilayers and in Natural Excitable Membranes

Type of Channel	Conductance (pS)
Molecules in BLM[a]	
EIM[b]	80, 400
Alamethin[c]	220, 930, 1900 2900, 3900
Hemocyanin	200
Gramicidin	17
K channels	
Squid axon	12
Frog SR	50–150
Frog node	3–5
Snail neuron	2.4
Na channels	
Squid axon	5.8
Tunicate egg	9
Myxicola axon	1.5
Frog node	8.8, 2.9
Rat node	15
Cultured rat muscle	18
Ca channels	
Cardiac cells	9–25

[a] BLM, Bilayer lipid membrane.
[b] EIM, Excitation inducing material.
[c] Shows five conductance state.

of 4–5 mV, it would imply a simultaneous movement of four or five charges. The equation given above predicts a sigmoidal dependence of the fraction of open channels, $n_o/(n_o + n_c)$, (proportional to the gross membrane conductance) on V. The midpoint of the conductance occurs at V_o, the point at which the energy for the movement of charges just balances the intrinsic energy difference between the two states.

How the voltage regulates the state of the conducting channel is an interesting question. At least three different mechanisms are compatible with the treatment just described.

1. A change in the conformation of the preformed channel could be triggered in response to a voltage-induced reorientation of dipoles at or around the channel.
2. A change in the membrane potential could alter the proportion of the channel-forming molecules in the membrane where they aggregate by lateral diffusion to form channels.

3. The change in membrane potential could alter the phase properties of the membrane such that the relative distribution of the components of the channel is altered.

Voltage-dependent behavior of BLM modified with alamethicin and monazomycin (Bauman and Muller, 1975) has been investigated. For example, BLM with monazomycin exhibits sigmoidal voltage versus conductance (g) curve with zero conductance at zero volts. The conductance increases with increasing V_m, with a slope of 5–8 mV for an e-fold change in conductance in the steepest part of the curve. The g versus V curve shifts along the V axis by divalent ions, presumably by changing the surface charge on the bilayer. The conductance measured at a fixed voltage increases with the nth power of monazomycin concentration, and value of n is between 5 and 10, depending upon the lipid composition. The transition from a low to a high conductance state in response to an applied potential proceeds along a sigmoidal time-course with an initial delay that depends upon the lipid composition, ionic composition and strength, membrane potential, and the monazomycin concentration. The membrane current also shows *inactivation*, i.e., the current increases and the declines even though the potential is held constant.

These observations are quantitatively consistent with a model in which monazomycin monomers are adsorbed on the bilayer interface. In response to the change in V the monomers reorient in the bilayer where they are aggregated. The kinetics of the conductance change represents the kinetics of aggregation of monomers in the bilayer, and rates depend only upon the concentration of the monazomycin monomers in the bilayer. Only the rate constants for the adsorption and desorption of the monomer are assumed to be voltage dependent. It is particularly striking to note that the monazomycin-doped BLM exhibits all the characteristics of an excitable membrane like the squid axon (Chapter 12), yet the molecular mechanism may be quite different (see also Menestrina et al., 1986).

Indeed, the elegant simplicity of ionophores and related peptides has provided us with many conceptual and experimental analogs for different modes of ion transport across membrane: carriers, channels, gates. If art was meant to imitate life, in the "Cunundrum of Workshop," one can almost hear the "Devil" of Rudyard Kipling whisper behind the leaves, "It's all pretty, but is it art?" Probably. Art is not meant to imitate life, but it could lead to a better appreciation of life. Models do the same. The analogies elaborated with the study of ionophores have begun to lead us toward such models as discussed in the following chapters.

10 Facilitated Transport

Did we really do it? And if we did it, what exactly did we do? And why did we do it? If we did it, and if we did do it, what will they do? I tell you, this spy scandal is a case for Solomon . . . or a commission of inquiry

Unknown

Transport of polar solutes across the hydrophobic barrier of biological membranes is facilitated by a variety of specific translocation mechanisms. An organism may contain transport systems for several solutes (for example, see Table 10-1). This is probably because translocation of solutes is not only necessary to provide nutrients, but translocation can also regulate levels of metabolites which in turn can modulate certain cellular processes. For example, ion transport across plasma membranes regulates cell volume, action potential, excitation-contraction coupling, fertilization, sensory transduction, coupled transport, energy transduction, lymphocyte activation, synaptic transmission, and cellular motility. Sometimes more than one transport system is found for the same solute; however such isotransport systems differ in substrate specificity and turnover rates. Several inherited disorders of transport in humans are known: renal glycosuria and intestinal malabsorption for sugars, cystinuria, Hartnup disease, and rickets. Some of these are molecular diseases arising from genetic defects. Indeed, not only the aspects related to genetic control of transport processes but also selectivity, specificity, saturation kinetics, asymmetry, susceptibility to protein reagents, and proteases demonstrate that most specific transport systems utilize proteins embedded in membranes.

Passive but facilitated transport systems have several general characteristics. Each transport system exhibits high structural specificity. The driving force for facilitated passive transport is the electrochemical gradient of the

TABLE 10-1 Mitochondrial Metabolite Transporters

Category	Name	Physiological Substrates	Inhibitors
Electroneutral proton compensated	Glutamate hydroxyl	Glutamate	*N*-Ethylmaleimide, avenaciolide
	Pyruvate	Monocarboxylic acids, ketone bodies, branched-chain ketoacids	α-Cyano-3-hydroxy-cinnamate, organic mercurials, *N*-ethylmaleimide
	Phosphate	Phosphate, arsenate	Organic mercurials, *N*-ethylmaleimide
	Ornithine	Ornithine, citrulline, lysine	
Electroneutral anion exchange	Dicarboxylate	Phosphate, malate, succinate, oxalacetate	Butylmalonate, bathophenanthroline, iodobenzylmalonate, phenylsuccinate, phthalonate, organic mercurials
	α-Ketoglutarate	Malate, α-ketoglutarate, succinate, oxalacetate	Phthalonate, bathophenanthroline, phenylsuccinate, butylmalonate
	Tricarboxylate	Citrate, isocitrate, phosphoenolpyruvate, malate, succinate	1,2,3-Benzenetricarboxylate, α-acetylcitrate, bathophenanthroline
Neutral	Carnitine	Carnitine, acylcarnitine	Organic mercurials, *N*-ethylmaleimide, sulfobetaines
	Neutral amino acids	Neutral amino acids	Organic mercurials
	Glutamine	Glutamine	Organic mercurials
Electrogenic	Adenine nucleotide	ADP, ATP	Atractyloside, carboxy atractyloside, bongkrekate, long-chain acyl CoA, α-acetylcitrate
	Glutamate/aspartate	Glutamate/aspartate	Glisoxepide (nonspecific)

solutes that ultimately leads to equalization of concentrations. Such transport processes, which occur through a limited number of sites, occur in both directions (influx and efflux), and the rate of transport exhibits saturation kinetics, i.e., the rate of transport is limited by the number of transport sites. Such characteristics give rise to a hyperbolic dependence of the initial rate of transport on the solute concentration (Fig. 10-1). Saturation kinetics of this type are usually described in terms of two parameters: K_t is related to the affinity of the solute for the transporter, and it is operationally defined as the

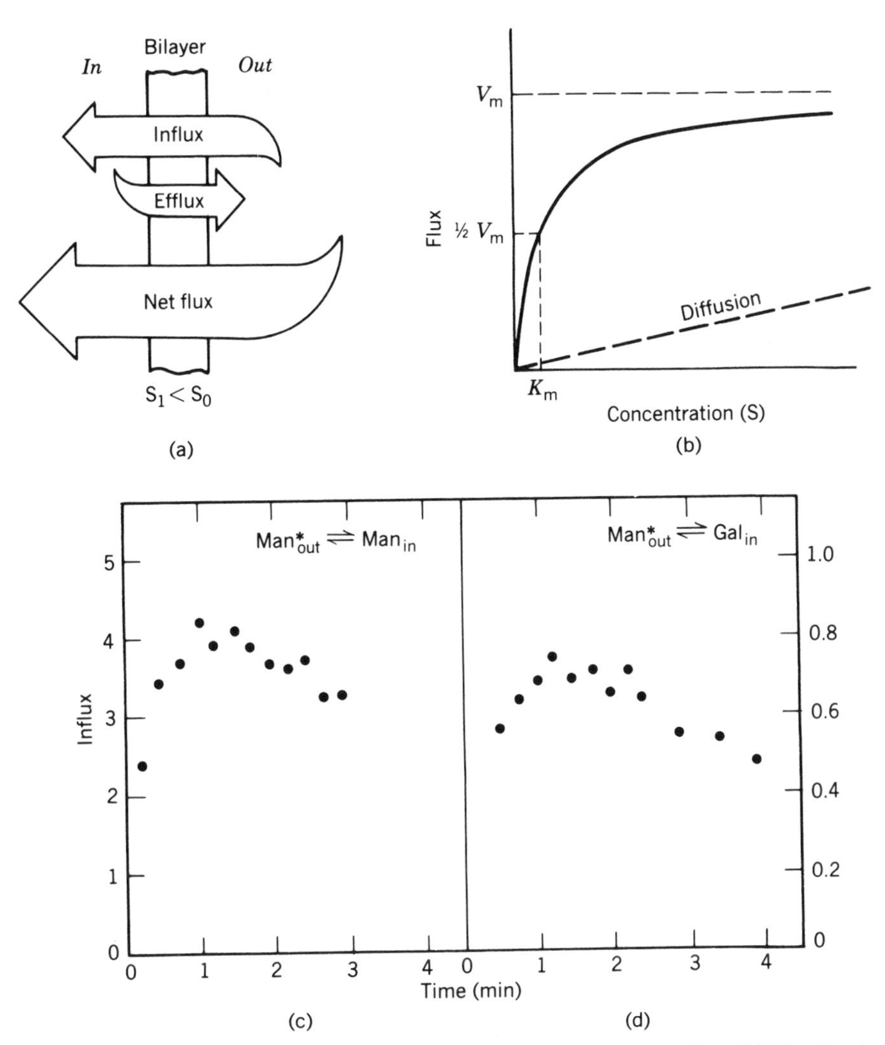

Fig. 10-1. (a) Net flux of solutes is the difference between influx and efflux. (b) The steady-state initial rate of transporter mediated flux of a solute increases hyperbolically as a function of the solute concentration (full line). Initially, the solute concentration in the other compartment is assumed to be zero. The rate of transport by the solubility diffusion mechanism and leakage is shown by the broken line. Interpretation of such a relationship requires several assumptions, however, these hyperbolic curves can be described by two parameters, K_t and V_m. (c, d) The influx rate of mannose or galactose against labeled mannose. Although the concentration of sugars is the same on both sides, countertransport (apparently uphill) is observed because the complexed transporter can move in both directions. It may be noted that the extent of stimulation of the flux of the two pairs is different. This suggests that the transporter is asymmetric, i.e., its affinity for sugars on the two interfaces is different.

concentration of the solute at which the rate of transport is half of the maximum; and V_m, the maximum rate of transport which is related to the turnover rate of the transporter. The relationship of these transport parameters to the kinetic (Michaelis–Menten) parameters for catalysis by enzymes is obvious. However, none of transporters are understood in details with which we understand the catalytic mechanisms of enzymes. The following treatment explicitly states several assumptions required for the interpretation of the kinetic characteristics of a transporter. As discussed in the last chapter (Fig. 9-1), at least four kinetically distinguishable steps are required for translocation of a solute S by a carrier C localized in the bilayer: (i) binding of S to C to form SC; (ii) diffusion of SC across the bilayer; (iii) dissociation of S from SC on side 2; and (iv) return of C to side 1. Microscopically all these steps are reversible, and many more kinetically distinguishable steps may be invoked in a real situation. Unfortunately, the methods used for ion translocation by ionophores cannot be used for characterizing transporters of neutral solutes in intact biomembranes. However, a complete kinetic description of this hypothetical process can be achieved in terms of two parameters (K_t and V_m) if the following assumptions are made:

1. Binding of S to C is reversible.
2. The free and complexed carriers are confined to the membrane phase but are accessible at both interfaces.
3. The rate of transfer of the free and complexed carrier between the two interfaces of the membrane is independent of the direction.
4. The rate of translocation of a carrier across the membrane is much greater than the rate of diffusion of the solute through the membrane.
5. The rate of translocation of the complex CS is much smaller than the rate of its formation and dissociation.
6. The affinity of the substrate for the carrier is the same at both interfaces.
7. The amount of carrier in the membrane remains constant, i.e., it is neither being synthesized, destroyed, solubilized, nor converted into a different form.
8. The rate at which C and CS cross the membrane is proportional to the difference in their concentrations at the two interfaces.
9. The contribution of the unstirred layer is negligible.
10. The translocation step does not necessarily imply diffusion or rotation of the free or bound carrier (C or CS). Any change that could alternately (but not simultaneously) present the binding site to the two sides would satisfy the intrinsic kinetic constraints.

Within the constraints of these assumptions, the net rate of transport of a solute from one interface of the bilayer to the other is given by:

$$v = D(\mathrm{CS}') - D(\mathrm{CS}'')$$

where D is the diffusion coefficient for the complex within the membrane. Since the concentration of the complex at the two interfaces CS′ and CS″ cannot be determined directly, it can however be related to bulk concentration by an equilibrium constant for the dissociation of the complex:

$$\mathrm{K} = \frac{(\mathrm{C})(\mathrm{S})}{(\mathrm{CS})}$$

The proportion of the total carrier present as CS at each interface is given by

$$\frac{[\mathrm{CS}]}{[\mathrm{C}] + [\mathrm{CS}]} = \frac{[\mathrm{S}]}{[\mathrm{S}] + \mathrm{K}}$$

If the equilibrium constant K at the two interfaces is the same, the net rate of transfer of the solute is

$$\text{net flux } v = D[(\mathrm{C}) + (\mathrm{CS})]\left[\frac{(\mathrm{S}')}{(\mathrm{S}') + \mathrm{K}} - \frac{(\mathrm{S}'')}{(\mathrm{S}'') + \mathrm{K}}\right]$$

The term D (CS′) − (CS″) expresses the maximum rate of transport, and therefore it can be replaced by a constant V_m. Thus, the rate of transport under the conditions of unidirectional flux (when S″ = 0) is

$$\mathrm{v} = \frac{V_m\,(\mathrm{S}')}{(\mathrm{S}') + K_t}$$

This relationship can be solved analytically. A diagram method described by Hill (1966) is often convenient to use for more complex mass–balance equations.

This exercise for modeling transport by carriers is meant to illustrate that under certain limiting conditions the rate of transport as a function of permeant concentration is determined by a hyperbola that resembles the Michaelis–Menten relationship. The mechanistic significance of K_t and V_m is very limited because these parameters represent a combination of primary rate constants that depend upon the specific kinetic model that is being examined. The rate parameter V_m is determined by the total population of the carrier as well as the turnover rate of individual carrier molecules. The constant K_t can be approximated as the equilibrium binding constant only if the rate of translocation is considerably slower than the rate of interfacial binding. From the preceding discussion, an analogy between the kinetic properties of transport proteins and catalytic proteins is obvious. The kinetic and molecular features of a carrier-type-facilitated transport of nonelectrolytes in biological membranes are summarized below:

1. Passive facilitated transport is driven only by the concentration gradient of the permeant. The initial steady-state rate of transport reaches a limiting rate at high permeant concentration (saturation kinetics). While the small size of the inner compartment would tend to promote a departure from the steady-state condition. Metabolism of the translocated solute often tends to maintain the steady-state concentration gradient for longer periods.
2. A transporter exhibits a high substrate specificity, i.e., K_t, which depends upon structure and conformation of the solute, tends to be in the micro- to millimolar range.
3. The overall rate of exchange of S between the two compartments is usually different from the rate of net flux. If the rate of translocation of CS is far greater than the rate of translocation of C, the rate of net transport will be considerably slower than the rate of exchange.
4. Transmembrane exchange of two different permeant species with different affinities can be mediated by a carrier in opposite directions. If the rate of translocation of C is slow, *countertransport* of the two solutes is promoted under exchange conditions.
5. Isotransporters with different K_t and V_m values can coexist or appear and disappear with age and metabolic state of an organism. The overall rates of transport would of course reflect fluxes through parallel pathways.
6. Temperature coefficients (Q_{10}) for most transporters are 2–3, that is, the rate of transport increases two- to three-fold for a 10° increase in the temperature. This corresponds to the activation energy greater than 10 kcal/mole, compared with a value of less than 5 kcal/mole for processes limited by diffusion in aqueous phase, and about 5 kcal/mole for lateral diffusion in bilayers.
7. Binding of the permeant and translocation of the complex are two distinct steps in the overall transport cycle. This implies that the CS complex must assume proper conformation before it can translocate. It also raises the possibility of regulation at the binding as well as at

the translocation level. Therefore, certain compounds may bind to the carrier, but the complex may not be translocated. In analogy with the behavior of catalytic proteins, it is possible to inhibit the binding or the translocation step by competitive, noncompetitive, uncompetitive, or allosteric mechanisms. A high degree of structural specificity for a solute in the translocation step would therefore imply that the translocation step for the CS complex is not a simple diffusion from one interface to the other.

8. Specific facilitated transport in almost all known cases is mediated by proteins, therefore such processes are inhibited by antibodies, protein reagents, and affinity labels.
9. Genetic regulation (by induction and repression) of transporters has been demonstrated by characterization of mutants and inherited disorders.
10. Transporters are subject to control by post-translational modification (glycosylation, phosphorylation), binding of regulatory ligands (hormones, messengers, transmitter, ions), or an electric field, as in voltage-gated channels (Chapter 12). Similarly, the translocation step can be coupled to the flux of other species (Chapter 11) or to a chemical reaction, as in active transport (Chapter 13).

Transporters occur in all organisms and in most tissue and cell types. A large number of transporters have been shown to satisfy many of the characteristics outlined above. As expected, most transporters are for essential metabolites like sugars, amino acids, cofactors, and ions. Some of the simple transporters are discussed in this chapter, and the more complex cases, where the translocation step is directly coupled to other regulatory functions, are discussed in subsequent chapters.

TRANSPORTER FOR GLUCOSE IN ERYTHROCYTES

Erythrocytes depend upon anaerobic glycolytic pathways for their energy supply. Uptake of glucose from the plasma therefore is the first step in the metabolism of glucose. The rate of uptake of glucose across the erythrocyte membrane is about a million times faster than the rate of its diffusion across a phospholipid bilayer. The concentration of glucose in plasma during fasting is 2–5 m*M*, and the rate of uptake of glucose by erythrocytes is about 250 times faster than the rate of its metabolism under physiological conditions. The rate of transport of glucose in erythrocytes is not affected by insulin, or by the rate of phosphorylation of glucose in the cytoplasm, or by inhibitors of glycolysis. A passive facilitated mode of transport of glucose is indicated by these observations.

The transport of glucose in erythrocytes is bidirectional, and the net rate of transport depends on the concentrations gradient (Carruthers, 1985). A rapid rate of exchange is observed in the absence of concentration gradient (Fig. 10-2). The half-time for efflux of tracer from cells under exchange conditions is about 10 sec at 37°C. The first-order rate constant for efflux is v/S, where S is the solute concentration and v is the initial rate. The S/v versus S plot (Eadie plots), based on the rearranged Michaelis–Menten relationship (Table 10-2), is often used for measuring K_t and V_m. Similarly, v/S values in the presence of inhibitors can be used to obtain K_i:

$$\frac{V}{v_i} = 1 + \frac{I}{K_i}$$

Type of Experiment	Sugar Concentration Out	Sugar Concentration In	K_t (mM)	V_m (mM/sec)
Zero-*trans* fluxes Uptake	Varied	Zero	1.6	0.6
Exit	Zero	Varied	2.5	2.15
Infinite-*trans* fluxes Uptake	Varied	Saturating		
Exit	Saturating	Varied		
Infinite-*cis* fluxes[a] Uptake	Saturating	Varied	2.8	–
Exit	Varied	Saturating	1.8	–
Equilibrium-exchange Internal [sugar] = external [sugar] uptake and exit	Varied	Varied	3.4	6

[a] Infinite-*cis* net fluxes were measured as the difference between unidirectional uptake and exit.

Fig. 10-2. Different protocols to study the transport of sugars across erythrocyte membranes. By varying the concentration of sugar in the two compartments (in and out), it is possible to obtain information about the asymmetry of the transporter. The transport parameters for different protocols are found to be different, which suggests that this transporter is asymmetric.

TABLE 10-2 Plots for Analysis of Initial Rate Data

Type of Plot	y-Intercept	x-Intercept	Slope
Lineweaver-Burk ($1/v$ vs $1/S$)			
$\frac{1}{v} = \frac{K}{V_m} \times \frac{1}{S} + \frac{1}{V_m}$	$\frac{1}{V_m}$	$\frac{-1}{K}$	K/V_m
Eadie (S/v vs S)			
$S/v = \frac{K}{V_m} + \frac{1}{V_m} \times S$	$\frac{K}{V_m}$	$-K$	$\frac{1}{V_m}$
Hofstee (v vs v/S)			
$v = V_m - \frac{v}{S} K$	V_m	V_m/K_1	K
Eisenthal Cornish-Bowden			
$\frac{V_m}{v} - \frac{K}{S} = 1$		(See Fig. 10-3)	

where v_i is the rate in the presence of the inhibitor concentration I. A plot of the type shown in Fig. 10-3 is best suited for obtaining inhibition constant K_i.

The transport parameters have also been measured under other conditions (Fig. 10-2):

a. *Zero-trans*, i.e., the solute concentration either inside or outside is zero.

b. *Infinite-cis*, i.e., the solute concentration on one side is varied, and the solute concentration on the other side is kept very high (more than 5

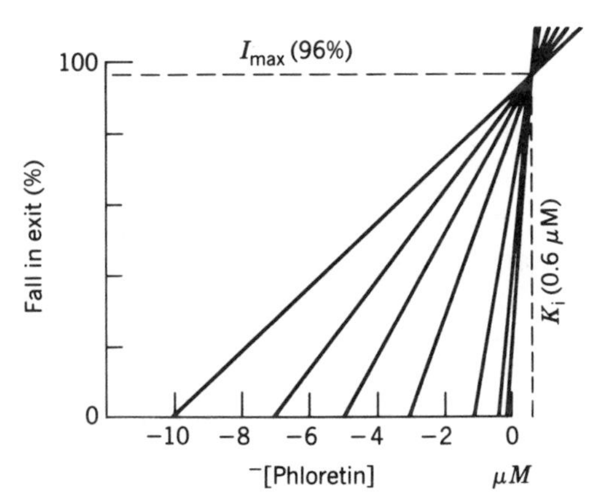

Fig. 10-3. An Eisenthal–Cornish–Bowden plot of the inhibition of efflux of D-glucose from human erythrocytes by phloretin. Data points for each −[phloretin] concentration are connected and extended into a positive quadrangle. The point of intersection of the lines provides the K_i and maximum inhibition. (From Carruthers, 1985.)

K_t). These fluxes are measured as the difference between unidirectional uptake and exit.

c. *Infinite-trans* i.e., countertransport. Under these conditions the flux of tracer from low concentration to the high concentration side is measured.

These protocols provide information about affinities of permeant at the two interfaces. For example, infinite-*cis* (influx) experiments measure K_t for outer interface, whereas infinite-*trans* and zero-*trans* relate to the affinity for the inner or the outer interface, depending upon the direction of flux (Lieb and Stein, 1972; Carruthers, 1985). These values for erythrocytes are compared in Fig. 10-2. Two classes of K_t are observed, suggesting that the carrier is asymmetrical.

The glucose transporter of human erythrocytes is specific for C1 conformation of D-glucose (K_t 3 mM) in which the hydroxyl groups at positions 2, 3, 4, and 5 are in equitorial conformation (Fig. 10-4). A slightly strained 1C conformation assumed by L-glucose makes it a poor substrate. K_t values for transport of the analogs are in the range of 2 to 3000 mM for 2-deoxy-D-glucose < D-glucose < D-mannose < D-galactose < D-xylose < 2-deoxy-D-galactose < L-arabinose < D-ribose < D-fucose < D-lyxose < L-glucose. On the other hand the V_m values are approximately the same, 300–600 mmoles/min per cell unit[1] or 13 μmole/min per mg membrane protein. This is in accord with the carrier model. In this series of sugars an increase in the affinity for the carrier involves hydroxyl groups at positions 1, 3, 4, and possibly 6 (Barnett et al., 1976). Different affinities of sugars at the two interfaces give rise to countertransport. This is also consistent with models invoking two different conformations of the transporters on the interface (Carruthers, 1985). As illustrated in Fig. 10-4, an interaction of different regions of the substrate are invoked in the two conformations of the transporter.

Another phenomenon consistent with the carrier mode of translocation is *countertransport or exchange diffusion* (Wilbrant, 1972; Widdas, 1980). This phenomenon arises due to an asymmetric competition for the transporter sites. If a solute S_1 is present at equal concentrations on both sides of the membrane, the carriers are equally distributed on the two interfaces, and no net flux occurs because there is no net concentration gradient. Addition of a second solute S_2 to one side reduces the number of sites available for the permeant S_1 on that side. Thus, a concentration gradient for the loaded

[1] One cell unit is the quantity of cells whose water volume is one liter under iso-osmotic conditions.

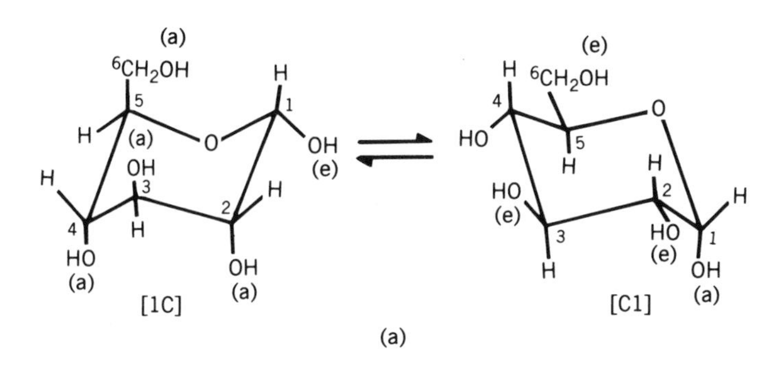

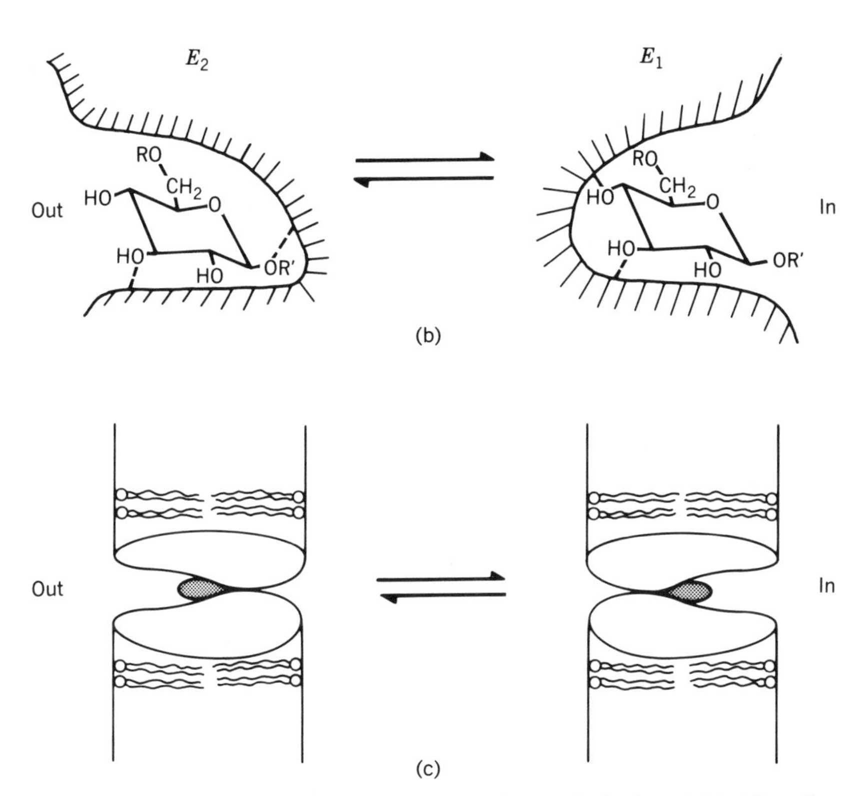

Fig. 10-4. (A) The two chair conformations of α-D-glucose. In the less stable 1C conformation the substituents at 2, 3, 4, and 5 positions are axially (*a*) oriented. In the more stable C1 conformation the substituents at 2, 3, 4, and 5 positions are equitorial (e). The C1 conformation is preferentially recognized by the glucose transporter of human erythrocytes. (B) Proposed model for the glucose transporter. 6-*O*-propyl-D-glucose (R = propyl; R′ = H) can bind to the E2 form of the transporter but cannot be transported for steric reasons. Propyl-β-D-glucopyranoside (R = H; R′ = propyl) can bind to the E1 form of the transporter but is not translocated. D-Glucose binds to both forms and is translocated by the transition SE1 to SE2 (From Carruthers, 1985). (C) A simplified minimal representation of asymmetric transporter which exposes the solute binding site alternately to the two sides on the membrane.

carriers (CS_1 and CS_2) is established which, as shown in Fig. 10-1 (C and D), leads to an apparently *uphill* movement of an equally distributed S_1. There is an alternative explanation for countertransport that is currently favored. If the complexed carrier moves faster than the free carrier, an increase in the concentration of the solute ($S_1 + S_2$) would stimulate the slower step, i.e., the return of C from the low permeant side. Such a countertransport mechanism is a direct consequence of the mobile carrier mechanism because it arises only when the carrier mediates a stoichiometric exchange and the movement of individual solute molecules occurs independently of one another. Such a mechanism is not possible for a simple channel-mediated process. Tests for a symmetrical carrier model have been found to be quantitatively inconsistent with a simple carrier model (Lieb and Stein, 1972; Le-Fevre, 1975).

To account for such departures from symmetrical transporter other possibilities have been considered. Glucose consists of α- and β-anomers which equilibrate slowly, and the transporter has higher affinity for the β-anomer. Other experimental factors that could influence experimental data are age of blood, proteolytic cleavage, and ATP depletion (see Wheeler and Hinkle, 1985). On the basis of studies with the reconstituted system, the following features of the glucose transporter emerge.

1. The affinity of the carrier at the two interfaces is different. Because the transport is passive, the ratio of K_m at the two interfaces should be equal to the ratio of the corresponding V_m. Indeed, erythrocytes show asymmetric kinetics with efflux having a three- to four-fold high V_m and a four- to 15-fold higher K_t than uptake.
2. Mobilities of the complexed and uncomplexed carriers are different.
3. The effect of unstirred aqueous layers at the two interfaces is asymetric.
4. The conformational change of the free transporter is rate-limiting because the rate of exchange is faster than the rate of net flux.
5. Inhibitors of transport (maltose, 4,6-*O*-ethyledeneglucose, 6-*O*-propylgalactose, propyl β-glucopyranoside, phloretin, diethylstilbesterol, and cytochalasin) apparently compete for the same substrate site although this site has different outward and inward conformation (Fig. 10-4) and, therefore, different affinities at the two interfaces.

Physicochemical characterization of the transport system also exhibits certain anomalies that remain to be explained. Temperature dependence of transport is a complex function: The plot of K_t versus $1/T$ is concave upward, whereas the plot of V_m versus $1/T$ is convex. Similarly, V_m shows a maximum at pH 7.1 and K_t shows a minimum at this pH. The transport system is also modified by treatment with proteolytic enzymes (from inside) as well as by protein reagents. Transport is inhibited by cytochalasin B (K_i <1 μM), phloretin (<1 μM), alloxan (1.8 μM), hydrocortisone (200 μM), prednisolone (400 μM), dipyridamole (100 μM), and diethylstilbesterol (2 μM). Of the several glucose analogs tested 6-deoxy-6-fluoroglucose (K_i 1.2 mM) is the most effective inhibitor. Transport is not inhibited by 1,3-dihydrocytochalasin, which inhibits cell motility like its parent cytochalasin. On the basis of inhibitor binding studies, it appears that the number of glucose transporters is about 100,000 per erythrocyte or about 6000 per square micron.

The glucose transport activity of erythrocyte is due to band 4.2 of 45 kD (Baldwin and Lienhard, 1981; Wheeler and Hinkle, 1985), whose amino acid sequence has been determined (Mueckler et al., 1985). Biochemical characterization and isolation of the glucose transport glycoprotein has been possible by using cytochalasin B binding for assay and phloretin as a ligand for affinity chromatography. These procedures have also been used to characterize glucose transporter in a variety of cells ranging from squid axon to adipocytes and cultured cells. They appear to contain cytochalasin binding protein of 45–60 kD, and all of these systems have different degrees of asymmetry (1.5- to 20-fold) and exchange acceleration (cf. Fig. 10-1 c and d).

Reconstitution procedures involving different phospholipids and protocols have been used successfully. The specific activity of the reconstituted transporter is about 5–15% of that expected for a fully active transporter on the basis of the number of band 4.5 monomers. There is still considerable disagreement about the effect of lipid composition on binding and transport. The degree of scrambling in the reconstituted preparations depends upon the method of reconstitution. Although not all the discrepancies between reconstituted and native transporters in erythrocyte are settled, Wheeler and Hinkle (1985) have tried to explain many of these. Finally, it should be emphasized that the model shown in Fig. 10-4 for glucose transporter is consistent with a carrier mechanism because the binding site presents alternately to the two interfaces and the osmotic separation of the two compartments is never violated. However, the translocation step in the transport cycle of the glucose transporter is a conformational change in contrast to the diffusion or rotation of ionophores. It is concievable that a high turnover by valinomycin (compared with that for other ionophores) is due to the symmetry of the bracelet complex, which can present itself alternately at the two interfaces by diffusion over a very short distance without significant rotation or *somersault*. Other less symmetrical ionophores have much lower turnover numbers because they must undergo a somersault within the bilayer so as to present the ion-binding site alternately to the two interfaces. In the glucose transporter the need for a somersault is obviated by exposing the site to the two interfaces only by a conformational change within the framework of a transmembrane passage (Fig. 10-4c). The exchange systems described below also have mechanistic similarities to this model.

ANION EXCHANGER OF ERYTHROCYTES

Several electrically silent ion exchangers across cell membranes have been reported. Of these the chloride–bicarbonate (1 : 1) exchanger in erythrocyte is probably the best characterized (Passow, 1986). It is involved in transport of carbon dioxide from tissues to the lung (Fig. 10-5). The pH drop associated with the hydration of CO_2 (mediated by carbonic anhydrase in the red cell) promotes release of oxygen (Bohr effect) and it is largely compensated by the high buffer capacity of hemoglobin. The exchange takes place in reverse in the pulmonary capillaries. The rate of exchange is rapid (half-time about 0.1 sec) so that it is almost 90% complete during the period a red cell spends in the capillary. The physiological importance of this transport system is further underscored by the fact that the anion exchange systems from different animals exhibit significant differences that are necessary for adaptation to their environments.

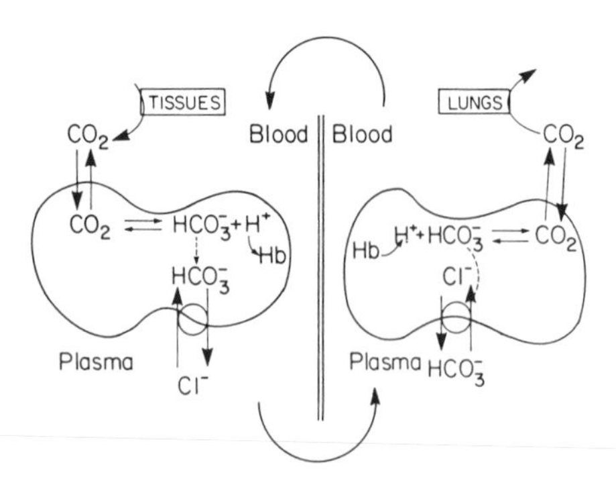

Fig. 10-5. The physiological role of the Cl-HCO_3 exchanger in erythrocytes is in transporting CO_2 from tissue to lungs. CO_2 produced in tissues diffuses into the plasma and from there to erythrocytes circulating in capillaries, where it is converted to bicarbonate by carbonic anhydrase. the proton released by this reaction is taken up by hemoglobin, which decreases its affinity for oxygen (Bohr effect). The release of CO_2 is facilitated in lungs because its partial pressure is low.

The anion exchange in erythrocytes is mediated by the band 3 protein (95 kD). A model of the transmembrane helices that could constitute the anion exchanger is shown in Fig. 10-6. It constitutes 25% of the total membrane protein with 0.55 million copies per cell. The permeability coefficient for the exchange of chloride P_{Cl} across erythrocyte membrane is 10^{-3} cm/sec, i.e., about 10^{10} ions per cell per second, and the half-time for the exchange of chloride ions in an erythrocyte is about 52 msec at 38°C. Net flow of chloride through the exchanger is about 10^{-5} of the exchange rate. This is considerably (about 100-fold) higher than $P_K = 10^{-10}$ cm/sec and $P_{Cl} = 10^{-12}$ cm/sec for lipid bilayers. The fact that this is an exchange system is further substantiated by an increase in ionophore-coupled flux of chloride. The activation energy for the exchange of chloride is about 30 kcal/mole and somewhat

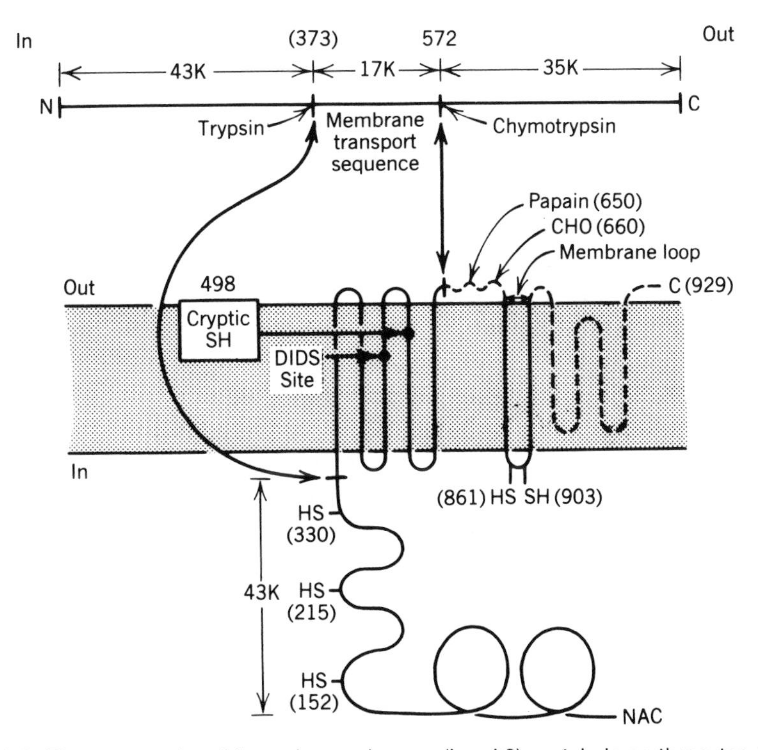

Fig. 10-6. The topography of the anion exchanger (band 3) protein in erythrocytes. Positions of the thiol groups, sites of cleavage, site of action of DIDS position of carbohydrate groups, number of loops, and other topological features are also shown. Intramembranous segments (around the DIDS binding site) flank the anion binding site, which can alternate between two nonidentical conformations. Only one anion can bind at a time, however the anions can reach the site from either surfaces. (For further details, see Jay and Cantley, 1986.)

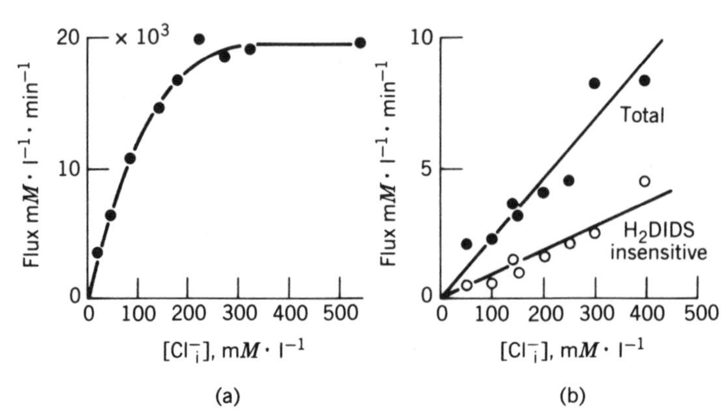

Fig. 10-7. The concentration dependence of chloride equilibrium exchange (a) and net flow of chloride (b). Note the 1000-fold difference between the ordinates. (From Passow, 1986.)

lower above 20°C. The break appears to depend upon the nature of the transported species, however the break occurs at about the same transport rate (about $4 \cdot 10^9$ ions per cell per second or about $5 \cdot 10^4$ ions per second per site).

The rate of chloride exchange shows saturation kinetics with $K_t = 30$ mM (Fig. 10-7). At high concentrations of certain anions, the rate of transport decreases, presumably because they interact with a modifier site. The selectivity for the various anions for the transport and for the modifier site is given in Table 10-3. Anions like superoxide, borohydride, phosphate, carbonate, and dithionite are also transported with rates that vary over a 10,000-fold range. The ability of anions to competitively inhibit the exchange of chloride is in the order malonate < lactate < acetate < propionate < Cl < Br < NO_3 < I < SCN. This suggests that anion binding to the transport system as reflected in the inhibition sequence is not the only factor that

TABLE 10-3 Apparent Dissociation Constants of Anion Binding to Transfer and Modifier Sites (mmol liter^{-1})

	Transfer Site	Modifier Site
Fluoride	88	337
Chloride	67	335
Bromide	32	160
Iodide	10	60
Thiocyanate	3	
Bicarbonate	10	600
Sulfate	30–40	350–600
Phosphate	68	200

determines the selectivity of the exchange pathway. This is further substantiated by the observation that the apparent dissociation constants for anions are not in the same order as the exchange rates. Interpretation of this data is complicated by the fact that each transport unit also has a modifier site. The transport site is the substrate binding site, and the low-affinity modifier site inhibits anion transport. Erythrocytes also contains a separate pathway for the transport of lactate, pyruvate, and other substituted carboxylates.

The anion exchanger of erythrocytes is noncompetitively inhibited by dipyridamole (I_{50} 5 μM), phloretin (1.5 μM), furosemide (100 μM), 1,8-anilinonapthalene sulfonic acid (25 μM), local anesthetics (about 1 mM), and long-chain alcohols (about 1 mM for octanol). Several compounds like DNDS (4,4′-dinitrostilbene-2,2′-disulfonate), which do not penetrate the membrane, competitively inhibit exchange by binding to a site (probably residue 561) accessible from the outside. Several other lipophilic stilbene derivatives have also been used as covalent modifiers. There exists a linear relationship between inhibition of transport and binding of inhibitors, and inhibition is complete when 10^6 molecules of inhibitor are bound per cell. Such studies suggest that irreversible inhibition of one subunit does not inhibit transport by the other subunit. Chemical modification of arginine (Zaki and Julien, 1985), as well as the amino groups of Lys-449 and -588 (Kampmann et al., 1982) suggest that side-chains of these residues are necessary for transport. The pH versus exchange rate profile for transport of chloride ions is bell shaped with a maximum at pH 6.5. This is also consistent with a role of the amino group in the translocation/binding step.

An obligatory exchange of anions by the exchanger implies that one transport cycle must involve two anion-binding steps. This could be simultaneous or alternate binding. Experimental data support an alternate or ping-pong mechanism which implies that the proportion of the two alternate transport sites depends on the concentration of the substrate and inhibitors, as well as on the intrinsic asymmetry of the protein. This conclusion is supported by the following experiments:

1. The *inward facing* and the *outward facing* conformations have been identified by using inhibitors. Stilbene disulfonate derivatives competitively inhibit the binding only from the extracellular medium, whereas another disulfonate (APMB) inhibits from inside.
2. The rate of exchange also depends on the relative concentrations of anions on the two sides, and as expected the apparent K_t is lower at smaller intracellular chloride concentration.
3. The rate of exchange of chloride with phosphate is about 10% of the rate of exchange of chloride with chloride.

4. Several inhibitors like H_2 4,4′-diisothiocyano-2,2′-stilbenedisulfonate (DIDS), niflumic acid, and N-(4-azido-2-nitrophenyl)-2-aminoethylsulfonate NAP-taurine are more potent in the presence of an outward chloride gradient.
5. At fixed intracellular chloride (0.1 *M*) extracellular chloride activates exchange (K_t about 2 m*M*). Similarly, at fixed extracellular chloride, intracellular chloride promotes exchange (K_t about 50 m*M*).

Band 3 protein of the erythrocyte membrane is labeled by reagents that modify amino group and inhibit anion exchange. The anion-exchange function of the erythrocyte membrane is due to the dimer or tetramer of band 3 (Jeannings, 1984). Each band 3 monomer contains a single carbohydrate chain (at residue 660) of variable length, and it has no known function in transport. The amino- and carboxy-terminii are localized in the cytoplasm, and a significant middle part of the protein is localized as 7–10 helices in the bilayer (Fig. 10-6). A large part of the amino-terminal region (about 43 kD, which can be cleaved off by intracellularly applied trypsin) is apparently not involved in the anion-exchange function. This region is attached to cytoplasmic skeleton and it also binds to enzymes of glycolytic pathway and to hemoglobin. Proteoliposomes containing band 3 exhibit many of the features of the exchange system (Kohne et al., 1983). Band 3 from mouse has been inserted biosynthetically in oocytes by transferring mRNA (Morgan et al., 1985).

Water Transport through Anion Channels

Water and small solutes show anomalously high permeability through the erythrocyte membrane. The osmotic flow is about 3.6 times the rate of exchange, and the activation energy for osmotic flow of water is about 6 kcal/mole compared with 4.5 kcal/mole for self-diffusion of bulk water and 12 kcal/mole for diffusion through the lipid phase. The osmotic flow is inhibited by thiol reagents. Such observations have been explained in terms of an equivalent pore radius of 4.5 Å in the human erythrocyte, and it has been suggested that band 3 could provide this pathway for water and other small polar solutes (Solomon et al., 1983).

ADP–ATP EXCHANGER OF MITOCHONDRIA

The inner mitochondrial membrane mediates exchange of a number of metabolites of the tricarboxylic cycle and oxidative phosphorylation (Table 10-

1). Stoichiometric (1 : 1) exchange of extramitochondrial ADP with intramitochondrial ATP is a key step in the overall energy balance. The rate of ADP–ATP exchange in rat liver mitochondria is about 100 μmole/min per gram of protein, which corresponds to a turnover number of 500/min with 0.2 μmole binding sites per gram of total membrane protein. Apparent K_t are 3 μM for ADP from outside, and 10 μM for ATP from inside. The inner surface has a low affinity for ADP while the outer surface has a low affinity for ATP and AMP. Other naturally occurring nucleotides do not have significant affinity for the exchanger. The activation energy for the transport system is 28 kcal/mole below 18°C and 11 kcal/mole from 18 to 30°C.

The relative affinity of the ADP–ATP exchanger for nucleotides as well as the rate of exchange depends upon the state of mitochondria. For example, at low ATP concentrations, energization of mitochondria inhibits translocation much more than at higher ATP concentrations. The 1 : 1 exchange of ATP for ADP involves the net transfer of one negative charge in the direction of ATP transfer. The potential across the mitochondrial membrane is inside-negative; this would prevent transfer of ATP from outside to inside, and at the same time it would facilitate the exchange of ATP(in) with ADP(ex). Indeed, a linear correlation between membrane potential and the $(ATP/ADP)_{ext}/(ATP/ADP)_{int}$ has been observed. It suggests that a net electrical charge movement accompanies the exchange of nucleotides, and that a change in membrane potential and pH difference during energization can regulate the exchange.

The ADP–ATP exchange protein has marked affinity for atractyloside and carboxyatractyloside (K_i 10 nm which prevents binding of ADP from outside), for bongkrekate (1 μM, which prevents dissociation of bound ADP from the matrix or inside), and for long-chain acyl-CoA. The exchanger has been isolated as an α_2 dimer of 60 kD with pI = 9.8. It is the most abundant protein in the inner mitochondrial membrane, where it constitutes about 6% of the total protein. There is evidence that this highly asymmetric exchanger functions by nucleotide-induced concerted interactions between subunits (Klingenberg, 1981).

IONIC CHANNELS IN EPITHELIAL CELLS

Epithelia protect an organism against external stress while providing pathways for exchange, transfer, and secretion of solutes. To fulfil these functions epithelia have special organization: Sheets of single or multilayered cells form a more or less tight barrier at the apical (mucosal, luminal) border by close apposition of the apical cell membranes at the outermost cell-to-cell

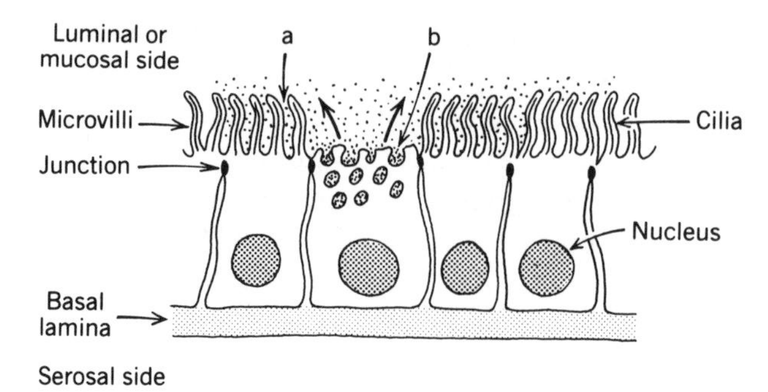

Fig. 10-8. Epithelial cells form coherent sheets called epithelia, which line the inner and outer surfaces of the body. Absorptive epithelial cells (a) have numerous hairlike microvilli projecting from their free surface (on the mucosal side) to increase the area for absorption; secretory cells (b) secrete substances into the surface. Adjacent epithelial cells can be of different types. Adjacent epithelial cells are bound together by junctions that give the sheet mechanical strength and also make it impermeable to small molecules. The sheet rests on basal lamina.

junction (Fig. 10-8). The outward membranes are markedly different from the serosal (basolateral) membranes both with respect to their microscopic structure and transport properties. These cells exhibit a large variety of transport and enzymic functions, and such structures were the first to be characterized for their bulk transport properties. However, progress in their biochemical characterization has been hindered by their morphological complexity. With the emergence of new techniques, interest in epithelial transport systems has been renewed. For example, patch-clamp techniques (Fig. 4-5) have shown the presence of separate Na and K channels in frog skin, rabbit colon, gall bladder, stomach, and locust hindgut. Only one Na permeability system is described in this section to illustrate the complexity of epithelial transport systems (van Driessche and Zeiske, 1985).

Pathways for Na appear to exist only on the apical cell membrane of epithelia whose physiological function is to control levels of Na^+, K^+, and to restore water balance between spaces with marked difference in osmolarity. The Na channels from such diverse sources as salivary glands, colon, urinary bladder, amphibian skin, trachea, and kidney tubules exhibit several common characteristics that are summarized below, although most of the information comes from studies on toad urinary bladder and frog skin.

1. Epithelial Na channels are permeable only to Na^+, Li^+, and possibly H^+. The conductance of these channels, γ, is about 3 pS and density is about 50–500 channels per μM^2. At high Na concentrations Na-permeability decreases as if these channels are closed at high sodium concen-

trations, and stimulated by anions like thiocyanate, iodide, nitrate, and chloride.

2. Pyrazine diuretics like amiloride (K_i about 1 μM) inhibit the channel, probably by directly blocking the ion binding site.
3. These channels are chemically modulated. Compounds containing strongly basic or cationic amino groups stimulate Na transport; however, considerable structural specificity is seen.
4. The total density of the channels is controlled by hormones like aldosterone, antidiuretic hormone like vasopressin, and prostaglandins. The effect of aldosterone is probably allosteric. Vasopressin seem to increase the density of channels.

PORINS

The wall of gram-negative bacteria consists of the outer membrane, the petidoglycan layer, and the inner membrane (Fig. 2-1). The inner membrane is the primary permeability barrier, and it contains the respiratory chain and a variety of other functional proteins (Di Rienzo et al., 1978). The peptidoglycan layer stabilizes the cell against lysis. The outer membrane of *Escherichia coli* and *Salmonella typhimurium* acts as a sieve through which hydrophilic solutes of up to MW 700 can pass, and this layer is impermeable to hydrophobic solutes. The sieving properties of the outer layer are attributed to porins which are found not only in bacterial walls but also in the outer membrane of mitochondria. Porin is a generic term, and porins from different organisms have very different sieving characteristics.

Porin from *S. typhimurium* (MW 40 kD) can be readily incorporated into vesicles and BLM (Benz et al., 1980). Properties of channels formed by porins exhibit considerable dependence on the lipid composition. Generally speaking, the conductance increase exhibits a first-order dependence on the protein concentration, which is probably present as a trimer in the detergent suspension. The *I-V* curve for the single channel is linear with γ about 90 pS and τ about 1 min. The diameter of the channel formed by porin is about 16 Å, and it is slightly (about three-fold) more selective for cations than for anions. The channel density in the bacteria is estimated to be about 10^{12} pores/cm^2.

Porins from other bacteria have similar properties, however, significant differences have been noted. For example, porin from *Pseudomonas aeruginosa* has shorter τ (about 100 msec), considerably greater anion selectivity (about 100-fold), and permits passage of solutes upto MW 6 kD (Benz et al.,

1983). Properties of porins induced in limiting growth media also differ, which reflects the role of porin in homeostasis. Several proteins that promote transport of nucleotides, phosphate, and sugars have been isolated from the outer membrane of gram-negative bacteria. The selectivity of these channels is generally seen if the molecular weight of solutes exceeds 600.

The primary sequence of porin is not particularly hydrophobic. The channel is formed largely from β-sheets oriented perpendicular to the plane of the bilayer (Garavito et al., 1983). Electron microscopy suggests that the pore is formed by a trimer in which three openings to the channel merge approximately in the middle of the membrane (Dorset et al., 1984).

CHANNELS FROM TOXINS

Proteins that damage membranes are produced by organisms ranging from prokaryotes to mammals. Two possible mechanisms appear to be involved. Many of these proteins (lipases, neuraminidases, glycosidases) have hydrolytic activity against many of the membrane components. Several toxins (e.g., pardaxin and mellitin) cause membrane damage by perturbing the organization of bilayers, by solubilizing lipids, or by creating nonspecific ion permeation pathways. On the other hand toxins like staphylococcal α-toxin, streptolysin-O, and complement $C_{5b\text{-}9}$ complex appear to form well-defined transmembrane multicomponent channels (Bhakdi and Tranum-Jensen, 1983). Several other toxins also use receptors on membranes for transport of their catalytically active subunit to the cytoplasm of the target cell. For example, the A fragment of diphtheria and *Pseudomonas* toxins is transferred to the cytoplasm by receptor-mediated endocytosis (see Chapter 15), where it ADP-ribosylates the diphthemide residue on the elongation factor EF-2. A single molecule of the A fragment is able to inactivate all EF-2 and thus cause cell death (Olsnes et al., 1985).

Immune cytolysis by complement occurs due to formation of channels. The complement system consists of 12 plasma proteins: C-1 thru C-9, and factors B, D, P, and a few regulatory proteins (I and H). The complement cascade is triggered via activation of C-1, C-4, and C-2 via B, D, P and C-3 + P which leads to the formation of a complex of C-5b to C-9 proteins, whose aggregate molecular weight is more than a million. Evidence for the formation of channels by this complex comes from the following experiments. The complex binds to detergents, liposomes, bilayer lipid membrane (BLM), and cells. On binding they release trapped markers and cause a discrete conductance change that is commonly seen with channels. Negative staining of membranes containing the complex shows hollow cylinders of 15 nm height

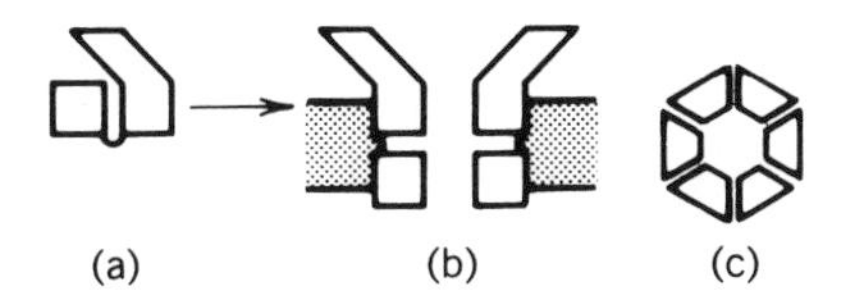

Fig. 10-9. A model for the assembly of α-toxin into a hexameric channel. (a) The monomer (side view) consists of two domains containing primary β-structure linked by a hinge. The top view (c) and side view (b) of the channel are also shown. On assembly the monomer opens up about the hinge, revealing an occluded hydrophobic surface. In the hexamer this surface is in contact with the lipid bilayer (or detergent), while the interior surface of the channel comprises residues that were originally on the surface of the monomer. (From Tobkes et al., 1985).

and a diameter of 10 nm. The complex also tends to solubilize membrane lipid, which depends upon the number of the complex molecules incorporated.

For the formation of channels many of the bacterial toxins undergo a transition from water-soluble form to an amphiphilic state. In many cases the protein channel can be isolated from membranes by immunoprecipitation or by mild detergent treatment. Thus for example, the α-toxin, produced by most pathogenic strains of *Staphylococcus aureus*, is readily incorporated

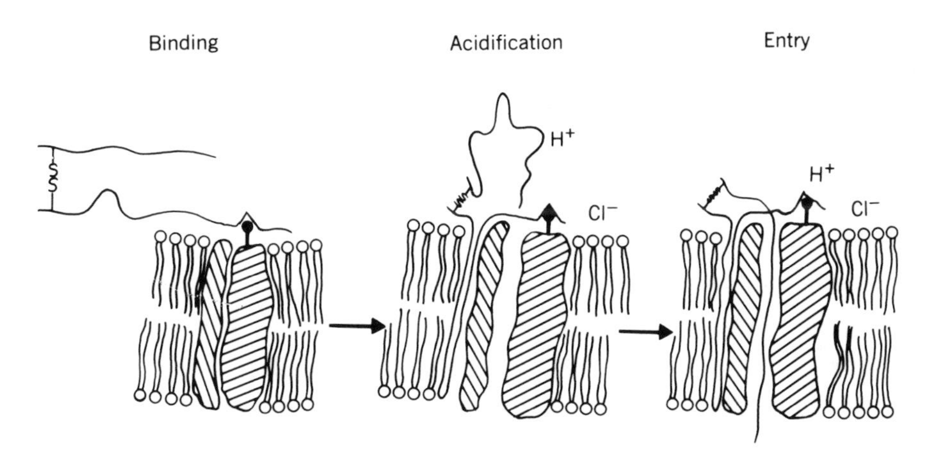

Fig. 10-10. A speculation about the entry of diphtheria toxin. The toxin binds to the receptor, which is probably the main anion transporter of the cells. When pH is reduced below 5.4 the hydrophobic domain of the B fragment inserts itself into the membrane in such a way that it inhibits the normal anion exchange. The inserted part of the B fragment could, together with the anion channel, form a pore which is sufficiently large to allow the A fragment to enter. A proton gradient across the membrane is required for entry. Ionic interactions between the peptide and bilayer could desolvate the microinterface and thus provide a hydrophobic site with local disorder.

into liposomes and erythrocytes and makes them leaky. It is a water-soluble monomeric protein (33 kD), and forms a transmembrane channel (annular diameter 10 nm, and the pore diameter 3 nm) by association of six subunits (Tobkes et al., 1985). Oligomerization of the toxin is triggered by detergents and lipids. The hexamer that is formed in detergent is not soluble in water, although it can be readily incorporated into preformed membranes. The toxin monomers have 6–10% α-helix content. The aggregated channel is apparently formed by domains of antiparallel β-sheet surrounding the central pore. A schematic model is shown in Fig. 10-9.

The mechanism of spontaneous insertion of peptides into preformed membranes is of considerable interest (Jain and Zakim, 1987) not only for the formation of pores but also for translocation of the signal sequence (Briggs and Gierash, 1986) and for the entry of picornaviruses and toxins like ricin, abrin, modaccin, viscumin, diphtheria toxin, and shigella toxin (Olsnes et al., 1985). One of the hypotheses is shown in Fig. 10-10. Two possible mechanisms are generally invoked. Binding of toxin to a protein receptor is followed by endocytotic uptake (Chapter 15). Under certain conditions the toxin bound to the surface could undergo a conformational change in which hydrophobic residues are exposed and thus insert into the bilayer, or the peptide chain could insert into a channel formed by the outer subunit of the toxin. A role of ion- or proton gradient has also been suggested.

11 Coupled Transport

More men are killed by overwork than the importance of the world justifies.

RUDYARD KIPLING
The Phantom Rikshaw

Most living organisms survive in an environment that has a low and fluctuating concentration of nutrients. This observation alone suggests that homeostatic mechanisms regulate the intracellular concentration of the energy-providing solutes, which is often higher than the extracellular concentration. Indeed, certain types of cells are able to establish more than a million-fold gradient of certain solutes. For example, the external pH of acid-secreting parietal cells of gastric mucosa is about 1, whereas the intracellular pH is maintained at 7. Such a transport of solutes against their concentration gradients is called *uphill* or *active* transport, i.e., uptake of solutes against their concentration gradients at the expense of metabolic energy.

The functional significance of the energy-driven uptake of nutrients by an organism may be appreciated by the fact that biosynthetically produced metabolites are energetically very expensive; about 40 moles of ATP are required for the synthesis of a mole of histidine or glucose. On a more subtle level, homeostasis of the internal environment in a cell is also necessary to maintain ionic and osmotic gradients, which in turn regulate turgor pressure, cell volume, constant pH and ionic composition, motility, reducing power, and ultimately the mass and energy balance that is necessary in the life of a cell.

The uphill transport processes exhibit many of the characteristics of the facilitated transport systems discussed in the last chapter: selectivity, substrate specificity, saturable rate of transport mediated by a few sites on the membrane, large temperature coefficient, modification by protein reagents, competitive inhibition, modulation by protein reagents, genetic inducibility, constitutive biosynthesis, and genetic impairment. In addition uphill transport systems are always unidirectional (*vectorial*) and require energy. The laws of thermodynamics dictate that the free energy required for uphill transport be greater than the steady-state concentration gradient that is produced. For an ionized solute with concentration C_1 and C_2 in the two compartments the free energy is:

$$\Delta\mu_{\text{ion}} = -RT \ln\frac{[C_1]}{[C_2]} + nF\,\Delta\psi$$

where n is the valence of the ion, F is the Faraday constant, and $\Delta\psi$ is the potential difference across the membrane. Thus, more than 1.36 kcal/mole of negative free energy is required for transfer of an un-ionized solute across a 10-fold gradient. This energy can be provided by the gradient of another solute (cotransport) or by an exergonic chemical reaction (active transport).

COTRANSPORT

A large class of membrane carriers is coupled obligatorily to cations for the translocation of the solute (Crane, 1977; Stevens et al., 1984). A carrier-mediated transport system requires formation of a binary CS complex in the interface as a prelude to the translocation step. As shown in Fig. 11-1, if the CS complex is not able to translocate or does not have enough driving force, a more mobile ternary complex CSX would translocate S along with X, as long as X is translocated down its concentration gradient. Cotransport systems have several constraints that regulate its efficiency (Heinz et al., 1975; Turner, 1983). These constraints relate to the order of binding and unbinding of S and X to C, their respective equilibrium constants, and the relative rates of translocation of C, CS, CX, and CSX from one interface to the other. It may be noted that the activator and the substrate binding sites are simultaneously or sequentially exposed to the aqueous phase on one side and then on the other side of the membrane. A cotransporter becomes an energy transducer as the electrochemical gradient of the activator or of the substrate drives the uphill transport of the other. This is because ultimately it is the gradient of the ternary complex, CSX, that determines the rate of translocation. The energy transduction occurs simply as a result of the stochastic cycling of the cotransporter, a process that can be accounted for and understood in terms of conventional reaction kinetics and equilibrium thermodynamics.

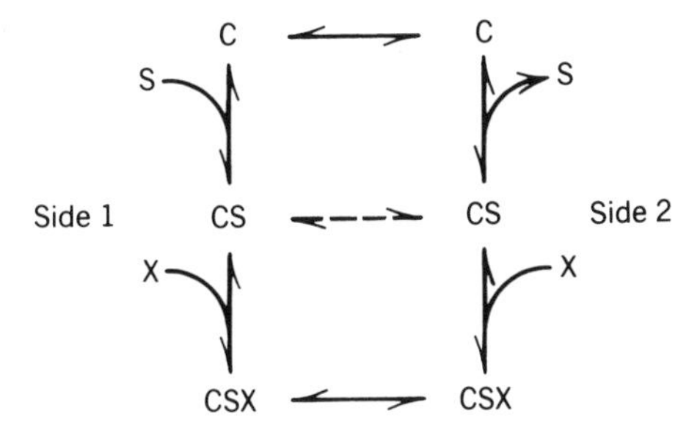

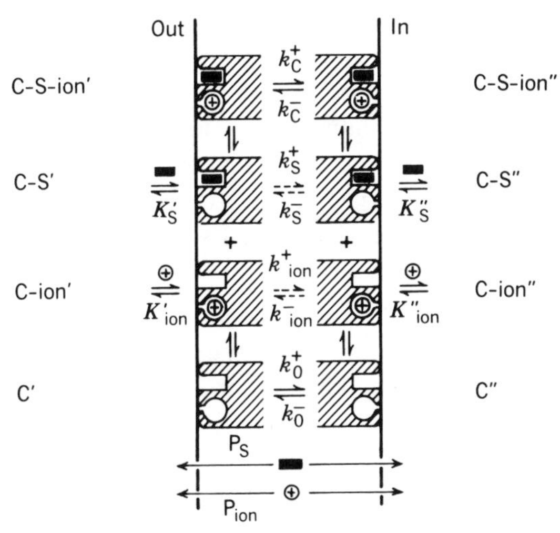

Fig. 11-1. A schematic representation of cotransport by simultaneous mechanism. (Top) The carrier (C) binds the substrate (S), and then the CS complex binds the ion X to form a ternary complex CSX. Normally only C or CSX are translocated, i.e., they undergo a transition between inward- and outward-facing sites. The strictly coupled translocation occurs when CS can not undergo such a transition. Under these conditions a gradient of X would drive S against its concentration gradient. The sequence of binding of S or X is not ordered. In the consecutive or "ping-pong" mechanism (not shown) S and X are translocated in mutually opposite directions. (Bottom) An elaborate representation of the cotransporter in terms of hypothetical binding sites.

Cotransport of Sugars

Several cotransporters driven by Na, K, or protons have been identified in a variety of organisms. Among these the cotransport of sugars and amino acids coupled to the gradient of Na ions is probably the best-characterized. The adsorptive endothelium (Fig. 10-8) of the small intestine is specialized in the uptake of nutrients from the lumen and deliver these to the blood and lymph vessels. The cotransporter operates in the brush borders (mucosal side), whereas transporters of low affinity are found on the serosal side across the lateral basal membrane. Na + glucose cotransporters in the choroid plexus, convoluted tubules of kidney, and intestinal brush borders have the following features (Crane, 1977; Turner, 1983; Semenza et al., 1984):

1. The cotransporter is localized in brush borders on the luminal side of small intestine. It translocates glucose from the luminal (mucosal) to the contraluminal (serosal) side with K_t of about 2–15 mM and an estimated turnover number of about 20/sec. There is indication that there are two Na + glucose transporters in intestinal brush border.
2. The cotransporter is specific for D-glucose, which is not chemically altered during the transport. It does not transport L-glucose or other monosaccharides. It is inhibited by phlorizin (K_i 5–20 μM), however, binding of phlorizin depends upon the concentration of sodium as well as the membrane potential, $\Delta\psi$. Cytochalasin B, which blocks the transporter of glucose on the serosal side, does not block the Na + glucose cotransporter.
3. Probably one Na ion is transported along with each glucose molecule. Transport is not observed in the absence of Na or in the presence of K, Rb, Cs, choline, Tris, guanidinium, or ammonium ions. Net uptake of glucose is completely blocked by ionophores such as monensin or gramicidin D, which dissipate the gradient of Na ions. The significance of many of these results is not easy to evaluate because the transport is also modulated by membrane potential.
4. At low Na ion concentration the gradient of glucose alone can drive the transport of glucose down its electrochemical gradient. Under physiological conditions the uphill cotransport of glucose is driven by a large downhill gradient of Na from the mucosal to the serosal side. Up to a 20-fold gradient of glucose can be achieved.
5. The rate of transport shows a Michaelis–Menten type of hyperbolic relationship for glucose concentration versus the rate of uptake. The effect of Na ions on K_t and V_m varies from species to species. In hamster intestine a decrease in the Na concentration increases K_t for the sugar without any effect on V_m. In rabbit ileum Na ions increase V_m without any significant effect on K_p. The Na : glucose stoichiometry appears to be 1 : 1 under most conditions. Also the order of binding and dissociation is probably C to CS to CSX (see Turner, 1983 for a discussion).

6. Translocation of the ternary complex requires a net movement of one positive charge in the direction of glucose uptake, i.e., the transport is electrogenic. The overall transport may appear electroneutral because translocation of cationic CSX is compensated by the resting membrane potential, or by the electrogenic Na pump operating in the opposite direction, or by leakage of anions in the direction of translocation. The inside negative membrane potential enhances the rate of Na-driven influx of D-glucose as well as the rate of binding of phlorizin from the outside.
7. The cotransporter is functionally asymmetric: The binding site for glucose is more easily accessible from the cytosolic side, it exhibits *trans*-inhibition by glucose or Na at $\Delta\psi = 0$.

The Na-glucose cotransporter is apparently coupled to or is in the vicinity of the digestive enzymes like sucrase–isomaltase complex (220 kD) localized at the outer (luminal) side of the brush border (Semenza, 1986). It can be detached from the brush border membranes by detergent solubilization or by treatment with papain. The complex has been resolved into two catalytically active subunits: a sucrase and an isomaltase. It is interesting to note that specific sugar intolerance, a genetic defect in humans, is associated with the lack of sucrase activity in the intestinal brush border.

Sucrase has several interesting properties. It hydrolyzes sucrose to glucose and fructose; it is activated by sodium; it is inhibited by phlorizin (K_i 8 μM) as well as alkali metal cations. The K_m and V_m for the catalytic activity and the corresponding values for transport are similar. Antibodies against sucrase (which do not inhibit the catalytic activity) inhibit sugar uptake. On the other hand Tris inhibits only sucrase but not the transport activity, presumably because the Tris binding site may not be available in intact cells. The transport and catalytic activities are blocked by conduritol β-epoxide. These observations suggest that the sodium-coupled cotransporter of glucose in intestinal mucosa is modulated by sucrase. The precise role of sucrase in the transport process is not understood. Most of the monosaccharides liberated by the action of disaccharidases are retained in the brush border region by its fuzzy coat and thus favor its absorption totally. No functional interaction has been detected between catalytic centers and the hydrophobic layer of the brush border membrane. The amino-terminal end of the isomaltase subunit contains a hydrophobic helical segment (residues 12–31), suggesting a possible role in anchoring to the membrane.

Attempts to characterize and purify the Na + glucose cotransporter protein have been largely unsuccessful, although considerable effort has been invested (Semenza et al., 1984). Studies with isolated vesicle preparations have yielded results that are largely consistent with those obtained with intact tissues.

Cotransport of Amino Acids

Several transport systems have been reported for amino acids (Christiansen, 1984) in avian and mammalian cells. Most cells absorb neutral amino acids via four Na-dependent and three Na-independent transport systems having partially overlapping substrate specificities. Anionic amino acids are similarly absorbed by a Na-dependent and a Na-independent system, whereas cationic amino acids are absorbed only by one Na-independent system. This classification is based mostly on the work with nonpolar and nonepithelial cells.

The Na-cotransporter for amino acids in epithelial cells exhibits essentially the same characteristics as the Na + glucose transporter. Jauch and Lauger (1986) have characterized the electrogenic properties of Na + alanine cotransporter of isolated and perfused pancreatic acinar cells by patch recording with microelectrodes. Results show that K_t for external alanine is 18 m*M*, and 64 m*M* for external Na. The coupling ratio is 1 : 1, the transport cycle is electrogenic, and the coupled transport involves a conformational transition of the ternary CSX complex between the inward- and outward-facing sites. These results are consistent with the simultaneous mechanism in which formation of the CSX complex is not ordered. With certain assumptions, the kinetic parameters could also be calculated. Detailed analysis of the *I-V* curves suggests that part of the potential drop for the ion being translocated occurs before arriving at the binding site, and a part of the drop occurs during translocation of the Na binding site within the membrane dielectric.

Na, K, 2Cl Cotransporter

Numerous types of animal cells have this quaternary electroneutral cotransporter which is inhibited by furosemide (Geck and Heinz, 1986). It is electrically silent and all the ions are transported in the same direction. The partial reactions of this cotransport could apparently account for a variety of other cotransporters and active transporters. The main function of this transporter is apparently in volume regulation. In avian erythrocytes this electroneutral cotransporter is activated by AMP as well as by catecholamines. The effect of cAMP could be on the volume sensor mechanism.

Lactose + Proton Cotransporter

This product of the *lac y* gene in the membrane of *Escherichia coli* cotranslocates galactosides at the expense of the proton gradient. As shown in Fig. 11-2, the proton gradient generated via the respiratory chain or through hydrolysis of ATP by proton-ATPase, drives the uptake of β-galactosides to generate a 100,000-fold gradient. The cotransporter behaves as if it were formally negatively charged and the transport of sugars is coupled with the transport of protons from outside to the cytoplasm of the cell. Since the flux of the substrate is coupled to the flux of protons, a downhill translocation of the substrate results in an uphill transport of protons and generation of protonmotive force, pmf.

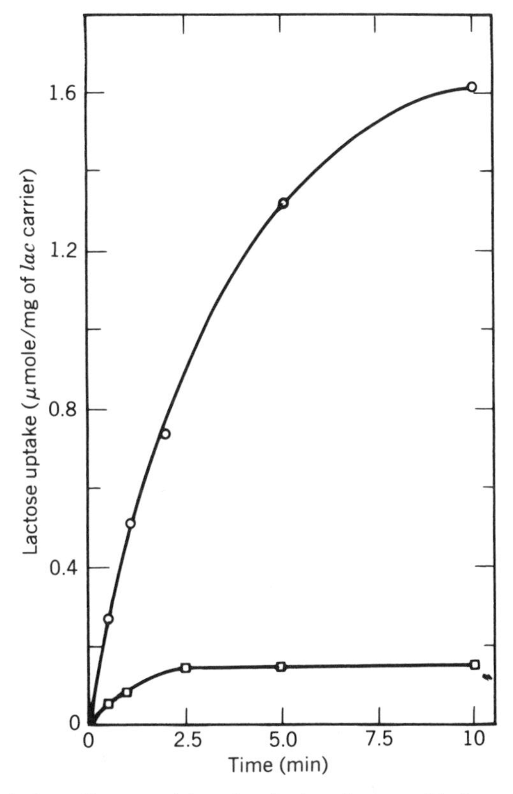

Fig. 11-2. Accumulation of lactose driven by electron transport in liposomes reconstituted with purified lactose cotransporter and by cytochrome oxidase. Vesicles were prepared by octylglucoside dilution followed by freeze-thaw/sonication. Lactose transport was measured in the absence (squares) and presence (circles) of reduced quinone as the substrate for electron transport. Values similar to those with squares were also obtained in the presence of valinomycin and nigericin. (From Kaback, 1983.)

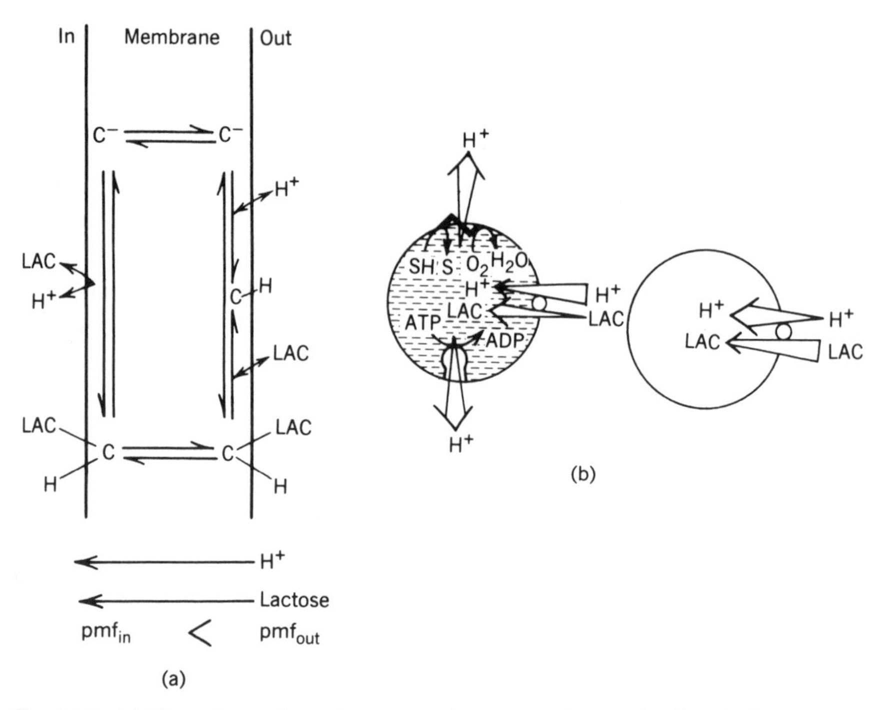

Fig. 11-3. (a) The scheme for cotransport of proton + lactose in *E. coli*. The order of substrate binding at the inner surface of the membrane is not implied. (b) In intact cells the proton gradient is generated by electron transport or by hydrolysis of ATP. (From Kaback, 1983.)

As implied in Fig. 11-3, translocation involves formation of a ternary complex and its reorientation to the cytoplasmic side where the complex dissociates, and then the empty carrier reorients for another turnover cycle. Because all the steps in the translocation cycle are reversible, the carrier also catalyzes the coupled movement of galactoside and proton from inside to outside. The turnover number of the carrier for both the facilitated diffusion and active transport is in the range 10–100/sec for different substrates (Table 11-1). The rate of of transport is not significantly affected by treatment of cells by energy poisons such as azide and iodoacetate. Thiol reagents inhibit transport, and other sugars do not offer any protection against transport. Nonmetabolizable sugars accumulate in sufficient quantity to swell osmotically and lyse protoplasts. The uptake of galactosides is significantly affected by lipid composition. Arrhenius plots show discontinuities and breaks, the temperature of which depends upon the acyl chain composition. The thermotropic transition of the bilayer is probably mirrored in these discontinuities, however interpretation of this data is not straightforward.

TABLE 11-1 Comparison of Turnover Numbers for the *lac* Carrier Protein: ML 308–225 Membrane Vesicles versus Proteoliposomes Reconstituted with Purified Carrier

	Turnover Numbers (sec^{-1})	
Reaction[a]	Membrane Vesicles[b]	Proteoliposomes
$\Delta\Psi$-driven influx ($\Delta\Psi$ = 100 mV)	16 (K_t = 0.2 mM)	16–21 (K_t = 0.5 mM)
Counterflow	16–39 (K_t = 0.45 mM)	28 (K_t = 0.6 mM)
Facilitated diffusion	8–15 ($K_t \simeq$ 20 mM)	8–9 ($K_t \simeq$ 3.1 mM)
Efflux	8 (K_t = 2.1 mM)	6–9 (K_t = 2.5 mM)

[a] All reactions were carried out at pH 7.5 and 25°C.

[b] Determination of the amount of *lac* carrier protein in ML 308–225 membrane vesicles is based on photolabeling experiments with [^{3}H]NPG, which indicate that the carrier represents about 0.5% of the membrane protein.

For example, the effect of the thermotropic phase change is primarily on K_t and not on V_m.

Experimental support for the cotransport model for the galactoside carrier is based on the following observations (Wright et al., 1981; Kaback, 1983). The carrier contains one galactoside binding site per carrier molecule in the membrane as well as in a detergent micelle. Both the binding and the translocation steps are inhibited by thiol reagents, histidine reagents, (diethylpyrocarbonate), and plumbagin—a lipophilic oxidizing agent. Binding of protons is inferred indirectly from the coupled movement of protons and galactosides. The pK_a of 8–10 has been inferred for the transported proton. Apparent K_t for translocation and the binding constants from either side of membrane is same in the absence of an electrochemical proton gradient.

The galactoside transport system mediates equilibration of the electrochemical potential gradients of galactose and protons (Fig. 11-3) according to the chemiosmotic energy conservation equation:

$$\Delta\mu = \Delta\psi \cdot F - 2.3\ RT\ \Delta\text{pH}$$

The stoichiometry for coupling is one under most experimental conditions. Thus, a 1000-fold gradient of galactoside is created in response to an electrochemical proton gradient of 180 mV directed from outside to inside. The electrochemical potential difference across the membrane also changes the affinity of the carrier. In the absence of the proton motive force, K_d and K_t for lactose is 14 mM. However, these constants decrease to 0.1 mM under the conditions of active transport. A mutant defective in coupled transport has been characterized, which implies that translocation of lactose can probably occur without proton gradients.

The purified lactose + proton cotransporter has been reconstituted into vesicles where it translocates lactose on the imposition of the proton gradient. The gene of lactose carrier (33 kD) has been sequenced. The gene product contains a single peptide of 417 residues, 70% of which are hydrophobic (Mieschendahl et al., 1981). Hydropathy plots and a high helix content suggests that eight or more helices span the membrane. Both the amino and carboxyl terminii are exposed to the cytoplasmic side. Direct interaction of the cotransporter on the cytoplasmic face with the factor III of phosphotransferase system has been observed: Transport of galactosides via the cotransporter is inhibited in the presence of sugars transported by the phosphotransferase system.

PHOSPHOTRANSFERASE SYSTEM

An uphill transport of solutes in certain organisms is accompanied by derivatizing the solute either during or after its translocation across the membrane. The modified solute does not leak out of the cell at the rate at which its parent solute is transported inside. Such uptake requires energy for the chemical modification of the solute; the derivatized solute accumulates; the transport is vectorial; and the reactions occur asymmetrically. One of the best characterized systems of this type is the phosphoenolpyruvate (PEP)-dependent phosphotransferase system (PTS) for the transport of sugars in anaerobes and facultative aerobes (Roseman and Postma, 1976). PTS is also intimately involved in regulating the utilization of other carbon sources by these organisms.

Studies with intact cells and isolated vesicles have demonstrated that the phosphorylation of sugar by PTS is intrinsically linked to the transport across the bacterial membrane (Fig. 11-4). The reaction is specific for PEP as a phosphate donor, although a variety of sugars can serve as acceptors. Isolated washed membranes catalyze the transfer of phosphate from phospho-HPr to sugars; however, in intact cells, the component III or IIA is phosphorylated first and then phosphate is transferred to the sugar. Most of the sugars are phosphorylated in position 6 and fructose is transported as 1-phosphate ester. The K_t for most sugars is about 0.1 mM. Several related transport systems with different sugar specificities can also be induced. The phosphotransferase system is also modulated by the proton gradient and by cyclic AMP (Dills et al., 1980). Since proton-coupled sugar transport system in bacteria also carry out a similar function, in such cases the phosphotransferase system takes precedence.

The PTS system consists of several components. The proteins such as

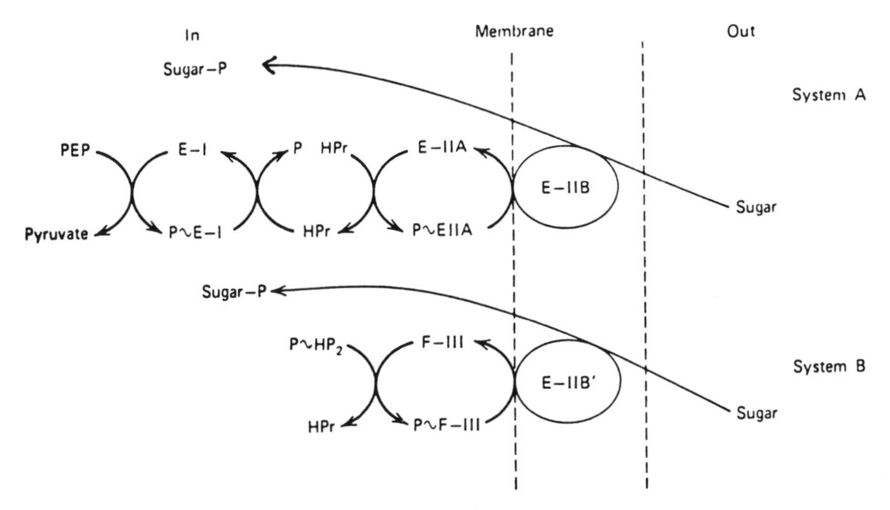

Fig. 11-4. Coupled reactions involved in the transfer of phosphate from phosphoenolpyruvate (PEP) to the sugar across the bacterial cell membranes by the various components of the phosphotransferase system (PTS) localized on the cytoplasmic face of a bacterial membrane. Two variations are observed: K_t for glucose is about 0.2 mM in the system A and about 5 μM in the system B.

enzyme I (E-I) and HPr are involved in the transfer of the phosphoryl moiety from PEP to the enzyme II (E-II) complex. E-I and HPr are common to all PTS systems in a given cell. Osmotically shocked cells lose their transport function, which can be restored by the addition of purified HPr. HPr from different cell types show considerable cross-reactivity. The sugar specificity of each PTS is determined by a group of constitutively synthesized or inducible proteins that make up the E-II complex. Solubilized and fractionated membranes yield E-IIA and E-IIB, and the A unit can be fractionated into sugar-specific components. Reconstitution of the enzyme activity requires a specific order of mixing of E-IIA, E-IIB, PG, and a divalent metal ion like Mg^{2+} or Ca^{2+}. When the order of mixing is E-IIB, ion, PG, and then E-IIA, a particulate active preparation is obtained. About 50 molecules of PG are required foı optimal activation of E-IIB, and PG cannot be replaced by most other lipids present in the bacterial membrane.

Histidine Transport

Salmonella typhimurium contains a periplasmic or shock-sensitive transporter that accumulates histidine against its concentration gradient. Some of the components of this transporter are located in the periplasm and can be released by osmotic shock (Fig. 11-5). It is one of the five histidine transport

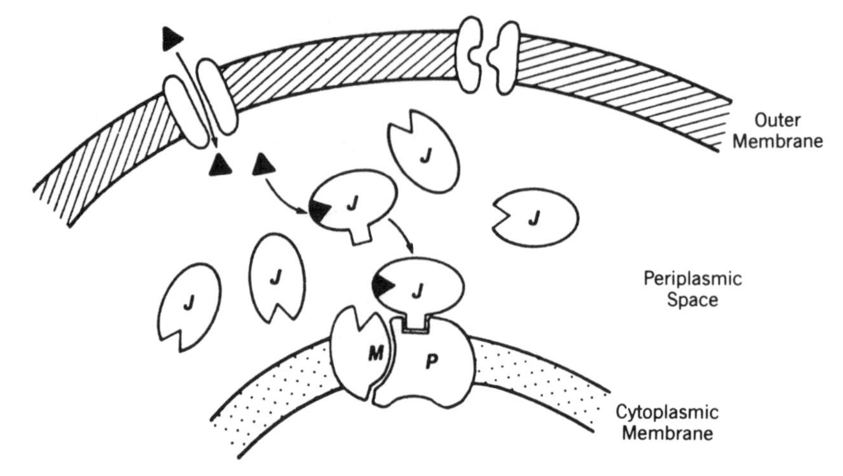

Fig. 11-5. Possible architecture of shock-sensitive periplasmic cotransporters. Several gene products are involved in the overall transport. For example three (M, P, J) membrane-bound proteins are represented in the translocation step. The role of the binding proteins (J) is implicated in accumulation of the solute in the periplasmic space, and also in the translocation step.

systems present in this organism, and a similar system has been identified in *Escherichia coli*. It consists of four gene products; three of these are localized in membranes and one (MW 25 kD) binds histidine (Higgins et al., 1982). Like other binding proteins (Hengge and Boos, 1983), the histidine-binding protein undergoes a change in conformation on binding its substrate. The complex of three proteins that is localized in the membrane can translocate histidine only when the binding protein with its substrate is also present. Very little is known about the function or the energy source of these cotransporters, analogs of which may also be present in eukaryotic cells. The periplasmic solute-binding proteins have also been implicated in chemotactic response of bacteria (Chapter 13).

Other Periplasmic Transporters

Many other periplasmic transporters, like the histidine transporter, have been described (Ames, 1986). Apparent K_t for solutes are typically 20–200 nM for the transporter as well as the binding proteins released during osmotic shock. The heat-stable binding proteins are water soluble, monomeric (MW 20–40 kD), and have one binding site. They appear to undergo a conformational change on binding of substrate. Their role in transport is implicated in the model shown in Fig. 11-5.

Regulation of Transport Systems

Transport of exogenous solutes into cells is the first step in the overlapping and interlocking cellular pathways for energy and metabolism. The diversity of energy coupling mechanisms operating on a wide range of transporters attests that they constitute alternative solutions to the problem of physiological regulation. At least five different mechanisms regulating uptake of sugars by bacteria have been demonstrated (Dills et al., 1980): regulation by phosphotransferase system of non-PTS sugar uptake; regulation by intracellular sugar phosphates; competition of the PTS-sugar uptake systems for the common phosphoryl donor (HPr); competition between external sugars for the substrate-binding site of a permease; regulation by proton electrochemical potential; expulsion of preaccumulated sugars.

Regulation by expulsion of sugars responds to stimuli in the form of variations in energy levels and metabolite concentrations by changing the direction and the rate of sugar transport. This response is achieved by covalent modification of a transport protein catalyzed by an ATP-dependent protein kinase.

12 Gated Channels

The mask of theory
over whole
face of nature . . .
WILLIAM WHEWELL

Intra- and intercellular transfer of information in a variety of cells occurs via channels that open and close in response to electrical and chemical signals. For example, transmission of nerve impulse, visual excitation, cell-to-cell communication through gap junctions, and synaptic transmission occur through gated channels of many different types present in a wide range of cells, especially in sensory organs (Latorre et al., 1985). The concept of voltage-gated channels was introduced in Chapter 9, and a schematic representation of channels gated by binding of a ligand or by an applied voltage is shown in Fig. 12-1.

Both, the voltage- and ligand-gated ionic conductance changes are used by nerve cells of central and peripheral nervous system to process sensory information. A schematic diagram of a nerve cell is shown in Fig. 12-2. It emphasizes several characteristic morphological features that are not normally seen in other types of cells. Besides the body (soma) of the cell it contains axons, dendrites, synapses, node of Ranvier, and nerve terminals. These morphological processes are a functionally integral part of the sensory information-processing network.

There are about 10^{11} nerve cells in the human body and they are functionally interconnected by a network of axons, dendrites, and synapses. The chemical and electrical communication between the nerve cells ultimately occurs through 10^{15} synaptic junctions between the nerve endings and body of the nerve cell or the muscle endplate. Many of the transport processes responsible for conduction and propagation of nerve signals through nerve

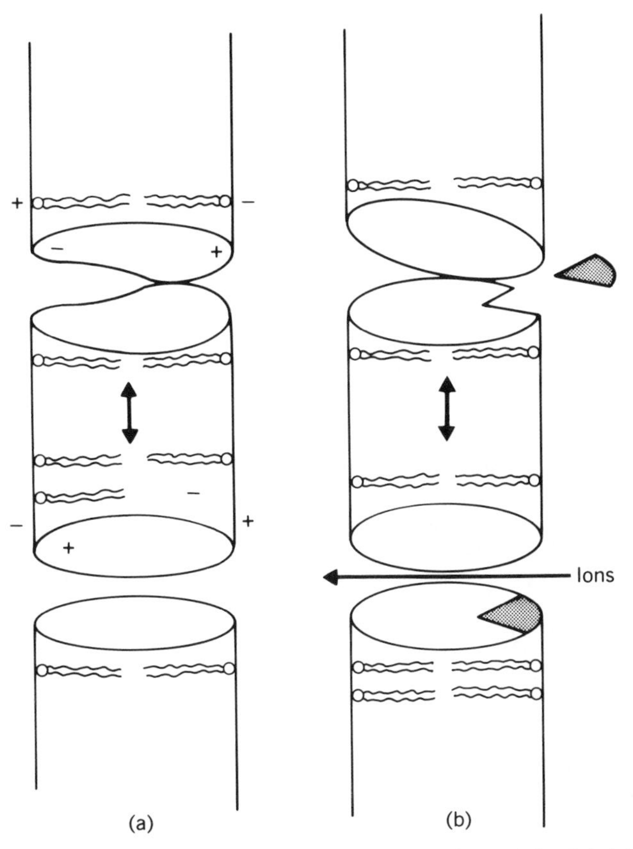

Fig. 12-1. A cartoon of a voltage-gated (a, left) and ligand-gated (b, right) ion channel. These are meant to illustrate the concept that an equilibrium between the open and closed conductance pathways can be shifted by changing the transmembrane potential or by binding of ligands, or by covalent modification (not shown). The gating mechanism of the molecule may be only a part of the transmembrane channel.

cells and axons are voltage gated (Table 9-6). Transfer of incoming electrical signals at the chemical synapses present at nerve terminals occurs by releasing neurotransmitters from their presynaptic terminal regions and inducing a change in the permeability to ions in the postsynaptic membranes of communicating cells. The channels in the postsynaptic membrane are often chemically gated, i.e., they open (excitatory synapse) or close (inhibitory synapse) in response to the binding of a transmitter. Due to the presence of such ligand-gated channels, chemical synapses act as efficient molecular amplification devices for sensory inputs as well as for the transmission of nerve signals over long distances. The activity of excitable cells ultimately generates a nonelectrical response, such as contraction, secretion, or a change in

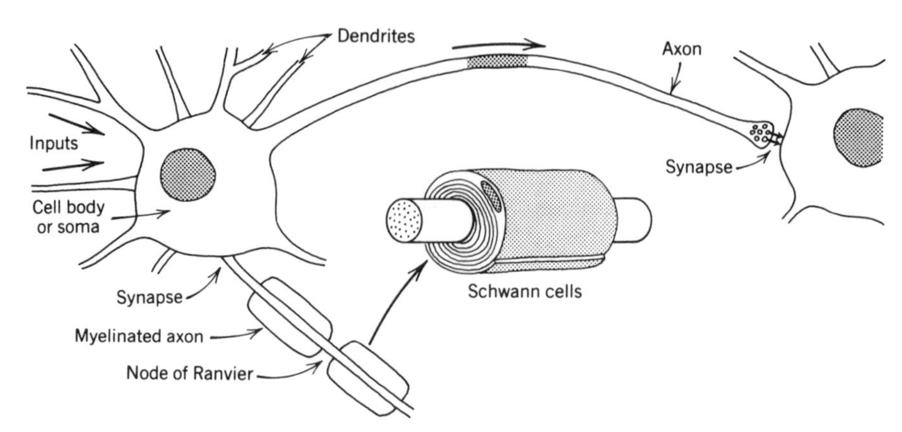

Fig. 12-2. Cartoon of a nerve cell. Nerve cells or neurons are specialized for communication. The brain and spinal cord, for example, are composed of a network of neurons among supporting glial cells. The functional parts of the neuron (their relative sizes are exaggerated) include soma or body, axon, dendrite, and synapse. Specialized cells, called Schwann cells or oligodendrocytes, wrap around some axons to form a multilayered membrane sheath called myelin. Gaps in this wrapping are nodes of Ranvier in myelinated axons. The sensory input from a neuron arrives to the nerve terminals where it is chemically or electrically transmitted to the next neuron via a synapse.

the intracellular metabolism. These are initiated via mechanisms that involve channels that open or close in response to the incoming electrical signals. The resulting change in the cytoplasmic levels of the ions like calcium leads to an appropriate biochemical and physiological response of the target cell. Mechanisms underlying the chemically and electrically gated processes are beginning to emerge, and salient characteristics of some are discussed in this chapter.

ACETYLCHOLINE RECEPTOR

There are several types of acetylcholine receptors (AChR). The nicotinic AChR has an ion channel that opens on binding of acetylcholine and its synthetic analogs (also called agonists) like carbamoylcholine or suberylcholine. One of the best characterized AChR is found on postsynaptic membranes of nerve terminals and neuromuscular endplates. Cholinergic synapses are also found in a large variety of organs and cells, and they have become a model for the chemical synapses in general. The soma and dendrites of a typical motoneuron of the spinal cord (one of the first systems to be examined in detail) are covered with synaptic boutons of the type schematically shown in Fig. 12-3. The number of such synaptic contacts on a

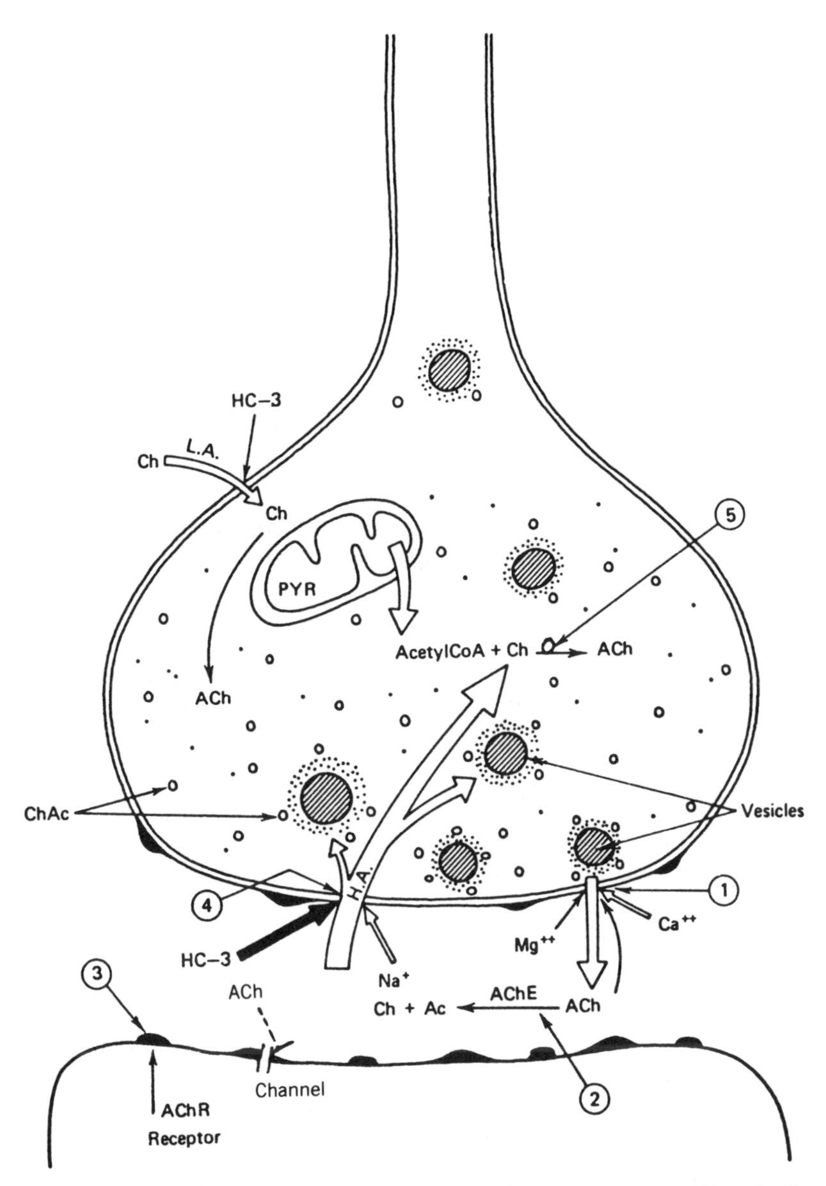

Fig. 12-3. A "road-map" for the synthesis, storage, release, response, and inactivation of acetylcholine at a cholinergic synapse. Choline (Ch) is taken up from the synaptic cleft by a high-affinity (H.A.) transporter, and from other parts by a low-affinity (L.A.) transporter. Inside the presynaptic terminal, choline is acetylated by Ch acyltransferase (ChAc) to acetylcholine (ACh), which is stored in the synaptic vesicles. On arrival of the presynaptic stimulus, the vesicles stored near the cleft fuse with the membrane and release their content in the cleft. ACh released in the cleft interacts with acetylcholine receptors (AChR) on the postsynaptic membrane to open the gated channel. Ultimately ACh is removed by hydrolysis catalyzed by acetylcholinesterase (AChE), which is present in the synaptic region. This cycle can be interrupted by a variety of inhibitors: (1) release of ACh by botulinum toxin (K_d 1 pM); (2) AChE by physostigmine (10 μM) and eserine; (3) AChR by curare, histrianicotoxin, venom toxins, clozapine, benztropine, and platyphyllin; (4) uptake of Ch by hemicholinium-3; (5) ChAc by oxotremorine and 4-(1-naphthylvinyl)pyridine. The release of ACh is activated by Ca, and the choline uptake is mediated by a Na cotransporter.

typical motoneuron may be upward of 2000 per cell and they are randomly distributed. Most of the boutons are about 100 nm in diameter and the spacing of the cleft is about 20–50 nm. The cleft is filled with extracellular fluid with which neurons are surrounded. This region also contains enzymes like acetylcholinesterase that hydrolyzes acetylcholine. Choline is taken up by the presynaptic membrane, where choline acetyltransferase mediates synthesis of acetylcholine. The space enclosed by the presynaptic membrane also contains mitochondria and neurofibrils along with about 300,000 synaptic vesicles (diameter 20–50 nm) containing about 1 *M* acetylcholine. The synaptic vesicles are localized near the cytoplasmic face of the presynaptic membrane, and about 50 vesicles fuse with it to release acetylcholine (in less than 1 msec) in the cleft in response to the presynaptic action potential. The mechanism by which acetylcholine is quantally released is not clear, however a role for Ca ions taken up during the action potential has been implicated.

Transmission of electrical signals across a chemical synapse involves several steps including synthesis, storage, release, and inactivation of the transmitter, as well as the interaction of the transmitter with a specific postsynaptic receptor. In response to a presynaptic electrical signal at a cholinergic synapse of the type just described, acetylcholine stored in vesicles in quantally (in packets of 6000–10,000 molecules) released in the synaptic cleft. The amount of the transmitter released in the cleft depends upon the frequency of the incoming signal. Typically 300 quanta are released synchronously in response to each incoming signal, which brings the concentration of acetylcholine to about 0.3 m*M* in the cleft. Acetylcholine released from the presynaptic membrane diffuses across the cleft to interact with specific receptors on the postsynaptic membrane. This interaction causes an increase in the conductance for cations (Fig. 12-4a). Since the resting membrane potential is about −65 mV, and each quanta of acetylcholine decreases the membrane potential by about 0.5–1 mV (called miniature endplate potential or mepp). The total depolarization (decrease in membrane potential that is called endplate potential) of about 5–10 mV is due to summation of mepp. The endplate potential triggers muscle contraction or initiates an action potential that travels along the membrane of the target cell. Although the action potential is an all-or-none response (described later in this chapter), the increase in membrane conductance caused by the acetylcholine is a graded response that depends upon the square of the concentration of the transmitter released in the cleft, i.e., two molecules of acetylcholine are required for opening the conductance pathway. The increase in the conductance of the postsynaptic membrane dissipates when the acetylcholine is hydrolyzed by acetylcholinesterase, which is also present in the cleft. The

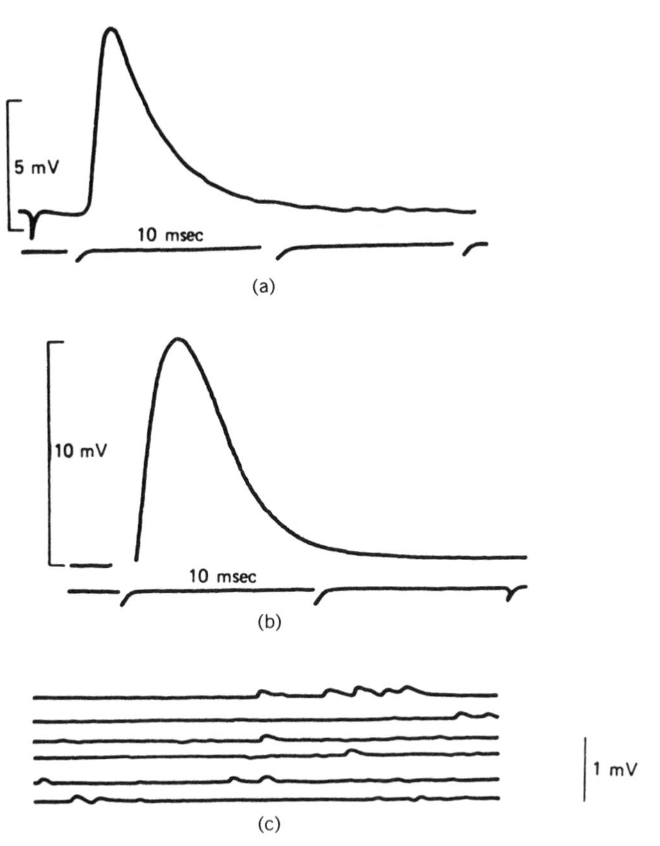

Fig. 12-4. (a) The time-course of the endplate potential (epp) evoked by stimulation of phrenic nerve in a rat diaphragm muscle fiber paralyzed by 15 m*M* Mg^{2+}; (b) epp evoked by an iontoporetic application of acetylcholine; (c) spontaneous miniature endplate potentials (mepp) of frog muscle endplate. The mepp arise from quantal release of acetylcholine from presynaptic vesicles.

transmission of an electrical signal across a chemical synapse in general, and across the cholinergic synapse in particular, exhibits the following characteristics:

1. A *delay* occurs during the passage of the information through the synaptic junction. Such a delay is intrinsic to the strength and the latency of the release mechanism, concentration of the transmitter in the cleft, size of the cleft, and the threshold level of the postsynaptic receptor.
2. The *duration* of the action of the transmitter is largely determined by the kinetics of its inactivation or reuptake.

3. *Specificity* of the synaptic transmission is seen at all levels ranging from release, action on receptor, and inactivation, as well as the structure of the agonist and antagonist.
4. Transmission of the signal through a synapse is *unidirectional.*
5. The postsynaptic conductance is slowly *desensitized* on prolonged exposure to agonists.

Both the number and density of the synapses varies extensively in different types of cells. There is only one synapse per muscle fiber. Higher density is found in cholinergic motor nerves, autonomic preganglionic trunks, and parasympathetic postganglionic fibers. The electroplax of electric fish (electric eel or torpedo fish) have more than 10,000 synapses on one face of the electrocyte, while the opposite noninnervated face is rich in Na,K-ATPase. The membranes from these two faces can be readily separated and provide a rich source of the ATPase and AChR. The receptor has been purified, cloned, and reconstituted in bilayer lipid membranes (BLM) and vesicles. Its behavior is in accord with its function as a channel that opens on interaction with acetylcholine, i.e., AChR has a channel that is gated by binding of acetylcholine.

Gating Properties of the AChR Channel

AChR channels have been studied at several levels. The increase in membrane permeability on binding of acetylcholine has been observed in intact tissues, in isolated membranes, and in reconstituted vesicles, as well as by the single-channel conductance and by the analysis of conductance fluctuations. Thus, endplate potential is evoked at neuromuscular junction on application of acetylcholine. In a patch-clamp on a section of membrane of about 10 μm^2, discrete step-like changes are seen on exposure to agonists (Fig. 12-5). For such experiments it is necessary to use agonists that are not readily hydrolyzed by acetylcholinesterase or to include its inhibitors in the medium. Direct resolution of single channels from synaptic junction or muscle endplate is difficult because of the high density of channels; they are also difficult to seal against the patch electrodes. Therefore most of the single-channel studies on AChR have been done of sparsely populated denervated or uninnervated cells. The conductance fluctuations of clustered synapses, from uninnervated fibers, or from the fibers where the receptors are diffusely distributed, appear to have the same channel parameters, however, a substate of 10 pS has been reported in uninnervated embryonic rat muscle (Hamill and Sakman, 1981). It may be noted here that channels of different conductance are often seen in most membranes.

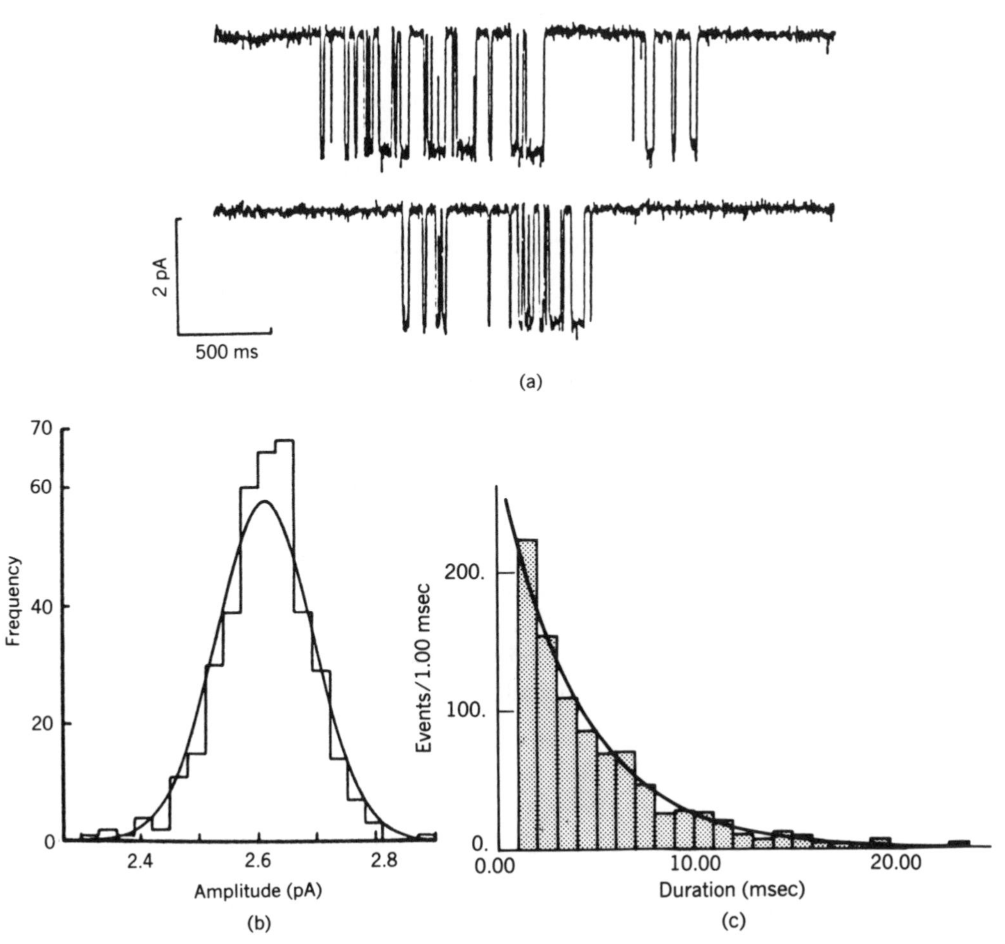

Fig. 12-5. (a) Recording of single-channel currents induced by acetylcholine. (b) Histogram of distribution of open-times (931 events) from single-channel (AChR) recordings; smooth curve shows a single exponential probability density function (normalized and weighted for the number of observations that lie under the fitted curve) with time constant of $\tau = 4.16$ msec or a rate constant of $1/\tau = 240\ s^{-1}$. (c) Histogram of the distribution of the amplitudes of single-channel currents; the smooth curve is a calculated Gaussian distribution. These three sets are not from the same experiment.

The channel parameters derived from such studies are similar to those obtained from the analysis of conductance fluctuations at constant low levels of agonists. The following assumptions are usually necessary for the analysis of the data:

1. The rectangular current pulses are due to random fluctuation of channels between open and closed states according to a Poisson process,

i.e., the behavior of channels is independent of each other and they do not have a memory of the past events.

2. All channels in a patch are homogeneous and have identical behavior.
3. The fraction of open channels at any given instance is low.

The spectral density of current fluctuations predicted from these assumptions suggests that the single-channel conductance is about 25 pS (corresponding to a depolarization of 0.3 μV) and the channel lifetime is about 1 msec. Thus, about 12,000 ions pass through a channel during its mean lifetime. The rate constants for ion translocation and for transition between open and closed states have been obtained from the single-channel conductance data, and these values correspond to those obtained from the relaxation kinetics (Hess et al., 1984). It appears that τ is weakly voltage-dependent, such that it is prolonged *e*-fold for each 100 mV of membrane hyperpolarization.

The channel parameters depend upon several factors (Neher and Sakman, 1976). The average open time depends upon the nature of the agonist (Fig. 12-5d). The number of open channels varies as the square of the agonist concentration, i.e., the channel exhibits a high probability of opening when the receptor has two agonist molecules bound to it, which is consistent with the following scheme:

$$\underset{(C)}{A + RR} \underset{4700S^{-1}}{\overset{1.5 \times 10^{8}\,M^{-1}S^{-1}}{\rightleftharpoons}} \underset{(C)}{ARR} \overset{9400S^{-1}}{\rightleftharpoons} \underset{(C)}{ARRA} \underset{500S^{-1}}{\overset{750S^{-1}}{\rightleftharpoons}} \underset{(O)}{AR^{*}R^{*}A}$$

The values of the rate constants given above are for acetylcholine as the agonist (Dionne and Liebowitz, 1982).

As suspected on the basis of the relaxation kinetics, lower efficacy of an agonist is consistent with shorter duration of fluctuations, while the single-channel conductance remains constant. The channel remains open as long as the agonist remains bound. The mean channel open time is apparently proportional to the affinity of the agonist. Channels exposed to agonist for long periods of time exhibit burst behavior, i.e., the channel remains in the desensitized state from which it occasionally recovers (Sakman et al., 1980). It appears that this desensitization consists of two or three components with time constants of about 2/sec and 0.2/sec. One of the features of desensitized channels is that they have several closed states. The analysis of channel closed times is complex because the closed state is actually the absence of information. If a channel has more than one closed state, then the distribution of all observed closed times should be multiexponential, each represented by a rate constant. When only one channel is present in the field of measurement and if individual components of a multicomponent exponential have sufficiently well-separated time constants, the individual components can be separated mathematically. Based on such studies the sequence of states of AChR shown in Fig. 12-6 has been proposed (Hess et al., 1984).

The physical significance of the closed states of a channel is not established yet, nor do the closed states represent a unique mode for the state of the channel. However, such modeling exercises and the underlying measurements do provide some useful alternatives. For example, the normal closed time and the open time are approximately comparable at 10 μM ACh (Sakman et al., 1980). The responses of a single channel often consist of bursts of activity separated by pauses. The length of such bursts corresponds to the time during which the channel is not in a desensitized state. The distribution of burst lengths would therefore provide information about the rate constants for the induction of the desensitized states.

A $\overset{K_1}{\rightleftharpoons}$ LA $\overset{K_1}{\rightleftharpoons}$ LAL $\overset{\Phi}{\rightleftharpoons}$ L(A)L $\underset{\text{(ion flux)}}{\overset{\bar{J}R_0}{\longrightarrow}}$

Open

K_{c_1} (LA ⇅ LI), K_{c_2} (LAL ⇅ LIL)

LI $\overset{K_2}{\rightleftharpoons}$ LIL

	Ach	CCh	SCh
K_1 (μM)	80	1900	4.5
K_2 (μM)	0.7	21	
K_{c2}	6×10^{-4}	10×10^{-4}	
Φ	1.5	2.8	1.0
$\bar{J}$ (M^{-1} s^{-1})	3×10^7	3×10^7	3×10^7

Parameter	Acetyl-choline	Carbamoyl-choline	Suberyl-dicholine
Average open time (msec)	11 ± 1.6	3.9 ± 0.4	19 ± 2.5
Single channel conductance (pmho)	15 ± 1.8	~15	~15

Fig. 12-6. A minimum kinetic model for the kinetics of AChR in vesicles (Hess et al., 1983). The model consists of six states: active closed (A, hatched circle), active open (A), and inactive (I) receptor conformations. In the table are given the values of the rate constants from this scheme for three different agonists for AChR from electric eel reconstituted into vesicles. The channel parameters for muscle endplate are tabulated for the three agonists. ACh, acetylcholine; CCh, carbamylcholine; SCh, subarylcholine.

The endplate current (Fig. 12-4) *in vivo* is carried by Na and K ions. AChR channels are permeable to all the alkali metal and alkaline earth cations of 7–8 Å diameter, but impermeable to anions. The permeability sequence for alkali metal cations through AChR channels parallels their mobility in the aqueous phase, whereas divalent cations have an opposite sequence. The reversal potential in the presence of varying Na or Ca concentration in the external medium is consistent with the Nernst electrodiffusion potential. The single-channel conductance increases from 10 to 30 pS with increasing Na and Ca concentrations. The single-channel conductance does not vary strongly over the range of membrane potential from −120 to −50 mV, however the current–voltage (*I-V*) curves have significant curvature. These results suggest that ion interactions occurring in AChR channel involve both surface charge effects and competition for an ion binding site. This data has been interpreted in terms of a two-barrier, one-well model (Eisenman et al., 1985).

Biochemical Characterization

A relatively unambiguous electrophysiological description of cholinergic receptor function, relative abundance of cholinergic synapses in a variety of tissues, and availability of specific agonists and antagonists have prompted considerable effort toward biochemical characterization of AChR (Anholt et al., 1985). It has been shown that AChR are localized on the outer surface of the postsynaptic cell or the muscle endplate. Their density ranges from about 100 to 100,000 per μm^2. Neurotoxins from snake venoms bind AChR with high affinity (K_D less than 1 nM), and they lock the channel in the open state. Antibodies against AChR produce neuromuscular block and other symptoms that suggest that at least one type of myasthenia gravis (characterized by abnormal fatigability of voluntary muscles) is an autoimmune response against the receptor.

AChR has been isolated by affinity chromatography using neurotoxins or analogs of an agonist conjugated to agarose. Purified AChR from electric organs as well as muscle endplates is a complex of five homologous glycopeptides (4–7% carbohydrate by weight) $\alpha_2\beta\gamma\delta$ of molecular weights between 50–60 kD. The amino acid sequences deduced from cDNAs indicate considerable homology between the subunits, suggesting that they are probably derived from a common ancestral gene (Noda et al., 1983, 1984; Popot and Changeux, 1984). Two α-subunits act as binding sites for acetylcholine as well as for α-bungarotoxin. Immunological cross-reactivity between subunits and their monoclonal antibodies has also been observed. Predictions based on the primary sequence data suggest a common motif of secondary structure for the portions of subunits inside the bilayer. It appears that five transmembrane helices are present in each functional channel (Fig. 12-7). AChR is expressed in oocytes not only when mRNA for all the four subunits is present (Mishina et al., 1984) but also when only three of the four subunits are present. Apparently, two of the largest subunits are interchangeable in the complex. Three-dimensional electron image reconstruction suggests that the subunits are disposed regularly around the channel over a large fraction of their length (Brisson and Unwin, 1985). The five subunits are symmetrically constrained within a 14-nm-long by 8-nm-diameter cylindrical shell.

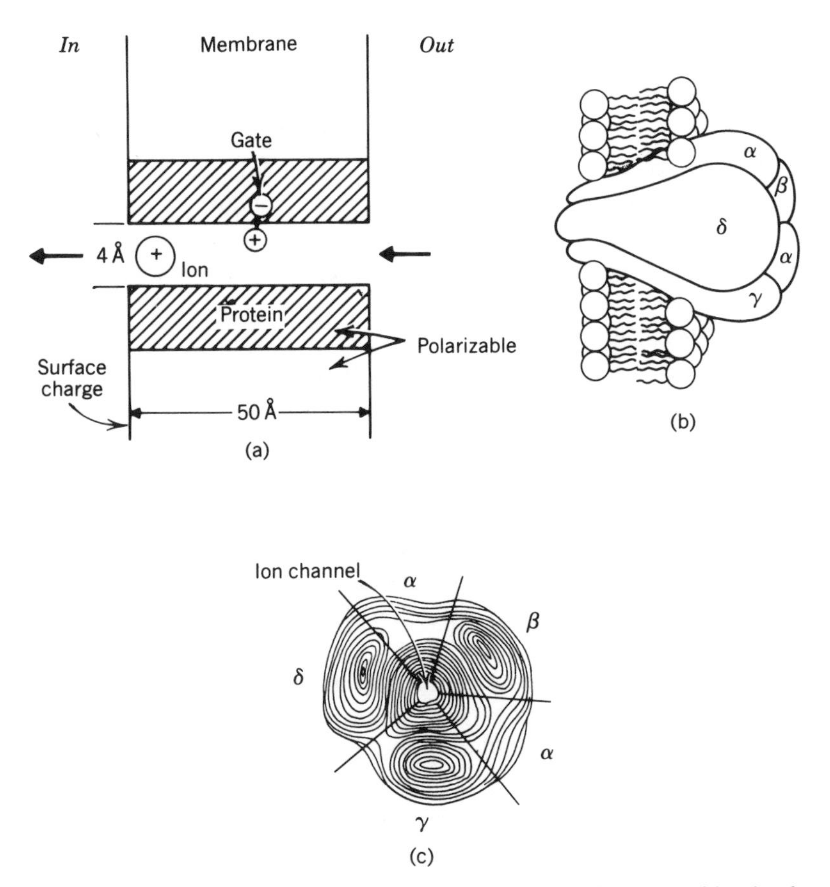

Fig. 12-7. A wishful reconstruction of the topography of AChR at several levels of complexity. (a) An idealized cross-section; (b) the transmembrane channel formed by subunits and all of them are accessible from both sides. The ACh-binding site is on the δ-subunit and is accessible only from the outside. Normally, cations move from outside to inside under their electrochemical gradient. (c) The cross-section of the channel from low-resolution X-ray diffraction profile.

The core of this cylinder delineates a water-filled opening along the axis. The opening of the channel is about 3 nm diameter near the entrance and it narrows down to about 1.1 nm.

In membranes of *Torpedo* electric organ, AChR occurs predominantly as a dimer bridged by a S—S bond between the δ-subunits on the cytoplasmic side. Dimer formation has not been demonstrated in the receptor from other sources and dimer form is not necessary for the function. The center-to-center separation between the pentamers is about 96 Å. These dimers apparently are immobilized at the synaptic junction through a 43-kD protein present on the cytoplasmic side as a cytoskeletal component.

AChR purified on an affinity column in detergent solution has been reconstituted by several methods, including dialysis and direct incorporation into BLM (Anholt et al., 1985). Reconstitution under optimal conditions requires a lipid/protein ratio of more than 16, and the reconstituted proteoliposomes apparently contain a constant lipid/protein ratio. About 70% of the reconstituted vesicles contain AChR with an agonist-binding site on the external surface. In reconstituted vesicles, leakage of Na is induced by agonists, and inhibited by antagonists like *d*-tubocurarine, α-bungarotoxin, and local anesthetics. Preexposure to an agonist could prevent subsequent leakage of ions, which would be expected if desensitization has occurred. Studies on reconstituted vesicles also demonstrate that the channels formed from monomer or dimer AChR are equally effective, and the receptors remain functionally active after extensive proteolysis with trypsin. Studies with dissociated complex suggest that the α-subunit contains the agonist binding sites, and occupancy of both of these sites is necessary for gating the channel to the open state. Results from reconstituted and intact AChR have been interpreted to analyze several models that relate ligand binding to opening of the channel (Hess et al., 1983; 1984). A minimum mechanism shown in Fig. 12-6 is based on the assumption that on binding of two ligands the protein isomerizes to an open conformation.

AChR reconstituted in BLM (Anholt et al., 1985) exhibit single-channel conductance of 20–25 pS and γ shows saturation at high Na^+ concentration. τ is 1–3 msec for carbamoylcholine applied to one side but not the other, and τ changes with the affinity of the agonist. Distribution of τ follows a double-exponential function, with the minor component of this distribution for each agonist about 5–10 times longer than the predominant short lifetime. Desensitization, burst kinetics, and inhibition by antagonists have also been observed. Desensitization is apparently induced by a conformational transition in which binding of a single agonist molecule to the AChR complex is sufficient.

ACTION POTENTIAL

Ionic gradients across a plasma membrane give rise to a resting potential. In all vertebrate cells, for example, the resting potential of about −70 mV is due to higher permeability for K ions and about 13-fold higher concentration of K in the cytoplasm compared to the external medium. Contribution from ionic gradients of Na and Cl ions is small because the permeability for these ions is relatively small, about 10% and 5% of the value for permeability of K ions. Small contributions to the resting potentials also come from electrogenic pump and the Donnon equilibrium potential. The functional significance of ionic gradients across cell membranes is far more subtle than that for the gradients of metabolites like sugars and amino acids. Inorganic ions and their gradients serve a variety of functions besides maintenance of the resting potential. Specific ions and their gradients also drive cotransport, regulate catalytic proteins and secretory processes, provide buffering action by cation–proton exchange, regulate cell volume through osmotic changes, and protect against the colloid osmotic pressure which develops during macromolecular synthesis.

Probably one of the most striking functions of ionic gradients is that they provide not only the driving force for the downhill flux of ions but also for the electrical excitability, which is ultimately responsible for most of the rapid information processing in sensory systems. The information in the nervous system is encoded in action potentials that arise due to transient perturbation of the resting potential by the voltage-gated opening of the sodium channels present in nerve membranes. Indeed, transmission, decoding, and storage of the receptor information by the nervous system is mediated by temporal fluctuations in the ionic permeabilities of the nerve cell membrane. Such time-dependent changes in the membrane potential have also been observed in a variety of plant and animal cells in response to the various stimuli, including electrical, chemical, light, and pressure. The following description of action potentials or controlled fluctuations in membrane potential is based primarily on the properties of the excitable membranes of neurons and their axons.

Neurons can be divided into four morphological compartments, and each plays a different role in signal transmission (Fig. 12-2). Dendrites receive synaptic input from numerous presynaptic elements and respond with graded or, in some cases, propagated changes in membrane potential. The cell body or soma also receives and integrates the graded synaptic inputs. Depolarization (less negative membrane potential) of the cell beyond a threshold value elicits one or a series of action potentials which are con-

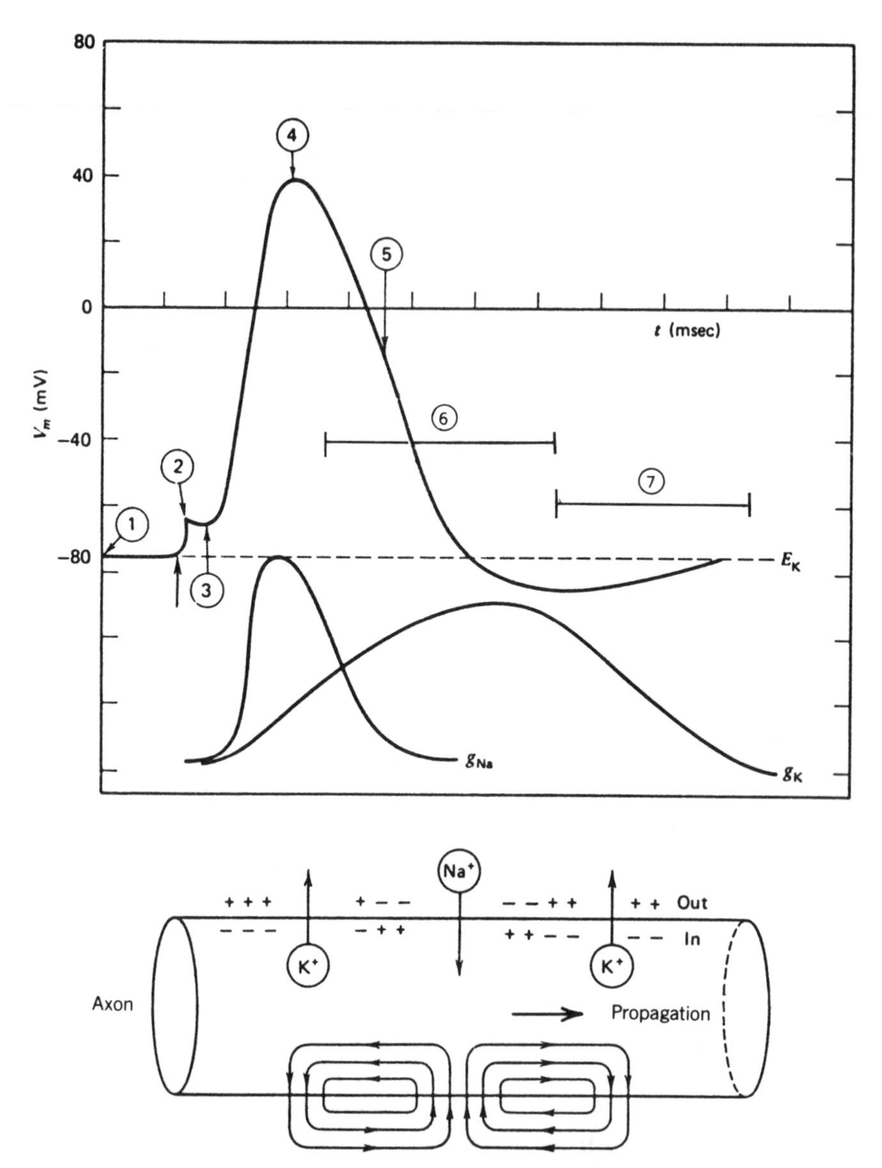

Fig. 12-8. (Top) An idealized representation of the time course of a "typical" action potential. A resting axon has a resting potential (1) close to the K equilibrium potential (E_K). When depolarized by a suprathreshold stimulus at the next arrow, one observes a stimulus artefact (2). After a latency period (3), the membrane potential swings positive to a peak value (4) and then declines (5) to the resting level. The rising and falling phases are observed even if the stimulus is withdrawn after the onset of the rising phase. After one action potential, the next cannot be induced even with a stronger stimulus during the absolute refractory period (6). After this in the relative refractory period (7), an action potential can be induced only with a stronger stimulus. The duration of the relative refractory period depends upon the magnitude of the suprathreshold stimulus. At the end of the

ducted down the axon to the nerve terminals. This depolarization of the terminals releases the transmitter into synaptic cleft, and the resulting excitation travels to succeeding neurons in the neural pathway of effector cells including skeletal muscle cells.

Conduction and Propagation

Operationally, axons, the hollow extensions from a nerve cell or neurons, are the communication lines of the nervous system in the sense that they transmit electrical impulses from one part of the nervous system to another. A unique feature of neurons is the brief transition in the membrane permeability that takes place when the resting transmembrane potential is reduced (*depolarized*) below a *threshold* value. Such a voltage-dependent change in the membrane permeability gives rise to the *action potential or spike* (Fig. 12-8). The threshold for excitation is always depolarizing. A hyperpolarizing stimulus, no matter how large, does not generate an action potential. If, however, a hyperpolarizing pulse of sufficient magnitude is suddenly withdrawn, an action potential is generated (*anode break excitation*). A slow withdrawal of the hyperpolarization does not lead to an action potential.

Action potential is initiated only when the resting membrane potential is depolarized from larger negative values. Typically, depolarization of the nerve membrane from −70 mV (normal resting value) to −55 mV initiates an action potential. The *threshold* current that produces an action potential can also be applied as an external electrical depolarization. In nerve cells containing appropriate synapses, such a threshold depolarization is attained by an increase in the membrane permeability induced by the transmitter. The conditions that determine the threshold are quite sensitive to the value of the resting potential and therefore on the K ionic gradient, to the rate of depolarization, to the history of the cell, and to the ionic composition and concentration of the medium. The threshold at the end of the hyperpolarization reaches its lowest value, and at low resting potentials the action potential cannot be generated no matter what the strength of the stimulus. Thus, all the factors that increase the membrane permeability nonselectively or de-

---

refractory period, when the resting potential has returned to the resting level, action potential can be reinitiated by the same stimulus. The peak value of the action potential is close to E_{Na} and it is independent of the magnitude of the suprathreshold stimulus. (Middle) The time-course of changes in conductance for Na (g_{Na}) and K (g_K) during the action potential. The scales are arbitrary but realistic. (Bottom) The propagated action potential along the axon generates local current-loops (shown for the flow of cations) because the direction of flow of ions changes outwards (resting) to inwards (during) to outwards (after).

plete ionic gradients block generation of the action potential because the resting membrane potential is also lowered. The duration over which depolarization is applied is also important. For example, a square depolarizing pulse of sufficient duration (about 10 mV for 1 msec) is suprathreshold, whereas the same depolarization in the form of a triangular pulse over 5 msec would not lead to an action potential (*accommodation*). These characteristics suggest that some kinetic factors control the threshold depolarization.

The height of the action potential generated by a suprathreshold stimulus depends on the transmembrane gradient of Na ions. The amplitude of the action potential (above 0 mV) can be predicted by Nernst equation (cf. Chapter 9), and at the peak of the action potential the permeability for Na ions is about 15 times larger than that for K ions. When a second depolarizing pulse is applied, the second action potential is generated only at the end of a *refractory period*. This period can be reduced by increasing the depolarization, however an absolute refractory period is reached when an action potential is not generated no matter how large is the depolarization. If the depolarization is maintained continuously, a series of action potentials or spikes are generated. For all practical purposes the response of a suprathreshold continuous stimulus is fixed in size shape, duration, and conduction speed over the length of the axon. Only the frequency of the action potentials changes as a function of the magnitude of the stimulus. This is a consequence of the refractory period that follows the peak of the action potential. At the beginning of the falling phase of the action potential, a new spike cannot be initiated no matter how strong a depolarizing stimulus is applied (absolute refractory period). Toward the end of the refractory period a spike can be initiated, however, the threshold value increases as one moves closer to the peak of the action potential. The maximum frequency at which a nerve fiber can transmit impulses depends on the magnitude of the absolute refractory period. The upper limit is usually $2 \cdot t$ impulses per second, where t is the absolute refractory period in seconds. A working range in human body is 5–500 impulses/sec. Values as high as 1500 impulses/sec have been observed in the discharge of the electric organ of electric fish.

In elongated excitable cells such as nerves, an impulse, once initiated by a stimulus, is propagated rapidly from the stimulating cathode to adjacent regions of the membrane and thus spreads as a wave over the membrane of the entire cell. Such self-propagation is described schematically in Fig. 12-8. During depolarization, the inside of the axon becomes more positive with respect to the adjacent section which still has a normal inside-negative resting potential. Therefore, a current flows from the positive region to the negative region, completing the circuit by returning to the positive region through the conducting solution outside the axon. The current arriving at the region of the normal resting potential triggers the generation of a spike similar to that in the region it has just left. In this manner, impulse (train of spikes) is regenerated from point to point along the length of an axon, and travels from one end of the axon to the other. The relay of impulses between neurons occurs via synapses. The velocity of conduction of an impulse depends upon the diameter D in microns) of the fiber; $v = 2.5\ D$ meters/sec for myelinated fibers, and v is proportional to the square root of D for nonmyelinated fibers. Usually, the conduction velocities are in the range of 1–100 meters/sec.

Clamping the Axon

The giant axon of squid (over 100 μM in diameter and several centimeters in length) is one of the best-characterized excitable membranes (cf. Baker, 1984). The features of *propagated* axonal conduction described above suggest that the overall process is all-or-none. This complex all-or-none uncontrolled behavior can be simplified (''tamed'') by the following ''clamping'' procedures:

1. If an axial metal electrode (usually a platinum or silver wire) is placed within the axon, it will short-circuit the longitudinal conduction (i_m = 0). In such a *space-clamped* axon the membrane potential is independent of the space coordinates along the long axis of the axon, and the membrane potential changes simultaneously over the whole length of the axon. In the space-clamped axon only one spike or action potential is generated after application of a square suprathreshold depolarizing pulse of sufficient duration. Most of the characteristics of the propagated and space-clamped action potentials are similar to those seen in unclamped axons, although some qualitative differences have been noted. For example, the threshold depolarization for space-clamped axons are higher. Similarly, the all-or-none or sharp response characteristics of the propagating impulse degenerate into an action potential response which is a smoothly continuous but steep function of the stimulus.
2. In a propagating or space-clamped axon, it is difficult to study the characteristics of the membrane at potentials between the resting (−70 mV) and the peak value (+30 mV) of the action potential. To study the ion-conductance characteristics of the membrane at any given time during the course of the action potential, it is necessary to clamp the

membrane potential such that $dV/dt = 0$. The basis of the *voltage-clamp* technique is to produce a sudden displacement of the membrane potential from the resting value to a desired value, and the potential is then held at the new level by means of a feedback amplifier. Thus, it is possible to follow the change in the membrane current as a function of the clamped voltage. Since the transmembrane current is directly proportional to the number and selectivity of the ion-conduction pathways, the time course of the change in the current under voltage-clamp conditions provides a direct measure of the underlying molecular processes. Under these conditions it is also possible to control other variables such as ionic strength and gradients (by perfusion) along with the applied membrane potential. As discussed later, through such experiments it has been established that a sudden sustained change in the membrane potential during an action potential is due to a transient change in the permeability for Na ions. The magnitude of this change depends upon the magnitude and sign of the membrane potential at which the membrane is held or clamped.

3. The whole range of patch-clamp techniques and correlation analysis of fluctuations in the steady-state conductance has been done on excitable membranes to obtain information about the channels and their affinity for the various antagonists. For patch-clamp the axon is first perfused with pronase and then with a solution of low K ion concentration. A fire-polished glass pipette is sealed against a small patch of membrane of about 1 μm^2 in area (Fig. 4-5). The few channels in that patch are studied on the basis of the fluctuations in the transmembrane current (Sackman and Neher, 1984). In favorable cases, the seal of the membrane to the pipette is so stable that the pipette may be withdrawn excising a patch of the membrane that can now be dipped into a variety of test solutions.

Voltage-Gated Na-Channels

Changes in the transmembrane current across a clamped axon, as shown in Fig. 12-9, have given considerable insight into the nature of the conductance changes that accompany an action potential. Following a clamped step depolarization there is a momentary surge of outward current due to the discharge of membrane capacity of about 1 $\mu F/cm^2$ (not shown in Fig. 12-9). This surge lasts about 10 μsec, and its relaxation is independent of the membrane potential. The charge transferred during the surge increases linearly with the potential, and the size of this linear component is not affected by the sign of the holding potential.

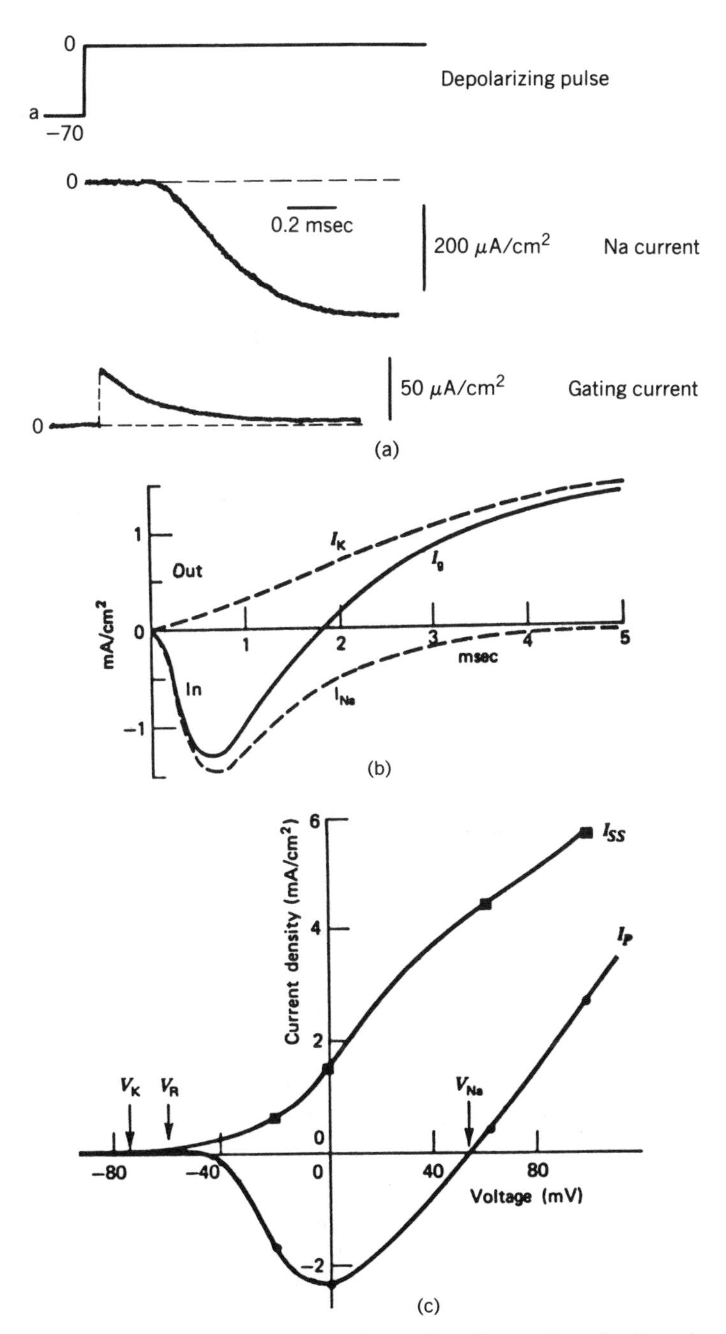

Fig. 12-9. (Top) Responses to a step depolizarization from −70 to 0 mV under voltage-clamp conditions. (a) Gating current (bottom) recorded in the absence of Na current and shown on an expanded scale. The Na current (middle) is shown for comparison. The direction of flow of charges is in the direction of the pulse (top), i.e., positive charges flow outwards. (b) The total ionic current (smooth curve, I_g) which can be resolved by perfusion or by use of appropriate inhibitors into fast inward Na current (I_{Na}) and slow outward K current (I_K). (c) The peak intensity of I_{Na} and I_K as a function of the clamp potential.

After the capacity charging there is a *gating current* of up to 100 pA that lasts about 100 μsec (Fig. 12-9a). Several characteristics of the gating current are of interest because its origin is probably in the mechanism that opens the Na-channel (Bezanilla, 1985). The time constant for the on response changes as a function of membrane potential. For brief pulses of the charge displaced during the on and off processes are equal. The steady-state distribution of charge displaced during the on response depends on the value of the membrane potential before the pulse. The plot of the charge displaced versus the depolarization is saturating, with a midpoint at about -25 mV. The total gating charge is about 1800 $e/\mu m^2$, and there are about 360 channels in a patch of this size in an unmyelinated axon.

At >100 μsec a slowly rising inward current appears (Fig. 12-9b) that peaks in about 1 msec and then declines and gives way to a prolonged outward current which declines only when the holding potential is withdrawn. The overall current has two components: the inward or fast current normally carried by Na ions (I_{Na}), and the outward slow current that is carried by K ions (I_K). The magnitude and the time course of both the inward and outward currents depend on the magnitude of the depolarizing pulse (Fig. 12-9b and c). A hyperpolarizing pulse does not produce such biphasic changes in the membrane current. If, however, the membrane is strongly hyperpolarized before depolarization, the conductance increases after a short but distinct delay. Additional features of I_{Na} and I_K (cf. Fig. 12-9) are summarized below:

1. After depolarization I_{Na} increases about fivefold more rapidly than I_K along a sigmoidal time-course.
2. Withdrawal of the depolarizing pulse (anywhere along its time-course) shuts off both I_K and I_{Na} along an exponential time course. Thus, repolarization of the membrane at the peak of the inward current does not give rise to the outward current before returning to the resting state.
3. The peak values (I_p) of the outward current vary as a function of the depolarizing holding potential. I_p reaches a peak value at about -30 mV and then decreases. The steady-state values of the inward current (I_{ss}) increase monotonically as a function of depolarized membrane potential. The intercepts of I_{ss} and I_p at the membrane potential axes (i.e., when the net current flow by each pathway is zero) correspond to the K and Na diffusion potentials. I_{Na} also increases with external Na ion concentration, and reaches a saturating value. Apparent dissociation constant for Na is about 370 mM at 0 mV, and the exact value depends on the membrane potential.

4. The rates of rise and fall of I_{Na} depend on the membrane potential; they are more rapid for large depolarizations than for small ones. Also the time constant for decay of gating transients is about 1.2 times slower than that for the decay of I_{Na}.
5. Both I_K and I_{Na} are blocked by separate groups of inhibitors (Table 12-1). For example, I_{Na} is blocked by tetrodotoxin and saxitoxin from outside; apparently they bind to separate sites. The outward I_K is blocked by tetraethylammonium ions applied only from the inside of the axon.
6. The total membrane current under certain conditions can exceed sum of I_K and I_{Na}. This suggests that the slow and the fast currents do not use a single pathway. However, these experiments do not necessarily rule out a voltage-dependent or sequential coupling between the two pathways.
7. The maximum conductance is 120 and 40 mS/cm^2 for the Na- and the K-selective pathways. The voltage dependence of the two conductances and their time constants have much in common. In both cases the conductance reaches a maximum and becomes independent of the voltage for positive membrane potentials, and both decrease exponentially for negative potentials. A change of 4 to 6 mV in the membrane potential produces an *e*-fold change in the conductance. The time constants for the change in conductance are also voltage-dependent with a maximum near the resting potential and decreasing on both sides of this maximum. These time constants are independent of the ionic concentrations in the medium.
8. The temperature dependence for the rate constants are found to be very high; 1.5- to 6-fold change for a 10°C change in the temperature (Q_{10}). However, Q_{10} is less than 1.5 for the conductance.
9. If Na ions in the external medium are replaced by organic cations such as choline, the early inward current is abolished but the outward I_K persists unchanged. Based on such experiments the selectivity sequence for I_{Na} decreases in the order Na^+, Li^+, hydroxylamine, hydrazine, ammonium, K^+, methylammonium, Rb^+, Cs^+; and for I_K the selectivity sequence in the decreasing order is K^+, NH^+, Rb^+, Cs^+, Na^+, Li^+. Thus, ammonium ions can pass through both the pathways.

The rapid and reversible changes in membrane permeability that underlie excitation and are dissected by the classical electrophysiological methods are caused by random and independent opening and closing of discrete Na and K channels gated by membrane potential. As discussed for other sys-

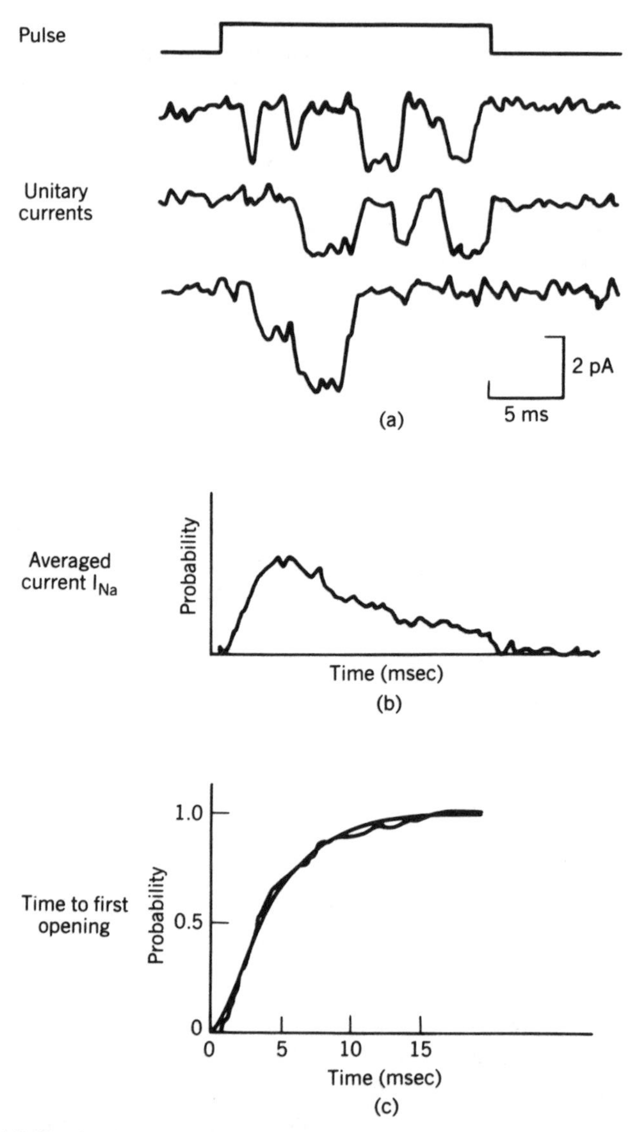

Fig. 12-10. (A) Single-channel recording from cultured rat myotubes on step depolarizations to −40 mV. The histograms (not shown) for the amplitudes and the open time for these channels are similar to those shown in Fig. 12-5 for AChR, which suggests that these channels represent a single open state. (B) However the probability of finding an open channel decreases with time. Averaged current from 144 recordings of the type shown above to produce the time course of I_{Na}. (C) Plot of the cumulative distribution of times after the start of the voltage step to the first opening for the same set of the voltage pulses. This can be fitted to a sequential model which requires at least two closed states leading to an open state. The amplitude histogram and the open time histograms (of the type shown in Fig. 12-5) of Na channels indicate the presence of only one type of channel. (From Patlak and Horn, 1982.)

tems, the kinetic states underlying gating are discrete and the open times are exponentially distributed for Na channels. As shown in Fig. 12-10, the time course of I_{Na} can be generated by summing conductance fluctuations obtained during a voltage-clamp pulse. Such patch-clamp experiments suggest that the probability of opening a Na channel decreases with time after depolarization. It appears that the channel also passes through more than one closed state. Thus, axonal excitation occurs because the rate of transitions necessary for opening the Na channel are modulated by the transmembrane potential. This is analogous to the opening of the AChR channels on binding of acetylcholine. Both of these processes are analogous to those that regulate functions of allosteric enzymes. The transmembrane potential acts not only as a driving force for ions across the Na channel but it also brings about two other changes:

1. The proportion of the open and closed channels depends on the magnitude and sign of V_m. Thus, the number of channels rather than the conductance of the open channels depends on V_m. This is elegantly demonstrated through the ensemble average of nonstationary single-channel bursts in response to step changes of membrane potential (Sigworth and Neher, 1980). From such data it is possible to reconstruct the time course of I_{Na}. This voltage dependence of the kinetics of opening and closing of the channels underlies the sigmoidal conductance-voltage curve as shown in Fig. 12-11. The midpoint slope of this curve (about an *e*-fold change in conductance for a shift of 4 mV) is determined by the work function or the activation energy of the process that brings about the two-state transition of channels.
2. For a voltage-dependent process of this type, movement of charges must occur either by dipole reorientation or by the release of bound

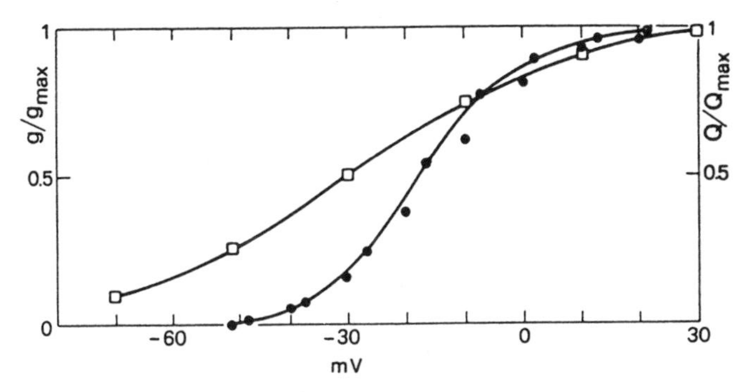

Fig. 12-11. The Na conductance versus voltage characteristics (circles) compared with gating current versus voltage characteristics (squares). (From Keynes et al., 1983.)

ions (Jain et al., 1970). Evidence for displacement of charges caused by the effect of the electric field on the gating mechanism for K and Na channels comes from the gating currents (Ulbrecht, 1977; White and Bezanilla, 1985). As shown in Fig. 12-9, the time-course of the gating currents is much faster than that of the ionic currents. This implies that there are some nonconducting states that precede the conducting states. Also the voltage-dependence of the gating current cannot be equated with the Na conductance curve (Keynes et al., 1983). As shown in Fig. 12-11, I_{Na} is insufficiently steep at its midpoint and is displaced laterally along the voltage axis. One of the possible explanations for this discrepancy is that two or more channels open in response to the movement of a single charge. Similarly, no charge movement corresponding to the closing of Na channels has been detected. This is probably because this process occurs in about 20 msec, and therefore the currents are too small.

The time-course of I_{Na} is obtained (Fig. 12-10) on a microscopic level via an ensemble average of nonstationary single-channel bursts in response to step changes of membrane potential (Sigworth and Neher, 1980; Patlak and Horn, 1982; Colquhoun and Hawkes in Sakman and Neher, 1984). Not only do these current–time curves resemble the voltage-clamp curves shown in Fig. 12-9, but they also provide molecular insights. The Na channel is nonstationary, i.e., the probability of opening changes with time. When current from such a pulse set is averaged, it gives mean current as a function of time; and the channels in a membrane function independently, and their currents add up to give the total mean current I_{Na}. Mean currents are directly proportional to the probability that any individual channel will be open as a function of time. This curve rises quickly after a slight delay, reaches a peak, and then returns to the zero level (inactivation). The probability of open channels is decreased when a depolarizing prepulse is applied. This implies that some channels are inactivated during the prepulse. Similarly, decay of inactivation of macroscopic Na-current can be explained by assuming that Na-channels might activate rapidly in response to a voltage stimulus, and the exponential decay of the current might be due to the voltage dependence of the closing rate constants. In such a process, the mean open-time of single channels will be the same as the time constant for the decay of macroscopic I_{Na}, will have only one open state, and will not reopen after closing. For Na-channels, these conditions are not simultaneously realized (Vandenberg and Horn, 1984). Thus, macroscopic inactivation is a complex process that has not been adequately modeled on the basis of the usual time-independent rate constants.

Blocking the Channel

Inhibition of conductance (Table 12-1) by specific molecules is taken as a strong indication for the presence of channels. However, inhibition could occur by a number of different mechanisms:

1. An open channel is occluded by inhibitors like pancuronium, local anesthetics, 9-aminoacridine, strychnine, and *N*-alkylguanidines.
2. A closed channel is blocked with gating kinetics intact. Under these conditions the inhibitor-bound channel can open and close but it can not translocate ions. For example, in the presence of tetrodotoxin the Na channel is not conducting, yet the gating current remains unaffected.
3. The gating kinetics is impaired. Under these conditions the channel looses its ability to open in the presence of the inhibitor. This is exemplified by the effect of polyarginine, which blocks both the Na current as well as the gating current.
4. Open channel is stabilized in the presence of inhibitors like batrachotoxin, grayanotoxin, and pyrathrins.

In neuroblastoma cells, tetrodotoxin-sensitive channels conduct 1.12 pA of current when the holding potential is −50 mV at 11°C. This corresponds to a flow of about 10^6 ion/second per channel, and the conductance of about

TABLE 12-1 Inhibitors of Excitable Membranes in the Various Phases

Inhibitor (Concentration)	Process Blocked	Sidedness
Tetrodotoxin (~1 n*M*)	Blocks Na channel	Out
Saxitoxin (~1 n*M*)	Blocks Na channel	Out
Batrachotoxin (~10 n*M*)	Leaves the Na channel open	Out
Grayanotoxin I (~1 μM)	Leaves the Na channel open	Out
α-Dihydrograyanotoxin	Leaves the Na channel open	Out
9-Aminoacridine	Slow outward current	In
Ba^{2+}	Slow outward current	In
Tetraethylammonium	Slow outward current	In
Aconitine	Leaves the Na channel open	Out
Veratridine	Leaves the Na channel open	Out
Condylactis toxin[2] (1 n*M*)	Prolongs the falling phase	Out
North African α-scorpion toxins	Slow inactivation	Out
American β-scorpion toxins	Enhance activation	Out
Pyrethroids	Modify closing of Na current	Out
Apamine	Ca-K channel is blocked	Out
Co(II)	Ca-K channel is blocked	Out
Tetraethylammonium	K current blocked	In

10–30 pS. The open time follows a Poisson distribution. The mean τ for channels is about 2.2 msec. If the membrane patch contains more than one Na channel, the probability of channel opening as a function of time during a step depolarization follows the same time course as the macroscopic I_{Na} recorded from the whole cell (Quandt and Narahashi, 1982). Application of tetrodotoxin decreases the number of conducting channels but the τ and the current amplitude remain unchanged. A dose-response curve indicates that tetrodotoxin blocks the channel in 1 : 1 stoichiometry, with the apparent dissociation constant of 2 n*M* (Narahashi, 1984). On the other hand in the presence of batrachotoxin τ reaches 100 msec (versus 2–3 msec) and the current amplitude decreases to about 50%. Although the probability of opening during step depolarization remains constant, the channels open at large negative membrane potential (about −80 mV) where normal channels do not open. These prolonged channels are blocked by tetrodotoxin. Two distinct channel populations are observed in the presence of batrachotoxin: normal and those having a reduced amplitude and longer τ. This behavior is consistent with the suggestion that batrachotoxin modifies the behavior of individual Na channels in an all-or-none manner, and both normal and modified channels are observed. Tetramethrin prolongs the τ for the Na channel, however, it has no effect on single-channel conductance and does not shift the conductance curve towards hyperpolarization.

Molecular Features

Biochemical identification and isolation of the Na channel has been facilitated by the use of toxins for assay and for affinity chromatography (Agnew, 1984; Catterall, 1984; Talvenheimo, 1985). Such studies suggest that the number of Na channels in most types of membranes (electric organ, skeletal muscle, brain, axon) is very small: about 10,000 per μm^2 in a node of Ranvier to 10–100 per μm^2 in unmyelinated axons, nerve terminals, and cell body. The Na-channels are not uniformly distributed in the plasma membranes of excitable cells. The glial and Schwann cells also express voltage-dependent K- and Na-channels, but to a lesser extent. Photoaffinity label on scorpion toxin has been used to identify a trimer of 270 + 39 + 37 kD peptides as the total complex of the Na channel in axons and rat brain synaptosomes. The smaller peptides are not part of all the preparations. The overall size is consistent with the target size for radiation inactivation. About 30% of the total mass of the large protein is carbohydrate. The α-subunit of the rat brain Na channel is also phosphorylated by cAMP-dependent protein kinase. There are also several Ca-binding sites (40–60) and the number of Ca ions bound to the channel complex appear to depend on its functional state.

Purified Na channels have been reconstituted in vesicles by the detergent dialysis method. The resulting vesicles are about 180 nm in diameter, each containing one to two channels. Functional activity of the channel has been assessed by toxin binding and by Na leakage in the presence of veratridine, which locks the channel in the open state. The ion selectivity is also similar to that seen with intact native Na channels. Interestingly, scorpion toxin-binding activity is observed only with the preparations reconstituted in phosphatidylcholine (PC) + brain lipids, whereas Na leakage is seen in channels reconstituted in PC vesicles. Proteoliposomes of Na channel in PC + phosphatidylethanolamine (PE) vesicles have been incorporated into BLM by fusion (Hartshorne et al., 1985). Single-channel conductance, γ, is 25 pS for Na^+, and 3.5 pS for K^+. The voltage at which the channel is open 50% of the time is -91 mV with an apparent gating charge of 4. K_i for tetrodotoxin is also voltage dependent.

The primary amino acid sequence of the Na channel (Noda et al., 1984) has been elucidated. There are four homologous regions between the Na channel and AChR (Greenblatt et al., 1985). Each homology region contains the sequence where positively charged arginine residues repeat with a period of 3, as if these charges could be part of a face of α-helix. Several other interesting features have emerged from the comparison of the theoretical tertiary conformation based on the sequences of these homologous proteins: the membrane-spanning regions are α-helical; carboxyl and amino terminii are cytoplasmic; the pseudo-symmetry generated by four homologous regions implies an even number of transmembrane segments per homologous region; the voltage-dependent change in the dipole moment of the "gate" implies displacement of charges buried within intramembranous hydrophobic interior of the protein. The water-filled pore (diameter 0.42 nm) lined with acidic groups can be predicted by aligning four helices of diameter 1.05 nm.

The gating process must involve a change that occurs spontaneously with a frequency that depends on the magnitude of the energy barriers encountered in the transition between the various conformations (for other possibilities see Hille, 1984; Honig et al., 1986). According to one model, voltage-controlled gating arises from the electric field-induced disruption of multiple ion pairs buried within the hydrophobic interior of the gate that regulates the channel. When a channel is closed and the membrane is hyperpolarized, the transmembrane potential pulls apart the ion pairs, thus stabilizing a relatively large transmembrane electrical dipole moment. In depolarized state, these pairs are allowed to relax toward one another. A change in dipole moment of 1000 Debye would require that 30 ion pairs each shift by approximately 0.8 nm in a direction normal to the plane of the membrane. This

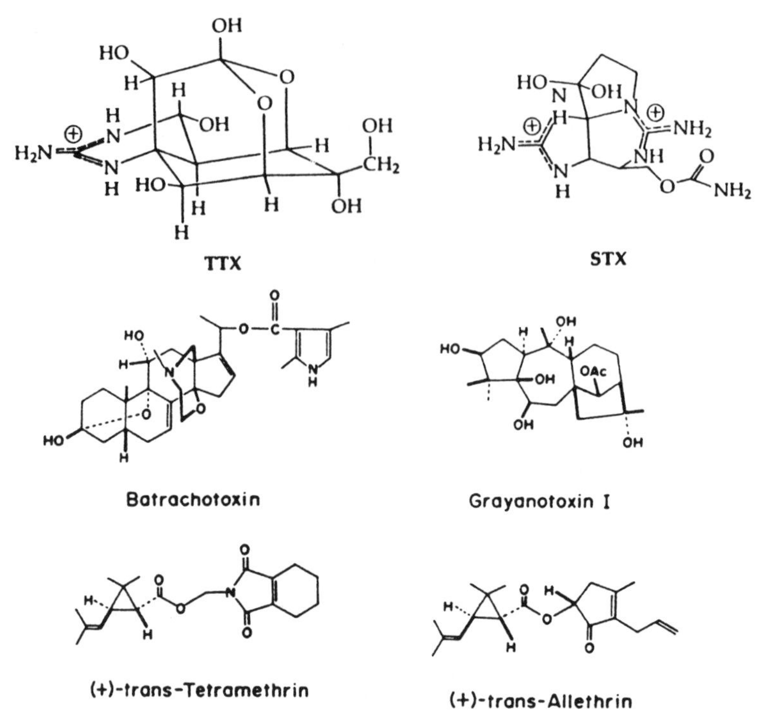

Fig. 12-12. Structures of inhibitors (left) and lipid-soluble activators (right) of the Na channel.

model is operationally similar to the one in which the binding equilibrium of ions (Ca or protons for example) in the membrane changes as a function of an applied electric field.

The biophysical observations summarized in this section are based on a wide range of sophisticated experiments. We are beginning to get a glimpse of the underlying molecular mechanisms involved in voltage-gated ion translocation processes. An understanding of this mechanism promises to provide insight into more complex functions of the nervous system.

Neurotoxins

A large variety of naturally occurring molecules (Figs. 12-12, 12-13, 12-14) from plants and venoms appear to interfere with the conduction of the nerve impulse. The mechanism of action of many of these is now fairly well settled (Table 12-1) while that of others is still being debated. Among the various naturally occurring toxic peptides such as snake neurotoxins, α-toxin, and

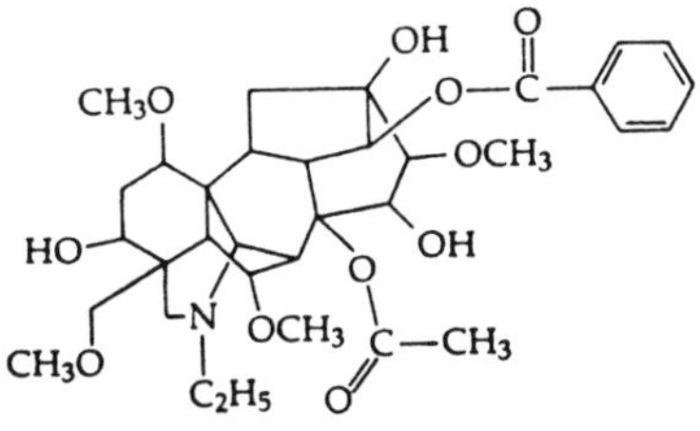

Aconitine

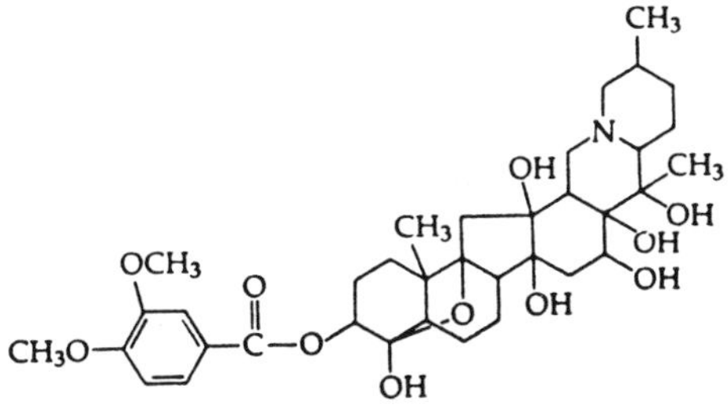

Veratridine

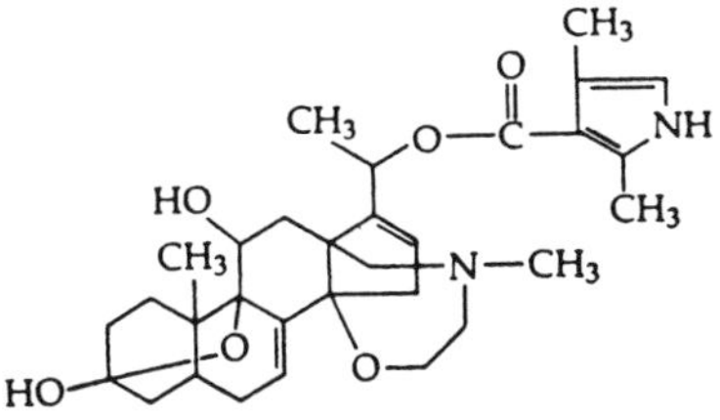

Batrachotoxin

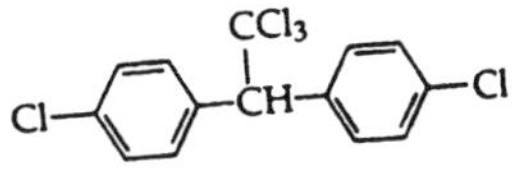

DDT

Fig. 12-12. (*Continued*)

Acetylcholine

Suberyldicholine

Carbamylcholine

d -Tubocurarine

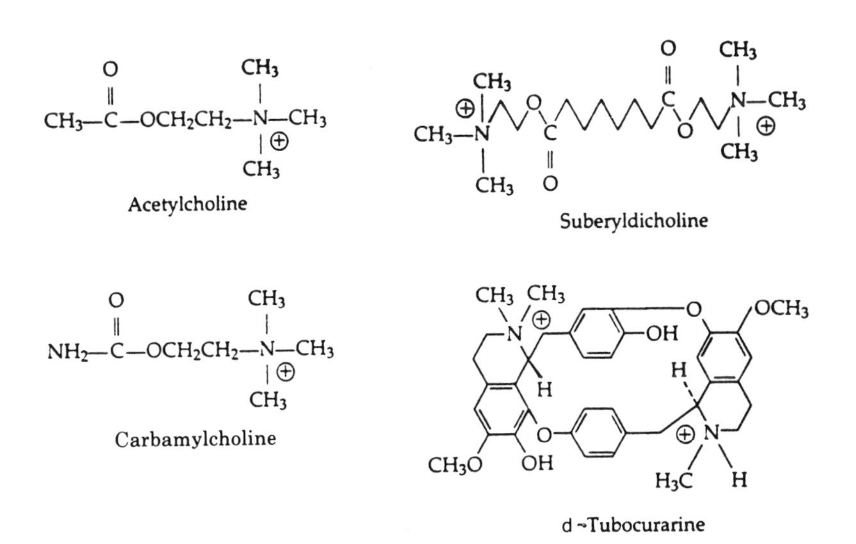

Fig. 12-13. Structures of three agonists and an agonist (*d*-tubocurarine) of AChR.

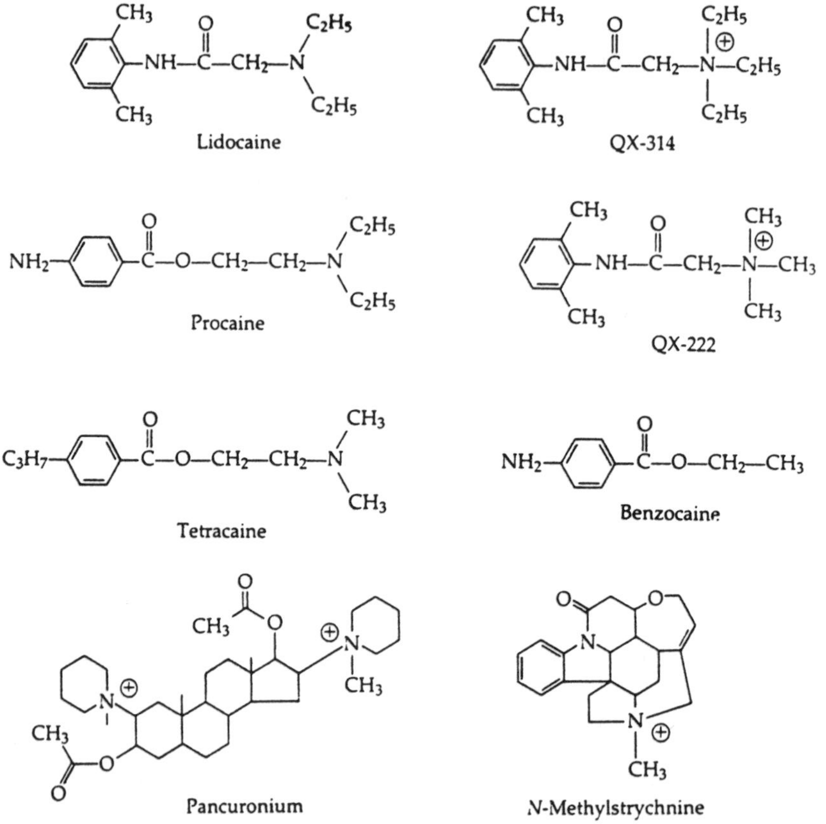

Lidocaine

QX-314

Procaine

QX-222

Tetracaine

Benzocaine

Pancuronium

N-Methylstrychnine

Fig. 12-14. Structures of some common local anesthetics that appear to block the AChR and the Na channel. Qx-222, Qx-314, pancuronium, *N*-methylstrychnine are membrane-impermeable quaternary ammonium compounds. Benzocaine is uncharged, whereas pK_a of lidocaine, procaine, and tetracaine is about 8, i.e., they are cationic under physiological conditions.

bungarotoxin) are among the true competitive antagonists that block the acetylcholine binding site. The receptor for α-toxin appears to be an AChR in a closed state. However, over a certain concentration range, binding of agonists is activated by the toxin. This is consistent with the two binding site model discussed earlier (Fig. 12-6). The hybrid species carrying a bound α-toxin and an agonist (carbamoylcholine) does not have an open channel (Sine and Taylor, 1980).

Classical antagonists like *d*-tubocurarine (Fig. 12-13) and gallamine block AChR by competing with the acetylcholine-binding site. Thus, tubocurarine reduces the frequency of the channel-opening events without altering the conductance characteristics of the open channel. This effect can be overcome by excess agonist. However, tubocurarine also induces desensitization of the receptor, and under certain conditions it also acts as a weak agonist causing the opening of very short-lived receptor channels (Trautman, 1982). The open time for channels was found to be 1–2 and 0.25–0.7 msec, and the mean open time for the channel is dependent on the concentration of curare. The fast component of the channel-open time distribution may arise from binding of a single molecule of curare to the high-affinity binding site, whereas the doubly liganded AChR may give rise to the slow component. Such studies suggest that the distinction between agonists and antagonists may not always be clear cut.

Local Anesthetics

Local anesthetics block propagated action potentials by blocking the Na channel or synaptic transmission. Structurally, local anesthetics are generally cationic amphipathic compounds (Fig. 12-14) that induce the loss of sensation only when placed in direct contact with nervous tissue (Adriani and Naraghi, 1977). Translocation of charged anesthetic across bilayers could be promoted by V_m or pH. Although anesthetics modulate bilayer organization and a variety of membrane functions, it appears that the primary site of action of at least some of the local anesthetics is on the voltage-gated Na channel (Hille, 1984; Haydon et al., 1984). The extent of inhibition of Na current by an internally applied anesthetic like QX-314 is enhanced in some cases by repetitive depolarizing pulses and is reduced in other cases by prolonged hyperpolarization. A number of amphipathic compounds, such as chlorpromazine, alcohols (Gage, 1978), benzocaine (Ogden et al., 1981), fatty acids, organic solvents, and detergents, act as voltage-independent noncompetitive blockers (Koblin and Lester, 1979; Anderson and McNamee, 1980).

Single-channel conductance records show that in the presence of quaternary lidocaine derivatives like QX222 single channels are fragmented into bursts of shorter current pulses, presumably due to repetitive blocking and unblocking of open channels (Nehr and Steinbach, 1978). Together with the voltage dependence and the requirement for the presence of the agonist, these observations have led to the hypothesis that the quaternary ammonium local anesthetics block the agonist-induced conductance by physically occluding the Na-channel from inside by voltage-dependent binding to a single site within the transmembrane electric field. It appears that the slow blocking observed with procaine and benzocaine could be due to a different mechanism, probably by binding to a closed state of the channel.

AChR has also been implicated as the site of action of anesthetics like procaine, lidocaine, and tetracaine, as well as proadifen, dimethisoquin, and trimethisoquin. The inhibitory effect of aminated local anesthetics on the endplate current is observed in the presence of agonists. Their effect is seen only when applied from outside. The decay of endplate currents shows two time constants in the presence of local anesthetics and the anesthetics may bind to AChR in the open state. Based on the sidedness of the effect of the anesthetics it appears that topological organization of AChR may be opposite to that of Na-channel.

Agonists enhance the binding of local anesthetics, and the local anesthetics enhance the binding of agonists to AChR. The net effect of some of the local anesthetics appear to be to stabilize the AChR in its high-affinity conformation and to increase the value of the allosteric constant for the second site. This enhances the rate and extent of desensitization resulting from exposure to agonists (Sine and Taylor, 1982). The affinity labeled local anesthetic trimethisoquin selectively labels δ-chain in AChR from *Torpedo* fish, and the labeling is enhanced by agonists. These observations show that anesthetics influence the distribution of receptors between different states. Thus, it is not always possible to draw a sharp distinction between the specific effects on a binding site and the general effects exerted through a change in the organization of the bilayer matrix. Indeed, in many cases macroscopic interpretations based on the pharmacological effects of antagonists and agonists have not yet found satisfactory interpretations based on the microscopic events.

MISCELLANEOUS GATED CHANNELS

Voltage-Gated K Channels

The voltage-clamp data (Fig. 12-9b) predict the presence of voltage-gated K channels. Such K channels are believed to be operationally separate from the Na channels, however, K channels have not been isolated. Single-channel conductance and K-gating currents have been recently recorded. The slow time-course of the gating current is the same as that of I_K (Bezanilla, 1985), suggesting the presence of many electrically silent slow states. Another complicating factor is that there may be many types of K channels and they are modified by ATP, probably via phosphorylation. One of the antagonists for the K channel is tetraethylammonium ion (K_i 0.1–100 mM depending upon the source of the channel). The inhibitory effect of the TEA ion from the cytoplasmic face is due to blocking of open K channels, and as expected its effect is potentiated by the membrane potential that favors open conformation (Stanfield, 1983).

Ca-Activated K Channels

There are two classes of activated K channels: those activated by changes in membrane potential, and those activated by Ca^{2+} and modulated by membrane potential. In nerve axons, only the voltage-sensitive K channels are found, however the soma of neurons and most other types of cells including erythrocytes contain K channels activated by Ca (Schwarz and Passow, 1983). The Ca-activated Na channels link metabolism of Ca to membrane polarization and apparently they are also responsible for the repetitive electrical activity in many excitable cells (Meech, 1978).

Ca-K channels have been detected in a variety of tissues including bovine chromaffin granules, rat myotubes, exterior pituitary cells, and acinar cells. They exhibit high ion selectivity and the single-channel conductance γ 200–300 pS. These channels are activated only by intracellular Ca, whose concentration under physiological conditions increases by the release of Ca from intracellular stores or by uptake from the external medium. An increase in K permeability causes hyperpolarization of membrane potential, i.e., it becomes more negative. The effect of Ca is half-maximal at about 3 μM in the presence of 2 mM K^+ outside. In the absence of external K the K channels are not activated. The Ca concentration dependence of K channels in erythrocytes and cultured muscle cells suggests that two or more Ca ions may be required for opening a channel. Single-channel conductance data have revealed that Ca increases the fraction of open channels with conductance (γ) of about 20 pS and mean open channel time, τ, of 5–6 msec (Grygorzyk and Schwarz, 1985). Inhibitors like quinine reduce the number of open channels without any effect on γ. Also at a given Ca concentration, the fraction of the open channels is reduced by hyperpolarization and increased by depolarization. At higher Ca concentrations, stronger hyperpolarizations are necessary to reduce the number of open channels. Channels can be activated by a variety of divalent cations including Pb^{2+}, whereas the channels are not permeable to Na (permeability ratio about 15 : 1). The behavior of K channels is also modulated in a yet uncharacterized manner by fluoride, other cations, energy poisons, and drugs like propanolol. The number of K channels per erythrocyte is probably well below 1000.

Voltage-Gated Ca Channels

Intracellular Ca (normally less than 10μM) plays an important role in a variety of processes, including stimulus-response coupling (Chapter 14). Influx of Ca via voltage-gated channels is directly controlled by the membrane potential, and is also modulated by hormones. Apparently, such channels are ubiquitous and their function is related to excitability. Channels from sarcolamellar membranes have been reconstituted in BLM: γ is 8 pS in 0.1 *M* Ba, and ion selectivity is Ba > Ca > Mg (1 : 0.45 : 0.08). These channels are competitively blocked by agonists like (−)D-600 (Ehrlich et al., 1986).

Ca-Gated Channels of Gap Junctions

A leakproof and continuous communication between cytoplasm of neighboring cells in eukaryotic tissues is often mediated by channels that span the membranes of two apposed cells. Due to the presence of such gap junctions, all cells in a communicating tissue share ions and metabolites. In excitable tissues like heart, gap junctions provide the basis for the rapid propagation of electrical signals between cells. Thus gap junctions are highly specialized structures that are clearly visible in electron micrographs. Low-resolution images (Brisson and Unwin, 1985) and the primary structure of the 32-kD peptide based on cDNA (Paul, 1986; Kumar and Gilula, 1986) that constitutes the sole component of gap junctions, exhibit remarkable similarity to the α-subunit of the AChR and the amino-terminal region of the Na channel. Their hydropathy profiles contain several hydrophobic sequences such that five to eight α-helices could orient through the bilayer to form a channel. The symmetrical organization of these helices around a central pore is also consistent with 40–60% homology in their sequence. The pores of these channels have different diameters at constrictions, however, in all these cases the constriction is on the cytoplasmic side.

A model for the Ca-gated opening of the pore of the gap junction is shown in Fig. 12-15. This cylindrical transmembrane channel is made up of six identical rod-shaped subunits. The channel traverses the plasma membranes of two apposed cells because hexamer in one membrane is joined symmetri-

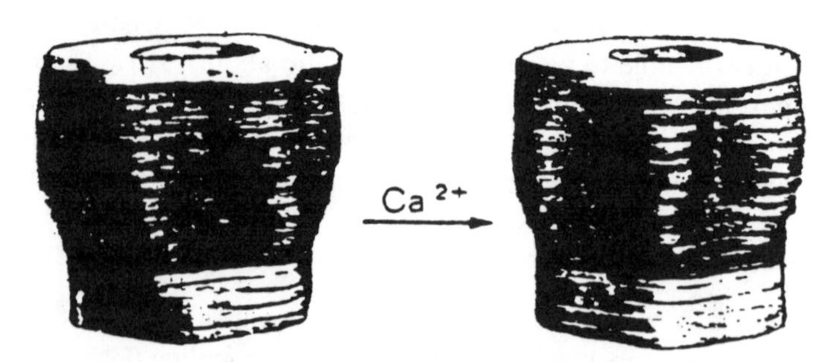

Fig. 12-15. An imaginative reconstruction of the Ca-gated channel from a gap junction. The configurations are related to each other by tilt axes passing radially through the bases of the subunits. Subunits on either side of the channel tilt in opposite directions about these axes. (From Brisson and Unwin, 1985).

cally to an oppositely facing hexamer in the other membrane. The assemblies made up of 12 subunits accumulate at gap junctions. They permit passage of small molecules, and the open–close transition is induced by Ca. Indeed, electron microscopy and X-ray studies on these membranes suggest that Ca induces a small coordinated rearrangement of the subunits around the channel (Unwin and Ennis, 1984). The switching between the two states concomitant with a large change in channel dimensions is accompanied by tilting in the opposite sense about a radial axis. Thus a displacement of 18 Å is accompanied by a large distortion of individual peptides. Energetically unfavorable interactions are apparently avoided because subunits move predominantly parallel to the plane of the bilayer. In this cooperative mechanism hydrophobic interactions remain essentially undisturbed.

Voltage-Gated Channels of Colicins

This proteins (63 kD with 522 residues) are bactericidal because they dissipate ion gradients in cell. The channel in colicin E1 is apparent in the carboxy-terminal tryptic fragment, as it shows single-channel conductance of 20 pS (Davidson et al., 1984). Although the channels of colicin E1 are not completely voltage-gated, γ for colicin A channels changes from 15 to 5 pS as a function of membrane potential.

13 Energy Transduction

. . . and the dumbfounded biologist found himself face to face with a scantily-clad young woman who was shamelessly gyrating the flagellalike tassels on her chest in an expert imitation of an Escherichia coli. Reflecting on the incident, at the end of his seminar Professor Howard Berg concluded: "These little bugs have shown everything they've got, although there are still a few tantalizing secrets they are keeping to themselves . . .

From *Smithsonian* (September, 1983)

Conversion of redox energy to other forms of energy in organisms occurs primarily via generation of ionic gradients. Thus, according to the chemiosmotic[1] or ion-gradient theory for energy transduction (Mitchell, 1979), the primary mode of energy transduction across bacterial, mitochondrial, and chloroplast membranes is generation of proton gradients. According to this theory, membranes provide a barrier to the free flow of ions; maintain transmembrane ionic gradients that give rise to diffusion potentials; provide a controlled internal milieu; and provide a matrix for asymmetric distribution of carrier and catalytic proteins.

The scheme shown in Fig. 13-1 points to the range and scope of membrane-linked energy-transducing systems. The primary energy-transducing membranes are the plasma membrane of prokaryotes and blue-green algae, the inner membrane of mitochondria, and the thylakoid membrane of chloroplasts. These membranes possess two classes of complexes which are together required for generating ionic gradients, and for the synthesis of ATP from the ionic gradients. There is an obligatory coupling between these processes. For example, NADH-ubiquinone reductase and oxaloacetate-pyruvate decarboxylase serve as the primary donor of the redox energy to the electron-transport chain. The energy released during the electron trans-

[1] Somehow the words "chemiosmosis" or "chemiosmotic coupling" evoke confusion amongst some people who are otherwise in general accord with the underlying concept. I have used the word "ion-gradient" theory as an inadequate compromise.

REDOX energy
Respiration
Light
↓
Ionic Gradient (Na, H)

$$\text{pmf} = \Delta\psi - \frac{2.3RT_{\Delta\text{pH}}}{nF}$$

↓
Work
ATP synthesis
Cotransport
Movement of flagella
NAD to NADH

Fig. 13-1. A scheme to illustrate the central role of ionic gradients in the overall energy transduction in living organisms.

port is used to generate electrochemical potential of protons or Na ions. The Na gradient appears to buffer the proton gradient across some bacterial membranes (Skulachev, 1985), whereas in others it is the primary mode of energy conservation. The free energy stored in these ionic gradients is used for chemical (ATP synthesis), osmotic (coupled metabolite transport), or mechanical (flagellar motion) work. Indeed, there is considerable circumstantial evidence suggesting that ionic gradients can also be used for thermoregulation, cotransport of Ca and other solutes, and hydrogen production (Skulachev and Hinkle, 1981). Some of the primary membrane-linked energy-transducing systems are described in this chapter.

The ion-gradient theory for energy coupling (Mitchell, 1979; Nicholls, 1982; Harold, 1986) emphasizes the equivalence and interconvertibility of the bond energy, redox energy, and transmembrane electrochemical potential of protons in the energy cycle of a cell. It is a very general thermodynamic statement about conservation and interconvertibility of the chemical energy in a bond with that of the concentration gradient. Since a coupling between the concentration gradient of ions and a chemical reaction requires vectorial fluxes between the compartments that lead to a charge separation, the theory emphasizes the unique role of the membrane not only as a barrier but also as a matrix to orient anisotropically the pumps, carriers, and catalysts that mediate electron transfer from a high-energy state to a low-energy state. The basic mechanistic principles of energy transduction in mitochondria, chloroplasts, and bacteria are the same. Also the structure and composition of the three sets of energy-transducing enzymes is similar. Both, the ATP synthase (proton-ATPase) and the quinol-cytochrome *c* (plastocyanin) oxidoreductase, are found in mitochondria, plasma membranes of prokaryotic cells, chromatophores of bacteria, and thylakoid membranes of chloroplasts.

The role of proton gradient in mediating energy transduction is based on a variety of observations. These include: energy coupling that requires topologically closed structures that retain proton gradients; an asymmetric proton gradient that is present in all energy transducing membranes; metabolic reactions that energize membranes also generate a proton gradient; all energy transducing molecules and complexes catalyze vectorial electrogenic translocation of protons; chemical and genetic manipulations that dissipate proton gradients also lead to a loss of energy transduction; and a direct coupling between proton gradient and electron and ion transport processes has been adequately demonstrated with uncouplers and by reconstitution studies.

The electrochemical driving force (also termed protonic potential or the proton motive force) for the movement of protons across energy transducing membranes is given by:

$$\text{pmf} = \frac{\Delta\mu\text{H}^+}{F} = \Delta\psi - 2.3\,\frac{RT}{nF}\,\Delta\text{pH}$$

and

$$-\Delta G\,(\text{ATP}) = nF \cdot \text{pmf}$$

where $\Delta\psi$ is the potential difference across the membrane and ΔpH is the pH difference across the membrane; z (= 2.303 RT/F) has a value of 59 mV at 25°C for monovalent ions like protons. The pmf to which an energy-transducing system localized in an asymmetric membrane would respond is $\mu H^+/z$. Thus, a total protonic transmembrane potential difference of 250 mV corresponds to a pH gradient of 4.2 units. Although a general consensus has developed in support of the coupling hypothesis, there appears to be considerable confusion about the contributions arising from the two terms (for example, see Nagle and Dilley, 1986). In different organisms the contribution of ΔpH and of $\Delta\psi$ may be appreciably different. In respiring mitochondria, $\Delta\psi$ is about 160 mV and ΔpH about 1. In thylakoid membranes virtually all the pmf is derived from the ΔpH term. In a system that is translocating protons, the pmf increases rapidly as protons are pumped: $\Delta\psi$ increases as the charge is thrown into the medium, and the work done in charging the membrane is conserved as the total pmf. However, other secondary processes and leaks could dissipate a significant part of the pmf. It may also be emphasized that only the transmembrane potential difference between the bulk media can contribute to pmf. Potentials arising from the ionization of surface groups cannot contribute significantly toward pmf because the energy of such systems can be used only by shifting surface p*K*a or by using up the transmembrane asymmetry of composition. Moreover, the capacity of such an energy storage device would be very small. Many other issues related to the mechanism and stoichiometry of coupling, as well as the role of local ionic gradients at the interface, still remain unresolved.

The pmf has been implicated as a driving force and also as a regulator for many different metabolic processes (Nicholls, 1982; Kaback, 1983; Hellingswerk and Konings, 1985) such as phosphorylation of ADP to form ATP, active and coupled-transport, reduction of NAD, transhydrogenation of NADP by NADH, modulation of enzymes by pH, bacterial motility, nitrogen fixation, transfer of genetic information, sensitivity and resistance to certain antibiotics, protein and cellulose synthesis, and processing and secretion of proteins. The key observations that have confirmed the validity of the chemiosmotic theory (Mitchell, 1979) are that: oxidative or light-induced electron-transport reactions produce proton gradients; the energy-transducing membrane is topologically closed and has low proton permeability; the movement of protons through ATP synthase is capable of producing ATP; these conservation reactions occur only in membranes that are impermeable to protons, and uncouplers abolish respiratory control and promote ATP hydrolysis; and many of the processes driven by pmf are reversible. Since the ATP synthase is reversible, any displacement from this equilibrium, which increases proton electrochemical potential or lowers the free energy of ATP, would cause the complex to reverse. Protons thus continuously circulate as the proton current across the membrane during oxidative phosphorylation, driven against their electrochemical potential by the electron-transport chain, and flowing down their electrochemical potential through the ATP synthase.

BACTERIORHODOPSIN

Halobacterium halobium grows in a medium of high salinity (up to 4 *M* NaCl) and low oxygen. Under such anaerobic growth conditions, purple patches (0.5 μm in diameter covering up to 50% of the total membrane area) are formed in its membrane. The fact that the content of these patches is not randomized suggests that their lateral diffusion coefficient is less than 10^{-12} cm^2 per sec.

The purple membrane has a single specialized function, and it contains only one major protein, bacteriorhodopsin (bR), which accounts for up to 75% of the membrane mass (Stoeckenius and Bogomolini, 1982). These membranes also contain two other minor pigments that function as a chloride pump and as a receptor for photoaxis. On the other hand, the rest of the membrane contains about 25 different proteins which carry out all the normal functions of the cell membrane. In the purple patches there are about 10 lipid molecules per bR. The lipids (25% by weight) of *Halobacterium* membrane contain phosphatidylglycerol phosphate (52%), phosphatidylglycerol (PG) (4%), phosphatidylglycerol sulfate (5%), glycolipid sulfate (10%), triglycosyl diether (19%), and squalene (9%). All these lipids contain dihydrophytol ether as the hydrophobic moiety. It is not known whether the lipids of purple patches are different from the lipids of the whole membrane.

In purple bacteria, light flashes cause a transient decrease in the 560-nm absorption and a corresponding increase in the 415-nm absorption. The primary event that follows photoexcitation is ejection of protons. The 560-nm light quantum leads to the release of one to two protons in the external medium. The rate of proton translocation (about 0.01–0.1 per sec) is apparently limited by the rate of transport (leak) of other ions. The original pigment bR_{570} is regenerated by reprotonation. Enhancment of net proton movements in the presence of cationic ionophores (uncouplers, valinomycin, gramicidin) suggests that the translocation of protons is uphill and electrogenic. Ultimately, the proton gradient generated from the primary photoexcitation event leads to phototaxis, synthesis of ATP, fixation of carbon dioxide, uptake of amino acids, uptake of K ions, ejection of Na ions via Na,H-exchange, a change in oxygen consumption, and membrane biosynthesis. The organism reverts to the native form within a few milliseconds in the dark.

The proton translocation function of purple membranes is due to bR (MW 26,000 kD, 248 residues; see Fig. 7-5), which has been purified to homogeneity. Delipidated preparations of bR have been reconstituted in liposomes. Vesicles prepared by cholate dialysis take up protons from the medium upon illumination. Fragments of purple membrane can be incorporated into vesicles where protons are pumped out because the protein has the opposite orientation. Experiments with intact and reconstituted bR systems have shown that the light-driven pump is regulated by a proton gradient.

The photocycle of bR has been characterized, as shown in Fig. 13-2. At least four different conformations of bR and five photointermediates have been associated with the photoevent that leads to proton translocation. bR can be considered an energy transducer that uses light of 0.9 eV to create a proton gradient of about 0.3 eV. The quantum yield of the cycle is 0.3 and up to two protons are translocated per cycle. The rate of the light-driven proton pump and the underlying photocycle of bR is lowered by the electrochemical gradient of protons that it develops or by an applied electrical potential. This implies that bR is under *feed-back* control from its own product.

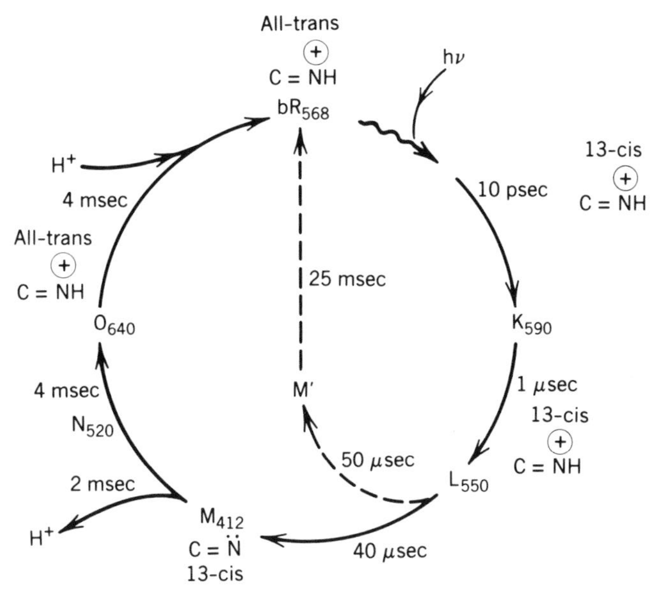

Fig. 13-2. Photocycle for light-driven bacteriorhodopsin (bR). All intermediates can decay rapidly back to bR when there is no light. The subscripts are maxima of the calculated absorption spectra.

The organization of bR in the purple patches is that of a two-dimensional crystal. As discussed in Chapter 7, the conformation of bR in membranes consists of seven transmembrane helical segments. These helical cylinders consist of several anionic residues that could act as a proton channel. The photopigment of bR, all-*trans* retinal is attached to Lys-216 via a protonated Shiff's base. The angle of absorbance vector of the chromophore is about 23° inclined from the membrane surface, and it is localized in the middle of the bilayer. Light causes isomerization about the C_{13}-C_{14} double bond of retinal, and a transient deprotonation of the Schiff's base.

Halorhodopsin

Most strains of *H. halobium* contain this minor component that mediates light-driven chloride influx. Its properties and light reactions are similar to those of bR except that deprotonation of the Schiff's base is not required (Lanyi, 1986). This pump compensates the uptake of K ions driven by the outside positive potential, and the K uptake required to balance the osmolarity due to high external Na concentration.

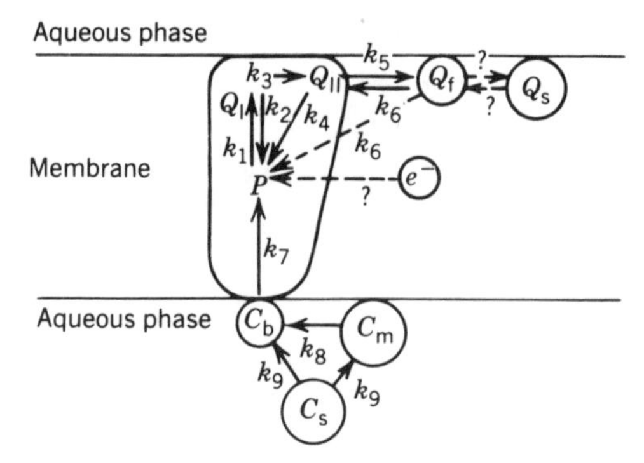

Fig. 13-3. Pathways for electron flow in a reaction center incorporated in BLM. P is the light-converting bacteriochlorophyll dimer; Q_1 the primary electron acceptor, a protein-bound ubiquinone; Q_{11}, a secondary acceptor; C_b, ferrocytochrome *c* which is bound to the reaction center; C_m, membrane-bound ferrocytochrome *c*; and C_s, ferrocytochrome in the aqueous phase. (From Apell et al., 1983.)

Slow Rhodopsin

About 5000 copies of this pigment per *Halobacterium* generates the attractant and repellant response. Action spectra show maxima at 580 nm and 370 nm, thus cells can avoid damaging blue light and move to regions where their light transducers are most effective. Its photocycle is slow (Spudich and Bogomolini, 1984).

Bacterial Reaction Center

The light-induced activity in reaction centers of bacteria causes an electric charge separation and thus generates a redox potential across their plasma membrane. The electron pathways of reaction center are shown in Fig. 13-3. The absorption of a photon leads to the cyclic electron flow that is coupled to a vectorial transport of protons from the cytoplasm to the periplasm across the membrane.

OXIDATIVE PHOSPHORYLATION

The inner and the outer membranes of mitochondria (Fig. 2-1b) have their evolutionary *roots* from bacterial *ancestors* that synergistically *colonized* almost all types of eukaryotic cells. The outer membrane of mitochondria regulates flow of solutes. The inner membrane specializes in energy transduction which involves two steps (Fig. 13-1): First, exergonic electron flow of a redox reaction produces a proton gradient due to a coupled translocation of protons to the space between the two membranes; second, the pmf of protons is used by ATP synthase on the inner membrane to phosphorylate ADP (Ernster, 1984).

Flow of electrons in mitochondria is coupled to the tricarboxylic acid (TCA) cycle which leads to the synthesis of ATP.

H_o^+

SH ⟶ S ; NAD^+ ⟵ NADH ; H_2O ⟵ $\frac{1}{2}O_2$

H_i

$$ADP + P_i + nH_i \rightarrow ATP + H_2O + nH_o$$

The energy released during the oxidation of a variety of substrates is used for the formation of NADH and FADH. Oxidation of these cofactors gives rise to the proton gradient across the inner membrane. During electron transport several bacteria also generate a Na gradient, while still others couple decarboxylation of oxaloacetate to pyruvate to generate a Na gradient (Dimroth, 1982).

In the electron transport chain (ETC) of mitochondria, transfer of electrons from NADH to oxygen is mediated by a series of coupled redox reactions catalyzed by a series of cofactors. The maximum value of pmf can be achieved only when the mitochondrial membrane is not permeable to protons as well as to anions. Both of these requirements are fulfilled by the inner mitochondrial membrane, which is relatively impermeable to these solutes. Moreover the convoluted topography of the inner mitochondrial membrane provides a large surface area. On the average, a complex of electron transport chain occupies a membrane area of approximately 900 nm^2 with approximately 1500 phospholipid molecules. The maximum velocity for electron transport through the complex is approximately 60 electrons per second. The intense activity of the oxidative pathways of metabolism is indicated by the fact that an active human adult uses and resynthesizes about its own weight in ATP every day via the electron-transport chain. The protons translocated in this process (about 500 moles) could acidify a lake of over 10 million liters of water from pH 7 to a pH of 5!

The inner membrane of mitochondria contains 75% protein by weight, and about 50% of these are involved in oxidative phosphorylation. The lipids of inner membrane are phosphatidylcholine (PC) (40%), phosphatidylethanolamine (PE) (30%), diphosphatidylglycerol (DPG) (20%), phosphatidylserine (PS) (10%), and there are about 2.8 double bonds in the two acyl chains of each phospholipid. Several proteins associated with the electron-transfer chain have been isolated and characterized. While some of these components catalyze specific reactions, others are devoid of any obvious catalytic function. ETC has been resolved by detergent treatment into five major complexes (I–V), each of which shows part of the overall function (Fig. 13-4). I is NADH : ubiquinone Q oxidoreductase; II is succinate : ubiquinone Q oxidoreductase; III is ubiquinol : ferricytochrome *c* oxidoreductase; IV is ferrocytochrome $c : O_2$ oxidoreductase; and V is ATP synthase. Functional interaction of these complexes occurs at different levels. Complexes I–IV make up the electron transport chain in which the redox energy (1.11 V) is conserved as the proton gradient. The pmf is used by complex V to synthesize ATP at separate locations in the inner membrane. Complex II is not directly involved in energy conservation. Complexes I, III, and IV are

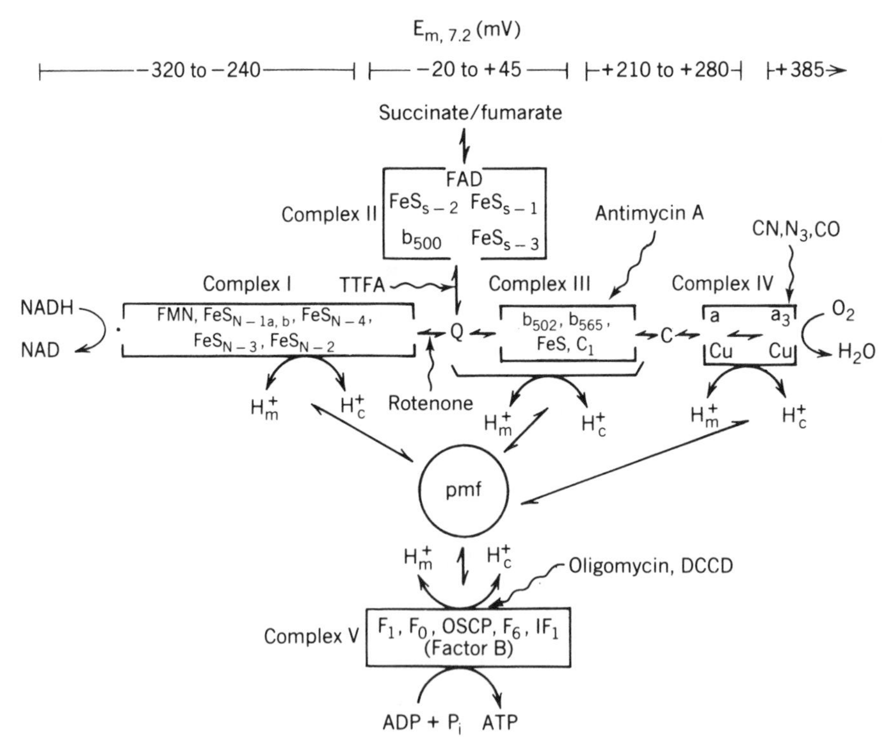

Fig. 13-4. Details of oxidative phosphorylation and electron transport coupled via pmf. (From Hatefi, 1985.)

embedded in the inner mitochondrial membrane. The catalytic functions of these complexes have been confirmed by reconstitution of the isolated complexes.

The subunit structure of each of the respiratory complexes is known (Table 13-1). Most of the subunits are small proteins of MW <20 kD. It is not known whether many of these subunits are actually required for the functional integrity of the complexes, or whether they are only copurified. By the criterion of immunoprecipitation and inhibition by antibodies, these subunits are apparently part of the respiratory complex. Most of the proteins in these complexes are synthesized on cytoplasmic ribosomes and coded by the nuclear genome (54 out of about 60); some subunits of cytochrome oxidase and the cytochrome bc_1 complex are coded in the mitochondrial DNA. Subunits synthesized in cytoplasm are transported as individual precursors into mitochondria.

ETC catalyzes transfer of reducing equivalents from NADH or FADH to an oxygen molecule. The energy from the flow of two electrons with a total redox span of 1.11 V (Fig. 13-4) is used to drive proton translocation from the matrix to the cytoplasmic side. The electron-transport carriers in the mitochondrial membrane are present in stoichiometric ratios with each other, although the stoichiometry is not 1 : 1. The maintenance of stoichiometry implies a lattice arrangement. It is also consistent with the observation that "dilution" of mitochondrial proteins in the bilayer by exogenous phospholipids reduces the rate of electron transport.

The mechanism of proton translocation mediated by ETC is not known; however, molecular details and the mechanism of electron transfer are beginning to emerge (Wickstrom and Saraste, 1984). Of the various complexes, the cytochrome c oxidase is probably best characterized. It functions as a

TABLE 13-1 Molecular Weights, Polypeptides, Prosthetic Groups, and Relative Abundance in the Bovine Heart Mitochondrial Inner Membrane of Complexes I, II, III, IV, and V

Complex	$M_r \times 10^6$ (Monomer)[a]	Polypeptides	mtDNA Encoded Polypeptides	Prosthetic Groups	Ratio in Mitochondria
I	0.7–0.9	25	—	FMN, Fe-S clusters	1
II	0.14	4–5	—	FAD, Fe-S clusters, b_{560} heme	2
III	0.25	9–10	1	b_{562}, b_{566}, c_1 hemes, [2Fe-2S] cluster	3
IV	0.16–0.17	8	3	aa_3 hemes, Cu_a Cu_{a3}	6–7
V	0.5	12–14	2	adenine nucleotides, Mg^{2+}	3–5

redox-linked proton pump: With the passage of two electrons, four to six protons are taken up from the M (matrix) side and transported to the space between the two membranes (C side). The precise stoichiometry of the electrons involved in the redox reactions and the protons translocated is not established. It is quite likely that the movement of charges and the change in pH may not have a linear relationship.

The reconstituted complexes or any of the individual components of ETC cannot synthesize ATP, although some of the complexes do generate a proton gradient and membrane potential. Experimental demonstration of the resulting potential is largely based on indirect techniques. However, a small (about 2 mV) transmembrane potential could be generated by incorporation of cytochrome O from *E. coli* in BLM (Hamamoto et al., 1985). As expected, this is reversed by a (about 150 mV) transmembrane potential or by cyanide.

Bacteriorhodopsin and FoF1 complex (see below) coincorporated into the vesicles containing the electron-transport complexes are able to synthesize ATP. Thus, a site III-specific oxidative phosphorylation system consisting of cytochrome oxidase, cytochrome *c*, and the ATPase complex or bR has been reconstituted. In all these preparations asymmetric distribution of reactants and products across an intact membrane is necessary for successful reconstitution. For example, reconstitution of a system that can catalyze oxidative phosphorylation requires not only phospholipids (highly unsaturated PE + PC are most effective) but also several factors (Fo, F1, F2, F6, oligomycin-sensitive coupling factor, and hydrophobic proteins) from mitochondria.

Thermogenesis

Shivering in response to cold is due to an uncoordinated muscular contraction that stimulates respiration, consumes ATP, and dissipates free energy as heat. In brown adipose tissues, mitochondria can switch from coupled ATP-producing respiration to an uncoupled mode, and thus generate heat directly (Nicholls and Locke, 1984). This loss of respiratory control is due to an interplay of the antagonistic effects of nucleotide triphosphates and free fatty acids. A higher concentration of free nucleotide triphosphates like GTP also increases the proton potential across the inner mitochondrial membrane.

THE CALCIUM PUMP

The cytoplasmic concentration of free calcium in living cells is about 100 n*M* compared with about 0.1 m*M* in the surrounding medium. This gradient is maintained by the Ca pump driven by Ca-ATPase found in sarcoplasmic reticulum (SR), endoplasmic reticulum, red cell, and plant cells. The Ca-uptake mechanism in mitochondria and several other organelles appears to be different and not very well understood. Most of these active Ca transporters have certain features in common: high affinity for Ca (K_d about 1 μM); they are driven by Mg-ATP with Ca/ATP ratio of 2; an acyl phosphate is formed as an intermediate; and they are inhibited by lanthanum and by ruthenium red. The relative contribution of these and other Ca-transport systems (like Na-Ca exchange in excitable cells, Ca-proton exchange in bacterial membranes, Ca channel) in homeostasis of Ca depends on the metabolic state and the type of cell. Some of the interesting examples include the effect of calmodulin on Ca-ATPase from erythrocytes, and phosphorylation of accessory protein on Ca-ATPase of cardiac sarcoplasmic reticulum (SR).

One of the best-characterized Ca pumps is from SR of muscles, where it is uniformly distributed in the free SR but not in junctional SR (Inesi, 1985). These membranes also contain two Ca-binding proteins (plus 12% calsequestrin and 4% "acidic protein") which are readily released by EGTA treatment, while Ca-ATPase (80%) and a proteolipid (probably a Ca channel protein, M55) remains attached to the membrane. The main lipids of SR are PC (65%), PE (20%), and PI/PS (10%). In membranes prepared by differential and density-gradient centrifugation, Ca-ATPase constitutes 50% of the total dry weight and about 70% of the total protein. On the basis of electron microscopy and X-ray diffraction evidence, a large portion of the ATPase polypeptide is located within the bilayer or protruded on the cytoplasmic surface of the bilayer. In isolated SR membranes the cytoplasmic side of the native Ca-ATPase is oriented toward the outside.

Purified Ca-ATPase consists of a single peptide chain of MW 115 kD. Both the amino and carboxyl terminus, as well as several loops, are exposed to the cytoplasmic phase, and there are four to eight transmembrane segments. The α-helix (31%) and β-sheet (21%) content of the protein appears to change in the presence of Ca. The protein in bilayers probably exists as an α_2 dimer, the functional significance of which is not clear. Under a variety of conditions two-dimensional arrays of Ca-ATPase have been noted. Thus, for example, vanadate promotes formation of two-dimensional crystalline arrays accompanied by inhibition of ATPase activity. Formation of arrays is prevented by 0.1 mM Ca or by inside negative membrane potential.

Muscle contraction results from depolarization of the surface membranes of T tubules, which triggers a charge movement in the junctional structures with SR, which in turn initiates the series of events that lead to the release of stored Ca from SR into cytoplasm. As the concentration of Ca in the cytoplasm increases from $<0.1\ \mu M$ to about $1\ \mu M$, binding of Ca to troponin changes the structure of the thin filaments. This facilitates the interaction of actin with cross-bridges of myosin, followed by activation of ATP hydrolysis and the development of contractile tension (Martonosi, 1984).

The Ca pump transports Ca from the cytoplasm into the SR lumen. It can lower the Ca concentration in the cytoplasm to about 10 nM even when the Ca concentration in the lumen is about 50 mM. Since SR membrane is freely permeable to anions (oxalate, phosphate, pyrophosphate), buffering and binding of Ca leads to its accumulation approaching saturation of the internal space in the inside-out vesicles. Precipitation and complexation of Ca would also lower the transmembrane diffusion potential arising from the gradient of Ca.

Native SR vesicles contain about 90 moles of phospholipid per mole of Ca-ATPase. Incubation with hydrophobic solutes (alkanols, diethyl ether, halothane) or depletion of phospholipids by detergent treatment, or with phospholipase A_2 or C depletes the ATPase activity. The catalytic activity can be restored by a variety of phospholipids and detergents, indicating a broad specificity. It appears that optimum activation of the enzymic activity depends upon the presence of a complete lipid annulus, i.e., about 35 moles of lipid per mole of enzyme. One of the best methods to reconstitute Ca-ATPase is by dilution of the detergent suspension of the protein with appropriate lipids. The reconstituted preparation consists of sheets of bilayer which are probably stabilized by cholate at the edges. There is virtually an absolute requirement for a fluid bilayer and the optimal chain length is 18–20 (Froud et al., 1987). Using brominated lipids East and Lee (1982) have shown that the relative binding of phospholipid species with different head groups or acyl chain length are very similar, and the exchange rate of the annular lipid with the bulk lipid is also about 100 nsec.

The reaction sequence for Ca-ATPase that is also consistent with the pump activity is shown in Fig. 13-5 and the corresponding rate constants are also given. Coupling of the catalytic activity with the transport cycle consists of several steps. The pump consists of Ca-ATPase oriented toward the cytoplasm and has high affinity for Ca. The Ca transport is initiated by a random-order interaction of Ca (K_d about $2\ \mu M$; in intact SR the affinity for Ca is about 100 times lower. Other di- or trivalent ions do not participate in the catalytic or transport functions. The K_m for ATP is about 2–3 μM; other nucleoside triphosphates like acetylphosphate and p-nitrophenylphosphate also participate in the catalytic and pump activity from the cytoplasmic side. Sequential binding of the second Ca ion occurs to this complex. The ternary complex ($ATP \cdot E \cdot Ca_2$) undergoes a conformational change that is reflected in intrinsic fluorescence, as well as in altered accessibility and mobility of

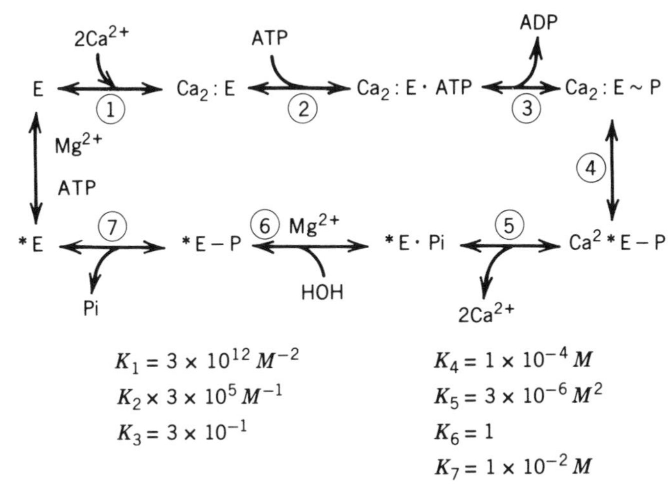

Fig. 13-5. A minimal kinetic scheme for Ca-ATPase with appropriate equilibrium constants. (From Fernandez et al., 1984.)

other functional groups on the protein. The bound ATP is cleaved, an aspartate residue is phosphorylated, and the bound Ca ions are occluded as they are no longer accessible to EGTA in the medium. The phosphorylated enzyme undergoes a slow conformational change, that exposes Ca to the opposite side of the membrane and the affinity for calcium decreases (K_d about 1 mM), and thus two Ca ions are released sequentially. The reaction sequence is completed by the hydrolysis of phosphoprotein.

Sarcoplasmic reticulum vesicles accumulate 2 Ca per ATP used. ATP is used to make a phosphorylated intermediate in which the terminal phosphate of ATP is transferred to an aspartate residue at the catalytic site. Each 115-kD subunit consists of one phosphorylation and two Ca-binding sites (apparent K_d is about 1 μM). The equilibrium binding isotherm for Ca displays a cooperative behavior and competition with respect to protons. The degree of cooperativity for Ca binding ($n > 3$) observed at low proton concentrations requires interaction of four binding domains corresponding to two 115-kD chains. This suggests that the dimeric form of the enzyme in the bilayer has a regulatory role. Binding of two Ca ions is a cooperative sequential process in which a conformational change induced by the first Ca ion promotes binding of the second Ca ion. The pump is electrogenic, which suggests that the charge compensation during the turnover cycle is at best partial.

According to the scheme shown in Fig. 13-6 and Table 13-2, the coupling of the catalytic activity with the transport cycle consists of several steps

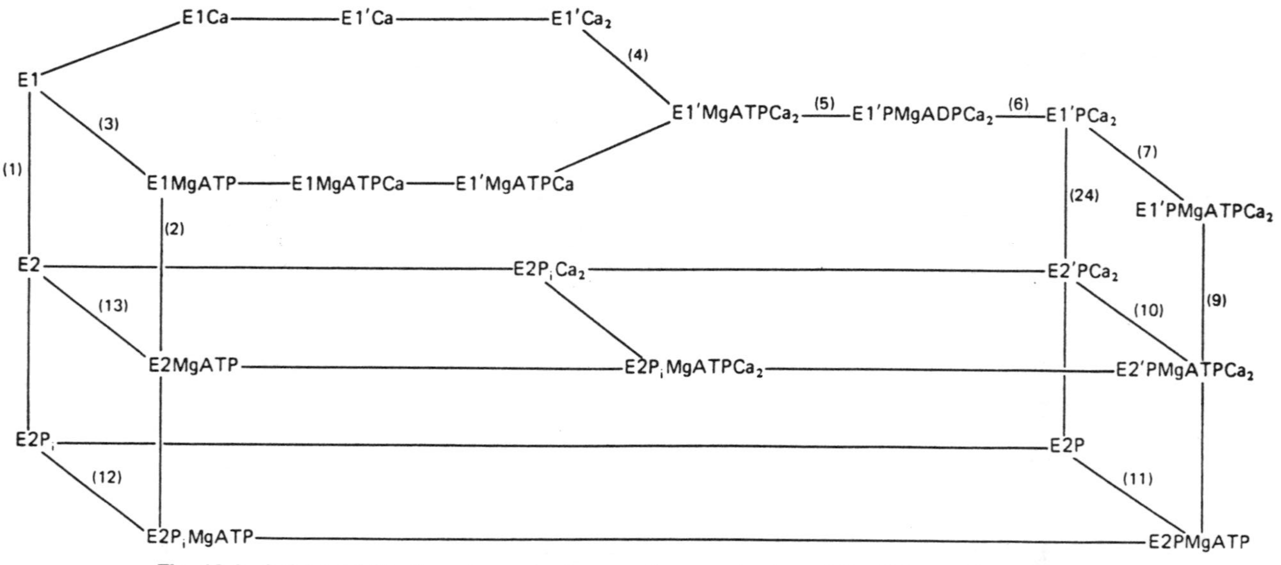

Fig. 13-6. A detailed kinetic scheme for the reaction mechanism of Ca-ATPase. All the equilibrium constants and the rate constants have been characterized as given in Table 13-2. (From Gould et al., 1986.)

TABLE 13-2 Kinetic Parameters for the Reaction Cycle of Ca-ATPase at 25°C

Reaction[a]	Equilibrium Constant[a] Symbol	Equilibrium Constant[a] Value	Forward Rate Constant (s^{-1})[a]
E1 + MgATP $\rightleftharpoons$ E1MgATP	K_3	1.25×10^5	18.5
E1 + Mg $\rightleftharpoons$ E1Mg	K_{16}	110.0	
E1 + MgADP $\rightleftharpoons$ E1MgADP	K_{14}	1.1×10^4	
$E1'Ca_2$ + MgATP $\rightleftharpoons$ $E1'MgATPCa_2$	K_4	1.25×10^5	45.5
$E1'Ca_2$ + Mg $\rightleftharpoons$ $E1'MgCa_2$	K_{16}	110.0	
$E1'Ca_2$ + MgADP $\rightleftharpoons$ $E1'MgADPCa_2$	K_{14}	1.1×10^4	
E1MgATP + Ca^{2+} $\rightleftharpoons$ E1MgATPCa	K_{C7}	5.0×10^4	1.0×10^8
E1MgATPH + Ca^{2+} $\rightleftharpoons$ E1MgATPHCa		*	1.0×10^8
$E1MgATPH_2$ + Ca^{2+} $\rightleftharpoons$ $E1MgATPH_2Ca$		*	1.0×10^8
E1MgATPCa $\rightleftharpoons$ E1'MgATPCa	K_{C8}	16.8	180.0
E1MgATPHCa $\rightleftharpoons$ E1'MgATPHCa		*	180.0
$E1MgATPH_2Ca$ $\rightleftharpoons$ $E1'MgATRPH_2Ca$		*	180.0
E1'MgATPCa + Ca^{2+} $\rightleftharpoons$ $E1'MgATPCa_2$	K_{C9}	1.44×10^8	1.0×10^8
E1'MgATPHCa + Ca^{2+} $\rightleftharpoons$ $E1'MgATPHCa_2$		*	1.0×10^8
$E1'MgATPH_2Ca$ + Ca^{2+} $\rightleftharpoons$ $E1'MgATPH_2Ca_2$		*	1.0×10^8
$E1_0^{Ca}MgATP$ + H^+ $\rightleftharpoons$ $E1_0^{HCa}MgATP$†	K_{H9}	1.5×10^8	
$E1'^{Ca}_0MgATP$ + H^+ $\rightleftharpoons$ $E1'^{HCa}_0MgATP$†	K_{H10}	1.0×10^9	
$E1'MgATPCa_2$ + H^+ $\rightleftharpoons$ $E1'MgATPHCa_2$	K_{H8}	1.67×10^7	
$E1'MgATPCa_2$ $\rightleftharpoons$ $E1'PMgADPCa_2$	K_5	0.5	500.0
$E1'PCa_2$ + MgADP $\rightleftharpoons$ $E1'PMgADPCa_2$	K_6	2.5×10^3	
$E1'PCa_2$ $\rightleftharpoons$ $E2'PCa_2$	K_{24}	0.03	0
$E1'PHCa_2$ $\rightleftharpoons$ $E2'PHCa_2$	K_{24}	0.03	0
$E1'PH_2Ca_2$ $\rightleftharpoons$ $E2'PH_2Ca_2$	K_{24}	0.03	11.5
$E1'PCa_2$ + H^+ $\rightleftharpoons$ $E1'PHCa_2$‡	K_{H8}	1.67×10^7	
$E1'PHCa_2$ + H^+ $\rightleftharpoons$ $E1'PH_2Ca_2$‡	K_{H8}	1.67×10^7	
$E2'PCa_2$ + H^+ $\rightleftharpoons$ $E2'PHCa_2$‡	K_{H8}	1.67×10^7	
$E2'PHCa_2$ + H^+ $\rightleftharpoons$ $E2'PH_2Ca_2$‡	K_{H8}	1.67×10^7	
$E1'PCa_2$ + MgATP $\rightleftharpoons$ $E1'PMgATPCa_2$	K_7	2.2×10^3	3.3×10^6
$E1'PMgATPCa_2$ $\rightleftharpoons$ $E2'PMgATPCa_2$	K_9	0.04	0
$E1'PMgATPHCa_2$ $\rightleftharpoons$ $E2'PMgATPHCa_2$	K_9	0.04	0
$E1'PMgATPH_2Ca_2$ $\rightleftharpoons$ $E2'PMgATPH_2Ca_2$	K_9	0.04	70.1§
$E2'PCa_2$ + MgATP $\rightleftharpoons$ $E2'PMgATPCa_2$	K_{10}	3.0×10^3‖	5.0×10^6
E2 + MgATP $\rightleftharpoons$ E2MgATP	K_{13}	2.0×10^4	
E2MgATP $\rightleftharpoons$ E1MgATP	K_2	15.6¶	2000.0
HE2MgATP $\rightleftharpoons$ HE1MgATP		**	221.4
$H_2E2MgATP$ $\rightleftharpoons$ $H_2E1MgATP$		**	24.3

[a] The reaction scheme is shown in Fig. 13-6. Rates of dephosphorylation of all Ca^{2+}-bound forms are assumed to be 0.05 times those of the Ca^{2+}-free form, and rates of dephosphorylation of all MgATP-free forms are assumed to be 1.5 times those of the MgATP-free forms. All unassigned rate constants are assumed to be fast. From Gould et al., 1986.

* Defined by K_{H3}, K_{H8}, K_{H9} and K_{H10}.

† 0 represents an unoccupied Ca^{2+}-binding site.

‡ Assumed to be unaffected by the binding of MgATP.

§ For a high-activity preparation; $k_{+9} = 41.4\ s^{-1}$ for a low-activity preparation.

‖ Defined by K_7, K_{24} and K_9.

¶ Defined by K_1, K_3 and K_{13}.

** Defined by K_2, K_{H1} and K_{H2}.

(Gould et al., 1986). The turnover for phosphorylation and dephosphorylation reactions are 150/sec and 60/sec, respectively. Since the overall turnover for the pump is about 10/sec (specific activity 5 IU), it implies that a rate-limiting step is or follows the dephosphorylation step. The *coupling* state of Ca pump appears to depend significantly on the lipid composition of proteoliposomes (Navarro et al., 1984).

Formation of the phosphoenzyme is observed with the delipidated enzyme, however, dephosphorylation requires lipids. The dephosphorylation step is also significantly inhibited by lipids in the gel phase, and the activity increases in the fluid-phase phospholipids. These observations suggest that the hydrolysis of the phosphoprotein or some step after dephosphorylation, which may also be the rate-limiting step in the overall turnover cycle, depends upon the lipid environment.

The Ca-transport process is reversible. ATP can be synthesized in vesicles loaded with Ca and suspended in a medium containing ADP and inorganic phosphate (P_i). Similarly, the enzyme can be phosphorylated from P_i under appropriate conditions. Affinity for Ca is reduced by binding of vanadate or phosphate ions. A mutual destabilization of the phosphate and Ca-binding site is apparently the central feature of the coupling mechanism. The energy derived from the interaction with phosphate or vanadate can be utilized to overcome the thermodynamic barriers within the enzyme cycle for the translocation of Ca.

THE SODIUM PUMP

Na,K-ATPase mediates active transport of 3Na and 2K ions in opposite directions against their electrochemical gradients (Fig. 13-6). It regulates ion and water balance, and acts as a receptor for cardioactive drugs like ouabain and their endogenous analogs (Anner, 1985). Na,K-ATPase has been isolated from a wide variety of plasma membranes. The active unit is either $\alpha\beta$ (about 265 kD) or $\alpha_2\beta_2$ (about 500 kD). Purified Na,K-ATPase in membrane fragments of 100–500 nm diameter contain 19000 particles /μ^2. The size of the particles (2.5 to 6 nm) depends upon the state of phosphorylation, and both $\alpha\beta$ and $\alpha_2\beta_2$ states can be crystallized in a two-dimensional array (Jorgenson, 1982). The topography of the α-subunit, which has the catalytic site and the ouabain-binding site, suggests that the carboxyl and amino termini are cytoplasmic, and probably six α-helices traverse the bilayer.

Phospholipids are absolutely necessary for the activity of Na,K-ATPase. A stoichiometry of about 66 lipids per 265-kD protein is found for motionally restricted spin-labeled lipids. Several lipid-soluble additives like 18 : 1, 18 : 2,

18:3, and 20:4 fatty acids, palmitoylcarnitine, and palmitoyl-CoA inhibit the ATPase activity, oubain binding, and Rb transport in intact cells. Stearic acid has little or no effect. Higher mole fractions of cholesterol inhibit, whereas a lower mole fraction of cholesterol increases Na,K-ATPase activity. These effects of lipid-soluble additives are probably due to a change in the lipid microenvironment induced by the additives. The lipid requirement for the ATPase activity can also be demonstrated by lipid extraction with detergents, organic solvents, or by treatment with lipolytic enzymes. Several studies suggest that about 250 lipid molecules per active $\alpha_2\beta_2$ molecule can reconstitute 100% of the ATPase activity. Preparations delipidated with detergents lose 100% of the catalytic activity, although ouabain-binding activity is retained with partial delipidation and partial inactivation of ATPase activity. Reactivation of completely delipidated Na,K-ATPase preparations has been achieved successfully in some cases, whereas only partial activation is achieved in others. The role of detergents used for solubilization of native Na,K-ATPase is apparently crucial (Anner, 1985). It appears that only those preparations in detergents can be reactivated in which the intrinsic lipid annulus is intact. Detergents that remove the native lipid annulus irreversibly inactivate. The reconstituted Na pump exhibits a chain length dependence for its activity, and the activation energy decreases with increasing length of the acyl chains containing one double bond (Marcus et al., 1986). The decrease in activity with chain length is apparently due to a different degree of "scrambling" in the orientation during reconstitution.

Orientation of Na,K-ATPase in reconstituted vesicles can be ascertained on the basis of several tests designed on the basis of the asymmetries intrinsic in the scheme shown in Fig. 13-7. These include test for correct orientation of vanadate and concanavalin A binding site; for trypsin or neuraminidase cleavage site; for ATP and ouabain sensitivity in intact and ruptured vesicles; and distribution of intramembrane particles in freeze-fracture replicas. In reconstitution of Triton-X-100-solubilized preparations, 85% of ATPase molecules appear to be inserted right side out. On the other hand, in cholate-solubilized preparations the right-side-out orientation is about 50%.

The role of the various partial reactions (Fig. 13-7) in reconstituted Na,K-ATPase systems has been reviewed (Anner, 1985). The ion translocation steps associated with the partial reactions of Na,K-ATPase are:

1. Passive but vectorial fluxes of Na and K ions are mediated apparently due to the ionophoric activity of the dephosphorylated enzyme.
2. Rb–Rb exchange mediated by dephospho-enzyme is inhibited by Na (I_{50} 1 mM) and Mg (I_{50} 1.3 mM), and is sensitive to vanadate. Apparent

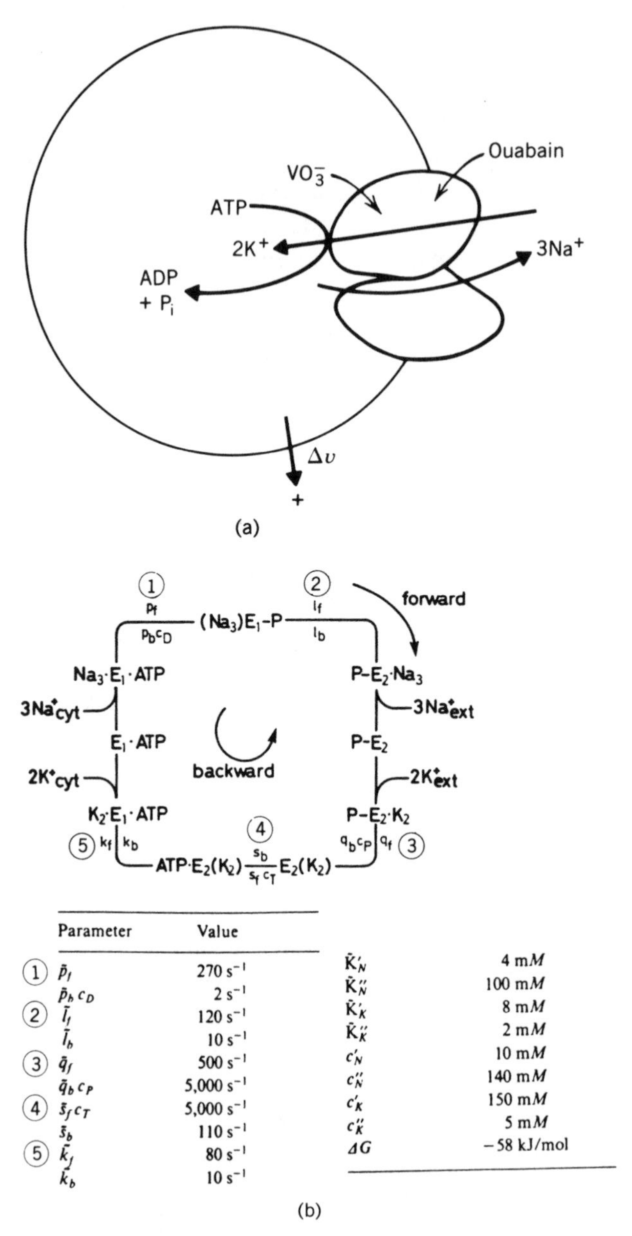

	Parameter	Value		
1	$\tilde{p}_f$	270 s^{-1}	$\tilde{K}'_N$	4 mM
	$\tilde{p}_b c_D$	2 s^{-1}	$\tilde{K}''_N$	100 mM
2	$\tilde{l}_f$	120 s^{-1}	$\tilde{K}'_K$	8 mM
	$\tilde{l}_b$	10 s^{-1}	$\tilde{K}''_K$	2 mM
3	$\tilde{q}_f$	500 s^{-1}	c'_N	10 mM
	$\tilde{q}_b c_P$	5,000 s^{-1}	c''_N	140 mM
4	$\tilde{s}_f c_T$	5,000 s^{-1}	c'_K	150 mM
	$\tilde{s}_b$	110 s^{-1}	c''_K	5 mM
5	$\tilde{k}_f$	80 s^{-1}	ΔG	−58 kJ/mol
	$\tilde{k}_b$	10 s^{-1}		

(b)

Fig. 13-7. (a) Minimum functional representation of Na pump. (b) The kinetic scheme for the pumping cycle with ping-pong mechanism and sequential translocation of cations. E_1 and E_2 are conformations of the enzyme with ion-binding sites exposed to the cytoplasm and to the extracellular medium, respectively. In the "occluded" states $(Na_3)E_1$ and $(K_2)E_2$ the bound ions cannot exchange with the aqueous phase. Dashes indicate covalent bonds and dots indicate noncovalent bonds. The rate constants are also summarized. (From Apell et al., 1986.)

affinity for Rb is 0.6 mM on the cytoplasmic side, and 0.2 mM on the extracellular side.

3. Rb–Rb or K–K exchange via phosphoenzyme is optimal above 150 mM internal and 16 mM external RbCl. E_1 to E_2 transition corresponding to an extensive α-helix to β-sheet conformational change is necessary for this exchange.
4. Ouabain sensitive Na–Na exchange and uncoupled Na transport is observed in the presence of MgATP. This ADP + ATP-stimulated Na–Na exchange occurs either in the partially poisoned cells or when Na replaces extracellular K in the binding site. Apparently Na ions are occluded in the ADP-sensitive E1, which is able to accept the terminal phosphate from ATP and transfer it to ADP.
5. Electrogenic Na–K exchange has been demonstrated in a variety of cell types in which the positive charge movement resulting from Na efflux is uncompensated by K influx. The most commonly observed stoichiometry is 3Na : 2K : 1ATP for the pump, and also 3Na-binding site : 2K-binding site : 1 phosphorylation sites on Na,K-ATPase. From the kinetic studies it seems that the Na discharge and the K-loading sites coexist rather than being a single set of sites alternating their affinity for Na or K ions. In reconstituted systems electrogenic potential of about 10 mV has been measured, and it is abolished by vanadate on the cytoplasmic side, or by oubain, palytoxin, or eryhtroplein on the extracellular side. Using indodicarbocyanine dye for measuring transmembrane potential, activation energy for the active transport process is found to be 27 kcal/mole.

The reaction and translocation mechanism for Na,K-ATPase (Fig. 13-7) is consistent with its function as an active transport system for Na and K (Albers et al., 1968; Post et al., 1969). The kinetic scheme can be viewed as a cycle composed of two sets of equilibria between a pair of phosphoenzymes and a pair of dephosphoenzymes, E_1 and E_2. The enzyme is phosphorylated by ATP in the presence of Na and Mg to give an acid-stable E_1-P, which can be readily cleaved by ADP and is insensitive to K. E_1-P is in equilibrium with E_2-P, which is insensitive to ADP and is readily cleaved by K. The dephosphoenzyme also exists in two forms, E_1 having a high affinity for Na ions, and E_2 having a high affinity for K ions. Since the pump can work in the absence of extracellular Na and in the absence of intracellular K, intracellular Na is responsible for activation of phosphorylation to E_1-P, and extracellular K for the activation of hydrolysis of E_2-P. The overall reaction appears to operate by a ping-pong mechanism with respect to Na and K. The partial catalytic reactions that are consistent with this mechanism include: Na-dependent phosphorylation, K-promoted dephosphorylation, Na-dependent ADP–ATP exchange, K-dependent exchange of O^{18} atoms between P_i and water, and the two conformations E_1 and E_2 that are promoted by Na and K. The partial transport reactions that also support this scheme are: the Na : Na exchange in the absence of K_o, and the K : K exchange in the absence of Na_i (Kaplan, 1985). An alternative model is gaining some support recently (Skou, 1982).

THE PROTON PUMP

The proton pump catalyzes ATP synthesis driven by pmf across membranes of bacteria, fungi, mitochondria, and chloroplasts (Fig. 13-8). Although the function of proton ATPase appears to be the synthesis of ATP at the expense of a proton gradient (1–3 protons per ATP), it can also generate a proton gradient concomitant with the hydrolysis of ATP. Basically, there are three different ATP-driven proton pumps. The one present in plasma membranes

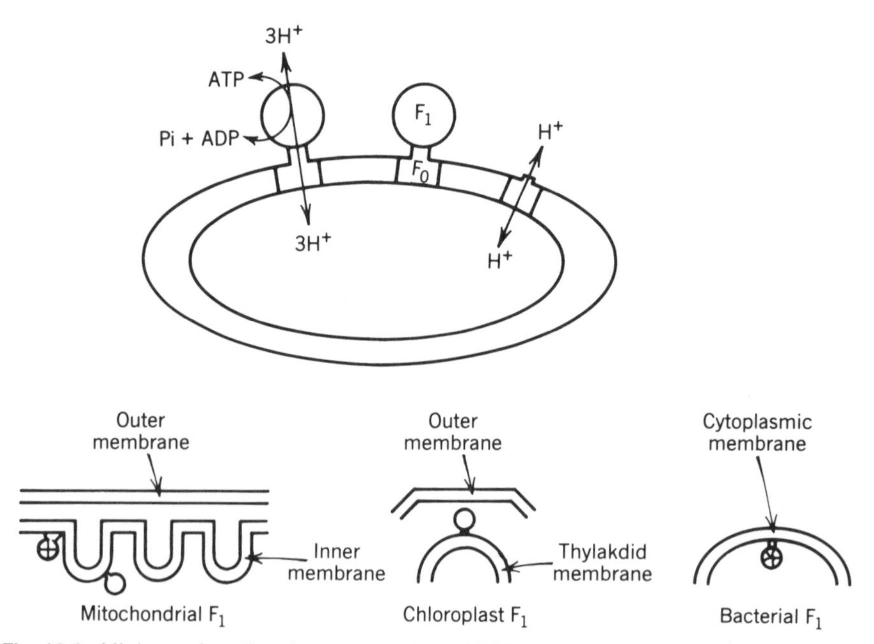

Fig. 13-8. Minimum functional representation of ATP synthase in submitochondrial particles. Topological localization of F_oF_1 is also shown for mitochondrial, chloroplast, and bacterial membranes.

of eukaryotic microorganisms and plants forms an acylated intermediate (usually aspartyl phosphoenzyme) and is inhibited by vanadate. Thus, it resembles Ca or Na pumps. The second type of proton pump, present in chloroplast, mitochondria, and plasma membranes of prokaryotes consists of multiple components, and it is also involved in the synthesis of ATP. The third type of proton pump is inhibited by *N*-ethylmaleimide (Xie et al., 1984). The bacterial proton pumps have several distinguishing characteristics (Senior, 1985). The following account is only for the proton pump from mitochondria and chloroplasts.

The proton-driven ATPase from mitochondria and prokaryotes consists of two parts which dissociate in the presence of urea. F1 contains the catalytic center, is localized outside the membrane, and it is bound to Fo which is thought to function as a transmembrane proton channel. Fo also confers oligomycin sensitivity to the catalytically active complex. When incorporated into liposomes FoF1 exhibits proton translocation, ATP-P_i exchange, ATP synthesis, and sensitivity to oligomycin. Fo in liposomes mediates translocation of protons and has a binding site for F1. The FoF1 complex differs from Ca-ATPase, Na,K-ATPase, and the proton-translocating ATPase from the fungal plasma membrane (Goffeau and Slayman, 1981). Unlike these transport ATPases, FoF1 synthesizes ATP under physiological conditions without forming a phosphorylated intermediate. In mitochondria ATPase-driven proton extrusion and reverse electron flow are also observed, but their physiological role is unknown but it could regulate oxidative phosphorylation.

The FoF1 complex from bacteria and chloroplast consists of eight dissimilar subunits, three of which belong to Fo. However, mitochondrial FoF1 is more complex: about 16 subunits in beef heart FoF1 and 10 in the yeast complex, along with several coupling factors. Relative stoichiometry of these subunits is not known, however the molecular weight of the complex is approximately 500 kD. The subunits of FoF1 are partly coded for by nuclear and partly by organellar genes.

The ATPase activity of FoF1 is tightly coupled to proton translocation, and a pmf of about 200 mV across the membrane would convert the ATPase into ATP synthase. Thus, ATPase activity in intact organelles is promoted by an uncoupler that makes membranes leaky to protons. Similarly, ATPase, ATP synthase, and P_i-ATP exchange reactions are sensitive to inhibitors like oligomycin that block translocation through Fo. The number of protons that are translocated downhill during the synthesis of ATP is probably three ($= n$). A small change in the apparent number of protons translocated during the synthesis of ATP would also occur at the p*K*a of ATP is different from that of ADP + P_i.

ATP synthesis mediated by a proton gradient has been observed in bacteria, chloroplast and mitochondria. However, interpretation of these results is not clear-cut because the energy required for ATP synthesis could have been stored in other components, rather than the proton gradient. Most reconstituted liposome preparations containing FoF1 become leaky during the acid–base transition, and, therefore, it is difficult to measure ATP synthesis. This difficulty is eliminated by using thermophilic FoF1 and thermophilic phospholipids (Sone et al., 1977). Under these conditions the maximal level of ATP synthesis is about 100 nmoles/mg of FoF1, whereas it is less than 2.5 nmole/mg of FoF1 in submitochondrial particles, bacterial membranes, and intact mitochondria.

In reconstituted vesicles, electrochemical potential in the form of a transmembrane potential or a pH gradient elicits equal rates of ATP synthesis. The extent of the electrochemical potential for protons that is available to FoF1 in a membrane is a matter of controversy. Direct measurement of pH during energization of mitochondria, and ATP synthesis driven by an imposed pH gradient in reconstituted vesicles, suggest that the bulk pH gradient can promote ATP synthesis. If, however, three protons are used to drive synthesis of one molecule of ATP, the overall reaction would be higher order. Therefore the rate of synthesis of ATP would be very slow at physiological pH, unless protons from the electron-transport chain circulate to the nearest FoF1 through the membrane surface.

Several features of the mechanism of the proton–ATPase reaction have emerged. The release of ADP from F1 is caused by ATP. There are probably six nucleotide binding sites on F1, three of which are on noncatalytic sites. While these sites give rise to positive cooperativity in V_m and negative cooperativity in the K_D, the exact role of the three catalytic sites in the overall turnover is not established. Several inhibitors (azide, aurovertin, tentoxin, oligomycin, quercetin) of FoF1 are known, however their mechanism of action is not known. The ATPase inhibitor protein isolated from mitochondria may prevent unnecessary ATP hydrolysis *in vivo*.

The proton gradient created by light-driven reactions in the chloroplast is used by FoF1 coupling factor that is similar to the mitochondrial coupling factor. F1 (MW 325 kD) has five subunits and at least three nucleotide-binding sites of different affinities. The nucleotides act as substrates as well as the regulators of the enzyme. The hydrophobic stalk, Fo (MW 100 kD) is made up of six to eight copies of a 8-kD peptide and two other subunits. F1 catalyzes hydrolysis that obeys Michaelis–Menten kinetics except at very low concentrations of ATP. Calcium is required for significant catalytic activity by F1, whereas Mg is required by membrane-bound FoF1. Kinetics of ATP hydrolysis and ATP synthesis as a function of the pH gradient shows that three protons are pumped per ATP synthesized (Dewey and Hammes, 1981; Tabake and Hammes, 1981). Binding of substrate shows no dependence on pH, but only on membrane potential. This suggests involvement of the catalytic step in the energy conservation step. The absolute values of the kinetic parameters depend on the external pH as well as on the pH gradient. However, dependence of the kinetic parameters on the external pH is independent of the pH gradient. Similarly, the dependence of the kinetic parameters on the pH gradient appears to be independent of the external pH. These observations are consistent with the suggestion that the catalytic site and the site for pumping of protons are not identical but they are coupled.

The direction of the reaction catalyzed by proton ATPases is determined by pmf. At low pmf it hydrolyzes ATP, and it generates ATP at high pmf. This could lead to a significant waste of energy under certain physiological conditions such as in chloroplasts in dark when pmf is low. Under such conditions the H–ATPase in chloroplast becomes inactive probably by tight binding of the ADP to the active site and by structural rearrangement of the FoF1 complex which is coupled with oxidation of a thiol group in the γ-subunit which is modulated by light intensity. In mitochondria, regulation occurs by tight binding of a natural protein inhibitor with ATPase, which is modulated by Ca.

SENSORS OF PROTON GRADIENTS

Chemotaxis, a migratory response of bacteria to a chemical gradient, serves a variety of purposes as attested by the diversity of chemotactic functions ranging from tumbling to the pheromone response. The means by which chemotactic response is expressed include rotation of flagella, beating of cilia, extension of pseudopods, or gliding over a surface. The specificity of mechanism, stimulus, and response reflect unique needs chemotaxis may serve for an organism. Thus bacteria like *E. coli* and *Salmonella* respond to gradients of nutrients like amino acids and sugars, whereas *Paramecium,* which feeds on bacteria, responds to the various excretion products of bacterial metabolism. Other functions associated with chemotactic response include homeostasis, defense, differentiation, contact for mating and aggregation.

Bacteria contain sensors that enable them to find their way toward attractants (nutrients and pheromones) and away from repellents like acetate and indole, which indicate overcrowding (Glagolev, 1984). For example, *E. coli* has 15 chemoreceptors as well as a temperature sensor and a pH sensor. The sensor for pmf guides *E. coli* away from uncouplers, and toward light, oxygen and certain substrates. The phosphotransferase system apparently acts as a pmf sensor. Proton gradient changes K_m for sugars, presumably by changing the redox state of the sulfhydryl groups. The rate of substrate transport by the phosphotransferase system increases in the presence of uncouplers, while oxygen and a respiratory substrate inhibit it.

Flagellar Motion

The organelles of bacterial motility are called flagella (latin word for whip). Under the control of sensory information from the environment, the flagella of bacteria enable them to migrate to more favorable environments. Many of the sensors on bacterial membranes are coupled to flagella for information processing. Bacteria use the rotary motion of flagella to propel themselves and attain speeds of about 20 cell lengths per second. *E. coli* contains 4–10 flagella, that are rigid helical filaments of 10–12 μm length and placed randomly on the cell surface. These close-to-equilibrium machines are made up of about 15 distinct proteins. The flagellum consists of a set of coaxial rings and a rod connected to a short structure (hook) in the membrane, which in turn is connected to the propulsive helical filament which is typically five times as long as the cell body and about 20 nm in diameter. Amino acid sequence and membrane orientation of the proteins involved in transduction have been proposed (Krikos et al., 1983).

During locomotion flagella are bundled in a rotary screw-like motion, so that the bundle acts as a propeller. A counterclockwise twirl (as seen from behind the cell) occurs when the cell is in the run mode, and a clockwise twirl occurs in the tumble mode. In an isotropic chemical environment, the random motion is made up of run and tumble modes. However, the movement is biased in the presence of a concentration gradient of a chemotactic attractant. The frequency of tumbles decreases when bacteria are swimming toward a higher concentration of attractants. The bacteria are also able to adapt to changes in the attractant concentration, both with respect to responsiveness and to threshold (Macnab and Aizawa, 1984).

There are several interesting features of this chemosensory system. The process is mediated by ligand-binding proteins present in the periplasmic space (Chapter 11). In the envelope of the gram-negative bacteria, the plasma membrane is surrounded by the cell wall or by the peptidoglycan, a thin, rigid layer that maintains the cell shape and prevents lysis in a hypotonic environment. Channels of porin in the outer membrane permit the passage of polar molecules of MW about 600. Between the inner membrane and the cell wall is an aqueous layer, known as periplasm, that contains soluble binding proteins. Together with some binding proteins located in the cytoplasmic membranes, the periplasmic binding proteins could act as chemotactic receptors. In *E. coli* there are about 25 chemoreceptors of this type. Quite a few of these have been purified and others have been identified by genetic techniques. The binding proteins (MW 30–40 kD) from periplasmic space (about 10,000 copies per cell, which corresponds to a concentration of about 1 m*M* in the periplasmic space), are water-soluble and are released by osmotic shock. These proteins apparently interact with other membrane proteins that ultimately lead to flagellar motion powered by the transmembrane proton gradient. Perhaps the best documented pmf-mediated response is aerotaxis, i.e., chemotaxis to oxygen. This is probably because increased binding of oxygen to the terminal oxidase of the respiratory chain stimulates electron transport and thereby generates a proton gradient.

14 Transbilayer Response of Signals

> **Those who have never seen two well-trained armies drawn up for battle can have no idea of the beauty and brilliance of the display. Bugles, fifes, oboes, drums, and salvoes of artillery produced such harmony as Hell itself could not rival. The opening barrage destroyed about six thousand men on each side. Rifle fire that followed rid this best of worlds of about nine or ten thousand villains who infested its surface. Finally, the bayonet provided "sufficient reason" for the death of several thousand more. The total casualties amounted to about thirty thousand. Candide trembled like a philosopher, and hid himself as best he could during this heroic butchery.**
>
> **When all was over, and the rival kings were celebrating their victory with Te Deums in their respective camps, Candide decided to find somewhere else to pursue his reasoning into cause and effect. He picked his way over piles of the dead and the dying, and reached a neighboring village on the Abar side of the border. It was now no more than a smoking ruin, for the Bulgars had burned it to the ground in accordance with the terms of international law.**
>
> **The following day, while creeping amongst the ruins Candide and Pangloss found something to eat. . . . Some of the citizens whom they had helped gave them as good a dinner as could be managed after such a disaster. The meal was certainly a sad affair, and the guests wept as they ate; but Pangloss consoled them with the assurance that things could not be otherwise:**
>
> **"For all this," said he, "is a manifestation of the rightness of things, since if there is a volcano at Lisbon it could not be anywhere else. For it is impossible for things not to be where they are, because everything is for the best."**
>
> **VOLTAIRE in *Candide***

Information transduction across membranes is an intriguing biological phenomenon. Indeed, the functional essence of an organism is its response to the environment. To carry out a variety of complex functions, cells undergo an impressive array of rather dramatic changes in response to stimuli. On the morphological level these include changes in shape, aggregation, and secretion, which are accompanied or followed by metabolic changes such as altered rates of transport, protein phosphorylation, arachidonate oxygenation, and many other less well-defined processes. The signal-response coupling across plasma membranes is taken axiomatically to account for the transbilayer response induced by a large variety of extracellular signals (hormones, transmitters, light). Typically, the response is dictated by the concentration of the signal, by the number of receptors, and by their mutual affinity. The occupied receptors in turn elicit a variety of secondary responses. The significance and extent of the secondary responses to the interaction of a signal with its receptor is the subject of intense investigation at several levels ranging from the biochemical characterization of receptors to the concepts related to receptor mobility, availability, biosynthesis, distribution, and the nature and extent of secondary responses.

ADENYLATE CYCLASE STIMULATED BY HORMONES

Formation of cAMP inside the cell represents the first link in a chain of biochemical reactions that ultimately result in the physiological expression of the interaction of a signal molecule. Such responses of cAMP range from effects on cell motility and morphology to specific regulation of intracellular metabolic pathways through modification of transporters and enzymes. The cAMP-mediated processes appear to be involved not only in regulation of transmembrane potential and permeability changes but also in transmitter synthesis, metabolism, cellular movements, and trophic and developmental processes. It may be emphasized that in many cases even now it is not certain which of these processes are a direct result of interaction with cAMP and which are secondary to the utilization of cAMP.

Formation of cAMP from ATP is catalyzed by membrane-bound adenylate cyclase with the catalytic site toward the cytoplasm (Fig. 14-1). Functionally, the cyclase activity is modulated by the interaction of a variety of signal molecules (hormones and transmitters) with their specific receptors on the external surface (Table 14-1). Thus, cAMP is formed as a cytoplasmic messenger (also called second messenger) of hormones and transmitters that interact with their receptors on the outer surface of the cell. There are several interesting characteristics of adenylate cyclase activated by hormones. The effect of the signal on adenylate cyclase is rapid (usually seen in less than 5 sec) and it is primarily on the turnover rate of the cyclase, rather than on its affinity for ATP. Since the signal is not chemically altered during its effect on the cyclase, it means that the effect of the signal molecule would be integrated over a time period to produce a large number of cAMP molecules (*amplification*). In a membrane, modulation of the same cyclase molecule may be coupled effectively to receptors of several hormones. Thus, integration, amplification, synergism, and antagonism of the responses of

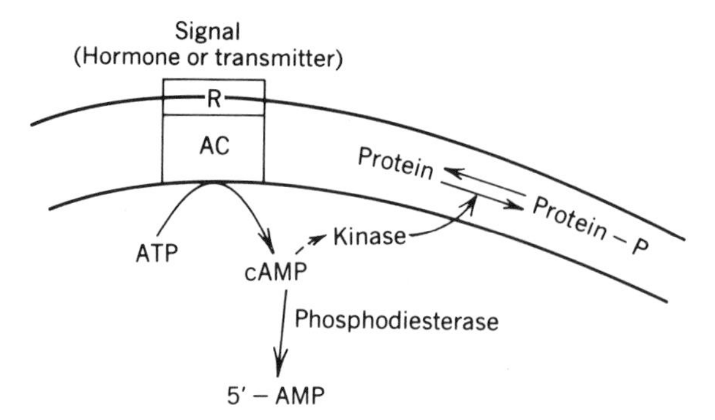

Fig. 14-1. An oversimplified sequence of events in cAMP-mediated transmembrane response to binding of a signal molecule (transmitters, hormones) to the receptor (R). Binding of the signal molecule to its extracellular receptor activates adenylate cyclase, which catalyzes formation of cAMP from ATP. Increased levels of cAMP may activate protein kinase resulting in increased levels of a phosphorylated protein. The effect of the signal is terminated by the hydrolysis of cAMP by phosphodiesterase (PDE) and by dephosphorylation of the phosphoprotein. The effect of the signal may be prolonged by inhibitors of phosphodiesterase (caffeine, theophylline, papavarine).

different signals can occur at the level of cytoplasmic adenylate cyclase activity. There is a conceptual analogy between the signal-activated adenylate cyclase and the acetylcholine receptor (AChR) channel. Both discriminate between stimuli and generate a response to a specific stimulus. Besides the usual criteria of stimulus specificity, selectivity of response, and a demonstrated variation in the levels of the signal and cAMP, a signal-stimulated adenylate cyclase also exhibits other similar characteristics: application of cAMP mimicks the action of the signal; inhibitors of phosphodiesterase, the only enzyme that inactivates cAMP, potentiate the effects of the signal; modulators of the cyclase (fluoride, GTP, prostaglandins) also modulate the action of the signal; specific antagonists and agonists of signal molecules show binding to their receptors, and the resulting modulation of adenylate cyclase correlates with their *in vivo* response; distribution of the signal-binding activity and the signal response in tissues is correlated; simultaneous changes in cAMP and the signal levels are demonstrable in response to external stimuli, such as firing of the presynaptic neuron. The coupling between the binding of signal and its response does not depend upon the membrane potential or the integrity of the membrane. However, the coupling between the binding of the signal and the adenylate cyclase is not obligatory and it can be bypassed.

Adrenergic Receptors

Stimulation of adenylate cyclase by adrenalin (epinephrine) is mediated by β-adrenergic receptors. The distinction between α- and β-adrenergic recep-

TABLE 14-1 Transmitter-Sensitive Adenylate Cyclase in Broken-Cell Preparations

Transmitter	Tissue	K_m (μM)	Percent Stimulation	Specific Inhibitors[a] K_1 (μM)
Dopamine	Cuadate nucleus	5–10	175–400	Thioxanthenes (0.001–0.04), phenothiazines (0.004–0.07), dibenzazepines (0.018–0.17), butyrophenones (0.09–0.22), tricyclic antidepressants (0.17–1.5).
	Olfactory tubercle	5–8	200	Similar to above
	Cerebral cortex	5–10	125	Haloperidol (0.5–1.0)
	Retina	1–3	200–900	Phenothiazines (0.01–0.08), haloperidol (0.06), ergotamine (1), phentolamine (5)
	Sympathetic ganglion	7	180	Phentolamine (25)
	Thoracic ganglion	2	250	Haloperidol (0.2)
	Neuroblastoma	1	200	Haloperidol (<10), phentolamine (<0.4)
Norepinephrine	Cerebellum	2–20	125–225	Propranolol (0.007)
	Cerebral cortex	10–50	120–180	DCI, propranolol (20–50)
	Glioma	0.4–5	200–700	
	Neuroblastoma	10–30	200	Propranolol (<1)
Serotonin	Colliculi	0.5–1	175	LSD (0.1–10), BOL, methysergide (<1)
	Thoracic ganglion	0.5	175	LSD (0.005), BOL (0.005), cyproheptadine (0.25)
Octopamine	Thoracic ganglion	1.5	400	Phentolamine (0.5)
Substance P	Cerebral cortex, cerebellum, hypothalamus, substantia nigra	0.2	140–200	
Prostaglandin E	Whole brain	2.5–25	160–250	Morphine, methadone (1–2), heroin (0.1–0.2)
	Neuroblastoma × glioma hybrids	1	2,000	Morphine (~2)
Morphine	Striatum	3	300	
Histamine	Hippocampus, neocortex, striatum	8	70–200	Metiamide (0.9)
Acetylcholine	Neuroblastoma	10	260	Atropine, nicotine (3), hexamethonium

[a] DCI, dichloroisoproterenol; LSD, lysergide; BOL, bromolysergide.

tors is based on the specificity for agonists and antagonists (Table 14-2). A typical α-adrenergic response is smooth muscle contraction. β-Adrenergic responses include smooth muscle relaxation, inotropic and chronotropic regulation in heart, and lipolysis. The cytoplasmic response linked to α-adrenergic agonists includes changes in levels of calcium and cAMP; in many cases, the response is not known. On the other hand activation of adenylate cyclase is always observed in response to binding of agonists to β-adrenergic receptors (Lefkowitz et al., 1985; Levitzki, 1986).

The use of radiolabeled agonists and antagonists of β-adrenergic receptors has provided a binding assay for monitoring distribution and fate of

TABLE 14-2 Characteristics of Adrenergic Receptors

	α	β
Agonist	Epinephrine > norepinephrine > isoproterenol	Isoproterenol > epinephrine > norepinephrine
Responses	Smooth muscle contraction	Smooth muscle relaxation, isotropic and chronotropic regulation in heart lipolysis
Linked response	Ca levels, cAMP levels	Activation of adenylate cyclase
Antagonists	Phentolamine Phenoxybenzamine Ergot alkaloids	Propranolol, alprenolol Pindolol

these receptors during fractionation. To be useful, specific binding of a radioligand should be of high affinity and saturable. The binding parameters and the kinetic parameters should reflect the bioresponse, and it should be possible to replace the radioligand with the natural hormone. Using these criteria, the competition binding curves of unlabeled antagonists for labeled antagonist have a slope of 1, indicating a uniform affinity of the receptors for antagonists. Use of suitable labeled agonists like hydroxybenzylisoproterenol has also been useful to characterize not only binding but also to establish the role of modulators like guanine nucleotides, stimulation by fluoride, and for solubilization and reconstitution of adrenergic receptors. Alprenolol immobilized on Sepharose 4B has been used for over 15,000-fold

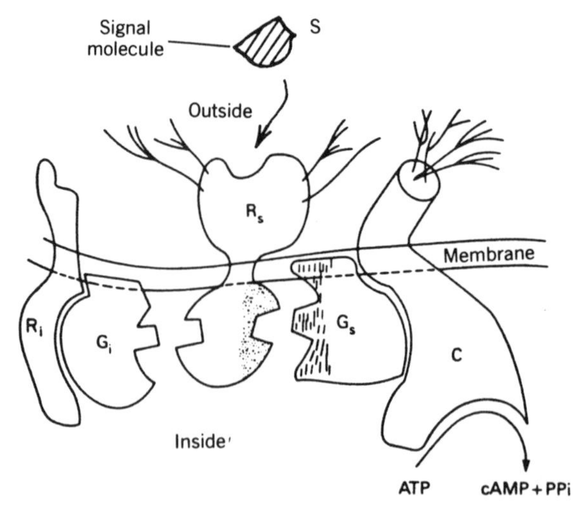

Fig. 14-2. A cartoon of adenylate cyclase complex that is activated by signal molecules. It contains the catalytic unit (C), receptor for the signal (R), GTP-binding protein (G_i and G_s for their inhibitory and stimulatory effects on the cyclase). The sequence of events mediated by S, G, R, and C is shown in Fig. 14-3.

purification by affinity chromatography of digitonin-solubilized β-adrenergic receptors. The binding protein of the β-adrenergic receptor from mammalian tissues is a glycoprotein of 64 kD, whereas the protein from turkey erythrocytes has two units of 40 and 50 kD.

Reconstitution studies have provided further information about other components necessary for the function of the receptor. As shown in Fig. 14-2, the high-affinity receptor with nucleotide-sensitive adenylate cyclase activity is a ternary complex, which consists of the agonist-binding protein or receptor (R), a guanine nucleotide regulatory protein (G), and the adenylate cyclase or catalytic (C, MW 150 kD) unit. Both R and G have been purified. The role of these three components is through a set of coupled reactions shown in Fig. 14-3. This model is consistent with most of the available data. Binding of the signal/agonist (S) to the receptor (R) promotes the binding of G to form the ternary complex SRG. This causes simultaneous release of GDP (normally bound to R at this stage). This promotes binding of GTP, which in turn releases R, then G(GTP) associates with the catalytic unit C to activate adenylate cyclase. The overall stimulation in intact cells is observed in less than 5 sec. The cyclase activity returns to the basal rate when a GTPase inactivates the CG(GTP) complex. Binding of the agonist to the receptor at this stage reinitiates the cycle. cAMP is inactivated by phosphodiesterase. Calmodulin has also been implicated in calcium-dependent stimulation of adenylate cyclase. Experimental evidence for the participation of the three components in the overall function of the complex is compelling.

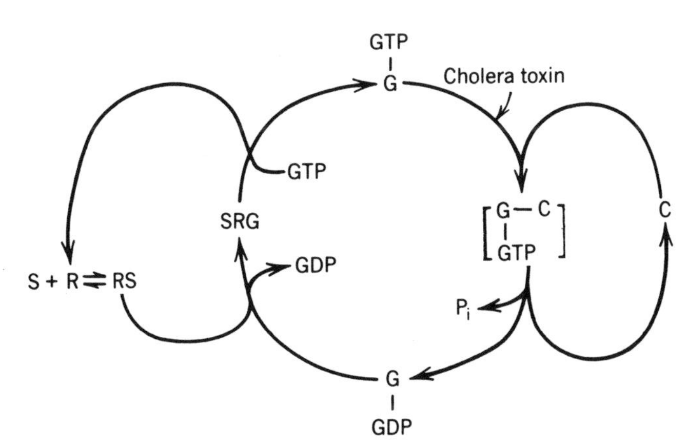

Fig. 14-3. A probable sequence of events during the activation of adenylate cyclase complex by signal molecules (S) like adrenalin (epinephrine). R is the receptor for S, C is adenylate cyclase, G is the G protein. Thus, the RS complex binds to G(GDP). This promotes exchange of bound GDP for GTP. The SRG(GTP) complex dissociates to give G(GTP), which can activate C. During this activation GTP is hydrolyzed and G(GDP) complex is formed.

However, biochemical characteristics of the functional complex are yet to be established. It is quite likely that as many as four other protein components also participate. Additional functional characteristics of the β-adrenergic receptor are given below.

Regulation by GTP

The chemical signal from the hormone is transmitted to the cyclase through the GTP regulatory protein (G). There are several different types of guanine nucleotide regulatory proteins that have been implicated not only in regulation of adenylate cyclase but also in transduction of photon signals in retinal rods and cones, ion channels, and the phosphoinositol cycle (Stryer, 1986). Two of the G proteins of immediate interest are: G_s refers to stimulatory and G_i to the inhibitory proteins that regulate adenylate cyclase. All G proteins are heterotrimers with virtually identical β- (35 kD) and γ- (8 kD) subunits. The α-chains differ (39–45 kD), and cDNA of several of these have been sequenced (Sullivan et al., 1986). The α-subunit has a GTP-binding site, GTPase activity, and the ADP-ribosylation sites by toxins (see below).

G proteins in the hormone-sensitive adenylate cyclase system have been linked to stimulatory and inhibitory responses in several eukaryotic cells. These proteins have a weak GTPase activity. The subunits dissociate in the presence of the nonhydrolyzable analogs of GTP and aluminium fluoride. According to the disaggregation theory proposed by Rodbell (1980), receptors are normally complexed with G proteins. On interaction with the hormone the α-subunit of the G protein dissociates and then associates with an effector like adenylate cyclase, and the receptor is transformed to a lower affinity state. In natural membranes as well as in reconstituted systems the R–G complex exhibits signal-induced binding and degradation of GTP; R can assume different affinity states, and a lower affinity is observed in the presence of GTP. As expected the molecular weight of the ground-state structure of receptors coupled to G exhibits a much higher molecular weight than the activated adenyl cyclase. GTP decreases the affinity of hormones for their receptors and synergistically amplifies hormonal modulation of adenylate cyclase activity. Phospholipids also potentiate the ability of G to stimulate the cyclase. In general nonhydrolyzable analogs like Gpp(NH)p promote the action more effectively than GTP. The fluoride binding site is also located on the G complex.

Regulation of Adenylate Cyclase

A variety of naturally occurring agents and hormones are also capable of inhibiting adenylate cyclase by interacting with their specific receptors on the plasma membrane. This increases the regulatory flexibility, attenuates the effect of activators, and reduces the basal level of cAMP. Above 1 μM GTP inhibits the cyclase activity; so do adenosine and prostaglandin E_1 (PGE_1) in adipocytes, or prostacyclin (prostaglandin I) in platelets. Several neurotransmitters and opiates and epinephrine also inhibit the cyclase activity in certain tissues but not in others. In the case of opiates, inhibition is potentiated by Na ions. These effects are much more pronounced in the presence of GTP, which shows that the inhibition is a receptor-mediated event. Treatment of membranes with *p*-hydroxymercuriphenylsulfonic acid inhibits the stimulation of the cyclase activity by GTP, but the inhibitory response to GTP and other agents is not lost.

Adenylate cyclase is also modulated by a direct interaction with several ligands. For example, it is activated by Ca and Mn, it is inhibited by forskolin and adenosine, and it is apparently down-regulated by cAMP. Modulation of lipid environment by lipid-soluble amphipathic solutes has a biphasic (stimulation followed by inhibition) effect on adenylate cyclase activity (Houslay and Gordon, 1983).

Cholera Toxin

Cholera toxin is an $\alpha\beta_5$ protein of MW 82 kD (β unit of 11.5 kD) produced by *Vibrio cholerae*. In brush border cells it increases the permeability for water by activating adenylate cyclase by ADP-ribosylation of the α-subunit of G_s, whereas the toxin from *Bordetella pertussin* acts on G_i. The action of cholera toxin is seen in all known eukaryotic cells with a lag phase of 60–90 min. The lag phase decreases with increasing concentration of ganglioside GM1 in the membrane, or by treatment with benzyl alcohol, or by depleting cholesterol. Unit A of cholera toxin catalyzes ADP-ribosylation of the G protein with NAD as substrate. This inactivates the GTPase activity. While unit A alone can catalyze this reaction in lysed cells, transfer of A into intact cells is mediated by unit B. Unit B binds to GM1 in cells and in vesicles (K_d about 1 nM). The mechanism for the transfer of unit A by binding of unit B to GM1 in a bilayer is not known, however, binding of the toxin to GM1 in vesicles has been shown to promote leakage of ions and glucose. Such effects are apparently due to a severe organizational change in the bilayer containing a complex of B unit with GM1. Other toxins also bind to gangliosides (Fishman, 1982) and several glycolipids have been implicated in receptor functions for hormones and transmitters (Loh and Law, 1980).

Desensitization or Down-Regulation

The adenylate cyclase activity is often reduced in the presence of excess signal. For example, the rate or extent of accumulation of cAMP to a given dose of agonist depends upon the history of the previous exposure to the agonist. Such ligand-induced desensitization and down-regulation (which probably represent two different underlying mechanisms) of receptors is a very general property of the receptors localized on membranes. It has been observed in receptors for catecholamines, insulin, glucagon, epidermal growth factor, prostaglandins, histamine, and acetylcholine. The underlying specific molecular mechanism for down-regulation has not been established in all cases, however, as discussed in the next chapter, endocytosis of occupied receptors could be a major pathway for down-regulation. Besides this, three other responses leading to down-regulation have been identified for β-adrenergic receptors: (i) a rapid uncoupling of the receptor from adenylate cyclase; (ii) a slower nonspecific desensitization of adenylate cyclase by cAMP; and (iii) a slow induction of phosphodiesterase activity mediated by cAMP.

The Role of Second Messenger

cAMP and GMP are formed from nucleotide triphosphates in many cells in response to specific signals on the cell surface. The second-messenger concept for cAMP proposes that the binding of a hormone (e.g., epinephrine, glucagon) to a specific receptor on the outer surface of the membrane activates adenylate cyclase on the inner surface of the plasma membrane. cAMPs are degraded by phosphodiesterase. Thus, cAMP levels are regulated primarily by these two enzymes. It appears that the steady-state turnover of cyclic nucleotides occurs rapidly, with half-time between 0.1 and 1 sec.

One of the well-established effects of cAMP is activation of cAMP-dependent protein kinases by dissociating the inactive holoenzyme into the dimeric cAMP-binding subunit and two monomeric catalytic subunits (Flockhart and Corbin, 1982). The catalytic unit phosphorylates and changes the functions of a variety of substrate proteins. Phosphorylation can change the activity of an enzyme not only by affecting its catalytic capacity but also by changing its sensitivity to effectors or by modifying its susceptibility toward proteolytic degradation. The effects of such post-translational phosphorylation can be more complex when the modified enzyme is a kinase or a phosphatase that in turn modifies yet another enzyme. The amplifying power of such a cascade is well exemplified in glycogen synthesis and degradation, where almost a millionfold amplification is achieved. In eurkaryotes, many of the target proteins of cAMP are not enzymes but proteins that control cell division, meiosis, transcription, translation, membrane permeability, hormone secretion, or muscle relaxation. Both in prokaryotes and in eukaryotes, cAMP is also known to regulate gene expression.

The site of phosphorylation on signal-transducing receptors from plasma membranes appears to have two specificities: some receptors phorphorylate tyrosine (tyrosine-specific kinase) while others phosphorylate (kinase C, 77 kD) serine and threonine on proteins. The tyrosine kinase activity appears to be associated with signal transduction (Cobb and Rosen, 1984), whereas the kinase C activity alters the affinity of receptors for their respective ligands and changes the absolute number of receptors per cell (Nishizuka, 1984). Kinase C is inactive in cytosol, however it is activated when bound to membranes containing phosphatidylserine (PS) or phosphatidylinositol (PI), presumably by an increase in its affinity for Ca. 1,2-Diacylglycerol, but not mono- and triacylglycerols, potentiate the action of Ca and increase the maximal enzyme activity. Phorbol esters, some of the most potent tumor-promoting agents, also activate protein kinase C in intact cells as well as in cell-free systems (Ashendel, 1985). Yet phorbol esters do not activate beyond the level maximally stimulated by diacylglycerol. Kinase C is also activated by 2-mercaptoethanol, lipid X, N^2,O^2—diacylglucosamine-1-phosphate, unsaturated fatty acids, and retinoic acid. On the other hand several amphipathic drugs, including anesthetics, inhibit protein kinase C.

cGMP also acts as a second messenger in certain cells. There are soluble and membrane-localized guanylate cyclase, cGMP phophodiesterase, and cGMP-specific protein kinases. However, there is no direct evidence for a direct coupling of guanylcyclase with a signal receptor function, or for the effect of cGMP on specific membrane processes. Several suggestions have been made. For example, the *Yin–Yang* hypothesis suggests that relative concentrations of cGMP and cAMP control processes like cell growth (Kaever and Resch, 1985). Certain peptides and vasodilators (azide, nitroglycerine, nitroprusside, hydroxylamine) increase the levels of cGMP in a variety of cells ranging from kidney to muscle to endothelial cells. This suggests that the activation of guanyl cyclase can occur via a receptor.

PHOSPHOINOSITOL CYCLE

Many receptors in response to a specific transmembrane signal trigger breakdown of phosphoinositols (Abdel-Latif, 1986). The PI cycle, as it is commonly known, putatively controls such diverse processes as contraction, secretion, sensory transduction, fertilization, cell division, and cell growth. The sustained activation of this pathway is believed to stimulate cell growth, and it is likely that certain oncogenes trigger this pathway for uncontrolled cellular proliferation. It is not certain, but a role of G proteins, protein kinases, cAMP, intermediates of phospholipid biosynthesis, and arachido-

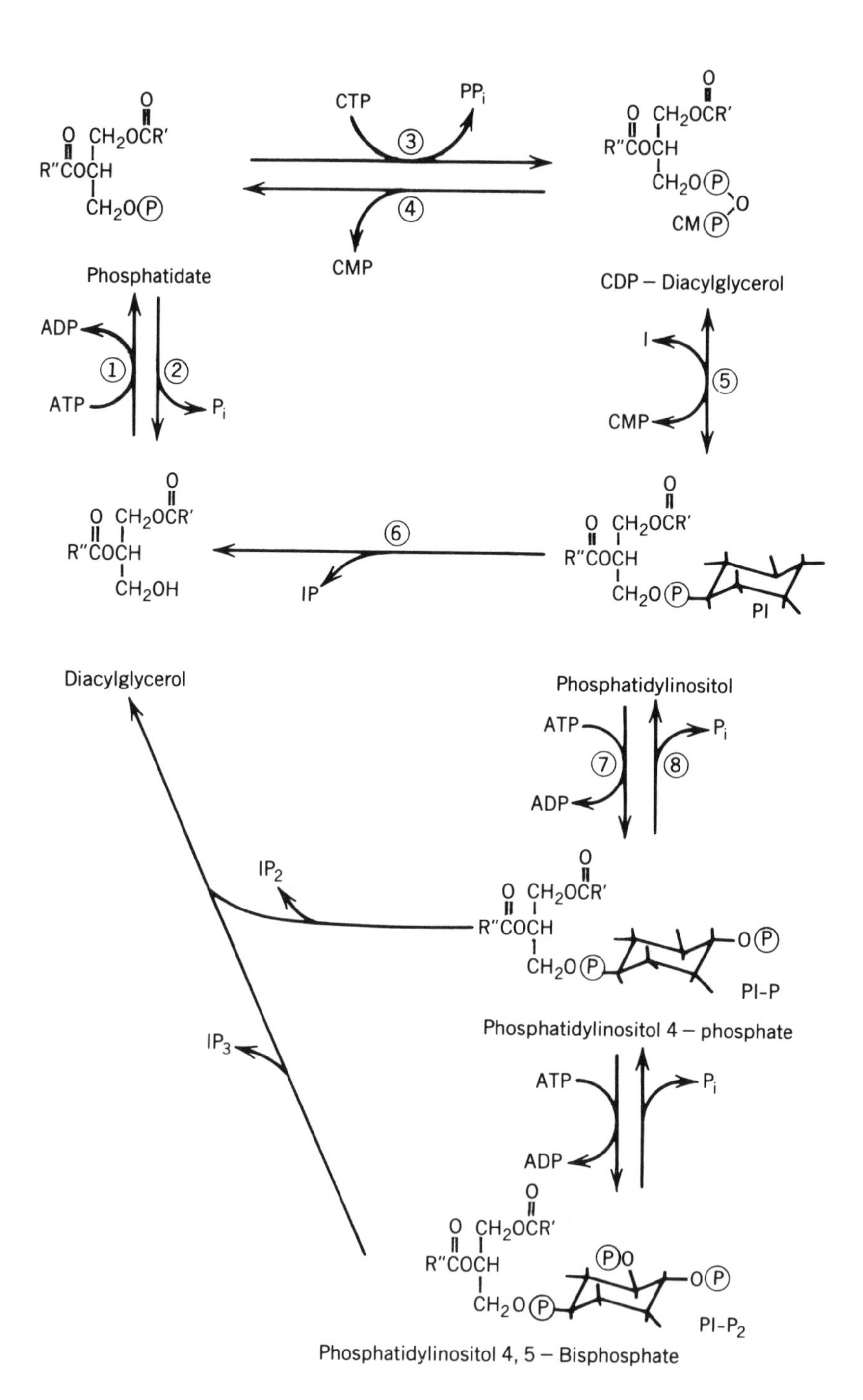

CTP
PP$_i$
③
④
CMP
Phosphatidate
CDP – Diacylglycerol
ADP
ATP
①
②
P$_i$
I
⑤
CMP
⑥
IP
PI
Diacylglycerol
Phosphatidylinositol
ATP
P$_i$
⑦
⑧
ADP
IP$_2$
PI-P
Phosphatidylinositol 4 – phosphate
IP$_3$
ATP
P$_i$
ADP
PI-P$_2$
Phosphatidylinositol 4, 5 – Bisphosphate

(a)

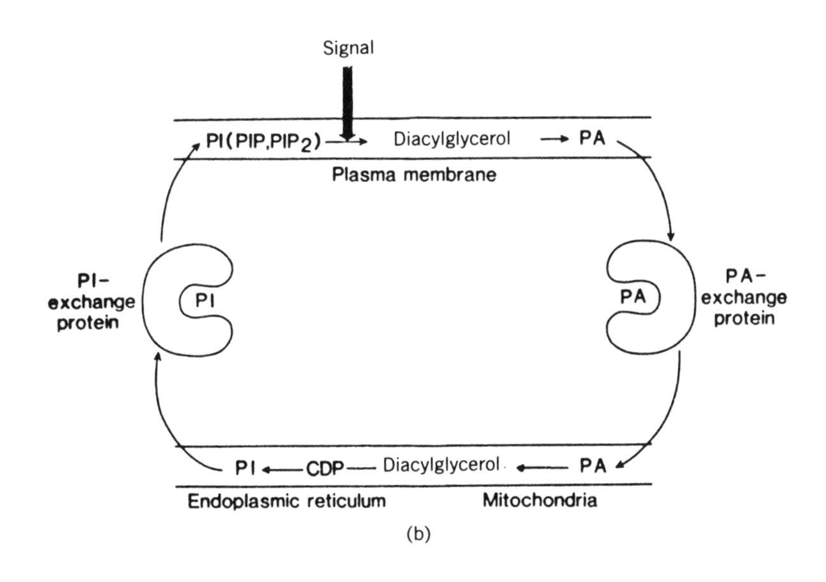

(b)

Subcellular Distribution	Enzyme
Plasma membrane	PA phosphatase
	Diacylglycerol kinase
	PI kinase
	PIP kinase
	PIP_2 phosphatase
Mitochondria	PA-cytidylyltransferase
Endoplasmic reticulum	CDP-diacylglycerol: inositol transferase
Lysosomes	Ca^{2+}-independent PI phosphodiesterase
	CDP-diacylglycerol hydrolase
	PIP phosphodiesterase
	PIP_2 phosphodiesterase
Cytosol	Ca^{2+}-dependent PI phosphodiesterase
	PIP phosphodiesterase
	PIP_2 phosphodiesterase

♦**Fig. 14-4.** Enzymatic interconversions of phosphoinositides. IP_n represents inositol phosphates. The enzymes involved are (1) diacylglycerol kinase, (2) PA phosphatase, (3) PA : CTP cytidyltransferase; (4) CDP-diacylglycerol pyrophosphatase; (5) CDPG : inositol phosphatidyltransferase; (6) PI phosphodiesterase; (7) PI kinase; (8) PIP phosphatase, etc. (b) Subcellular loci implicated in the stimulated turnover of phospholipids. Formation and turnover of diacylglycerol and phosphatidic acid occurs in the plasma membrane. In nerve tissue PA is apparently transported to mitochondria, where it is converted to CDP-diacylglycerol, and then transferred to the endoplasmic reticulum (ER). Several other enzymes also are probably involved. (c) Localization of the enzymes involved in the PI cycle.

nate metabolites is often implicated with this pathway. A complex series of interconnected events are likely to emerge when the dust of current enthusiasm in this area settles.

Phosphoinositides are minor (less than 5% typically) membrane components that turnover more rapidly than other membrane lipids. PI is the major inositol phospholipid, although PI-4-phosphate, PI-4,5-bisphosphate, and PI-1,4,5-trisphosphate are also present. In response to stimulus the overall turnover of inositides increases (Fig. 14-4). The first step, hydrolysis of PI by phospholipase C or a phosphoinositidase, is believed to be rate limiting. Resulting diglyceride and inositolphosphates act as putative second messengers. Inositol-1,4,5-triphosphate modulates intracellular calcium fluxes (Berridge and Irvine, 1984). The diglyceride, among other things, activates protein kinase C (Nishizuka, 1984). In the normal course, diglyceride is phosphorylated to phosphatidic acid which is converted back to PI. An accelerated turnover of phosphoinositides is apparently an early response (within a minute in most cases) to a signal in many cells.

SODIUM–PROTON EXCHANGE SYSTEM

Transmembrane exchange of Na for protons is ubiquitous, as it has been identified in virtually every type of cell ranging from ovum and erythrocyte to muscle and endothelial cells (Aronson, 1985). Under physiological conditions it mediates uphill efflux of protons coupled to downhill influx of Na into the cell. This process plays a role in regulating intracellular pH and cell volume, and it is probably also involved in modulating metabolic responses. Activation of Na/H exchange appears to be a universal response of quiescent vertebrate cells (human, mouse, platelets, fibroblasts, neutrophils, hepatocytes) to growth-promoting agents like serum, epidermal growth factor, platelet-derived growth factor, vasopressin, bradykinin, concanavalin, and α-thrombin (Moolenaar, 1986). Such an exchange of extracellular Na for intracellular protons in sea urchin eggs is, for example, observed following fertilization. The exchange system is mediated by a protein, and mutants defective in the exchange have been isolated. The kinetics of exchange is tightly coupled, concentration dependent, asymmetric, reversible, and electroneutral; it is driven by ionic gradients but insensitive to membrane potential; it has stoichiometry of 1 : 1; and it is inhibited by amiloride which is not transported. Kinetic studies with renal brush border membrane vesicles show that external protons interact at a site where Na, Li, NH_4, and amiloride compete for binding (Aronson et al., 1983). In contrast, internal H^+ interacts at the transport site and at an activator site. Thus, the exchange system behaves as a pH sensor and a regulator of cytoplasmic pH. In response to low pH, the exchange system is activated, and it slows down below a threshold pH of 7.2.

Growth-promoting agents also appear to modulate the kinetics of the exchange system (Moolenaar, 1986). Synthetic diacylglycerol and phorbol esters activate Na/H exchange in human fibroblasts. This suggests that the protein kinase C could transduce the effect of growth factors, and these stimulants of protein kinase C have no effect on cytoplasmic free Ca. It appears that this effect is via phosphorylation of Na/H-exchanger. A hypothetical scheme integrating the various cellular effects of ion translocation and catalytic steps is shown in Fig. 14-5.

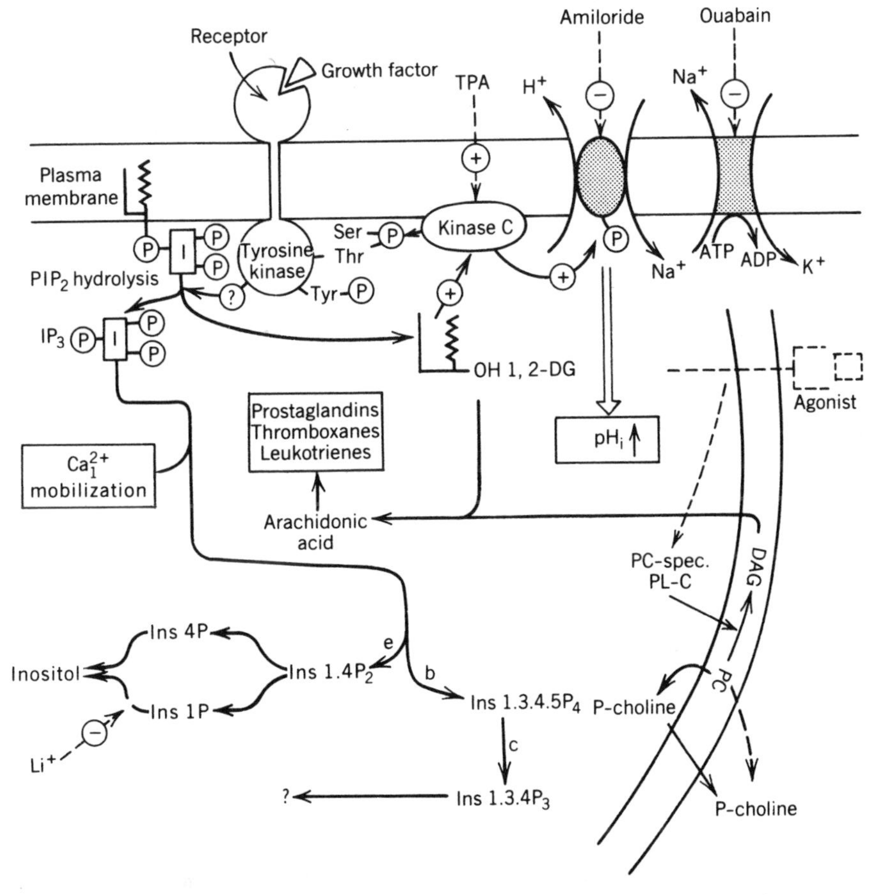

Fig. 14-5. A possible scheme for the integration of the events that may occur in response to certain agonists like growth factors. This is to illustrate the possibility that the protein kinases are coupled to PI-cycle intermediates. (See also Moolenaar, 1986, and Berridge, 1986.)

PHOTOTRANSDUCTION IN VISION

Pigmented sensory cells in retina (rods and cones) transform a light stimulus into an electrical signal (Fig. 14-6). The outer segment of rods and cones contains a long stack of photosensitive flattened discs covered by the plasma membrane, which contains channels (about 2500 per disc) that permit transport of Na in dark. Single-channel conductance of these channels is estimated to be 20 fS (Schwartz, 1985). Embedded within the disc membrane in rods are rhodopsin molecules. Absorption of visible light by rhodopsin ultimately leads to a decrease in the conductance of the outer plasma membrane. One photon apparently closes about 100–300 sodium channels in

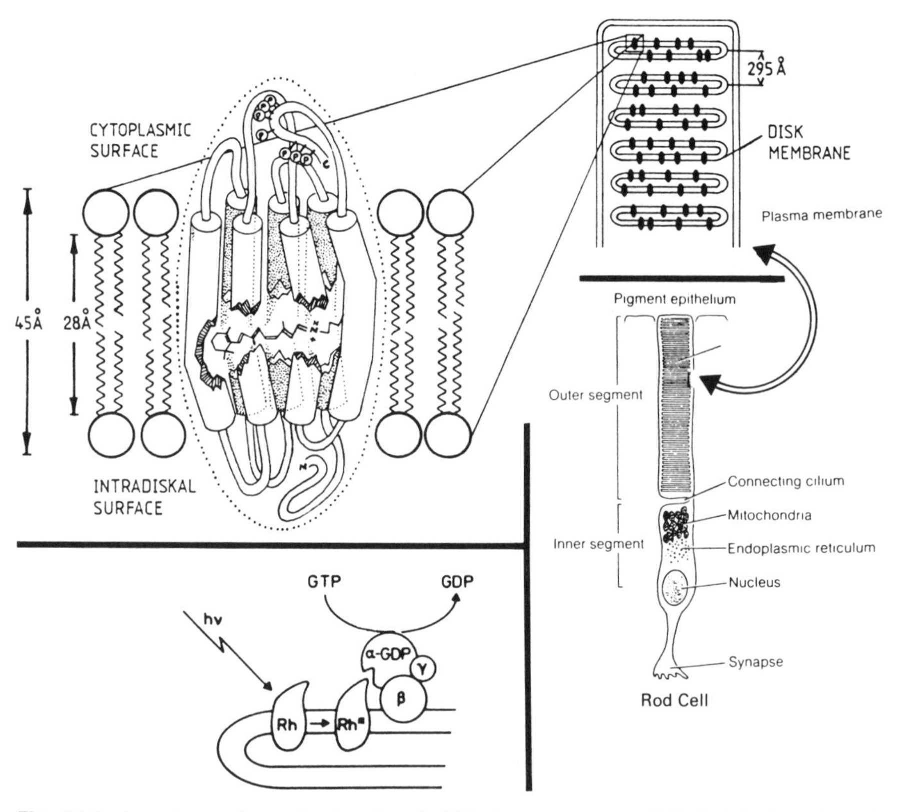

Fig. 14-6. A cartoon of a retinal rod cell. This is a sensory cell that detects external photostimuli. It is a modified epithelial cell that carries specialized photosensitive regions containing membranous discs enclosed with a plasma membrane. Activation of rhodopsin in discs hyperpolarizes the plasma membrane, and thus sends information to neighboring cells. An exploded view of rhodopsin is also shown along with the putative rhodopsin-activated GTPase (transducin).

about 0.1 sec. The exact mechanism of the intervening events is not known. However, two major events have been characterized (Chabre, 1985). First, absorption of a quanta of light by rhodopsin causes a rapid photoisomerization; second, the final electrophysiological response results from a reduction of inward sodium current that flows in the dark through the rod outer segment membrane. Two working hypotheses are under active consideration. According to the Ca channel hypothesis, rhodopsin or a protein coupled to it causes Ca influx which inhibits the Na channel. According to the cGMP hypothesis, binding of cGMP opens Na-channels and rhodopsin regulates the cytoplasmic cGMP levels through a cascade of reactions involving transducin (a G protein), GTPase, or a phosphodiesterase.

Rhodopsin, the primary photoreceptor in discs, is a transmembrane glycoprotein (MW 41 kD) with about half of its mass embedded in the bilayer. In membrane it is present as a monomer at 3.3 m*M* concentration (of the rod volume) and occupies 25% of the membrane area (34 nm^2 per protein). The average intermolecular separation is about 56 Å. There are about 65 phospholipid molecules, 2 diglycerides, 5 fatty acids, and 9 cholesterol molecules per rhodopsin. About 60% of the rhodopsin chain is α-helical, which could make seven to eight transmembrane helices. Bovine rhodopsin contains two oligosaccharide chains at Asn-2 and Asn-15, each with six to eight residues. The amino terminus is in the intradiscal surface away from the cytoplasm, and the carboxyl terminus is in the cytoplasm. The carboxyl region is rich in hydrophilic residues including several serines and threonines, which can be phosphorylated by rhodopsin kinase when rhodopsin is exposed to light. It is probably relevant to note that there is considerable homology between rhodopsin, muscarinic AChR (not described in this book), and β-adrenergic receptor (Kubo et al., 1986). This implies a similarity in their transmembrane topographies, which is also substantiated by their ability to interact with G proteins.

The initial rapid photoevents (Fig. 14-7) in rhodopsin occur on 11-*cis*-retinal, covalently attached to Lys-297 and lies toward the center of the rhodopsin molecule. The chromophore is buried in the center of the membrane and lies nearly in the plane of the membrane. The visible absorption spectrum of retinal is red-shifted 50–140 nm, depending upon the state of protonation of the charged residues (Asp-83, Glu-122, Glu-134) which lie near the Schiff base linkage. On absorption of photons, 11-*cis*-retinal isomerizes to 11-*trans*-retinal. According to one of the theories, metarhodopsin II (Rh*) formed within a few milliseconds after photoactivation of rhodopsin interacts with transducin (a G protein) on the surface of the disc membrane. During this brief encounter Rh* catalyzes the exchange of bound GDP for GTP. This GTP-containing transducin in turn activates a phosphodiesterase

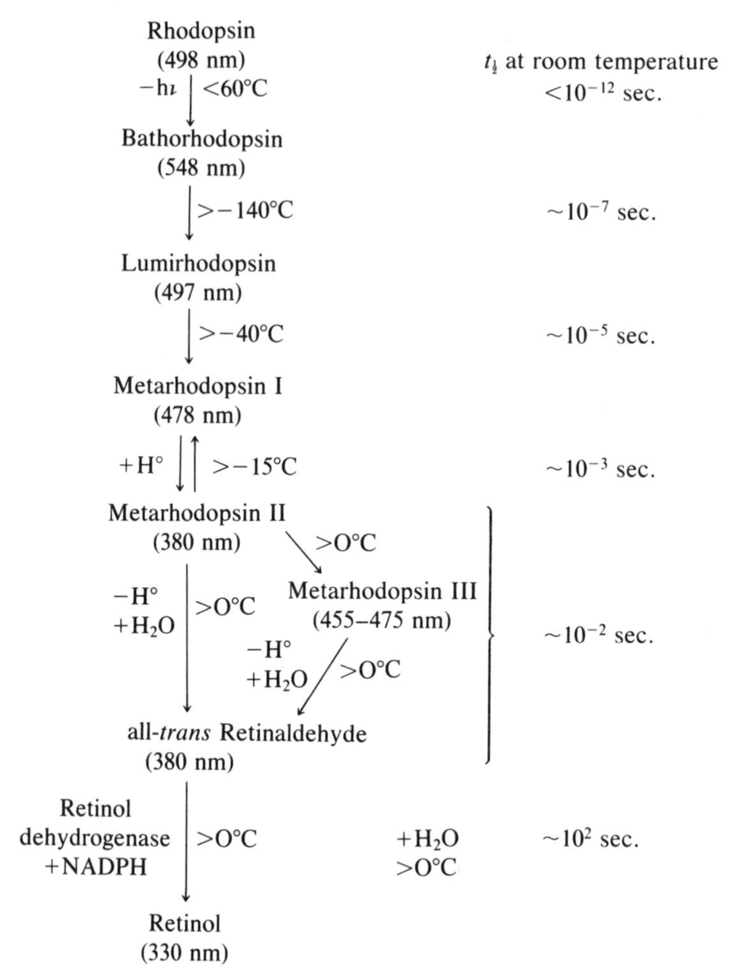

Fig. 14-7. The sequence of photoevents following photon incidence on rhodopsin. Rhodopsin is localized in the stack of membranes bounded by the plasma membrane of the rod outer segment. The cation channel in the plasma membrane is kept open by GTP, and the incidence light on rhodopsin somehow causes the channel to close.

specific for cGMP. Thus, one Rh* can activate as many as 500 transducin molecules and therefore as many PDE molecules. Activation of PDE molecules is terminated by a slow hydrolysis of GTP bound to G protein. This high cGMP level keeps the channel normally open in the plasma membrane, whereas light inhibits this. Thus, the hyperpolarizing response of light is due to a reduction in the cytosolic concentration of cGMP. Calcium ions are also known to reduce the extent of the dark current, and therefore could be responsible for the photoresponse.

Patch-clamp experiments with the plasma membrane of rods suggest that two molecules of cGMP (apparent K_d 50 μM) activate the conductance of the rod outer segment membrane (Yau and Nakatani, 1985), and the channel is inhibited by 1-*cis*-diltizem. It appears that the effect of cGMP is cooperative. The unit conductance of these channels is 50–100 fS, and they exhibit weak selectivity for alkali metal cations (about threefold in favor of Na and Li). These channels also appear to be permeable to divalent ions. The presence of Na-channels that remain open in the presence of cGMP has also been shown in excised patch of rod membrane (Matthews, 1987). Purified Na channels of the rod outer segment membrane have also been incorporated into bilayer lipid membranes (BLM) where they exhibit similar conductance characteristics, however, the effect of cGMP is not observed. Apparently a crucial component is missing!

Modulation of Channels by cAMP

Circumstantial evidence suggests that cAMP modifies ion channels in many types of cells (Bittar, 1983; Levitan, 1985). AChR and the Na channel have been shown to be phosphorylated by protein kinase, which presumably is activated by cAMP. The physiological significance of this phosphorylation is not known. One of the best-characterized channels that is modulated by cAMP is the K channel from the abdominal ganglion neuron of *Aplysia*, where the duration of action potential is increased by serotonin. Voltage-clamp studies suggest that this is accompanied by a decrease in K current. The same effect can be induced by intracellular injection of cAMP or by the catalytic unit of adenylate cyclase. The effect of serotonin is counteracted by the injection of a naturally occurring inhibitor. This reversal by the inhibitor is rapid, which suggests that the phosphorylation is rather labile. Most of these effects of serotonin can be simulated in patch-clamp studies.

15 Bulk Transport by Fusion and Secretion

It is an old and consistent tradition with us to be concerned with words we use and with their purification.

ROBERT OPPENHEIMER

In a living cell some of the membrane-bound compartments are continuously in a state of flux through fusion reactions. The formation of vesicles from plasma membrane and cellular organelles is apparently very common. A wide range of intra- and intercellular events involve a step where specific recognition of membranes is followed by secretion of vesicles, involving fusion of membranes. Fusion and secretion of membrane-bound structures is a process for bulk transfer of aqueous phase as well of the membrane components. Such a process appears to be involved not only in the release of secretory vesicles but also in uptake of enveloped viruses and parasites, fertilization, myogenesis, membrane flow, protein secretions, post-translational transfer and "sorting," transfer of solutes between organelles, and other manifestations of endo- and exocytosis (Evered and Whelan, 1984). These diverse phenomena have been shown to occur in a variety of organelles and cell types ranging from erythrocytes, yeast, and amoeba to muscle endplates, adrenal medulla, and parotid acinar cells (White et al., 1983; Duzguenes, 1985). As a technique, fusion of vesicles is an important process for incorporation of integral membrane proteins, formation of hybrid cells, and transfer of the vesicle contents into cells (Margolis, 1984).

Operationally fusion involves three stages. First, cell–cell contact brought about by the agglutination or cross-linking or modulation of the interface. The next step requires modification of the bilayer organization such that hydrophobic groups of apposed membranes somehow come in contact. Finally, there is intermixing of the components of membrane and the aqueous phase. Fusion is a triggered process and in cells it displays a remarkable degree of specificity and selectivity in terms of the type of cell, type of inductor, and the life cycle of the cell. Calcium ions and proteins have been implicated as modulators or triggers of fusion in many instances. Fusion can also be induced by brief electrical pulses applied to closely apposed cells. As a rule fusion reactions are rapid, local, and nonleaky and do not result in detectable changes in the composition of the membranes. The symmetry of membranes is maintained. No general pattern has emerged regarding the mechanism of fusion. It is obvious that a controlled violation of the topography of the bilayer occurs during the fusion event. The mechanism underlying the rate-limiting step probably involves formation of transient defects or perturbations in the organization of the bilayer: Calcium-induced intermembrane bridging, Ca-induced changes in the properties of the bilayer, involvement of specific proteins, and modification and activation of specific effector molecules and enzymes like lipases and kinases could facilitate one or more steps in the overall process.

Fusion of membrane-bound structures is a complex process, as it involves interactions of polar and nonpolar regions of a bilayer, and the integrity of the bilayer is violated transiently at the contact site. In some cases, it appears that direct mixing of lipid monolayers in the two interacting vesicles can occur without mixing of their contents in the aqueous compartments. Similarly, the origin of the thermodynamic driving force and the precise nature of the bilayer at the site of fusion is still under active consideration. For such reasons, interpretation of fusion processes in biological membranes is not possible yet. A role of fusogens like Ca, specific surface proteins, and the bilayer-perturbing agents has been invoked. While the intermembrane point contact promoted by some of these agents may facilitate fusion, it is the perturbation of the bilayer organization that is believed to be the rate-limiting step. While the nature of such defects in the organization of the bilayer is still being debated, a role of lipidic particles and hexagonal II phase can be ruled out because such polymorphs are seen only after the fusion process has occurred.

Fusion is a macroscopic event that leads to the formation of larger particles, which can be distinguished from the species that are fusing on the basis of one or more techniques including light scattering, ultracentrifugation, gel filtration, electron microscopy, nuclear magnetic resonance (NMR), electron spin resonance (ESR), and differential scanning calorimetry. The continuous methods for measuring fusion are based on monitoring mixing of the aqueous phase or mixing of the lipid phase (Fig. 15-1). Thus, chelation of turbium ions by picolinic acid on mixing the contents of the two aqueous

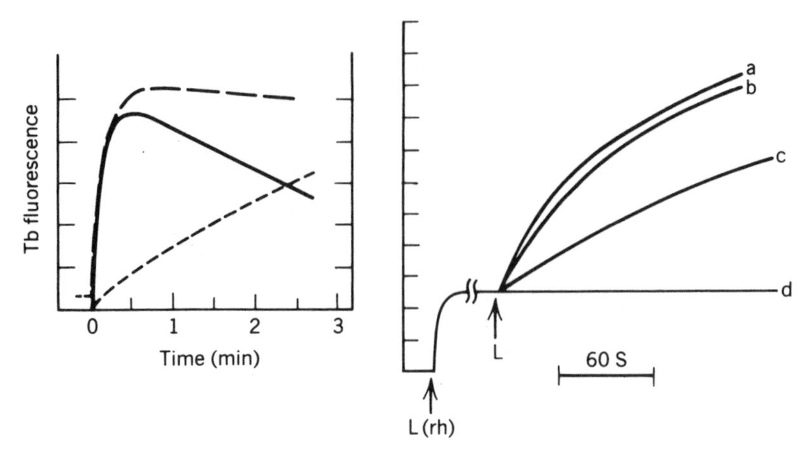

Fig. 15-1. Kinetics of Ca-induced fusion of vesicles (Left) of PS by (from top) scattering measurement, by mixing of turbium and picolinate in separate vesicles, and by release of trapped carboxyfluoresceine. (From Wilschut et al., 1980.) (Right) Increase in fluorescence intensity of self-quenched octadecylrhodamine (rh) in vesicles on fusion (from top) of DMPC (rh) + DMPMe(a), DMPMe + DMPMe(rh)(b), DMPMe(rh) + DMPC(c), and DMPC + DMPC(rh)(d). Fusion was induced by adding the nonfluorescent vesicles. DMPMe is dimyristoylphosphatidylmethanol. (From Jain et al., 1986a.)

compartments of the fusing vesicles leads to an increase in the fluorescence intensity. In this assay system each one of the pair of reactants is encapsulated in different populations of vesicles. It is also necessary to include in the external aqueous medium an inhibitor of the reaction so that a leakage does not contribute to the signal from the fusion event. Similarly, release of self-quenching due to surface dilution or the resonance energy transfer of fluorophores occurs when the appropriate nonexchangeable components in the bilayer are mixed. In this class of methods, it is necessary to differentiate between mixing of bilayers from the intervesicle exchange of the probe by adhesion or aggregation, or through the aqueous phase.

Fusion of Vesicles

Vesicles of well-defined composition have proven to be excellent models for fusion (Nir et al., 1983). The rate of fusion of most phospholipid vesicles is very slow. However, spontaneous fusion of vesicles is accelerated under a variety of conditions. For example, vesicles (25 nm diameter) of zwitterionic lipids like dimyristoylphosphatidylcholine (DMPC) are stable for days, however these vesicles fuse readily in the presence of as little as 1% of the products of degradation. Similarly, fusion of the vesicles of anionic phospholipids like phosphatidylserine (PS) is promoted by divalent ions like Ca, and the rate of fusion is further accelerated by certain membrane proteins or by physiological concentrations of phosphate. Under a variety of conditions, it has been shown that aggregation and fusion of PS vesicles are concurrent processes, whereas the process leading to leakage is slower. Kinetics of fusion depends upon the vesicle concentration, fusogen concentration, the nature of fusogen and phospholipid, temperature, as well as pH (Klappe et al., 1986). Thus, the rate of fusion of the various anionic phospholipid vesicles decreases in the order phosphatidic acid (PA), diphosphatidylglycerol (DPG), PS, phosphatidylglycerol (PG), phosphatidylinositol (PI). Codispersions of PS with PC do not fuse as readily as the dispersions of PS with PE. Heterofusion of PC vesicles with vesicles of dimyristoylphosphatidyl methanol (DMPMe) occurs readily in the presence of Ca (Jain et al., 1986a). Such heterofusion is seen only when DMPMe membranes are phase-separated. Apparently such a process has not been reported for PS vesicles.

Calcium ions are the most commonly studied fusogens for vesicles of anionic phospholipids, although other divalent cations are also fusogenic. Subtle difference in the relative rate of fusion, aggregation, and leakage are noted in the presence of other divalent cations. For example, Mg-induced fusion occurs at a slower rate, there is some delay in the rate of fusion and aggregation, fusion is accompanied by a pronounced leakage, and vesicles do not collapse even at a higher concentration and retain most of their aqueous contents. The rate of fusion increases above a threshold concentration of Ca, which depends upon the nature of the lipid. For 25-nm vesicles of PA, PS, and PG, these concentrations are 0.2, 1.5, and 5 m*M*, respectively. However, the threshold concentrations depend upon the size of the vesicles as well as the presence of other components that interact with vesicles. Synexin, a protein isolated from brain, liver, adrenal medulla, and platelets, increases the initial rate of fusion and reduces the threshold concentration of Ca required for fusion of PS vesicles at pH 6, and the effect is considerably less pronounced at pH 7.2. A similar effect is seen with mellitin, polymyxin, myelin basic proteins, fragment of albumin, alamethicin, clatharin, and glycophorin. Lectins containing appropriate ligands in the vesicles also promote fusion. It is interesting to note that most of these peptides and proteins remain anchored in the interface of the bilayer.

Several suggestions have been made to describe the molecular interactions involved in fusion of bilayer vesicles. Calcium-induced heterofusion of PC vesicles with DMPMe vesicles (Jain et al., 1986a) rules out the role of interfacial charge repulsion, a role of the water layer of the "hydration force," and a role of intervesicle bridging by Ca in the rate-limiting step of the overall process. Since the rate of fusion shows a second-order dependence on the vesicle concentration, the rate-limiting step must involve collision of two vesicles. The fact that a Ca-induced change in the bilayer of only one of the two interacting vesicles is enough for heterofusion suggests that the site of contact of the fusing vesicles is probably a Ca-induced change in the organization of the bilayer.

One of the most likely Ca-induced changes in the bilayer that promotes fusion of vesicles is lateral-phase separation. A maximal rate of fusion is often observed in the vicinity of the phase-transition temperature. However, it has been difficult to quantitate the changes that accompany lateral-phase separation and cause fusion. Often these are vaguely described as defects or instabilities, and possibly a nonlamellar state like hexagonal II phase in some cases.

Fusion of Erythrocytes

Hen erythrocytes have been studied as models for fusion because they do not divide nor do they normally fuse. Moreover fusion can be followed by phase-contrast microscopy by monitoring the appearance of the multinucleated cells (Lucy, 1980). In these cells fusion can be induced by a variety of agents including polyethylene glycol, lysophospholipids, dimethylsulfoxide, sorbitol, calcium phosphate at high pH, and viruses. Similar results have been obtained with nonnucleated erythrocytes. It appears that the fusion event in these cases involves perturbation of membrane proteins. This raises the possibility that the cytoskeletal network is perturbed before or during fusion.

Endocytosis

The term "endocytosis" was originally coined to cover all forms of internalization of the surrounding medium and particles adsorbed on the external surface of a cell. One of the modes of bulk transfer is via vacuoles generated from surface ruffles and micropinosomes which fuse with lysosomes (Pastan and Willingham, 1981b; Schneider et al., 1985). It is the major mode of transport for large particles not only in macrophages but also in most eukaryotic cells. As visualized by electron microscopy, endocytosis starts with invagination of the plasma membrane around substances to be captured. The endocytic vesicles are pinched off into the cytoplasm where they migrate and ultimately fuse with lysosomes. In other cases the endocytic vesicles may appear in the Golgi region, while in epithelial cells the vesicles cross the cytoplasm and release their contents at another pole of the cell. In endothelial cells, there is a continuous bidirectional exchange of macromolecules between the plasma and the interstitial fluid. In cultured cells, as many as 3000 vesicles may form in a minute. Formation of membrane vesicles accounts for the turnover of phospholipids. For such reasons, in some cases the half-life of membrane phospholipids may be less than one hour.

Physiologically, the nonspecific endocytic mechanism of bulk uptake provides the cell with an energy-efficient mode of uptake of nutrients, including amino acids, carbohydrates, and lipids as products of action of hydrolases in lysosomes. In parallel to this endocytic mechanism, many substances like hormones and proteins are internalized by binding to specific receptors present at the cell surface. Such an uptake plays a fundamental role in growth, nutrition, and differentiation of animal cells. Endocytic processes are also involved in selective secretion, delivery, and clearance of substances including toxins, viruses, bacteria, antigens, and antibodies. Endocytic processes in conjunction with exocytic events serve to recycle membranes which fuse with the plasma membrane at the presynaptic endings or in secretory cells.

Nonspecific fluid endocytosis (*pinocytosis*) or ingestion of invaginated large particles (*phagocytosis*) have been operationally distinguished. Pinocytosis is a ubiquitous process that occurs at very different rates in different cells, thus permitting some degree of selectivity. On the other hand, phagocytosis requires some yet uncharacterized interaction with the plasma membrane, appears to require energy, and is much more dependent on the cytoskeleton. Nonreceptor-mediated adsorptive endocytosis has also been invoked for the uptake of low-density lipoproteins. The fate of the material taken up by these different processes may not necessarily be the same.

Receptor-Mediated Endocytosis

During receptor-mediated endocytosis (RME), solutes bind to specific receptors on the cell surface and the complex is internalized. Many large molecules and aggregates like viruses, transport proteins, serum lipoproteins, transferrin, lectins, macroglobulin-protease complex, antibodies, immunoglobulins, toxins, hormones, and plasma proteins enter cells by RME (Pastan and Willingham, 1985). Although these ligands share a common endocytic pathway, their rates of uptake are often very different because the overall rate is determined by a complex set of interacting processes, which include binding of the ligand to the receptor, lateral diffusion of the complex, trapping of the complex in coated pits, internalization and processing of the coated pits, and the endocytic process that completes recycling of the receptor.

In general, during endocytosis the macromolecular ligands bind to specific cell-surface receptors, which move to a cluster within specialized regions (coated pits) of all eukaryotic cells. Each type of extracellular solute is recognized by specific surface receptors synthesized by the cell. The types of receptors expressed on any given cell and the level of expression depends on the tissue or cell type and the metabolic state of the cell. Some receptors are found only in certain types of cells, whereas others are virtually ubiquitous. All the receptors examined so far appear to be transmembrane glycoproteins synthesized and processed by Golgi and rough endoplasmic reticulum (ER). Isolation, reconstitution (Braell et al., 1984) and genetic studies of the components involved in RME (Sege et al., 1984) have been reported.

RME has been studied extensively by physiological and ultrastructural methods in cultured and isolated cells. With the exception of mature erythrocytes, all mammalian cells carry out endocytosis (except during mitosis). RME also occurs in liver, brain, and endothelial tissues. All RME pathways share common features. The molecules that are taken up are rapidly transferred from coated pits (diameter about 0.2 μm and 50 to 2000 per cell) on the cell surface to coated vesicles (receptosomes) and then, after uncoating, directly to lysosomes (Pearse and Bretscher, 1981) via the Golgi system. In a fibroblast, there are about 1000–2000 coated pits, each with a diameter of about 140 nm. Coated pits appear as indentations or invaginations in plasma membrane and they occupy 1–2% of the membrane area. Each coated pit gives rise to a receptosome about every 20 sec, irrespective of the occupancy of the receptor. The cytoplasmic face of the pits is covered with clatharin (180 kD) and two or three other proteins, all of which together form a skeleton called the triskelion. When viewed under the scanning electron microscope, the coated pits and vesicles appear as cage-like structures composed of polygonal lattice of hexagons and pentagons.

Prior to their uptake the bound ligands appear to be concentrated in coated pits. On the average, each receptor encounters a coated pit every 3–4 sec, and the coated pit transfers ligands into intracellular uncoated vesicles once every 20 sec. The processes involved in receptor-mediated uptake of asialoglycoprotein by coated pits are discussed later on. Trapping of ligands by receptors appears to occur in different modes. For example, receptors for epidermal growth factors are randomly distributed on the cell surface and only become clustered in coated pits in the presence of ligands. On the other hand, receptors for low-density lipoprotein are found clustered in coated pits in the absence of added ligand. This suggests that under certain conditions coated pits can distinguish occupied receptors from free receptors.

Once receptors are bound with their appropriate ligands within the coated pits, the pits pinch off from the membrane. Resulting coated vesicles (diameter 50–150 nm) contain both receptors and structural proteins. For all practical purposes these vesicles contain a domain distinct from either the extracellular environment or the cytoplasm. Receptosomes have low pH and do not contain significant amounts of functional hydrolytic enzymes. The vesicles move along tracks of microtubules to the trans-Golgi system, where they deliver their content and mix their membranes. Estimates of the lifetime of receptosomes range from 5 to 60 min in different types of cells. It appears that the receptosomes containing occupied receptors are internalized without intake of clatharin-coated pits. The mechanism responsible for sorting receptors from ligands and from other receptors is not known. Also, the mechanism for recycling of receptors and the membrane is unknown. However, it appears that lysosomes and microtubules comprise the exocytic link of the pathway. Additional features of RME are illustrated by the two examples discussed below.

Insulin Receptors

Insulin regulates cellular metabolism and development by promoting anabolic functions and by slowing down catabolic pathways (Denton et al., 1981). Although liver, muscle, and fat cells are the main physiological targets of insulin for overall metabolic homeostasis, many other types of cells also exhibit regulatory effects of insulin. The effects of insulin include activation of transport (within a few minutes) of sugars, amino acids, and ions; activation (in several minutes) of enzymes like glycogen synthase and pyruvate dehydrogenase; inhibition of lipase in adipocytes; and modulation of transcription in about 20 min. The primary site of action of insulin is its receptor (heterodimer 130 + 90 kD; K_d less than 100 pM) on the plasma membrane on a variety of target cells (Gammeltoft, 1984; Czech, 1985). The larger subunit of the receptor has the binding site, and the smaller subunit appears to be a tyrosine kinase, which autophosphorylates both the subunits. The kinase is regulated by β-adrenergic agonists and other reagents that elevate cAMP. They decrease affinity of the receptor for insulin and thus partially uncouple the receptors from tyrosine kinase activity. The kinase is also inhibited by phorbol esters.

One of the events that follows binding of insulin is that it is internalized by an energy-requiring process. It is not certain whether the internalized receptors are destroyed by association with lysosomes, or they are recycled back to the plasma membrane. Either of these possibilities is capable of accounting for down-regulation, i.e., prolonged exposure of cells to insulin leads to a net loss of receptors from the target membranes. The internalized receptors and insulin could function as a signal and modulator of specific cytoplasmic events. Thus, bivalent anti-insulin receptor antibodies mimic a variety of effects of insulin on adipocytes. This implies that clustering and internalization of receptors without insulin can elicit insulin-like response.

Effect on Glucose Uptake

In adipocytes, glucose metabolism is necessary for the production of glycerol-3-phosphate, which is used for the synthesis of glycerides. For metabolism of glucose the uptake of glucose is rate-limiting in the absence of insulin, and the hexokinase reaction is rate-limiting in the presence of insulin. As summarized in Table 15-1, the transport system for glucose in adipocytes (Fig. 15-2) is similar to that in erythrocytes, however several differences have been noted. K_m for glucose under physiological conditions is about 8 m*M*. Half-time for efflux with tracer exchange is about 20 sec at 37°C, and insulin stimulates it by a factor of about 10. The rate of uptake of glucose is also stimulated by epinephrine and inhibited by glucocorticoids. The K_m for the various analogs increases in the order 2-deoxyglucose, 3-*O*-methylglucose, D-glucose, L-arabinose, D-allose. Thus, the effect of insulin on glucose transporter in adipocytes shows only a weak requirement for hydrogen-bonding at the 4 and 6 positions, but a strong requirement for β-OH at position 1 and equatorial OH at 3.

It has been suggested that endocytosis of insulin receptors triggers the transfer of inactive glucose carriers from an internal vesicle pool to the plasma membrane where they are activated (Cushman et al., 1983). In obese rats the increase in receptor population induced by insulin is significantly smaller than it is in normal rats. At physiological concentrations of insulin,

TABLE 15-1 Kinetic Parameters for 3-*O*-Methyl-D-Glucose Transport in the Rat Adipocyte

	K_m(mM)		V_{max}(mM sec^{-1})	
Experiment	Basal	Plus Insulin	Basal	Plus Insulin
Equilibrium exchange	About 5	About 5	0.07–0.2	1.6–1.9
	2.5–5	2.5–5	0.058	0.8
	4.22 ± 1.24	4.45 ± 0.26	0.058 ± 0.001	0.84 ± 0.002
Zero *trans* entry (measures outside site)	2.5–5	2.5–5	—	—
	5.41 ± 0.98	6.10 ± 1.65	0.034 ± 0.034	1.20 ± 0.19
Zero *trans* exit (measures inside site)	4.09 ± 1.05	2.66 ± 0.26	0.153 ± 0.023	1.19 ± 0.07
	—	5.65 ± 2.05	—	—
Infinite *cis* entry (measures inside site)	9.03 ± 3.28	6.51 ± 0.83	0.066 ± 0.013	0.98 ± 0.09
Infinite *cis* exit (measures outside site)	4.54 ± 1.32	3.60 ± 1.33	0.106 ± 0.026	1.76 ± 0.63

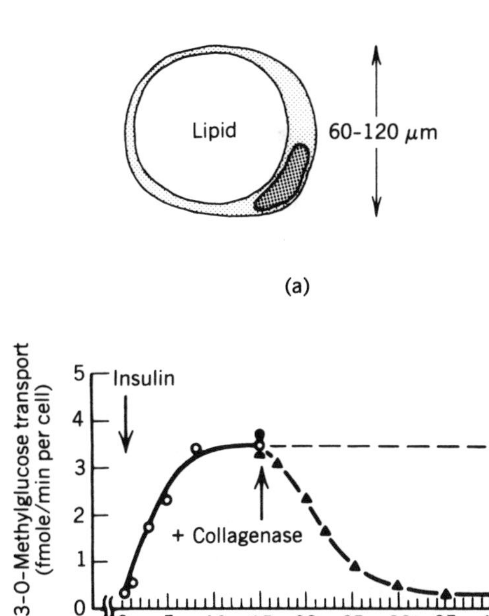

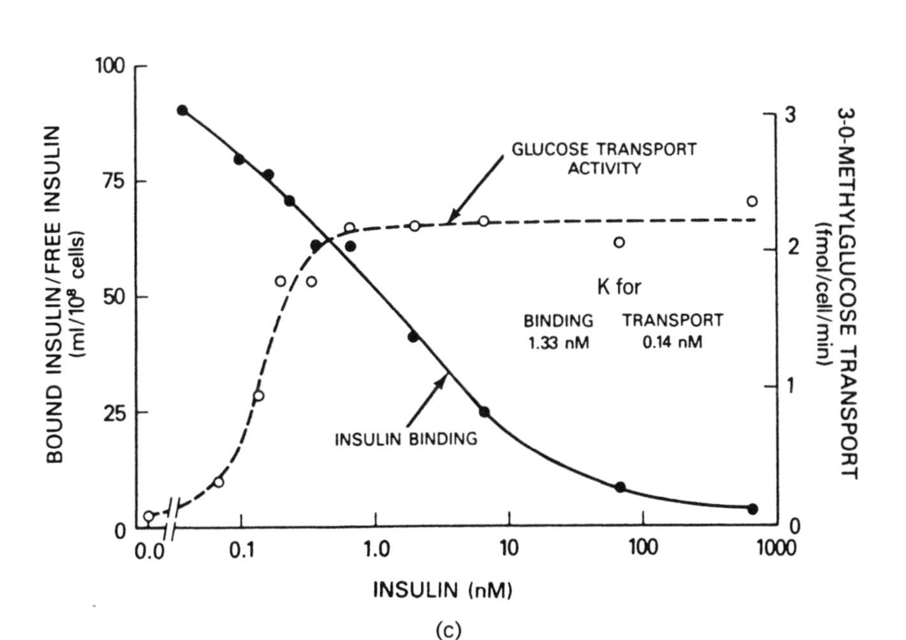

Fig. 15-2. (a) Adipose cells are among the largest cells in the body. These cells are responsible for the production and storage of fat. The nucleus and cytoplasm are squeezed to the cell periphery by a large lipid droplet. (b) Stimulation of glucose transport by insulin in the isolated rat adipose cells. Time-course in response to insulin and its reversal by removing insulin by treatment with collagenase. (c) Effect of insulin on binding of insulin (solid line) and on glucose transport (broken line) activity. A decrease in the binding with increasing concentration of insulin is due to down-regulation.

the rate-limiting step for activation or deactivation appears to be the association or dissociation of insulin from the receptors. The insulin receptor is an $\alpha_2\beta_2$ tetrameric protein. The transmembrane β-subunit of insulin receptor is apparently phosphorylated by a tyrosine-specific kinase. Specific guanine nucleotide regulatory protein and associated protein kinases are also activated by the action of insulin on target cells. Similarly, action of insulin on intact hepatocytes as well as on broken membranes activates a high-affinity cAMP phosphodiesterase. However, this activation is completely abolished if the cells are pretreated with glucagon (Heyworth et al., 1983).

One of the main functions of insulin is to stimulate glucose metabolism in adipose tissues and muscle cells. The stimulatory effect is due to an increase in the number of glucose transporters in the plasma membrane derived from a large intracellular pool of transporters (Fig. 15-3); intracellular distribution of insulin receptors and of growth factors also changes in response to insulin treatment of adipocytes (Simpson and Cushman, 1985). Uptake of glucose in adipocytes is best demonstrated by using nonmetabolizable analogs of glucose such as 2-deoxyglucose, L-xylose, and 3-*O*-methylglucose. The onset

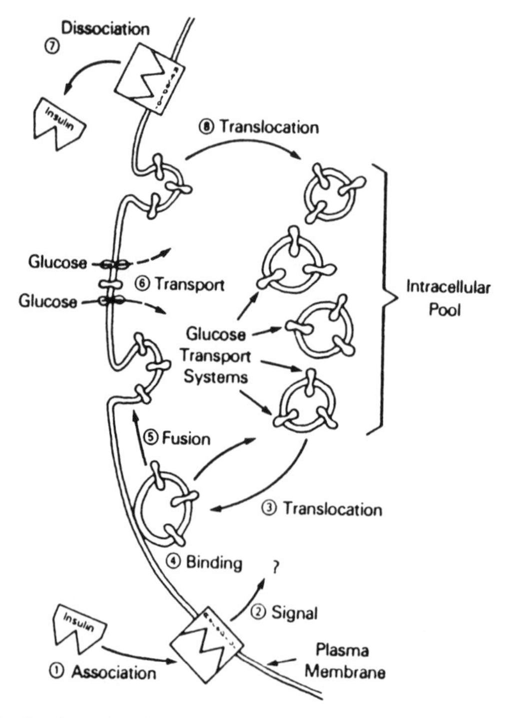

Fig. 15-3. A hypothetical mechanism to account for stimulation of glucose transporter by insulin. (From Karnieli et al., 1981b.)

of the action of insulin is rapid, and a 10- to 40-fold increase in the rate of uptake is seen in about 10 min. (Fig. 15-2B). The concentration of insulin that elicits half-maximal response (about 0.3 n*M*) (Fig. 15-2C) corresponds to a receptor occupancy of 10%. The insulin-mediated transport is inhibited by cytochalasin B, K_i 100 n*M*. The glucose transporter from adipose cell membranes has been reconstituted in egg phosphatidylcholine (PC) vesicles by the detergent dialysis method. These studies also show that the number of cytochalasin-binding sites is higher in preparations obtained from insulin-treated adipocytes.

Based on this and other experimental data, the model illustrated in Fig. 15-3 has been proposed for translocation of glucose transporters by insulin. In the basal state, most of the glucose transporter in adipocytes reside in a large intracellular pool. These fully functional transporters are associated with a Golgi-enriched fraction during the fractionation. In response to binding of insulin with its receptors, intracellular transporters are translocated to the plasma membrane. Although this system is exocytotic, it is not associated with secretion of any ligand or protein. On dissociation of insulin, the transporters cycle back into the intracellular pool. It is quite likely that the recycling of the transporter is a continuous process, and that the steady-state rate is somehow stimulated by binding of insulin to its receptors. It has several unknown steps, however it does illustrate the complexity of the overall process involving binding and fusion events. Apparently energy is required because energy poisons like cyanide or azide can effectively freeze the transporters at any stage by reducing the ATP levels. Similarly, compounds like hydrogen peroxide, vanadate, *p*-chloromercuriphenylsulfonate, or trypsin, known to stimulate glucose transport, do so by the same mechanism as with insulin. The state or site of storage of the latent receptors is not known.

Purification of the glucose transporter from adipocytes has not been possible, however indirect evidence suggests that it is a transmembrane protein of about 54 kD.

Asialoglycoprotein Receptor

The asiologlycoprotein receptor (ASGP-R) mediates the specific recognition and receptor-mediated endocytosis of glycoproteins in liver. The receptor has been purified. The detergent-depleted glycoprotein aggregates to a significant extent, and the receptor from human hepatic parenchymal membranes is a single peptide of 41 kD, whereas the receptor from rabbit liver has two peptides of 48 and 40 kD in 1 : 2 ratio. The human receptor peptide appears to have only 38 amino acids of the amino terminus on the cytoplasmic side, 27 residues in the membrane, and over 200 residues with three glycosidic chains on the outside. The lipid-free protein does not bind ligands, but the receptor function is restored in the reconstituted proteoliposomes.

The ASGP receptor, which may be considered a lectin, specifically recognizes ligands with terminal galactose or *N*-acetylgalactosamine residues. The K_d for glycosides decrease in the order: mono- (m*M* range), di- (μM), tri- and tetra- (n*M*). There is an absolute requirement for Ca binding, and other divalent ions cannot substitute. The dissociation rate constant for the complex is 0.0015/min) for the rabbit and 0.0017/sec for the human receptor. After ligand binding the receptors are concentrated in the coated pits and then internalized with a rate constant of about 0.2/min at 37°C in hepatocytes. Although most of the internalized ligand is metabolized inside the cell, under some conditions receptor recycling has been noted. Distribution of

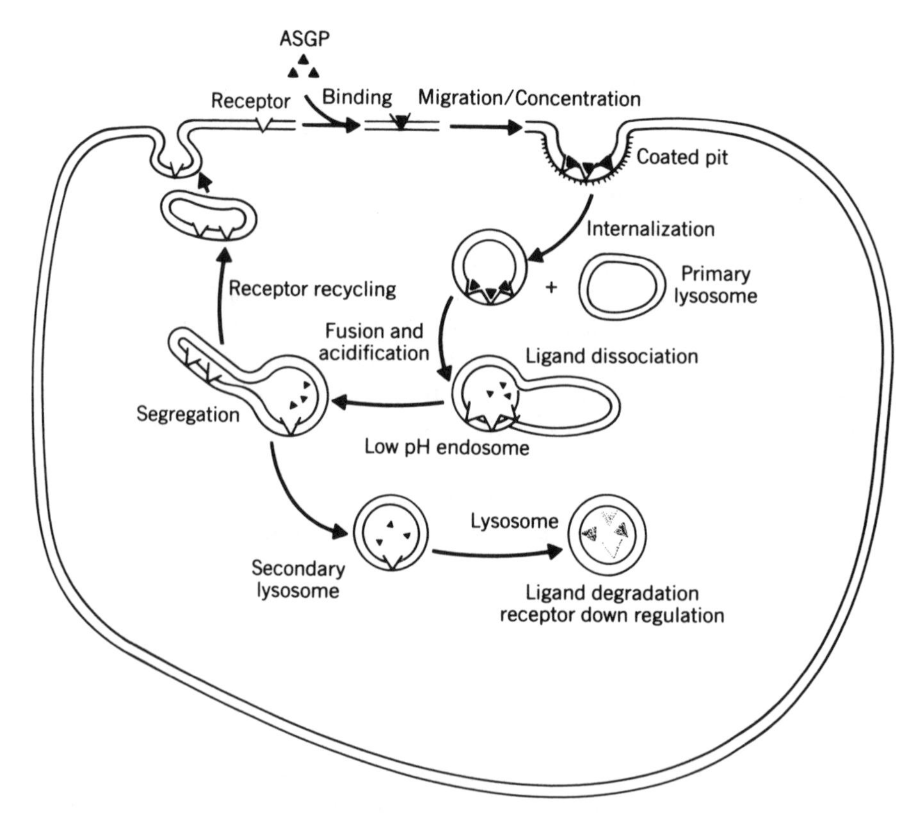

Fig. 15-4. Pathways for receptor mediated endocytosis of asialoglycoprotein (ASGP) and ASGP-receptor. The ligand binds to the receptors on the surface and segregate to coated pits. The pits pinch off to form coated vesicles and then loose their coats. The smooth vesicles are then processed and their contents recycled.

internalized vesicles apparently requires considerable processing and uncoupling of the receptor from ligands. The products of degradation of the glycoprotein ligand appear in about 30 min. Some of the steps for internalization and recycling of ASGP-R are summarized in Fig. 15-4. It is also important to note that the cycle time for different receptors are also very different.

LIPOSOMES AS DRUG CARRIERS

Liposomes are attractive vehicles for controlled bulk delivery of drugs, hormones, enzymes, and DNA into cells. They are nontoxic, weakly immunogenic, biodegradable, capable of encapsulating a wide range of solutes, and thus protecting these solutes from the environment. Their bilayers can be modified with molecules that can specifically recognize and therefore be targeted to specific cells. Drugs and enzymes encapsulated in liposomes have been tested as carriers in several systems. A variety of drugs, both water-soluble and lipid-soluble, have been found to be more effective in liposomes (Alving, 1983; Poznansky and Juliano, 1984) and several are in an advanced stage of a commercial feasibility study. One of the significant success stories is the use of amphotericin B encapsulated in liposomes for treatment of *Candida* and *Aspergillus* infections. Liposomes offer interesting possibilities in the diagnosis and treatment of disease by increasing the therapeutic index (the ratio of effect/toxicity) of drugs by modifying biodistribution and by improving the interaction of drugs with cells. However, a successful use of liposomes for drug delivery would ultimately entail understanding the pharmacodynamic factors that regulate the fate of larger particles in the body rather than that of small solutes that are molecularly dispersed.

The principle underlying the use of liposomes for targeting drugs is based on the premise that liposomes bearing a high concentration of a drug can deliver it to a specific target cell (Gregoriadis et al., 1985). The overall process is rather complex. Liposomes are typically constrained by endothelial barriers to remain in circulation. In most cases, the accumulation of liposomal drugs in the reticuloendothelial cells of liver and spleen represents an undesirable sink for the drug, diverting it from possible action at other sites. The liposomes are cleared by phagocytic cells, thus they are ideally suited for delivery of drugs to phagocytic cells, for immune modulation, or for treatment of infectious disease and for targeting into specific body spaces. Possible steps for delivery of drugs to parasites in macrophages would involve: entry of liposomes into the cell via phagosomes, fusion of phagosomes to lysosomes, and fusion of liposomes to lysosomes where the encapsulated drug may come into contact with the parasite. Since liposomes deliver their contents via an endocytotic pathway, for effective delivery the target cells must have an active endocytosis mode, and the drug must be stable in the acidic environment of lysosomes and capable of escape from lysosomes under acidic conditions.

In a multicellular organism there are many factors that prevent liposomes from reaching their target. Because parenterally injected liposomes naturally travel to liver, spleen, lung, and bone marrow, these organs show highest uptake. Targeting of drugs to other organs must be faciliated by induction of fusion and by specific binding of liposomes conjugated with specific glycoconjugates, antigens, and antibodies capable of interaction with the target cells (Huang, 1983). Interaction of liposomes with cells and with serum components is known to modulate properties of liposomes (stability, size, permeability, surface charge, composition, aggregation) that create additional difficulties.

Mechanism of liposome–cell interaction could involve fusion, adsorption, endocytosis, and lipid exchange (Huang, 1983). Initially, liposomes are adsorbed on the cell surface, and they will be endocytosed if the cells were capable of endocytosis. The relative contribution of other processes probably depends on the specific system under investigation, and such processes are not mutually exclusive. Uptake of protein-free liposomes with cells is best quantitated by monitoring transfer of one or more nonmetabolizable and nonexchangeable markers in the aqueous phase (chromate, inulin, carboxyfluoresceine) as well as the bilayer (radiolabeled hexadecylcholesteryl ether, lipids attached to a fluorescent label) of liposomes.

Bibliography

I dashed to the library at the first opportunity; I refer to the venerable library of the University of . . . , at that time, like Mecca, impenetrable to infidels and even hard to penetrate for such faithful as I. One had to think that the administration followed the wise principle according to which it is good to discourage the arts and sciences: only someone impelled by absolute necessity, or by an overwhelming passion, would willingly subject himself to the trials of abnegation that were demanded for him in order to consult the volumes. The library's schedule was brief and irrational, the lighting dim, the file cards in disorder; in the winter, no heat; no chairs . . . ; and finally, the librarian was an incompetent, insolent boor of exceeding ugliness. . . .

PRIMO LEVI

in *The Periodic Table* (in search of a method "to prepare alloxan from chicken and snake shit for the cosmetic industry.")

Abdel-Latif, A. A. (1986). Pharmacol. Rev. 38, 227–272.

Adamany, A. M., Blumenfeld, O. O., Sabo, B., McCreary, J. (1983). J. Biol. Chem. 258, 11537–11545.

Adriani, J., and Naraghi, M. (1977). Ann. Rev. Pharmacol. Toxicol. 17, 223–242.

Agnew, W. S. (1984). Ann. Rev. Physiol. 46, 517–30.

Albers, R. W., Koval, G. J., Siegel, G. J. (1968). Mol. Pharmacol. 4, 324–36.

Alberts, B., Bray, D., Lewis, J., Raff, M., Roberts, K., Watson, J. D. (1983). *Molecular Biology of the Cell,* Garland Publ. Inc., New York.

Albrecht, O., Gruler, H., Sackman, E. (1978). J. Physique 39, 301–313.

Allegrini, P. R. Van Scharrenburg, G., DeHaas, G. H., Seelig, J. (1983). Biochim. Biophys. Acta 731, 448–455.

Allegrini, P. R., Pluschke, G., Seelig, J. (1984). Biochem. 23, 6452–6458.

Almers, W., Stirling, C. (1984). J. Membrane Biol. 77, 169–186.

Alving, C. R. (1983). Pharm. Theraup. 22, 407–424.

Alving, C. R. (1983). Biochem. Soc. Trans. 12, 342–344.

Ames, G. F. (1986). Ann. Rev. Biochem. 55, 397–425.

Andersen, D. S. (1983). Biophys. J. 41, 119–133; 147–165.

Anderson, C. R., Stevens, C. F. (1972). J. Physiol. 235, 655–699.

Anderson, R., Hanson, K. (1985). J. Biol. Chem. 260, 12219–23.

Andreasen, T. J., McNamee, M. (1980). Biochem. 19, 4719–4726.

Anholt, R., Lindstrom, J., Montal, M. (1985). In *The Enzymes of Biological Membranes,* ed. A. N. Martonosi, Plenum, New York, pp. 335–401.

Anner, B. M. (1985). Biochim. Biophys. Acta 822, 319–334; 335–354.

Apell, H. J., Snozzi, M., Bachofen, R. (1983). Biochim. Biophys. Acta 724, 258–277.

Araujo, P. S., Rosseneu, M. Y., Kremer, J. M. H., Van Zoelen, E. J. J., DeHaas, G. H. (1979). Biochemistry 18, 580–586.

Aronson, P. S. (1985). Ann. Rev. Physiol. 47, 545–560.

Aronson, P. S., Suhm, M. A., Nee, J. (1983). J. Biol. Chem. 258, 6767–6771.

Ashendel, C. L. (1985). Biochim. Biophys. Acta 822, 219–242.

Avigad, G., Amaral, D., Asensio, C., Horeker, B. L. (1962). J. Biol. Chem. 237, 2736–2743.

Axelrod, D. (1983). J. Membrane Biol. 75, 1–10.

Babbitt, B. P., Huang, L. (1985). Biochemistry 24, 2186–94.

Baker, P. F. (1984). Curr. Top. Memb. Transport 22, several chapters in this volume.

Bakker, R., Dobbelmann, J., Borst-Pauwels, G. F. W. (1986). Biochim. Biophys. Acta 861, 205–209.

Balasubraminan, A., McLaughlin, S. (1982). Biochim. Biophys. Acta 685, 1–5.

Baldwin, S. A., Lienhard, G. E. (1981). J. Biol. Chem. 256, 3685–89.

Bamberg, E., Lauger, P. (1974). Biochim. Biophys. Acta 367, 127–133.

Bangham, A. D., Hill, M. W., and Mason, W. T. (1980). Progress in Anesthesiology 2, 69–77.

Barchfeld, G. L., Deamer, D. W. (1985). Biochim. Biophys. Acta 819, 161–169.

Barenholz, Y. (1983). In *Physiology of Membrane Fluidity,* ed. M. Shinitzky, CRC Press, pp. 131–173.

Barnett, J. E., Holman, G. D., Chalkley, R. A., Munday, K. A. (1976). Biochem. J. 145, 417–429.

Barry, P. H., Diamond, J. M. (1984). Physiol. Rev. 64, 763–872.

Barsukov, L. I., Victorov, A. V., Vasilenko, I. A., Evstigneeva, R. P., Bergelson, L. D. (1980). Biochim. Biophys. Acta 598, 153–168.

Barton, P. G., Gunstone, F. D. (1975). J. Biol. Chem. 250, 4470–4476.

Bashford, C. L., Smith, J. C. (1979). Methods Enzymol. 55, 569–586.

Bauman, G., Muller, P. (1975). J. Supramol. Struct. 2, 538–557.

Bayley, H., Staros, J. V. (1984). In *Azides and Nitrenes: Reactivity and Utility,* Academic Press, pp. 433–490.

Beechem, J. M., Brand, L. (1985). Annu. Rev. Biochem. 54, 43–71.

Bellini, F., Phillips, M. C., Pickell, C., Rothblatt, G. H. (1984).

Biochim. Biophys. Acta 777, 209–215.

Ben-Shaul, A., Gelbart, W. M. (1985). Annu. Rev. Physical Chem. 36, 179–211.

Benz, R., McLaughlin, S. (1983). Biophys. J. 41, 381–398.

Benz, R., Ishii, J., Nakae, T. (1980). J. Membr. Biol. 56, 19–29.

Benz, R., Gimple, M., Poole, K., Hancock, R. E. W. (1983). Biochim. Biophys. Acta 730, 387–390.

Berde, C. B., Anderson, H. C., Hudson, B. S. (1980). Biochemistry 19, 4279–93.

Berg, O. G., Von Hippel, P. H. (1985). Annu. Rev. Biophys. Bioeng. 14, 131–160.

Bergelson, L. D., Molotkowsky, J. G., Manevich, Y. M. (1985). Chem. Phys. Lipids 37, 165–195.

Bergmann, W. L., Dressler, V., Haest, C. W. M., Deuticke, B. (1984). Biochim. Biophys. Acta 772, 328–336.

Berridge, M. J. (1986). J. Exp. Biol. 124, 323–325.

Berridge, M. J., Irvine, R. F. (1984). Nature 312, 315–321.

Bezanilla, F. (1985). J. Membr. Biol. 88, 97–111.

Bhakdi, S., Tranum-Jensen, J. (1983). Biochim. Biophys. Acta 737, 343–372.

Billah, M. M., Finnean, J., Colman, R., Mitchell, R. H. (1977).

Biochim. Biophys. Acta 433, 45–54; 465, 515–526.

Bittar, E. E. (1983). Progr. Neurobiol. 20, 1–54.

Blake, R., Hager, L. P., Gennis, R. B. (1978). J. Biol. Chem. 253, 1963–1971.

Blaurock, A. E. (1982). Biochim. Biophys. Acta 650, 167–207.

Bloch, K. (1983). Crit, Rev. Biochem. 14, 47–92.

Blocher, D., Six, L, Guterman, R., Henkel, B., Ring, K. (1985).

Biochim. Biophys. Acta 818, 333–342.

Blume, A. (1983). Biochemistry 22, 5436–42.

Blume, A., Eibl, H. (1981). Biochim. Biophys. Acta 640, 609–618.

Blume, A., Griffin, R. G. (1982). Biochemistry 24, 6230–42.

Blume, A., Rice, D. M., Wittebort, R. J., Griffin, R. G. (1982a). Biochemistry 24, 6220–30.

Blume, A., Wittebort, R. J. DasGupta, S. K., Griffin, R. G. (1982b). Biochemistry 24, 6243–53.

Boggs, J. M. (1984). In *Membrane Fluidity,* ed. M. Kates and L. A. Mason, Plenum Publ., New York, pp. 3–53.

Boggs, J. M., Rangaraj, G., Koshy, K. M. (1986). Chem. Phys. Lipids 40, 23–34.

Bolard, J. (1986). Biochim. Biophys. Acta 864, 257–304.

Born, M. (1920). Z. Phys. 1, 45–48.

Boron, W. F. (1983). J. Membr. Biol. 72, 1–16.

Bouma, S. R., Drislane, F. R., Huestis, W. H. (1977). J. Biol. Chem. 252, 6759–6763.

Bradley, R. J., Prasad, K. U., Urry, D. W. (1981). Biochim. Biophys. Acta 649, 281–285.

Brady, R. O., ed. (1986). Glycolipids. Chem. Phys. Lipids 42, 1–248.

Braell, W. A., Schlossman, D. M., Schmid, S. L., Rothman, J. E. (1984). J. Cell Biol. 99, 734–741.

Briggs, M. S., Gierasch, L. M. (1986). Adv. Prot. Chem. 38. 109–179.

Brisson, J. R., Carver, J. P. (1983). Biochemistry 22, 1362–68; 3671–80; 3680–83.

Brisson, A., Unwin, P. N. T. (1985). Nature 315, 474–477.

Brito, R. M. M., Vaz, W. L. C. (1986). Anal. Biochem. 152, 250–55.

Brown J. C., Hunt, R. C. (1978). Int. Rev. Cytol 52, 277–349.

Brown, M. F., Williams, G. D. (1985). J. Biochem. Biophys. Methods 11, 71–81.

Buldt, G., Gally, H. U., Seelig, J., Zaccai, G. (1979). J. Molec. Biol. 134, 673–691.

Burke, T. G., Tritton, T. R. (1984). Anal. Biochem. 143, 135–140.

Burke, T. G., Tritton, T. R. (1985a). Biochemistry 24, 5972–5980.

Burke, T. G., Tritton, T. R. (1985b). Biochemistry 24, 1768–1776.

Busath, D., Szabo, G. (1981). Nature 294, 391–393.

Cadenhead, D. A., Muller-Landau, F. (1984). Can. J. Biochem. Cell Biol. 62, 732–737.

Caffrey, M. (1985). Biochemistry 24, 4826–44.

Cafiso, D. S., Hubbell, W. L. (1981). Biophys. J. 33, 144a.

Cantor, R. S., Dill, K. A. (1984). Maromolecules 17, 384–388.

Capaldi, R. A., Marshall, F. A., Staples, S. J. (1983). Commun. Mol. Cell. Biophys. 1, 365–379.

Capaldi, R. A., Vanderkooi, G. (1972). Proc. Natl. Acad. Sci. 69, 930–932.

Carnie, S., McLaughlin, S. (1983). Biophys. J. 44, 325–332.

Carlson, R. M. K., Tarchini, C., Djerassi, C. (1980). In *Frontiers of Bioorganic Chemistry and Molecular Biology,* ed. A. N. Ananchenko, Pergamon Press, New York, pp. 211–224.

Carruthers, A. (1985). Progr. Biophys. Mol. Biol. 43, 33–69.

Carruthers, A., Melchior, D. L. (1983). Biochemistry 22, 5797–5807.

Carruthers, A., Melchior, D. L. (1984). Biochemistry 23, 27 12–18; 6901–11.

Casal, H. L., Mantsch, H. H. (1984). Biochim. Biophys. Acta 779, 381–401.

Casal, H. L., Cameron, D. G., Mantsch, H. H. (1983). J. Phys. Chem. 87, 5354–57.

Catterall, W. A. (1984). Annu. Rev. Biochem. 55, 953–85.

Cevc, G., Marsh, D. (1985). Biophys. J. 47, 21–31.

Cevc, G., Marsh, D. (1987). *Phospholipid Bilayer: Physical Principles and Models.* Wiley, New York.

Chabre, M. (1985). Annu. Rev. Biophys. Biophys. Chem. 14, 331–360.

Chadha, J. S. (1970). Chem. Phys. Lipids 4, 104–108.

Chou, P., Fasman, G. (1978). Adv. Enzymol. Mol. Biol. 47, 45–148.

Chin, J. H., Goldstein, D. B. (1985). Membrane Fluidity in Biol. 3, 1–38.

Chapman, D., Williams, R. M., Ladbrook, D. B. (1973). Chem. Phys. Lipids *3,* 304–367.

Chrezcszczyk, A., Wishnia, A., Springer, C. S. (1977). Biochim. Biophys. Acta 470, 161–169.

Christiansen, H. N. (1984). Biochim. Biophys. Acta 779, 255–269.

Christie, W. W. (1985). J. Lipid Res. 26, 507.

Ciani, S., Eisenman, G., Szabo, G. (1969). J. Membrane Biol. 1, 1–36.

Clegg, R. M., Vaz, W. L. C. (1985). In *Progress in Protein-Lipid Interactions,* ed. A. Wats, p. 173–229. Elsevier, Amsterdam.

Cobb, M. H., Rosen, O. M. (1984). Biochim. Biophys. Acta 738, 1–8.

Collin, K. D., Washabaugh, M. W. (1985). Quart. Rev. Biophys. 18, 323–422.

Colquhoun, D., Hawkes, A. G. (1981). Proc. Roy. Soc. Lond. (Biol) 211, 205–235.

Comfurius, P., Zwaal, R. F. A. (1977). Biochim. Biophys. Acta 488, 36–42.

Coolbar, K. P., Berde, C. B., Keough, K. M. W. (1983). Biochemistry 22, 1466–73.

Cooper, K., Jakobsson, E., Wolynes, P. (1985). Progr. Biophys. Mol. Biol. 46, 51–96.

Cornell, B. A., Separovic, F. (1983). Biochim. Biophys. Acta 733, 189–193.

Corti, M., Degregorio, V., Ghidoni, R., Sonnino, S., Tettamanti, G. (1980). Chem. Phys. Lipids 26, 225–238.

Crane, R. K. (1977). Rev. Physiol. Biochem. Pharm. 78, 99–159.

Crook, S. J., Boggs, J. M., Vistnes, A. I., Koshy, K. M. (1986). Biochem. 25, 7488–7494.

Cross, G. A. M. (1987). Cell 48, 179–181.

Cullis, P. R., Hope, M. J., DeKruijff, B., Verkleij, A. J., Tilcock, C. P. S. (1985). In *Phospholipids and Cellular Regulation,* ed. J. F. Kuo, CRC Press, Boca Raton, Florida, pp. 1–59.

Cunningham, C. C., Hager, L. P. (1971). J. Biol. Chem. 246, 1575–1582, 1583–1589.

Cunningham, B. A., Lis, L. J. (1986). Biochim. Biophys. Acta 861, 237–242.

Cushman, S. W., Wardzala, L. J., Simpson, I. A. Karneilli, E., Hissin, P. J., Wheeler, T. J., Hinkle, P. C., Salans, L. D. (1983). In *Hormones and Cell Regulation,* Vol. 7, Elsevier, pp. 73–84. Amsterdam.

Czech, M. P. (1985). Annu. Rev. Physiol. 47, 357–381.

Dahl, J., Dahl, C. (1983). Proc. Natl. Acad. Sci. U. S. 80, 692–696.

Danielli, J. F., Davson, H. (1935). J. Cell. Comp. Physiol. 5, 595–610.

Das, S., Rand, R. P. (1984). Biochem. Biophys. Res. Commun. 124, 491–496.

DasGupta, S. K., Rice, D. M., Griffin, R. G. (1982). J. Lipid Res. 23, 197–200.

Davidson, V. L., Brunden, K. R., Cramer, W. A., Cohen, F. S. (1984). J. Mol. Biol. 79, 105–118.

Davis, J. H. (1983). Biochim. Biophys. Acta 737, 117–171.

Davis, P. J., Keough, K. M. W. (1983). Biochemistry 22, 6334–40.

Davis, P. J., Fleming, B. D., Coolbear, K. P., Keough, K. M. W. (1981). Biochemistry 20, 3633–3637.

Davis, P. J., Kariel, N., Keough, K. M. W. (1986). Biochim. Biophys. Acta 856, 395–398.

Dawidowicz, E. A. (1987). Curr. Top. Membranes and Transp. 29, 175–202.

Deamer, D. W. (1986). In *Origins of Life*.

Deamer, D. W., Nichols, J. W. (1985). In *Water and Ions in Biological Systems,* ed. A. Pullman, V. Vasilescu, L. Packer, Plenum, New York, pp. 469–481.

Deamer, D. W., Uster P. S. (1983). In *Liposomes,* ed. M. J. Ostro, Dekker, New York, pp. 27–51.

DeDuve, C. (1964). J. Theoret. Biol. 6, 33–59.

Defay, R., Prigogine, I., Bellemans, A., Everett, D. H. (1966). *Surface Tensions and Adsorption,* Wiley, New York.

Denton, R. M., Brownsey, R. W., Belsham, G. J. (1981). Diabetologia 21, 347–362.

De Rosa, M., Gambeacorta, A., Gliozzi, A. (1986). Microbiol. Rev. 50, 70–80.

Devaux, P. F., Seigneuret, M. (1985). Biochim. Biophys. Acta 822, 63–125.

Dewey, T. G., Hammes, G. G. (1981). J. Biol. Chem. 256, 8941–8946.

Diamond, J. M., Katz, Y. (1974). J. Membr. Biol. 17, 121–154.

Diamond, J. M., Wright, E. M. (1969). Annu. Rev. Physiol. 31, 581–646.

Dilger, J. P., McLaughlin, S. (1979). J. Membrane Biol. 46, 359–384.

Dill, K. A., Koppel, D. E., Cantor, R. S., Dill, J. D., Bendedouch, D., Chen, S. H. (1984). Nature 309, 42–45.

Dills, S. S., Aperson, A., Schmidt, M. R., and Saier, M. H. (1980). Microbiol. Rev. 44, 385–418.

Dimitrov, D. S., Li, J., Angelova, M., Jain, R. K. (1984). FEBS Lett. 176, 398–402.

Dimroth, P. (1982). Eur. J. Biochem. 121, 435–441.

Dionne, V., Liebowitz, M. (1982). Biophys. J. 39, 253–261.

DiRienzo, J. M., Nakamura, K., Inouye, M. (1978). Annu. Rev. Biochem. 47, 481–532.

Dluhy, R. A., Cameron, D. G., Mantsch, H. H., Mendelsohn, R. (1983). Biochemistry 22, 6318–25.

Doody, M. C., Pownall, H. J., Kao, Y. J., Smith, L. C. (1980). Biochem. 19, 108–116.

Dorset, D. L. Engel, A., Massalski, A., Rosenbusch, J. P. (1984). Biophys. J. 45, 128–129.

Dufourc, E. J., Parish, E. J., Chitrakorn, S., Smith, I. C. P. (1984). Biochem. J. 23, 6062–6071.

Dufourcq, J., Faucon, J. F. (1978). Biochemistry 17, 1170–1176.

Duzguenes, N. (1985). *Subcellular Biochemistry,* vol. 11, ed. D. B. Roodyn Plenum Press. New York, 195–186.

Duzguenes, N., Wilschut, J., Hong, K., Fraley, R., Perry, C., Friend, D. S., James, T. L., Papahadjopoulos, D. (1983). Biochim. Biophys. Acta 732, 289–299.

East, J. M., Lee, A. G. (1982). Biochemistry 21, 4144–4151.

Easwaran, K. R. K. (1985). In *Metal Ions in Biological Systems: Antibiotics and their Complexes,* ed. H. Sigel, vol. 19, pp. 109–137. Mercel Decker, New York.

Easwaran, K. R. K. (1987). *Ion Transport through Membranes,* ed. B. Pullman and K. Yagi. Academic Press, in press.

Edelman, G. (1984). Trends in Neurosci. 7, 78–84.

Ehrlich, B. E., Schen, C. R., Garcia, M. L., Kaczorowski, G. J. (1986). Proc. Natl. Acad. Sci. 83, 193–197.

Eibl, H. (1980). Chem. Phys. Lipids 26, 405–428.

Eibl, H. (1983). Membrane Fluidity in Biol. 2, 217–236.

Eibl, H., Woolley, P. (1986). Chem. Phys. Lipids 41, 53–63; other papers in preparation.

Eisenberg, R. S., Frank, M., Stevens, C. F. (1984). *Membranes, Channels and Noise,* Plenum Press, New York.

Eisenman, G. (1962). Biophys. J. 2, 259–323.

Eisenman, G., Szabo, G., McLaughlin, S. G. A., Ciani, S. M. (1973). Bioenegetics 4, 93–148.

Eisenman, G. (1983). In *Mass Transfer and Kinetics of Ion Exchange,* ed. L. Liberti and F. G. Helfferich, NATO ASI Series, Martinus Nijhoff Publ., The Hague, pp. 121–155.

Eisenman, G., Dani, J. A., Sandblom, J. (1985). In *Ion Measurements in Physiology and Medicine,* eds. M. Kessler, D. K. Harrison, J. Hoper, pp. 54–66. Springer-Verlag, New York.

Eisenman, G., Horn, R. (1983). J. Membrane Biol. 76, 197–225.

Eisenman, G., Ciani, S., Szabo, G. (1969). J. Membrane Biol. 1, 294–345.

Elamrani, K., Blume, A. (1984). Biochim. Biophys. Acta 769, 578–584.

Elder, M., Hitchcock, P. B., Mason, R., Shipley, G. G. (1977). Proc. Roy. Soc. Ser. A 354, 157–170.

Eklund, K. K., Virtanen, J. A., Kinnunen, P. K. J. (1984). Biochim. Biophys. Acta 793, 310–312.

Elliot, J. R., Needham, D., Dilger, J. P., Haydon, D. A. (1983). Biochim. Biophys. Acta 735, 95–103.

Engelbert, H. P., Lawaczeck, R. (1985). Ber. Bunsenges. Phys. Chem. 89, 754–759.

Engelman, D. M., Steitz, T. A., Goldman, A. (1986). Annu. Rev. Biophys. Biophys. Chem. 15, 321–353.

Ernster, L., ed. (1984). *Bioenergetics,* Elsevier, Amsterdam.

Ermishkin, L. N., Kasumov, K. M., Potseluyev, V. M. (1977). Biochim. Biophys. Acta 470, 357–367.

Esfahani, M., Rudkin, B. B., Cutler, C. J., Waldron, P. E. (1977). J. Biol. Chem. 252, 3194–3198.

Estep, T. N., Mountcastle, D. B., Barenholz, Y., Biltonen, R. L., Thompson, T. E. (1979). Biochemistry 10, 2112–2117.

Etemadi, A. (1985). Adv. Lipid Res. 21, 281–428.

Ethier, M. F., Wolf, D. E., Melchior, D. L. (1983). Biochemistry 22, 1178–1182.

Evered, D., Whelan, J., eds. (1984). *Cell Fusion,* Ciba Foundation Symp., vol. 103, Pittman Publ., London.

Eytan, G. D. (1982). Biochim. Biophys. Acta 694, 185–202.

Eze, M. O., McElhaney, R. N. (1987). Biochim. Biophys. Acta 897, 159–168.

Fato, R., Battino, M., Esposti, M. D., Castelli, G. P., Lenaz, G. (1986). Biochemistry 25, 3378–90.

Feigenson, G. W. (1986). Biochem. 25, 5819–25.

Felgner, P. L., Thompson, T. E., Barenholz, Y., Lichtenberg, D. (1983). Biochemistry 22, 1670–1674.

Fendler, J. H. (1982). *Membrane Mimetic Chemistry,* Wiley, New York.

Fendler, K., Grell, E., Haubs, M., Bamberg, E. (1985). EMBO J. 4, 3079–3085.

Fenwick, E. M., Marty, A., Neher, E. (1982). J. Physiol. (Lond.) 331, 559–635.

Ferguson, M. A. J., Low, M. G., Cross, G. A. M. (1985). J. Biol. Chem. 260, 14547–55.

Fernandez, F., Kurzmack, M., Inesi, G. (1984). J. Biol. Chem. 259, 9687–98.

Finegold, L., Melnick, S., Singer, M. A. (1985). Chem. Phys. Lipids 38, 387–390.

Finegold, L., Singer, M. A. (1986). Biochim. Biophys. Acta 855, 417–420.

Fisher, A., Sackman, E. (1984). J. Physique 45, 517–527.

Fishman, P. H. (1982). J. Membrane Biol. 69, 85–97.

Fleishcher, S., Packer, L. (eds.) (1974). Methods Enzymol. 31, several articles of interest.

Fleischer, S., McIntyre, J. O., Churchill, P., Fleer, E., Maurer, A. (1983). In *Structure and Function of Membrane Proteins,* ed. E. Quagliarielli and F. Palmieri, Elsevier, Amsterdam, pp. 283–300.

Flockart, D. A., Corbin, J. D. (1982). Crit. Rev. Biochem. *12,* 133–186.

Flory, P. J. (1969). *Statistical Mechanics of Chain Molecules,* Wiley, New York.

Fontell, K. (1982). Mol. Cryst. Liquid Cryst. 63, 59–82.

French, R. J., Horn, R. (1983). Annu. Rev. Biophys. Bioengg. 12, 319–56.

Frezzatti, W. A., Toselli, W. R., Schreier, S. (1986). Biochim. Biophys. Acta 860, 531–538.

Froud, R. J., East, J. M., Rooney, E. K. Lee, A. G. (1987). Biochemistry 25, 7535–44; 7544–52.

Gabriel, N. E., Roberts, M. F. (1984). Biochem. 23, 4011–4015.

Gaffney, B. J. (1985). Biochim. Biophys. Acta 822, 289–317.

Gage, P. W. (1978). J. Physiol. 274, 279–298.

Gahmberg, C. G., Kiehn, D., Hakamori, S. (1974). Nature 248, 413–415.

Galla, H. J., Hartman, W. (1980). Chem. Phys. Lipids 27, 199–219.

Gammeltoft, S. (1984). Physiol. Rev. 64, 1321–1378.

Garavito, R. M., Jenkins, J. A., Jansonius, J. N. Karlsson, R., Rosenbusch, J. P. (1983). J. Mol. Biol. 164, 313–327.

Gaub, H., Sackman, E., Buschl, R., Ringsdorf, H. (1984). Biophys. J. 45, 725–731.

Gaub, H., Buschl, R., Ringsdorf, H., Sackman, E. (1985). Chem. Phys. Lipids 37, 19–43.

Geck, P., Heinz, E. (1986). J. Membr. Biol. 91, 97–105.

Geiger, B. (1983). Biochim. Biophys. Acta 737, 305–341.

Gennis, R. B., Shinenski, B., L. Strominger, J. L. (1976). J. Biol. Chem. 251, 1264–1269; 1277–1282.

Gershfeld, N. L. (1976). Annu. Rev. Phys. Chem. 27, 349–368.

Gershfeld, N. L., Tajima, K. (1977). J. Coll. Interf. Sci. 59, 597–604.

Glagolev, A. N. (1984). *Motility and Taxis in Prokaryotes,* Harwood Academic Publishers, New York.

Glossman, H., Neville, D. M. (1971). J. Biol. Chem. 246, 6339–6346.

Goffeau, A., Slayman, C. W. (1981). Biochim. Biophys. Acta 639, 197–223.

Goldfine, H. (1983). Curr. Top. Membr. Trans. 17, 1–43.

Goldstein, D. B. (1984). Annu. Rev. Toxicol. 24, 43–64.

Goldstein, J., Brown, M. (1983). Annu. Rev. Biochem. 52, 223–261.

Gorter, E., Grendel, F. (1925). J. Exp. Med. 41, 439–443.

Gould, G. W., East, J. M., Froud, R. J., McWhirter, J. M., Stefanova, H. I., Lee, A. G. (1986). Biochem. J. 237, 217–227.

Grant, C. W. M., Peters, M. W. (1984). Biochim. Biophys. Acta 779, 403–422.

Greenblatt R. E., Blatt, Y., Montal, M. (1985). FEBS Lett. 193, 125–134.

Gregoriadis, G., Poste, G., Senior, J., Trouet, A. (1985). *Receptor-Mediated Targeting of Drugs,* Plenum, New York.

Griffin, R. G. (1981). Methods Enzymol. 72, 108–174.

Grinstein, S., Rothstein, A., Sarkadi, Gelfand, E. W. (1984). Am. J. Physiol. 246, c204–c215.

Gruen, D. W. R. (1985). J. Phys. Chem. 89, 146–153; 153–163.

Grunner S. M., Cullis, P. R., Hope, M. J., Tilcock, C. P. S. (1985a). Ann. Rev. Biophys. Bioeng. 14, 211–238.

Grunner, S. M., Jain, M. K. (1985). Biochim. Biophys. Acta 818, 352–355.

Grunner, S. M., Lenk, R. P., Janoff, A. S., Ostro, M. J. (1985b). Biochem. 24, 2833–2842.

Grygorzyk, R., Schwarz, W. (1985). Eur. Biophys. J. 12, 57–65.

Gulik, A., Luzatti, V., DeRosa, M., Gambacorta, A. (1985). J. Mol. Biol. 182, 131–149.

Gupta, C. M., Radhakrishnan, R., Khorana, H. (1977) Proc. Natl. Acad. Sci. 74, 4315–4319.

Gutknecht, J., Walter, A. (1981). Biochim. Biophys. Acta 641, 183–188.

Hainesworth, A. H., Hladky, S. B. (1987). Biophys. J. 51, 27–36.

Hamamoto, T., Carrasco, N. Matsushita, K., Kaback, H. R., Montal, M. (1985). Proc. Natl. Acad. Sci. 82, 2570–2573.

Hamill, O. P., Sakman, B. (1981). Nature 294, 462–464.

Hamilton, J. A., Miller, K. W., Small, D. M. (1983). J. Biol. Chem. 258, 12821–12826.

Hanahan, D. J. (1986). Annu. Rev. Biochem. 55, 483–509.

Harbison, G. S., Griffin, R. G. (1984). J. Lipid. Res. 25, 1140–41.

Hargreaves, W. R., Deamer, D. W. (1978). Biochemistry 17, 3759–65.

Harlos, K., Eibl, H., Pascher, I., Sundell, S. (1984). Chem. Phys. Lipids 34, 115–126.

Harold, F. M. (1986). *The Vital Force: A Study of Bioenergetics,* W. H. Freeman, New York.

Haydon, D. A., Kimura, J. E. (1981). J. Physiol. Lond. 312, 57–70.

Hartley, G. S. (1936). *Aqueous Solutions of Paraffin-chain Salts,* Hermann & Cie, Paris.

Hartshorne, R. P., Keller, B. U., Talvenhe, J. A., Catteral, W. A., Montal, M. (1985). Proc. Natl. Acad. Sci. 82, 240–244.

Hatefi, Y. (1985). Annu. Rev. Biochem. 54, 1015–1069.

Hauser, H. (1984). Biochim. Biophys. Acta 772, 37–50.

Hauser, H. (1985). Chimia 39, 252–264.

Hauser, H., Shipley, G. G. (1983). Biochemistry, 22, 2171–2178.

Hauser, H., Shipley, G. G. (1984). Biochemistry 23, 34–41.

Hauser, H., Shipley, G. G. (1985). Biochim. Biophys. Acta 813, 343–346.

Hauser, H., Hinckley, C. C., Krebs, J., Levine, B. A., Phillips, M. C., Williams, R. J. P. (1977). Biochim. Biophys. Acta 468, 364–377.

Hauser, H., Pascher, I., Sundell, S. (1980). J. Mol. Biol. 137, 249–264.

Hauser, H., Pascher, L., Pearson, R. H., Sundell, S. (1981). Biochim. Biophys. Acta 650, 21–51.

Hawco, M. W., Davis, P. J., Keough, K. M. W. (1981). J. Appl. Physiol. 51, 509–515.

Haydon, D. A., Kimura, J. E. (1981). J. Physiol. 312, 57–70.

Haydon, D. A., Elliott, J. R., Hendry, B. M. (1984). Curr. Top. Membrane Transport 22, 445–482.

Heast, C. W. M. (1982). Biochim. Biophys. Acta 694, 331–352.

Heinz, E., Geck, P., Pietrzyk, C. (1975). Ann. N. Y. Acad. Sci. 264, 428–441.

Helenius, A., Simon, K. (1975). Biochim. Biophys. Acta 415, 29–79.

Hellingswerf, K. J., Konings, W. N. (1985). Adv. Microbial. Physiol. 26, 125–154.

Henderson, R., Unwin, P. N. T. (1975). Nature 257, 28–32.

Hengge, R., Boos, W. (1983). Biochim. Biophys. Acta 737, 443–478.

Hentke, K., Braun, V. (1973). Eur. J. Biochem. 34, 284–296.

Hess, G. P., Cash, D. J., Aoshima, H. (1983). Annu. Rev. Biophys. Bioeng. 12, 443–73.

Hess, G. P., Kolb, H., Lauger, P., Schoffeniels, E., Schwarze, W. (1984). Proc. Natl. Acad. Sci. 81. 5281–5285.

Heyworth, C. M., Wallace, A. V., Housley, M. D. (1983). Biochem. J. 214, 99–110.

Houslay, M. D., Gordon (1983). Curr. Top. Mem. Trans. *18,* 179–231.

Higgins, C. F., Haag, P. D., Nikaido, K., Ardeshir, F., Garcia, G., Ames, G. F. L. (1982). Nature 298, 723–27.

Hill, M. W. (1974). Biochim. Biophys. Acta 356, 117–124.

Hill, T. L. (1966). J. Theoret. Biol. 10, 442-459.

Hille, B. (1975). J. Gen. Physiol. 66, 535–560.

Hille, B. (1984). *Ion Channels of Excitable Membranes,* Sinauer, Sutherland, Massachusetts.

Hinz, H. J., Corner, O., Nicolau, C. (1981). Biochim. Biophys. Acta 643, 557–571.

Hinz, H. J., Six, L., Ruess, K. P., Leiflander, M. (1985). Biochemistry 24, 806–13.

Hladky, S. B. (1979). Curr. Top. Membr. Trans. 12, 53–164.

Hladky, S. B., Haydon, D. A. (1984). Curr. Top. Membr. Transp. 21, 327–372.

Hochman, Y., Zakim, D. (1984). J. Biol. Chem. 259, 5521–5525.

Hochman, Y., Zakim, D., Vessey, D. (1981). J. Biol. Chem. 256, 4783–4788.

Hodgkin, A. L., Huxley, A. F. (1952). J. Physiol. 116, 449–472; 473–496; 497–515; 117, 500–544.

Honig, B. H., Hubbell, W. L., Flewelling, R. F. (1986). Annu. Rev. Biophys. Chem. 15, 163–193.

Hope, M. J., Bally, M. B., Webb, G., Cullis, P. R. (1985). Biochim. Biophys. Acta 812, 55–65.

Hopfer, V., Lehninger, A. L., Lennarz, W. I. (1970). J. Membr. Biol. 3, 142–155.

Hopp, T. P., Wood, K. R. (1981). Proc. Natl. Acad. Sci. 78, 3824–3828.

Horie, T., Vanderkooi, J. M. (1984). Life Chemistry Reports 2, 141–178.

Horn, R., Patlak, J., Stevens, C. (1981). Nature 291, 426–428.

Houslay, M. D., Gordon, L. M. (1983). Curr. Top. Memb. Transp. 18, 179–231.

Huang, C., and Mason, J. T. (1978). Proc. Natl. Acad. Sci. 75, 308–310.

Huang, C., Mason, J. T. (1986). Biochim. Biophys. Acta 864, 423–470.

Huang, L. (1983). In *Liposome,* ed. Ostro, p. 87–124, Marcel Dekker, New York.

Hughes, R. C. (1976). *Membrane Glycoproteins: A Review of Structure and Function,* Butterworths, London.

Hui, S. W., Huang, C. (1986). Biochemistry 25, 1330–1335.

Hui, S. W., Coden, M., Papahadjopoulos, D., Parsons, D. F. (1975). Biochem. Biophys. Acta 382, 265–275.

Hui, S. W., Mason, J. J., Huang, C. (1984). Biochemistry 23, 5570–77.

Illsley, N. P., Verkman, A. S. (1987). Biochemistry 26, 1215–1219.

Imai, T. (1978). Biochim. Biophys. Acta 523, 37–182.

Inesi, G. (1985). Annu. Rev. Physiol. 47, 573–601.

Israelachvili, J. N. (1985). *Physics of Amphiphiles: Micelles, Vesicles and Microemulsions,* North Holland, Amsterdam.

Israelachvili, J. N., Marcelja, S., Horn, R. G. (1980). Q. Rev. Biophys. 13, 121–229.

Issacson, Y. A., Deroo, P. W., Rosanthal, A. F., Bittman, R., McIntyre, J. O., Bock, H. G., Gazzotti, P., Fleischer, S. (1979). J. Biol. Chem. 254, 117–126.

Ivanov, V. T., Schev, S. V. C. (1982). In *Biopolymer Complexes,* ed. G. Snatzke and W. Bartmann, Wiley, New York, pp. 107–125.

Jacobs, R. E., Hudson, B. S., Anderson, H. C. Biochemistry 16, 4349–4359.

Jain, M. K. (1972). *The Bimolecular Lipid Membrane,* Van Nostrand, New York.

Jain, M. K. (1975). Curr. Top. Membranes Transp. 6, 1–57.

Jain, M. K. (1983). Membrane Fluidity in Biol. 1, 1–36.

Jain, M. K., Wray, L. V., Wu, N. M. (1975). Nature 255, 494–495.

Jain, M. K., White, H. B. (1977). Adv. Lipid Res. 15, 1–60.

Jain, M. K., Wu, N. M. (1977). J. Membrane Biol. 34, 157–201.

Jain, M. K., Wary, L. V. (1978). Biochem. Pharm. 27, 1294–96.

Jain, M. K., Zakim, D. (1987). Biochim. Biophys. Acta 906, 33–68.

Jain, M. K., DeHaas, G. H. (1981). Biochim. Biophys. Acta 642, 203–211.

Jain, M. K., Jahagirdar, D. V. (1985). Biochim. Biophys. Acta 814, 319–326.

Jain, M. K., Marks, R. H. L., Cordes, E. H. (1970). Proc. Natl. Acad. Sci. U. S. 67, 799–805.

Jain, M. K., White, F. P., Williams, E., Strickholm, A., Cordes, E. H. (1972). J. Membr. Biol. 363–385.

Jain, M. K., Van Echteld, C. J. A., Ramirez, F., DeGier, J., DeHaas, G. H., Van Deenen, L. L. M. (1980). Nature 284, 486–487.

Jain, M. K., Singer, M. A., Creceley, R. W., Pajouhesh, H., Hancock, A. J. (1984a). Biochim. Biophys. Acta 774, 199–205.

Jain, M. K., Singer, M. A., Ramirez, F., Marecek, J. F., He, N. B., Hui, S. W. (1984b). Biochim. Biophys. Acta 775, 426–434.

Jain, M. K., Rogers, J., Simpson, L., Gierasch, L. (1985a). Biochim. Biophys. Acta 816, 153–162.

Jain, M. K., Crecely, R., Hille, J. D. R., DeHaas, G. H., and Grunner, S. (1985b). Biochim. Biophys. Acta 813, 68–76.

Jain, M. K., Jahagirdar, D. V., Van Linde, M., Roelofsen, B., Eibl, H. (1985). Biochim. Biophys. Acta 818, 356–364.

Jain, M. K., Rogers, J., Jahagirdar, D. V., Marecek, J. F., Ramirez, F. (1986a). Biochim. Biophys. Acta 860, 435–447.

Jain, M. K., Maliwal, B. P., DeHaas, G. H., Slotboom, A. J. (1986b). Biochim. Biophys. Acta 860, 448–461.

Jain, M. K., Rogers, J., Marecek, J. F., Ramirez, F., Eibl, H. (1986c). Biochim. Biophys. Acta 860, 462–474.

Jain, M. K., DeHaas, G. H., Marecek, J. F., Ramirez, F. (1986d). Biochim. Biophys. Acta 860, 475–483.

Janoff, A. S., Miller, K. W. (1982) *Biological Membranes,* ed. D. Chapman, vol. 4, Academic Press, New York, pp. 417–476.

Jannings, M. L. (1984). J. Membr. Biol. 80, 105–117.

Jauch, P., Lauger, P. (1986). J. Membr. Biol. 97, 117–127.

Jaworski, M., Mendelsohn, R. (1985). Biochemistry 24, 4322–28.

Jay, D., Cantley, L. (1986). Annu. Rev. Biochem. 55, 511–538.

Jiang, R. T., Shyy, Y., Tsai, M. (1984). Biochemistry 23, 1661–1667.

Jones, O. T., Lee, A. G. (1985). Biochim. Biophys. Acta. 812, 731–739.

Jordan, P. C. (1984). Biochemistry. J. 45, 1101–1107.

Jorgenson, P. L. (1982). Biochim. Biophys. Acta 694, 27–68.

Kaback, H. R. (1983). J. Membr. Biol. 76, 95–112.

Kaever, V., Resch, K. (1985). Biochim. Biophys. Acta 846, 216–225.

Kampmann, L., Lepke, S., Fasold, H., Fritzsch, G., Passow, H. (1982). J. Membr. Biol. 70, 199–216.

Kannenberg, E., Blume, A., Poralla, K. (1984). FEBS Lett. 172, 331–334.

Kannenberg, E., Blume, A., Geckeler, K., Povalla, K. (1985). Biochim. Biophys. Acta 814, 179–185.

Kapitza, H. G., Sackman, E. (1980). Biochim. Biophys. Acta 595, 56–64.

Kaplan, J. H. (1985). Annu. Rev. Physiol. 47, 535–44.

Karnovsky, M. J., Kleinfeld, A. M., Hoover, R. L., Kausner, R. D. (1982). J. Cell. Biol. 94, 1–6.

Kasianowicz, J., Benz, R., McLaughlin, S. (1984). J. Membr. Biol. 82, 179–190.

Katz, Y., Simon, S. A. (1977). Biochim. Biophys. Acta 471, 1–15.

Kauzman, W. (1959). Adv. Protein Chem. 14, 1–63.

Keough, K. M. W. (1985). Membrane Fluidity in Biol. 3, 39–84.

Keough, K. M. W., Davis, P. J. (1984). In *Membrane Fluidity,* ed. M. Kates and L. A. Mason, Plenum Press, New York, pp. 55–97.

Kerry, C. J., Kits, K. S., Ramsey, R. L., Sansom, M. S. P., Usherwood, P. N. R. (1986). Biophys. J. 50, 367–374.

Keynes, R. D., Greef, N. G., Van Holden, D. F. (1983). Proc. Roy Soc. Lond. 220, 1–30.

Khan, M. I., Sastry, M. V. K., Surolia, A. (1986). J. Biol. Chem. 261, 3013–19.

Killian, J. A., DeKruijff, B. (1986). Chem. Phys. Lipids 40, 259–284.

Kirk, R. G. (1968). Proc. Natl. Acad. Sci. 60, 614–621.

Kirschner, D. A., Casper, D. L. D. (1979). In *Myelin,* ed. P. Morell, Plenum Press, New York. pp. 51–89.

Klein, R. A. (1982). Q. Rev. Biophys. 15, 667–757.

Klingenberg, M. (1981). Nature 290, 449–454.

Knoll, W., Apell, H. J., Eibl, H., Miller, A. (1986). Eur. J. Biophys. 13, 187–193.

Knoll, W., Schmidt, G., Ibel, K., Sackman, E. (1985). Biochemistry 24, 5240–5246.

Koblin, D. D., Lester, H. A. (1979). Mol. Pharmacol. 15, 559–580.

Kodama, M., Hashigami, H., Seki, S. (1985). Biochim. Biophys. Acta 814, 300–306.

Koehler, L. S., Fossel, E. T., Koehler, K. A. (1980). Progr. Anesthesiol. 2, 447–455.

Kohn, A. B., Schullery, S. E. (1985). Chem. Phys. Lipids 37, 143–53.

Kohne, W., Deuticke, B., Haest, C. W. M. (1983). Biochim. Biophys. Acta 730, 139–150.

Krag, S. S. (1985). Curr. Top. Memb. Transp. 24, 181–249.

Kramp, W., Pieroni, G., Pinkard, R. N., Hanahan, D. J. (1984). Chem. Phys. Lipids 35, 49–62.

Krikos, A., Mutoh, N., Boyd, A., Simon, M. (1983). Cell 33, 615–622.

Kubo, T., Fukuda, K., Mikami, A., Maeda, A., Takahashi, H., Mishima, M., Haga, T., Haga, K., Ichiyama, A., Kaugawa, K., Kojima, M., Matsuo, H., Hirose, T., Numa, S. (1986). Nature 323, 411–416.

Kumar, N. M., Gilula, N. B. (1986). J. Cell Biol. 103, 767–776.

Kuypers, F. A., Roelofsen, B., Op den Kamp, J. A. F., Van Deenen, L. L. M. (1984). Biochim. Biophys. Acta 769, 337–347.

Kyte, J., Doolittle, R. F. (1982). J. Mol. Biol. 157, 527–37.

Lange, Y., Gary-Bobo, C. M. (1974). J. Gen. Physiol. 63, 690–706.

Langmuir, I. (1917). J. Am. Chem. Soc. 39, 1848–1860.

Lackowicz, J. (1983). *Principles of Fluorescence Spectroscopy,* Plenum, New York.

Lanyi, J. K. (1986). Annu. Rev. Biophys. Bioengg. 15, 11–28.

Lasic, D. D. (1982). Biochim. Biophys. Acta. 692, 501–502.

Latorre, R., Alvarez, O., Cecchi, X., Vergara, C. (1985). Annu. Rev. Biophys. Biophys. Chem. 14, 79–111.

Lau, A., McLaughlin, A., McLaughlin, S. (1981). Biochim. Biophys. Acta 645, 279–292.

Lauger, P., Benz, R., Stark, G., Bamberg, E., Jordan, P. C., Fahr, A., Brock, W. (1981). Quart. Rev. Biophys. 14, 513–598.

Lauger, P. (1985). Angew. Chem. (Engl.) 24, 905–923.

Laycock, J. D. (1953). Anesthesia 8, 15–20.

Lee, A. G. (1976). Nature 262, 545–548.

Lee, A. G. (1977). Biochim. Biophys. Acta 472, 237–281; 285–344.

LeFevre, P. G. (1975). Ann. N. Y. Acad. Sci. 264, 398–413.

Lefkowitz, R. J., Cerione, R. A., Codima, J., Birnbaumer, L., Caron, M. G. (1985). J. Membr. Biol. 87, 1–12.

Levin, I., Thompson, T. E., Barenholz, Y., Huang, C. (1985). Biochemistry 24, 6282–6286.

Levitan, I. B. (1985). J. Membr. Biol. 87, 177–190.

Levitt, D. G. (1986). Annu. Rev. Biophys. Biochem. 15, 29–57.

Levitzky, A. (1986). Physiol. Rev. 66, 819–854.

Lewis, B. A., Das Gupta, S. K., Griffin, R. G. (1984). Biochemistry 23, 1988–1993.

Lewis, R. N., and McElhaney, R. N. (1985). Biochemistry 24, 4903–11; 2431–39.

Li, W., Haines, T. H. (1986). Biochemistry 25, 7477–7483.

Liao, T. H., Gallop, P. M., Blumenfeld, O. O. (1973). J. Biol. Chem. 248, 8247–8253.

Lichtenberg, D., Rosen, R. J., Dennis, E. A. (1983). Biochim. Biophys. Acta 737, 285–304.

Lieb, W. R., Stein, W. D. (1972). Biochim. Biophys. Acta 265, 187–207.

Lieb, W. R., Stein, W. D. (1986). J. Membr. Biol. 92, 111–119.

Liebovitch, L. S., Fishbarg, J., Koniarck, J. P., Todorova, I., Wang, M. (1987). Biochim. Biophys. Acta 896, 173–180.

Ling, G. N. (1962). *A Physical Theory of the Living State,* Grim (Blaisdell), Boston.

Lis, L. J., McAlister, M., Fuller, N., Rand, R. P., Parsegian, V. A. (1982). Biophys. J. 37, 667–672.

Loesche, M., Mohwald, H. (1985). Eur. Biophys. J. 11, 35–42.

Loh, H. H., Law, P. Y. (1980). Annu. Rev. Pharm. Tox. 20, 201–234.

Longmuir, K. J. (1985). Curr. Top. Membr. Transport 22.

Low, M. G., Kinkade, P. W. (1985). Nature 318, 62–64.

Lucy, J. A. (1980). In *Membrane–Membrane Interactions,* ed. N. B. Gilula, Raven Press, pp. 81–98.

Lucy, J. A., Ahkong, Q. F. (1986). FEBS Lett. 199, 1–11.

Luxnat, M., Galla, H. J. (1986). Biochim. Biophys. Acta 856, 274–282.

Luzatti, V., Tardieu, A. (1974). Annu. Rev. Phys. Chem. 25, 79–82.

Lytz, R. K., Reinert, J. C., Church, S. E., Wickman, H. H. (1984). Chem. Phys. Lipids 35, 63–76.

MacDonald, P. M., Sykes, B. D., McElhaney, R. N. (1984). Can. J. Biochem. 62, 1134–1150.

MacDonald, P. M., Sykes, B. D., McElhaney, R. N., Gunstone, F. D. (1985). Biochemistry 24, 177–184.

Macnab, R. M., Aizawa, S. I. (1984). Annu. Rev. Biophys. Bioeng. 13, 51–83.

Maddy, A. H., ed. (1976). *Biochemical Analysis of Membranes,* Wiley, New York.

Magee, A. I., Courtneidge, S. A. (1985). EMBO J. 4, 1137–1144.

Magee, A. I., Schlesinger, M. J. (1982). Biochim. Biophys. Acta 694, 279–289.

Maggio, B., Ariga, T., Sturtevant, J., Yu, R. K. (1985). Biochemistry 24, 1084–1092.

Mangold, H. K., Paltauf, F. (1983). *Ether Lipids: Biochemical and Biomedical Aspects,* Academic Press, New York.

Mantsch, H. H., Modec, C. (1985). Biochemistry 24, 2440–2446.

Marcus, M. M., Apell, H. J., Roudna, M., Schwendener, R. A., Weder, H. G., Lauger, P. (1986). Biochim. Biophys. Acta 854, 270–278.

Margolis, L. B. (1984). Biochim. Biophys. Acta 779, 161–189.

Marinetti, G. V., Crain, R. C. (1978). J. Supramol. Struct. 8, 191–213.

Marinetti, G. V., Cattieu, K. (1982). Biochim. Biophys. Acta 685, 109–116.

Marsh, D., Seddon, J. M. (1982). Biochim. Biophys. Acta 690, 117–123.

Marsh, J. W., Dennis, J., Wriston, J. C. (1977). J. Biol. Chem. 252, 7678–7684.

Martin, R. B., Yeagle, P. L. (1978). Lipids 13, 594–597.

Martinek, K., Levashov, A. V., Klyachko, N., Khmelnitski, Y. L., Berezin, I. V. (1986). Eur. J. Biochem. 155, 453–468.

Martonosi, A. N. (1984). Physiol. Rev. 64, 1240–1320.

McClosky, M., Poo, M. (1984). Int. Rev. Cytol. 87, 19–81.

McDaniel, R. V., McIntosh, T. J., Simon, S. A. (1983). Biochim. Biophys. Acta 731, 97–108.

McElhaney, R. N. (1982). Chem. Phys. Lipids 30, 229–259.

McElhaney, R. N. (1984). Biochim. Biophys. Acta 779, 1–42.

McElhaney, R. N. (1986). Biochim. Biophys. Acta 864, 361–421.

McIntosh, T. J., Simon, S. A. (1986). Biochemistry, 25, 4058–4066.

McIntosh, T. J., McDaniel, R. V., Simon, S. A. (1983). Biochim. Biophys. Acta 731, 109–114.

McIntosh, T. J., Simon, S. A., Ellington, J. C., Porter, N. A. (1984). Biochemistry 23, 4038–4044.

McLaughlin, S., Murine, N., Greslfi, T., Vaio, G., and McLaughlin, A. (1981). J. Gen. Physiol. 77, 445–473.

McLean, L. R., Phillips, M. C. (1984). Biochemistry 23, 4624–4630.

Meech, R. W. (1978). Annu. Rev. Biophys. Bioengg. 7, 1–18.

Meier, P., Ohmes, E., Kothe, G. (1986). J. Chem. Phys. 85, 3598–3614.

Melchior, D. (1982). Curr. Top. Membr. Transport 17, 263–316.

Mena, P. L., Djerassi, C. (1985). Chem. Phys. Lipids 37, 257–270.

Menestrina, G., Voges, K., Jung, G., Boheim, G. (1986). J. Membr. Biol. 93, 111–132.

Menger, F. M., Doll, D. W. (1984). J. Am. Chem. Soc. 106, 1109–13.

Middlekoop, E., Lubin, B. H., Op den Kamp, J. A. F., Roelofsen, B. (1986). Biochim. Biophys. Acta 855, 421–424.

Mild, K. H., Lovtrup, S. (1985). Biochim. Biophys. Acta 822, 155–167.

Mieschendahl, M., Buchel, D., Bocklage, H., Muller-Hill, B. (1981). Proc. Natl. Acad. Sci. 78, 7652–56.

Miller, C. (1984). Annu. Rev. Physiol. 46, 549–58.

Miller, C., Arvan, P., Telford, J. N., Racker, E. (1976). J. Mol. Biol. 30, 271–282.

Miller, C., ed. (1986). *Ion Channel Reconstitution,* Plenum Press, New York.

Mills, J. T., Furlong, S. T., Dawidowicz, E. A. (1984). Proc. Natl. Acad. Sci. 81, 1385–1388.

Mishina, M., Kurosaki, T. Tobimatsu, T., Morimoto, Y., Noda, M., Yamamoto, T., Terao, M., Lindstrom, J., Takahashi, T., Kuno, M., Numa, S. (1984). Nature 307, 604–608.

Mitchell, P. (1979). Eur. J. Biochem. 95, 1–20.

Mittal, K. L., Bothorel, P., eds. (1985). *Surfactants in Solution,* Plenum Press, New York. (also earlier volumes).

Moczydlowski, E., Lattore, R. (1983). J. Gen. Physiol. 82, 511–542.

Moddy, A. H., ed. (1976). *Biochemical Analysis of Membranes,* Wiley, New York.

Montal, M., Mueller, P. (1972). Proc. Natl. Acad. Sci. 69, 3561–66.

Moolenaar, W. (1986). Annu. Rev. Physiol. 48, 363–376.

Morariu, V. V., Pop, I. V., Popescu, O., Benga, G. H. (1981). J. Membrane Biol. 62, 1–5.

Morgan, M., Hanke, P., Grygorczyk, R., Tinschl, A., Fasold, H., Passow, H. (1985). EMBO J., 4, 1927–1931.

Moss, C. W. (1981). J. Chromatogr. 203, 337–347.

Mrsny, R. J., Wolwerk, J. J., Griggith, O. H. (1986). Biochim. Biophys. Acta 39, 185–191.

Mueckler, M., Caruso, C., Baldwin, S. A., Panico, N., Blench, I., Morris, H. R., Allard, W. F., Lienhard, G. E., Lodish, H. F. (1985). Science 229, 941–945.

Mulders, F., Van Lange, H., Van Ginkel, G., Levine, Y. K. (1986). Biochim. Biophys. Acta 859, 209–218.

Muller, D., Churchill, P., Fleischer, S., Guengerrich, F. P. (1984). J. Biol. Chem. 259, 8174–8182.

Muller, H., Luxnat, M., Galla, H. (1986). Biochim. Biophys. Acta 856, 283–289.

Murakami, Y., Kikuchi, J., Takaki, T., Uchimura, K., Nakano, A. (1985). J. Am. Chem. Soc. 107, 2161–2167.

Murari, R., Murai, M. P., Bauman, W. J. (1986). Biochemistry 25, 1062–1067.

Myher, J. J., Kuksis, A. (1982). Can. J. Biochem. 60, 638–650.

Nagle, J. F. (1980). Annu. Rev. Phys. Chem. 13, 157–195.

Nagle, J. F., Wilkinson, D. A. (1982). Biochemistry 21, 3817–3821.

Nagle, J. F., Dilley, R. A. (1986). J. Bioenerg. Biomemb. 18, 55–64.

Narahashi, T. (1984). Annu. Rev. Neurobiol. 16, S39–S51.

Navarro, J., Kinnucan, M. T., Racker, E. (1984). Biochemistry 23, 130–35.

Navarro, J., Chabot, J., Sherrill, K., Aneja, R., Zahler, S. A., Racker, E. (1985). Biochemistry 24, 4645–50.

Neher, E., Sakman, B. (1976). Nature 260, 799–802.

Neher, E., Steinbach, J. H. (1978). J. Physiol. 277, 153–176.

Neher, E., Strühmer, W. (1985). Phys. Bl. 41, 329–334.

Neufeld, E. F., Ashwell, G. (1980). In *The Biochemistry of Glycoproteins and Proteoglycans,* ed. W. J. Lennarz, Plenum Press, New York, pp. 241–266.

Nicholls, D. G. (1982). *Bioenergetics,* Academic Press, London.

Nicholls, D. G., Locke, R. M. (1984) Physiol. Rev. 64, 1–64.

Nichols, J. W. (1985). Biochemistry 24, 6390–98.

Nir, S., Bentz, J., Wilschut, J., Duzgunes, N. (1983). Progr. Surface Sci. 13, 1–124.

Nishizuka, Y. (1984). Nature 308, 693–698.

Noda, M., Furutami, Y., Takahashi, H., Toyosato, M., Tanabe, T., Shimizu, S., Kikyotani, S., Koyano, T., Hirose, T., Inayama, S., Numa, S. (1983). Nature 305, 818–823.

Noda, M., Shimizu, S., Tanabe, T., Takai, T., Kayano, T., Ikeda, T., Takahashi, H., Nakayama, H., Kanaoka, Y., Minamino, N., Kanagawa, K., Matsuo, H., Raftery, M. A., Hirose, T., Notake, M., Inayama, S., Hayashida, H., Mixata, T., Muma, S. (1984). Nature 312, 121–127.

Noggle, J. H. (1985). *Physical Chemistry,* Little Brown Boston.

Nuhn, P., Brezesinski, B., Dobner, B., Forster, G., Gutheil, M., Dorfler, H. (1986). Chem. Phys. Lipids 39, 221–236.

Ogden, D. C., Siegelbaum, S. A., Colquhoun, D. (1981). Nature 289, 596–598.

Ohki, S., Ohshima, H. (1984). Biochim. Biophys. Acta 776, 177–182.

Ohki, S., Ohshima, H. (1985). Biochim. Biophys. Acta 812, 147–154.

Ohki, S., Roy, S., Ohshima, H., Leonards, K. (1984). Biochemistry 23, 6126–6132.

Ohshima, H., Ohki, S. (1985a). Biophys. J. 47, 673–678.

Ohshima, H., Ohki, S. (1985b). J. Coll. Interface Sci. 103, 85–94.

Okahata, Y. (1986). Acc. Chem. Res. 19, 57–63.

Oku, N., MacDonald, R. C. (1983a). J. Biol. Chem. 258, 8733–8738.

Oku, N., MacDonald, R. C. (1983b). Ciochim. Biophys. Acta 734, 54–61.

Okubo, T., Kitano, H., Ishiwatari, T., Isem, N. (1979). Proc. Roy. Soc. Lond. A366, 81–90.

Olden, K., Parent, J. B., White, S. L. (1982). Biochim. Biophys. Acta 650, 209–232.

Olsnes, S., Sandvig, K., Madshus, J. H., Sundan, A. (1985). Biochemistry Soc. Symp. 50, 171–191.

Olson, E. N., Towler, D. A., Glaser, L. (1985). J. Biol. Chem. 260, 3784–3790.

Op den Kamp, J. A. F. (1979). Annu. Rev. Biochem. 48, 47–71.

Op den Kamp, J. A. F., Roelofsen, B., Van Deenen, L. L. M. (1985). Trends Biochem. Sci. 9, 320–323.

Overton, E. (1899). Vjschr. Naturforsch Ges. Zurich 44, 88–98.

Owen, P., Kaback, H. R. (1978). Proc. Natl. Acad. Sci. U. S. 75, 3148–3152.

Pagano, R. E., Sleight, R. G. (1985). Science 229, 1051–1057.

Parente, R. A., Lentz, B. (1986). Biochem 25, 1021–1026.

Pascher, I., Sundell, S. (1977). Chem. Phys. Lipids 20, 175–191.

Pascher, I., Sundell, S. (1986). Biochim. Biophys. Acta 855, 68–78.

Pascher, I., Sundell, S., Hauser, H. (1981). J. Mol. Biol. 153, 791–806; 807–824.

Pascher, I., Sundell, Eibl, H., Harlos, K. (1986). Chem. Phys. Lipids 39, 53–64.

Passow, M. (1986). Rev. Physiol. Biochem. Pharmacol. 103, 61–203.

Pastan, I., Willingham, M. C. (1981). Annu. Rev. Physiol. 43, 239–250.

Pastan, I., Willingham, M. C. (1985). *Endocytosis,* Plenum Press, New York.

Patlak, J. B. (1984). In *Membrane Channels and Noise,* eds. R. S. Eisenberg, M. Frank, Stevens, C. F., pp. 197–234. Plenum Press, New York.

Patlak, J., Horn, R. (1982). J. Gen. Physiol. 79, 333–351.

Paul, C., Rosenbusch, J. P. (1985). EMBO J. 4, 1593–97.

Paul, D. L. (1986). J. Cell Biol. 103, 123–134.

Pearse, B. M. F., Bretscher, M. S. (1981). Annu. Rev. Biochem. 50, 85–101.

Pearson, R., Pascher, I. (1979). Nature 281, 499–501.

Perly, B., Smith I. C. P., Jarrell, H. C. (1985), Biochemistry 24, 1055–1063; 4659–65.

Peters, M. W., Grant, C. W. M. (1984). Biochim. Biophys. Acta 775, 273–282.

Peters, R., Beck, K. (1983). Proc. Natl. Acad. Sci. 80, 7183–7187.

Peters, R. (1986). Biochim. Biophys. Acta 864, 305–359.

Pierotti, R. A. (1976). Chem. Rev. 76, 717–726.

Pigor, T., Lawaczeck, R. (1983). Z. Naturforsch 38C, 207–312.

Pink, D. (1984). Cand. J. Biochem. 62, 760–777.

Pjura, W. J., Kleinfeld, A. M., Karnovsky, M. J. (1984). Biochemistry 23, 2039–43.

Pond, J. L., Langworthy, T. A., Holzer, G. (1986). Science 231, 1134–1136.

Pope, C. G., Urban, B. W., Haydon, D. A. (1982). Biochim. Biophys. Acta 688, 279–283.

Popot, J. L., Changeux, J. P. (1984). Physiol. Rev. 64, 1162–1239.

Post, R. L., Kume, S., Tobin, T., Orcutt, B., Sen, A. K. (1969). J. Gen. Physiol. 54, 306s–326s.

Powell, G. L., Marsh, D. (1985). Biochemistry 24, 2902–08.

Pownall, H. J., Hickson, D. L., Smith, L. C. (1983). J. Amer. Chem. Soc. 105, 2440–2445.

Poznansky, M. J., Juliano, R. L. (1984). Pharm. Rev. 36, 277–336.

Prasad, K. U., Trapane, T. L., Busath, D., Szabo, G., Urry, D. W. (1982). Intern. J. Peptide Protein Res. 19, 162–171.

Pratt, L. R. (1985). Annu. Rev. Physical Chem. 36, 433–449.

Presti, F. T. (1985). Membrane Fluidity in Biology 4, 97–146.

Pryde, J. G. (1986). Trends Biochem. Sci. 11, 160–163.

Quandt, F. N., Narahashi, T. (1982). Proc. Natl. Acad. Sci. 79, 6732–6736.

Rainier, S., Jain, M. K., Ramirez, F., Soannou, P. V., Marecek, J. F., Wagner, R. (1979). Biochim. Biophys. Acta 558, 187–198.

Ramirez, F., Marecek, J. F. (1985). Synthesis 5, 449–488.

Ramasammy, L. S., Brockerhoff, H. (1982). J. Biol. Chem. 257, 3570–74.

Rand, R. P., Parsegian, V. A. (1984). Can. J. Biochem. 62, 752–759.

Ranck, J. L. (1983). Chem. Phys. Lipids 32, 251–67.

Ranck, J. L., Tocanne, J. F. (1982). FEBS Lett. 143, 171–174; 175–178.

Rando, R. R., Bangerter, F. W. (1979). J. Supramol. Struct. 11, 295–309.

Rao, J. K. M., Argos, P. (1985). Biochim. Biophys. Acta 869, 197–214.

Rauvala, H., Finne, J. (1979). FEBS Lett. 97, 1–7.

Recktenwald, D. J., McConnell, H. M. (1981). Biochemistry 20, 4505–10.

Reusch, R. N., Sadoff, H. L. (1983). Nature 302, 268–270.

Reynolds, J. A., McCaslin, D. R. (1985). Methods Enzymol. 117, 41–53.

Richards, C. D., Martin, K., Gregory, S., Keightley, C. A., Hesketh, T. R., Smith, G. A., Warren, G. B., Metcalfe (1978). Nature 276, 775–779.

Rilfors, L., Eriksson, P., Arvidson, G., Lindblom, G. (1986). Biochemistry 25, 7702–7711.

Ritchie, R. J. (1984). Progr. Biophys. Mol. Biol. 43, 1–32.

Robins, S. J., Patton, G. M. (1986). J. Lipid Res. 27, 131–139.

Rodbell, M. (1980). Nature 284, 17–22; Trends Biochem. Sci. 10, 461–464.

Rogers, J., Lee, A. G., Wilton, D. C. (1979). Biochim. Biophys. Acta 552, 23–30.

Romstead, L. S. (1984). In *Surfactants in Solution,* ed. Mittal, K. L., Lindman, B., vol. 2, Plenum Press, New York, pp. 1015–1035.

Roseman, S., Postma, P. W. (1976). Biochim. Biophys. Acta 57, 213–257.

Rothgeb, T. M., Oldfield, E. (1981). J. Biol. Chem. 256, 6004–6009.

Rothgeb, P. F., Janoff, A. S., Miller, K. W. (1983). J. Lipid Res. 24, 47–51.

Rowe, E. S. (1985). Biochim. Biophys. Acta. 813, 321–330.

Ruocco, M. J., Atkinson, D., Small, D. M., Skarjune, R. P., Oldfield, E., Shipley, G. G. (1981). Biochem. 20, 5957–5966.

Ruocco, M. J., Siminovitch, D. J., Griffin, R. G. (1985a). Biochemistry 24, 2406–2411.

Ruocco, M. J., Makriyannis, A., Siminovitch, D. J., Griggin, R. G. (1985b). Biochemistry 24, 4844–4851.

Saffman, P. G. (1976). J. Fluid. Mech. 73, 593–602.

Sakman, B., Neher, E. (1983). *Single Channel Recording,* Plenum Press, New York.

Sakman, B., Neher, E. (1984). Annu. Rev. Physiol. 46, 455–72.

Sakman, B., Patlak, J., Nehr, E. (1980). Nature 286, 71–73.

Sandblom, J., Eisenman, G., Hagglund, J. (1983). J. Membr. Biol. 71, 61–78.

Sauter, J. F., Braswell, L. M., Miller, K. W. (1980). In *Progress in Anesthesiology,* vol. 2, ed. R. A. Fink, Raven Press, New York, pp. 119–207.

Schauer, R. (1982). Adv. Carbohydrate Chem. Biochem. 40, 131–234.

Schindler, H. (1979). Biochim. Biophys. Acta 555, 316–336.

Schindler, H. (1980). FEBS Lett. 122, 77–79.

Schmidt, R. B. (1986). Angew. Chem. (Engl.) 23, 212–235.

Schneider, Y., Octave, J., Trouet, A. (1985). Curr. Top. Membr. Transport 24, 413–458.

Schurtenberger, P., Hauser, H. (1984). Biochim. Biophys. Acta 778, 470–480.

Schwartz, E. A. (1985). Annu. Rev. Neurosci. 8, 339–67.

Schwartz, M. A., McConell, H. M. (1978). Biochemistry 17, 837–840.

Schwarz, W., Passow, H. (1983). Annu. Rev. Physiol. 45, 359–74.

Scotto, A. W., Zakim, D. (1985). Biochemistry 24, 4066–75.

Seelig, J., Seelig, A. (1980). Quart. Rev. Biophys. 13, 19–61.

Seeman, P. (1972). Pharmacol. Rev. 24, 583–655.

Seddon, J. M., Cevc, G., Marsh, D. (1983). Biochemistry 22, 1280–89.

Seddon, J. M., Cevc, G., Kaye, R. D., Marsh, D. (1984). Biochemistry 23, 2634–2644.

Sege, R. D., Kozarsky, K., Nelson, D. L., Kreiger, M. (1984). Nature 307, 742–5.

Segrest, J. P., Jackson, R. L., Andrews, E. P., Marchesi, V. T. (1971). Biochem. Biophys. Res. Commun. 44, 390–395.

Segrest, J. P., Kahane, I., Jackson, R. L., Marchesi, V. T. (1973). Arch. Biochem. Biophys. 155, 167–183.

Seigneuret, M., Devaux, P. F. (1984). Proc. Natl. Acad. Sci. 81, 3751–55.

Semenza, G. (1986). Annu. Rev. Cell. Biol. 2, 255–313.

Semenza, G., Kessler, M., Hosang, M., Weber, J., Schmidt, U. (1984). Biochim. Biophys. Acta 779, 343–79.

Senior, A. E. (1985). Curr. Top. Membr. Trans. 23, 135–151.

Severs, N. J., and Robenek, H. (1983). Biochim. Biophys. Acta 737, 373–408.

Shew, R. L., Deamer, D. W. (1985). Biochim. Biophys. Acta 816, 1–8.

Shinitzky, M., Barenholz, Y. (1978). Biochim. Biophys. Acta 515, 367–394.

Sigworth, F. J., Neher, E. (1980). Nature 287, 447–49.

Silvius, J. R. (1982). In *Lipid Protein Interactions* vol. 2, ed P. C. Jost, O. H. Griffith, John Wiley, New York, pp. 239–281.

Simon, S. A., McIntosh, T. J. (1984). Biochim. Biophys. Acta 773, 169–172.

Simon, S. A., Stone, W. L., Busto-Lattorre (1977). Biochim. Biophys. Acta 468, 378–388.

Simpson, I. A., Cushman, S. W. (1985). Annu. Rev. Biochem. 55, 1037–1057.

Sine, S. M., Taylor, P. (1980). J. Biol. Chem. 255, 10144–10156.

Sine, S. M., Taylor, P. (1982). J. Biol. Chem. 257, 8106–8114.

Singer, M. A., Finegold, L. (1985). Biochim. Biophys. Acta 816, 303–312.

Singer, S. J., Nicholson, G. (1972). Science 175, 720–731.

Singh, A. P., Nicholls, P. (1985). J. Biochem. Biophys. Methods 11, 95–108.

Sixl, F., Watts, A. (1983). Proc. Natl. Acad. Sci. 80, 1613–15.

Skou, J. C. (1982). Ann. N. Y. Acad. Sci. 402, 169–84.

Skulachev, V. P. (1985). Eur. J. Biochem. 151, 199–208.

Skulachev, V. P., Hinkle, P. C. eds. (1981). *Chemiosmotic Proton Circuits in Biological Membranes,* Addison-Wesley, Reading, Massachusetts.

Small, D. M. (1985). *Physical Chemistry of Lipids: From Alkanes to Phospholipids,* Plenum Press, New York.

Smith, I. C. P. (1985). In *Nuclear Magnetic Resonance of Liquid Crystals,* ed. J. W. Emsley. D. Reidel Publ. Co., New York. pp. 533–566.

Smith, I. C. P., Ekiel, I. H. (1984). In *Phosphorus-31 NMR,* Academic Press, New York, pp. 447–475.

Smith, R. L., Oldfield, E. (1984). Science 225, 280–88.

Solomon, A. K., Chasen, B., Dix, J. A., Lukacovic, M. F., Toon, M. R., Verkman, A. S. (1983). Ann. N. Y. Acad. Sci. 414, 97–124.

Somerharju, P. J., Virtanen, J. A., Eklund, K. K., Vainio, P., Kinnunen, P. K. J. (1985). Biochemistry 24, 2773–81.

Sone, N., Yoshida, M., Hirata, H., Kagawa, Y. (1977). J. Biol. Chem. 252, 2956–2960.

Spatz, L., Strittmatter, P. (1971). Proc. Natl. Acad. Sci. U. S. (1971). 68, 1042–1046.

Spudich, J. L., Bogomolni, R. A. (1984). Nature 312, 509–513.

Stanfield, P. R. (1983). Rev. Physiol. Biochem. 97, 1–67.

Steck, T. L., Weinstein, R. S., Straus, J. S., Wallach, D. F. H. (1970). Science 168, 255–257.

Stepanov, A. E., Shvets, V. I. (1979). Chem. Phys. Lipids 25, 247–263.

Stevens, B. R., Kaunitz, J. D., Wright, E. M. (1984). Annu. Rev. Physiol. 46, 417–433.

Stevens, V. L., Lambeth, J. D., Merrill, A. H. (1986). Biochemistry 25, 4287–92.

Stewart, T. P., Hui, S. W., Portis, A. R., Papahadjopoulos, D. (1979). Biochim. Biophys. Acta 556, 1–16.

Stilbs, P., Arvidson, G., Lindblom, G. (1984). Chem. Phys. Lipids 35, 309–314.

Stoeckenius, W., Bogomolini, R. (1982). Annu. Rev. Biochem. 51, 587–616.

Stryer, L. (1986). Annu. Rev. Neurosci. 9, 87–119.

Stubbs, C. S., Smith, A. D. (1984). Biochim. Biophys. Acta 779, 89–137.

Stumpel, J., Nicksch, A., Eibl, H. (1980). Biochemistry 20, 662–666.

Stumpel, J., Nicksch, A., Eibl, H. (1983). Biochim. Biophys. Acta 727, 246–254.

Sullivan, K. A., Liao, Yu-Chien, Aborzi, A., Beiderman, B., Chang, F., Masters, S. B., Levinson, A. D., Bourne, H. R. (1986). Proc. Natl. Acad. Sci. 83, 6687–6691.

Surolia, A., Bhachawat, B. K. (1978). Biochem. Biophys. Res. Commun. 83, 779–785.

Surolia, A., Bhachawat, B. K., Podder, S. K. (1975). Nature 257, 802–804.

Szabo, G., Eisenman, G., Ciani, S. (1969). J. Membr. Biol. 7, 346–382.

Szoka, F., Papahadjopoulos, D. (1980). Annu. Rev. Biophys. Bioeng. 9, 467–508.

Tabake, T., Hammes, G. G. (1981). Biochemistry 20, 6859–6864.

Tall, A. R. (1986). J. Lipid. Res. 27, 361–367.

Talvenheimo, J. A. (1985). J. Membr. Biol. 87, 77–91.

Tamm, L. K., McConnell, H. M. (1985). Biophys. J. 47, 105–113.

Tamura-Lis, W., Reber, E. J., Cunningham, B. A., Collins, J. M., Lis, L. J. (1986). Chem. Phys. Lipids 39, 119–124.

Tanford, C., Reynolds, J. A. (1976). Biochim. Biophys. Acta 457, 133–170.

Tanford, C. (1980). *Hydrophobic Effect,* Wiley, New York.

Tatulian, S. A. (1983). Biochim. Biophys. Acta 736, 189–195.

Tausk, R. J. M., Karmiggelt, J., Oudshorrn, C., Overbeek, J. Th. G. (1974a). Biophys. Chem. 1, 175–183.

Tausk, R. J. M., Van Esch, J., Karmiggelt, J., Voordouw, G., Overbeek, J. Th. G. (1974b). Biophys. Chem. 2, 53–63.

Tausk, R. J. M., Oudshoorn, C., Overbeek, J. Th. G. (1974c). Biophys. Chem. 2, 53–63.

Taylor, J. A. G., Minguis, J., Pethica, B. A. (1976). J. Chem. Soc. Farady Trans. 1, 2694–2702.

Tenenbaum, D., Folch-pi, J. (1963). Biochim. Biophys. Acta 115, 141–147.

Thompson, T. E., Tillack, T. W. (1985). Annu. Rev. Biophys. Biophys. Chem. 14, 361–86.

Tien, H. T. (1985). Progr. Surface Sci. 19, 169–274.

Tilcock, C. P. S., Cullis, P. R., Grunner, S. M. (1986). Chem. Phys. Lipids 40, 47–56.

Tobkes, N., Wallace, B. A., Bayley, H. (1985). Biochemistry 24, 1915–1920.

Toko, K., Yamafuji, K. (1980). Chem. Phys. Lipids 26, 79–99.

Tomasi, M., Montecucco, C. (1981). J. Biol. Chem. 256, 11177–11181.

Towler, D., Glaser, L. (1986). Biochemistry 25, 878–884.

Trauble, H. (1971). J. Membr. Biol. 4, 193–208.

Trautmann, A. (1982). Nature 298, 272–275.

Tritton, T. R., Hickman, J. A. (1985). In *Experimental and Clinical Progress in Cancer Chemotherapy,* ed. F. M. Muggia, Martinus Nijhoff Publishers, Boston, pp. 81–131.

Trudell, J. R. (1977). Anesthesiology 46, 5–10.

Tse, M. Y., Singer, M. (1984). Can. J. Biochem. Cell. Biol. 62, 72–77.

Tsien, R. Y., Hladky,S. B. (1982). Biophys. J. 39, 49–56.

Turner, R. J. (1983). J. Membr. Biol. 76, 1–15.

Ulbrecht, W. (1977). Annu. Rev. Biophys. Bioeng. 6, 7–31.

Unwin, P. N. T., Ennis, P. D. (1984). Nature 307, 609–613.

Urry, D. W. (1984). In *Spectroscopy of Biological Molecules,* ed. C. Sandorfy and T. Theophanides, D. Reidel Publ. Co., Norwell, Mass, pp. 487–510; 511–538.

Urry, D. W., Prapane, T. K., Prasad, K. U. (1983). Science 221, 1064–1067.

Urry, D. W., Venkatachalam, C. M., Wood, S. A., Prasad, K. U. (1985). In *Structure and Motion: Membranes, Nucleic acids and Proteins,* eds. E. Clementi, G. Corongiu, M. H. Sarma, R. H. Sarma, Adenine Press, Guilderland, NY, pp. 185–203.

Vassort, G., Whittembury, J., Mullins, L. J. (1986). Biophys. J. 50, 11–19.

Van Dael, H., Centerickx, P. (1984). Chem. Phys. Lipids 35, 171–181.

Van den Berg, C. A., Horn, R. (1984). J. Gen. Physiol. 84, 535–564.

Van Dijk, P. W. M., Kaper, A. J., Oonk, H. A. J., De Gier, J. (1977). Biochim. Biophys. Acta 470, 58–69.

Van Driessche, W., Zeiske, W. (1985). Physiol. Rev. 65, 833–903.

Van Duijn, G., Valtersson, C., Chojnacki, T., Verkleij, A. J., Dallner, G., DeKruijf, B. (1986). Biochim. Biophys. Acta 861, 211–223.

Van Hoogewest, P., DeGier, J., De Kruijff, B. (1984). FEBS Lett. 171, 160–164.

Vance, D., Vance, J., eds. (1985). *Biochemistry of Lipids and Membranes,* Benjamin Publishing Co., San Francisco.

Vaver, V. A., Todria, K. G., Prokazova, N. V., Rozynov, B. V., Bergelson, L. D. (1977). Biochim. Biophys. Acta 486, 60–69.

Venter, J. C., Schmidt, U., Eddy, B., Semenza, G., Fraser, C. M. (1985). In *Tissue Culture of Epithelial Cells,* ed. M. Taub, Plenum, New York, pp. 179–201.

Vaz, W. L. C., Goodsaid, F., Jacobson, K. (1984). FEBS Lett. 174, 199–207.

Verheij, H. M., Slotboom, A. J., DeHaas, G. H. (1981). Rev. Physiol. Biochem. Parmacol. 91, 91–203.

Von Schulthess, G. K., Cohen, R. K., Sakato, N., Benedek, G. B. (1976). Immunochemistry 13, 955–962.

Wallach, D. F. H., Lin, P. S. (1973). Biochim. Biophys. Acta 300, 211–254.

Watts, A., Harlos, K., Marsh, D. (1981). Biochim. Biophys. Acta 645, 91–96.

Watts, A., Marsh, D., Knowles, P. F. (1978). Biochemistry 17, 1792–1801.

Weber, J., Warden, D. A., Semenza, G., Diedrich, D. F. (1985). J. Cell. Biol. 27, 83–96.

Weiss, R. M., McConnell, H. M. (1984). Nature 310, 47–49.

Weiss, R. M., Roberts, W. M., Strühmer, W., Almers, W. (1986). J. Gen. Physiol. 87, 955–983.

Welti, R., Mullikin, L. J., Yoshimura, T., Helmkamp, G. M. (1984). Biochem. 23, 6086–6091.

Wennerstrom, H., Lindman, B. (1980). Phys. Rep. 52, 1–86.

Wheeler, T. J., Hinkle, P. C. (1985). Annu. Rev. Physiol. 47, 503–17.

White, M. M., Bezanilla, F. (1985). J. Gen. Physiol. 85, 539–554.

White, J., Kielian, M., Helenius, A. (1983). Q. Rev. Biophys. 16, 151–195.

Wickstrom, M., Saraste, M. (1984). *Bioenergetics,* ed. L. Ernster, Elsevier, Amsterdam, pp. 49–94.

Widdas, W. F. (1980). Curr. Top. Membr. Transp. 14, 166–223.

Wiegandt, H., ed. (1985). *Glycolipids,* Elsevier, Amsterdam.

Wilbrandt, W. (1972). J. Membr. Biol. 10, 357–366.

Wilcox, C. A., Olson, E. N. (1987). Biochemistry 26, 1029–1036.

Wilshut, J., Duzgunes, N., Fraley, R., Papahadjopoulos, D. (1980). 19, 6011–6021.

Wislocki, P. G., Miwa, G. T., Lu, A. Y. H. (1980). Enzymic Basis of Detoxification, 1, 135–182.

Wisenieski, B. J., Bramhall, J. S. (1981). Nature 289, 319–321.

Wolber, P. K., Hudson, B. S. (1981). Biochemistry 20, 2800–2810.

Wolfensen, R., Anderson, L., Cullis, P. M., Southgate, C. C. B. (1981). Biochemistry 20, 849–855.

Womack, M. D., Kendall, D. A., MacDonald, R. C. (1983). Biochim. Biophys. Acta 733, 210–215.

Wong, M., Brown, R. E., Barenholtz, Y., Thompson, T. E. (1984). Biochemistry 23, 6498–6505.

Wong, P. T. T. (1984). Annu. Rev. Biophys. Bioengg. 13, 1–24.

Wong, P. T. T., Mantsch, H. H. (1985). Biochemistry 24, 4091–4096.

Wright, E. M., Diamond, J. M. (1977). Physiol. Rev. 57, 109–156.

Wright, J. K., Riede, I., Overath, P. (1981). Biochemistry 20, 6404–6415.

Wu, S. H., McConnell, H. M. (1975). Biochemistry 14, 847–854.

Xie, X., Stone, D. K., Racker, E. (1984). J. Biol. Chem. 259, 11676–11678.

Yau, K., Nakatani, K. (1985). Nature 317, 252–255.

Yavin, Z., Yavin, E., Kohn, L. D. (1982). J. Neurosci. Res. 7, 267–278.

Yeagle, P. L. (1985). Biochim. Biophys. Acta 822, 267–287.

Yguerabide, J., Foster, M. C. (1981). In *Membrane Spectroscopy,* ed. E. Grell, Springer-Verlag, New York.

Yogeeswaran, G. (1980). In *Cancer Markers, Diagnostic and Developmental Significance,* ed. S. Snell, Humana Press, New York, pp. 371–401.

Yue, B. Y., Jackson, C. M., Taylor, J. A. G., Minguis, J., Pethica, B. A. (1976). J. Chem. Soc. Faraday Trans. 1, 2685–2693.

Zaki, L., Julien, T. (1985). Biochim. Biophys. Acta 818, 325–332.

Zulauf, M. (1985). In *Physics of Amphiphiles, Micelles, Vesicles and Microemulsions,* North-Holland, Amsterdam, pp. 663–676.

Zwolinsky, B. J., Eyring, H., Reese, C. E. (1949). J. Phys. Colloid Chem. 53, 1426–1453.

Index

A-23187, *see* Calcimycin
Acetylcholine, 296, 303, 359
Acetylcholine receptor, 296–306
Acetylcholinesterase, 297
Actin, 170
Active transport, 281
Acyl chain:
  conformation, 123
  structure, 27–29
Adenylate cyclase, 357–365
Adipocyte, 382
ADP–ATP exchanger, 274–275
Adrenergic receptors, 358–365
Alamethicin, 252, 254, 255, 377
Aldosterone, 277
Allethrin, 319, 322
Alloprenolol, 360
Alloxan, 269
Amiloride, 277
Amino acid transporter, 286
Aminopeptidase, 188
Amphotericin B, 253
Anesthetics:
  general, 163–165
  local, 162, 273, 324–326
Anion exchanger, 170, 171, 180, 270–274
Ankyrin, 171, 183
Antamanide, 225
Arabinose, 197
Arachidonic acid, 39
Asialoglycoprotein receptor, 385
ATPase, Ca-, 186
ATPase, Na, K-, 186
ATP synthesis, 331, 336
Atractyloside, 275
Atropine, 359
Axon, 296

Bacterial reaction center, 336
Bacteriohopane, 30
Bacteriorhodopsin, 178, 179, 180, 192, 333–335
Band, 3. *See also* Anion exchanger
Beauvericin, 222
Benzocaine, 324
Benztropine, 297
Bilayer:
  anesthetics, 163–165
  binding of solutes, 149–153
  carriers, 215–238
  channels, 238–256
  conductance, 230
  diffusion potential, 228
  drugs, 162
  effects of solutes, 157–162
  fluidity, 125
  ion translocation, 227

Bilayer (*Continued*):
molecular areas, 135
order and dynamics, 122–146
order parameter, 124–125
packing, *see* Bilayer, order and dynamics
partition coefficient, 148–149, 154–157
phase change, 126–135
phase properties, 137–146
planar, 94–96
properties, 86–121
subphases, 81
thermotropic changes, 126–135
thickness, 135
viscosity, 125
Bipolar lipids, 35
BLM, 94–96, 229, 240–243
Bongkrekate, 275
Born equation, 53, 215
Bubble, air, 52

Calcimycin, 222, 226
Calcium pump, 340
Carbamylcholine, 303, 323
Carbocyanin dye, 237
Carbohydrates, 6, 11, 194–212
Carboxyatractyloside, 275
Cardiolipin, *see* Diphosphatidylglycerol
Cation carriers, 215–233
CCP, 235, 237
CDP-Diacylglycerol, 40
Cells, 1–3
adipose, 383
animal, 2
epithelial, 276
nerve, 296
plant, 296
Centrioles, 2
Cephalin, *see* Phosphatidylethanolamine
Ceramide, 36
Ceramide lactoside, 37
Ceramide trihexoside, 37
Channels, 238–256
calcium, 255
colicin, 329
epithelial, 275
gap junction, 328
gated, 294–329
porins, 277
potassium, 241, 255, 373
sodium, 256, 312–322, 370
toxins, 278
voltage-gated, 254–256, 312–322
Chemotaxis, 353
Chloroplast, 2, 12
Cholera toxin, 361, 363
Cholesterol, 30, 66, 74, 141–146
Choline, 297, 298
Chromosomes, 2
Clatharin, 377, 380
Clozapine, 297
Coat, 12
Coated pits, 380
Complement system, 278
Complexons, *see* Ionophores
Concanavalin, 209, 347, 368
Conductance, 228
single channel, 240–247, 301, 316
Conduritol β-epoxide, 285
Cord factor, 198
Cotransport, 282–290
Countertransport, 259, 264, 266
Coupled transport, 281
Critical micelle concentration, 70
Cubic phase, 79
Cyclic AMP, 357–368
Cyclic GMP, 370–373
Cytochalasin, 183, 209, 268, 284, 385
Cytochrome b_5, 180, 188

DDT, 323
Diabolic acid, 198
Diacylglycerol, 39, 40
Dianemycin, 222, 226
DIDS, 271, 274
Diethylstilbesterol, 268
Diffusion:
lateral, 7, 68, 108–112
transbilayer, 115–121
transverse, 115–121
Diffusion potential, 228
Dinactin, 222
Diphosphatidylglycerol, 32, 38, 40, 132
Dipicrylamine, 237

Diplopterol, 30
Dipyridamole, 269, 273
DNDS, 273
Dopamine, 359
Down-regulation, 363, 383
DTFB, 235
Duramycin, 186

EIM, 255
Electrical double layer, 98–105
Electron transport chain, 337–339
Emulsion, 52, 79
Endocytosis, 209, 278, 378–387
Endoglycosidase H, 205
Endoplasmic reticulum, 2, 13, 167
Endplate potential, 299
  miniature, 299
Energy transduction, 330–355
Enniatin, 222, 224, 225
Epinephrine, 358–364
Erythrocyte, 106, 167, 169, 180, 260, 378
Eserine, 297
Etruscomycin, 253
Exchange diffusion, *see* Countertransport

Facilitated transport, 257–280
Fatty acids, 18
  attached to proteins, 184
FCCP, 235
Filipin, 252
Flagellar motion, 354
Flip-flop, *see* Transbilayer movement
Fucose, 197
Furosamide, 273
Fusion, 209, 373–379

Galactocerebroside, 37
Galactosamine, 197
Galactose, 197
Galactosidase, 41
Ganglioside, 37, 38, 71, 200
Gating current, 313–317
Globoside, 37, 200
Glucocerebroside, 37, 200
Glucosamine, 197
Glucose, 197
Glucose transporter, 170, 263–270, 282–286, 382–385
Glycerophospholipids, 31
Glycocalyx, 12, 195
Glycoconjugates, 194–212
  antigens, 203, 206–207
  identification, 195
  isolation, 196
  receptor, 204
  structure, 196
Glycolipids, 194
Glycophorin A, 170, 180, 188, 207, 377
Glycoproteins, 194–212
Glycosphingolipids, 36, 37, 199–201
Glycosyltransferase, 206
Golgi complex, 2, 12, 200
Gouy–Chapman theory, 100–103
G protein, 362, 365, 371–373
Gramicidin, 220, 241–251, 256
Grisoryxin, 222

Halorhodopsin, 335
Hematoside, 37
Hemicholinium-3, 297
Hemocyanin, 255
Hexagonal (H_{II}) phase, 76–78
Hexamethonium, 359
Histamine, 359
Histidine transporter, 291–292
HLA antigen, 180
Hormones on adenylate cyclase, 357
HPETE, 39
Hydrocortisone, 269
Hydrophobic effect, 5, 51–54
β-Hydroxybutyrate dehydrogenase, 190–192

Influenza virus receptor, 188
Ion gradients, 228–230, 330
Ionophores, 213–256
Ion selectivity, 216–221, 244, 250, 272, 277, 315, 328, 346–348, 368
C_{55}-isoprenoid alcohol phosphokinase, 192
Isoproterenol, 360

Kinase, 293, 358, 365, 370, 384

Lactose transporter, 287
Lamellar phase, 79–84
Lasalocid, 222
Lectins, 203, 209–212
Leucoplast, 2
Leucotriene, 39
Lidocaine, 324
Lipid A, 199
Lipidic particles, 78
Lipid–protein interaction, 166–193
Lipid–solute interaction, 147–165
Lipid storage diseases, 49
Lipid–water interface, 51–85
  hydration, 84–85
Lipid X, 365
Lipolytic enzymes, 38–41, 50
  genetic defects, 49
Lipophilic ions, 236
Liposomes, 86–94, 387–388
Lipoteichoic acid, 198
Lipoxygenase, 39
LSD, 359
Lung surfactant, 68–69
Lysophospholipase, 40
Lysosomes, 12, 386

Malonaldehyde, 39
Maltose, 268
Mannose, 197
Mellitin, 278, 377
Membrane, asymmetry, 93, 106–108
  components, 25–44
  composition, 27, 334, 338, 341
  functions, 3–7
  ghosts, 11
  isolation, 10
  markers, 18–24
  molecular motions, 7, 8
  morphology, 10, 12, 17
  organelles, 11, 27
  organization, 7
  plasma, 1, 12
  receptors, *see* Receptor
  surface coat, 12, 195
Membrane potential:
  action potential, 307–326
  Donnon equilibrium, 99
  endplate potential, 299
  surface, 98
  transmembrane, 98
Membrane proteins, 27, 166–193
  characterization, 171
  composition, 27
  conformation, 174–181
  covalently attached fatty acids, 184
  isolation, 168
  reconstitution, 172–174
  rotation, 182
  solubilization, 167
  topography, 181
  translation, 182
Micelles, 52, 69–76
  inverted, 52
Mitochondrion, 2, 12, 167
MN blood group determinant, 170, 180, 207
MN protein, 170, 188
Monactin, 222, 225
Monamycin, 225
Monazomycin, 253–256
Monensin, 222, 224
Monogalctosyl diglyceride, 33
Monolayer, 52, 59
  phase diagram, 66
  pressure-area curves, 61–63
  surface potential, 65
Morphine, 359
Mucopolysaccharides, 12, 195
Mycolic acid, 34
Myelin, organization, 83

NANA, *see* Neuraminic acid, *N*-acetyl-
Neuraminic acid, 36, 37, 197
  N-acetyl-, 197
Neuraminidase, 41, 180
Nicotine, 359
Niflumic acid, 274
Nigericin, 222, 224, 226
Nonactin, 222, 224
Norepinephrine, 359
Nucleolus, 2
Nucleus, 2, 12
Nystatin, 252

Octopamine, 359
Oligosaccharides, *see* Glycoconjugates
Order parameters, 124–125
Organelles, 2, 23
Ornithine lipids, 34
Ouabain, 348
Oxanol dye, 237
Oxidative phosphorylation, 336–340
Oxotremorine, 297

PAF, *see* Platelet activating factor
Pancuronium, 324
Pardaxin, 278
Partition coefficient, 148–149
Patch-clamp, 96, 312
Perfluoropinacol, 237
Periplasm, 12, 292, 354
Permeability, 115–121, 213–256
Peroxisomes, 12
Phagocytosis, 379
Phase:
  hexagonal I, 53
  hexagonal II, 53
  lamellar, 53
Phase diagram, 58, 66, 87, 138–144
Phase properties, effect of solutes, 159–162
Phentolamine, 360
Phloretin, 268, 273
Phlorizin, 285
Phorbol esters, 365
Phosphatidic acid, 32, 39, 40, 98, 132
Phosphatidylcholine, 32, 39, 40, 43, 122, 127, 130, 139, 157, 186, 250, 338, 340
Phosphatidylethanolamine, 32, 39, 40, 132, 134, 139, 186, 338
Phosphatidylglycerol, 32, 40
Phosphatidylinositol, 32, 40, 365–369
Phosphatidylserine, 32, 40, 97, 103–105
Phosphodiesterase, 358, 364
Phosphoinositol cycle, 365–368
Phospholipase A1, 31
Phospholipase A2, 31, 39, 40, 188–190
Phospholipase B, 40
Phospholipase C, 31, 40, 368
Phospholipase D, 40
Phospholipid exchange proteins, *see* Phospholipid transfer proteins
Phospholipids:
  conformation, 43–48
  dispersions, 97–98
  effect on catalysis, 187
  exchange, 112–115
  intracellular translocation, 115
  molecular area, 64
  phase properties, 72
  polymorphism, 54–59
  self-association, 51
  structure, 31
  synthesis, 48
Phospholipid transfer proteins, 113–115
Phosphonolipids, 31
Phosphotransferase system, 290
Phototransduction, 370
Physostigmine, 297
Phytohemagglutinin receptor, 188
Phytosphingosine, 36
PI cycle, 365–368
Pimaricin, 252
Pinocytosis, *see* Endocytosis
Platelet activating factor, 40
Platyphyllin, 297
PLEP, *see* Phospholipid transfer proteins
Polyene antibiotics, 252
Polyiodide, 237
Polymyxin, 377
Polysaccharides, 12
Porins, 277–278
Prednisolone, 269
Procaine, 324
Propranolol, 360
Prostacyclin, 39
Prostaglandin, 39, 277, 359
Proton carriers, 234–236
Proton gradient, 234–236, 330–340, 354, 368
Proton motive force, 330–340, 354, 368
Proton pump, 333, 337, 350
PTS, 290
Pump:
  calcium, 340
  proton, 333, 337, 350
  sodium, 346

Pyrazine, 277
Pyruvate oxidase, 193

Qx-222, 324
Qx-314, 324

Receptor:
  acetylcholine, 296–306, 359
  antigen, 206–207
  asialoglycoprotein, 385–387
  influenza virus, 188
  insulin, 381
  lectins, 209
  phytohemagglutinin, 188
Receptor mediated endocytosis, 379–387
Retinaldehyde, 372
Retinoic acid, 365
Rhodopsin, 370–373
Ribosome, 2
Rod cell, 370
Rod outer segment membrane, 370

Sarcoplasmic reticulum, 167, 340–343
Second messenger, 364
Secretion, 373
Seminolipid, 33
Serotonin, 359
Sitosterol, 30
Sodium–proton exchange, 368
Sodium pump, 346
Space-clamp, 311
Sphinganine, 36
Sphingolipid, 38
Sphingomyelin, 36, 132
Sterols, 29
Stigmaterol, 30
Streptolysin-O, 278
Strychnine, *N*-methyl-, 324
Suberyldicholine, 323
Substance P, 359
Sucrase–isomaltase, 180, 188, 285
Sulfatide, 37, 200
Sulfoglycolipid, 198
Surfactants, 68, 71
Synapse, 296

Tay-Sachs ganglioside, 37, 200
Tetracaine, 324
Tetrahymenol, 30
Tetramethrin, 319, 322
Tetranactin, 222
Tetraphenylarsonium, 237
Tetraphenylboron, 237
Thermogenesis, 340
Thromboxanes, 39
Toxin:
  abrin, 280
  α-toxin, 279
  apamine, 319
  batrachotoxin, 319, 322
  botulinium, 297
  channels, 278
  cholera, 361, 363
  condylactis, 319
  curare, 297
  diptheria, 278, 280
  grayanotoxin I, 319, 322
  histrianicotoxin, 297
  modaccin, 280
  neurotoxins, 322
  ricin, 280
  saxitoxin, 319, 322
  shigellin, 280
  streptolysin-O, 278
  tetrodotoxin, 319, 322
  tubocurare, 323
  venom, 297, 304
  viscumin, 280
Tranducin, 370
Transbilayer movement, 105–108
Transport, *see* Active transport; Coupled transport; Facilitated transport; Permeability
Transport defects, 257
Transporters:
  ADP–ATP exchanger, 274–275
  amino acid, 286
  anion, 270–274
  glucose, 259, 263–270, 282, 382
  histidine, 291
  kinetics, 264–265
  lactose, 287
  metabolite, 258

Trinactin, 222
TTFB, 235
Tunicamycin, 207

UDP-glucuronosyltransferase, 193
Uncoupler, *see* Proton carriers
Uncoupler-1799, 235
Uphill transport, *see* Active transport

Valinomycin, 221–233
Vanadate, 348
Vasopressin, 277
Veratridine, 319, 323
Vesicles, 11, 86–94, 297, 321, 342, 376, 379
Vision, 370
Voltage-clamp, 311–312

Water:
- hydration layer, 84
- permeability, 119–121
- physical properties, 52

X-206, 222, 224, 226
X-537 A, *see* Lasalocid
Xylose, 197